2026 ElCon Conference of Young Researchers in Micro and Nanoelectronics, Electrical Engineering Materials, Plasmas and Fields, Photonics and Electro-Optics (ElCon-MN 2026)

Saint Petersburg, Russia
3-5 February 2026

IEEE Catalog Number: CFP267BZ-POD
ISBN: 979-8-3315-4784-4

**Copyright © 2026 by the Institute of Electrical and Electronics Engineers, Inc.
All Rights Reserved**

Copyright and Reprint Permissions: Abstracting is permitted with credit to the source. Libraries are permitted to photocopy beyond the limit of U.S. copyright law for private use of patrons those articles in this volume that carry a code at the bottom of the first page, provided the per-copy fee indicated in the code is paid through Copyright Clearance Center, 222 Rosewood Drive, Danvers, MA 01923.

For other copying, reprint or republication permission, write to IEEE Copyrights Manager, IEEE Service Center, 445 Hoes Lane, Piscataway, NJ 08854. All rights reserved.

*** This is a print representation of what appears in the IEEE Digital Library. Some format issues inherent in the e-media version may also appear in this print version.

IEEE Catalog Number:	CFP267BZ-POD
ISBN (Print-On-Demand):	979-8-3315-4784-4
ISBN (Online):	979-8-3315-4783-7

Additional Copies of This Publication Are Available From:

Curran Associates, Inc
57 Morehouse Lane
Red Hook, NY 12571 USA
Phone: (845) 758-0400
Fax: (845) 758-2633
E-mail: curran@proceedings.com
Web: www.proceedings.com

TABLE OF CONTENTS

A Comparative Analysis of Laser Drilling Methods for Microholes in Transparent Media.................................. 1
Selemani A. Chemchem, Salhab Mhmad, Anastasia Vasilieva, Vladimir Ivanov

Development and Investigation of a Reduced Graphene Oxide-Based Impedimetric Immunosensor
for the Detection of FGF21 Protein... 5
Elisavata I. Epifanova, Andrey A. Ryabko, Alexey A. Kolobov, Ivan K. Khmelnitskiy, Yuri V. Cheburkin, Nikita O. Sitkov

Fabrication of Non-Lithographic Metal Masks for Metal-Assisted Chemical Etching of Silicon 10
Ekaterina Glushets

Direct Laser Writing As An Alternative To Existing Methods Of Silicon Microstructuring 14
Olga S. Gribovskaya

Understanding and Predicting Signal Degradation in Printed Circuit Board Transmission Lines
Using Neural Networks 18
Alexey V. Moskvin, Ivan Yu. Gritskevich

Features of Formation and New Areas of Application of Composite Structures Based on Aluminum
Oxide 23
Ekaterina N. Muratova

Modeling and Analysis of the Design of Sensitive Element of a Capacitive Type Micromechanical
Accelerometer 27
Paing Soe Thu, Viktor V. Kalugin, Elena S. Kochurina, Ye Min Hlaing, Satt Naing Moe

Application of an Autoencoder for Improving the Accuracy of OTDR Measurements 33
Pavel Petrov, Nikolay Kolybelnikov

Study of Sensing Elements with Various Resonator Designs on a Square Membrane for High-
Precision MEMS Pressure Sensors... 38
Phyo Win Tun, Sergey P. Timoshenkov, Boris M. Simonov

Modeling of a SiC trench gate MOSFET 43
Mikhail A. Potapov

Investigation of the Formation of an Impedance Biosensor Response for a Structure Based on Zinc
Oxide Nanorods and Peptide Recognition Elements.. 46
Alina S. Printseva, Andrey A. Ryabko, Alexey A. Kolobov, Valentina V. Trushlyakova, Nikita O. Sitkov

AFM Investigation of Surface Morphological Changes in Microplastics under Photocatalytic
Decomposition.. 50
Dmitry G. Radaykin

Nanostructuring by Local Anodic Oxidation Using Atomic Force Microscopy (AFM) for the
Formation of Mask Structures 54
Matvey Repkin

Study of the Influence of DX-Centers on the Characteristics of a Pseudomorphic High-Electron-
Mobility Transistor (pHEMT) 57
Alexander V. Sapozhnikov, Roman S. Kryukov, Vadim V. Perepelovsky, Anatoly L. Dudin, Artyom I. Baranov

Controlled Hot-Injection synthesis of Lead Halide Perovskite Quantum Dots for Enhanced Photoluminescence and Sensor Application 61
  Sohail Amir

Design and Analysis of a Stacked Silicon Wafer-Based Technology for Multilayer Multi-Chip Modules 64
  Ilya A. Solovyov, Andrey Yu. Titov, Denis V. Vertyanov

Thermoelectric Numerical Simulation of a Memristor-based Microchip 68
  Maxim V. Sozonov

A Low-Power 12-bit SAR ADC with a Switching Energy Minimization Algorithm in 180-nm CMOS 72
  Anastasiia Tsepilova, Vladimir Losev, Aleksandr Timoshenko

Quantum Dots in Nanoelectronics: From Fundamental Principles to Revolutionary Applications 76
  Ivan Y. Yaroshenko, Sergey R. Galimov, Alexander V. Petruhanov, Dmitry M. Kuznetsov

A Mini Review: Perovskite Solar Cells 79
  Al Walo Walo

A Comparative Study of ZnO and TiO2 as ETL Performance in Cs2SnI6-Based Perovskite Solar Cells Using SCAPS-1D 83
  Al Walo Walo

MACl-Assisted Improvement of Morphology in Mixed-Halide FAPbBr2Cl Perovskite Films 89
  Maksim A. Balutin, Yuriy E. Isaev, Aleksandr S. Tarasov, Aleksander E. Degterev, Mariya M. Degtereva, Ivan A. Lamkin

Acetone Detection via ZnO/Zn2SnO4 Impedance Spectroscopy Analysis 93
  Cong Doan Bui, Svetlana S. Nalimova

Characterization of Nanostructured Ternary Oxide Multisystems for Gas Sensors 96
  Sergey Buzovkin, Arina Rybina, Roman Kryukov, Svetlana Nalimova, Vyatcheslav Moshnikov

Technology of Vacuum Deposition of Combined Thin-film Precision Strain Gauges with Temperature Self-compensation 100
  Sergey A. Gurin, Anastasia E. Shepeleva, Maksim D. Novichkov, Ekaterina A. Pecherskaya, Dmitry V. Agafonov, Vadim S. Volkov

Remote Temperature Monitoring Method Based on Tapered Fiber Sensors 106
  Vyacheslav G. Nesterov, Sergei A. Shagako, Diana D. Shagako

Effect of Micro- and Nanostructure of Carbon Anode on the Performance of Biophotovoltaic Cell 110
  Rakshaev Bair Ts., Konditerov Pavel B., Snarskaya Dina D., Kononova Svetlana V., Fallahzade Peyman Z., Zimina Tatiana M.

Resent Advances in the Production and Research of Composite Materials Based on Graphene and Polymers 115
  Pavel F. Samsygin, Svetlana S. Nalimova

Formation of Low-Dimensional Halide Perovskite Crystals in Radiation-Resistant Phosphate Glass 119
  Tatyana Sedegova

Formation of a Contact Pad in the Ceramic Structure of an Electrostatic Chuck 124
  Viktor S. Traktirshchikov, Viktor V. Kalugin, Maksim E. Shiryaev, Ivan A. Korotkevich

Investigation of the Effect of Laser Annealing on the Structural Properties of BaSnTiO3 Thin Films............ 127
    Oleg E. Zaytsev, Igor N. Zakasovsky

Study of a Resistive-Type Superconducting Fault Current Limiter (SFCL) for Fault Protection in a
Microgrid... 131
    Qusai K. Al Naimi, Abdulhadi Haj Ahmad, Hayder J. Mohammed

Plasma-Enhanced Atomic Layer Deposition of WO3 Using Tungsten Hexacarbonyl and Activated
Oxygen 136
    Vladislav E. Zamoshets

Biconical Antenna for Use in Systems with High Radiated Power 141
    *Alexander A. Borisov, Valeriy V. Demshevsky, Igor A. Bogachev, Andrey D. Bazhenov,
    Vyacheslav V. Lobodin, Stanislav S. Sidorenko*

Clustered Antenna Array for AESA with Wide Angles of Electronic Scanning 145
    *Valeriy V. Demshevsky, Grigoriy S. Anikin, Stanislav S. Sidorenko, Ilya A. Tsygitsa, Dmitri V.
    Bagno, Eugene V. Iliin*

Numerical Modelling of Spin Wave Logic Gates... 150
    Roman Haponchyk, Alexey Ustinov

Tunable Ring Resonator Based on Ferroelectric Capacitors 154
    Kamil Karymsakov, Kirill Dmitriev, Svetlana Zubko, Danyar Alzhanov, Valeriia Bobrovskaia

Application of X-Ray Fluorescence Analysis for Identification of Paint Pigments in Easel and Wall
Paintings 157
    Laleh M. Moazzami Lavasani

Development of an Automated Reflectometer Control System for Fiber Optic Communication
Lines.. 160
    *Evgeny S. Minin, Sergey S. Gryzulev, Bogdan K. Reznikov, Timofey A. Kotov, Nikolay Yu.
    Kolybelnikov, Dmitry I. Isaenko.*

Development of a Simulator for a Transceiver Device for Long-Distance Fiber Optic
Communication Lines 166
    *Evgeny S. Minin, Sergey S. Gryzulev, Bogdan K. Reznikov, Timofey A. Kotov, Nikolay Yu.
    Kolybelnikov, Dmitry I. Isaenko*

A Mobile Multifunctional Sensor for Monitoring the Spectral Composition of Radiation Sources................ 170
    Andrey Novikov, Gleb Sozykin, Natalya Burenkova, Vladimir V. Skryabin, Anna S. Klyuchnik

Fractal Antenna Array for a Compact Active Electronically Scanned Array 175
    *Stanislav S. Sidorenko, Valeriy V. Demshevsky, Grigory S. Anikin, Aleksander A. Borisov, Ilya
    A. Tsygitsa, Vyacheslav V. Lobodin*

Optimization of a Grow Light Design for Uniform Plant Illumination 180
    *Gulnaz Galina, Roman Kurenkov, Mariya Degtereva, Yevgeniy Levin, Alexander Degterev,
    Ivan Lamkin*

Optical Properties Study of Metal Halide Perovskites after Polymer Modification 183
    *Yuriy E. Isaev, Maksim A. Balutin, Aleksandr S. Tarasov, Aleksander E. Degterev, Marina D.
    Pavlova, Sergey A. Tarasov*

Study of Technology for Forming Arrays of Polymer Microlenses for Fluorimetric Biosensor
Systems.. 187
    Diana I. Khasanova, Anastasia D. Tarasenko, Nikita O. Sitkov

Development of an Energy-Efficient Phytotron 191

Roman A. Kurenkov, Gulnaz M. Galina, Mariya M. Degtereva, Yevgeniy Levin, Alexander E. Degterev, Ivan A. Lamkin

Laser Polishing for Creating Microfluidic Chips on Stainless Steel 195

Valery V. Lavrinenko

Laser Cleaning of Parchment: Assessment of Treatment Results... 199

Taniana K. Lepekhina, Angelina D. Neelova, Yan O. Guttovskiy

Peculiarities of the Influence of Laser Radiation Parameters on the Measurement Error of the Refractive Index of Liquid Media in a Differential Refractometer 204

Daniil S. Provodin, Alexandra D. Kurkova

Compact Cross-Pumped 2-μm Laser with Passive Q-Switching by Glass:PbS Nanocrystals.......... 209

Mhmad Salhab, Tatiana Zotova, Vladimir Ivanov, Aleksei Onushchenko, Aleksandr Titov, Anastasia Vasilieva

Functional Automatic Laser Beam Alignment System Capabilities.. 212

Sergey A. Shagako, Diana D. Shagako, Vyacheslav G. Nesterov

Modeling of the GaAs Optical Waveguide Phase Shifter.. 216

Alexander Shevtsov, Vitalii Vitko, Alexey Ustinov

Development of Organic Photosensitive Structures with a Bulk Heterojunction Based on P3HT and PBDTT-DPP 220

Gordey A. Shutkin, Marina D. Pavlova, Nikita A. Khorshev, Aleksandr E. Degterev, Aleksander S. Tarasov, Ivan A. Lamkin

Integrated Optical Delay Line with Electronic Control... 224

Vitalii Vitko, Alexander Shevtsov, Andrey Nikitin, Alexey Ustinov

Author Index

Proceedings
of the 2026 ElCon Conference of Young Researchers in Micro and
Nanoelectronics, Electrical Engineering Materials, Plasmas and Fields,
Photonics and Electro-Optics
(2026 ElCon-MN)

February 3-5, 2026

St. Petersburg
Russia
2026

Preface

IEEE Russia North West Section and Saint Petersburg Electrotechnical University "LETI" are pleased to present the Proceedings of the 2026 ElCon Conference of Young Researchers in Micro and Nanoelectronics, Electrical Engineering Materials, Plasmas and Fields, Photonics and Electro-Optics (2026 ElCon-MN) held in St. Petersburg, Russia, on February 3-5, 2026. This conference was hosted by Saint Petersburg Electrotechnical University "LETI". The Organising and Technical Program Committees believe and trust that we have been true to the spirit of collegiality that members of IEEE value whilst also maintaining a high standard as we reviewed papers, provided feedback and now present a strong body of published work in this collection of proceedings.

The themes for this year's conference were chosen as means of bringing together the many orientations of micro and nano-electronics, plasmas and fields, photonics and electro-optics research and teaching, and providing a basis for discussion of issues arising across the young engineering community in these fields.

The aim in these proceedings has been to present high quality work in an accessible medium, for use in the teaching and further research of all people associated with micro and nano-electronics, plasmas and fields, photonics and electro-optics, electrical engineering materials studies. To achieve this aim, all abstracts were double blind reviewed, and full papers submitted for publication in this volume of proceedings were subjected to a rigorous reviewing process.

Prof. Victor A. Tupik, Conference Chair
Dr. Sergey Shaposhnikov, Conference Publication Chair

A Comparative Analysis of Laser Drilling Methods for Microholes in Transparent Media

Selemani A. Chemchem
Department of Photonics.
Saint Petersburg Electrotechnical Univerity *"LETI"*
Saint Petersburg, Russia
selemanichemchem@gmailcom

Salhab Mhmad
Department of Photonics.
Saint Petersburg Electrotechnical Univerity *"LETI"*
Saint Petersburg, Russia
mhmadsalhab@gmail.com

Anastasia Vasilieva
Department of Photonics.
Saint Petersburg Electrotechnical University "LETI"
Saint Petersburg, Russia
anastasiastru@mail.ru

Vladimir Ivanov
Department of Photonics
Saint Petersburg Electrotechnical University "LETI"
Saint Petersburg, Russia
vnivan@mail.ru

Abstract— Laser drilling of micro-holes in transparent materials presents a technological challenge due to the low thermal conductivity and fragility of optical media. This technique is critical for uses in the fields of micro-optics, microfluidics, biomedical devices, and photonic integrated circuits. Laser drilling has become the preferred non-contact technique owing to its precision, flexibility and ability to process hard and brittle substrates. This study compares nanosecond laser drilling methods for glass, focusing on the effect of an absorbing coolant on drilling quality under dry, top-surface, and bottom-surface cooling conditions. The results indicate that localized top-surface cooling provides superior thermal management, reducing heat-affected zones, cracking and debris while improving hole geometry. These findings demonstrate the effectiveness of controlled cooling strategies for enhancing precision and surface integrity in laser drilling of transparent materials.

Keywords: Laser Drilling, Transparent Materials, Prism, Vintage microscope MBS-9, Cooling solution.

I. INTRODUCTION

The advancement of drilling technologies in micro-optics, photonics and biomedical engineering is fundamentally driven by the ability to create precision micro-holes in transparent substrates such as glass [4,7]. These micro-holes act as important light guides, flood channels and surgical tools [8]. Their precision directly determines the performance and reliability of the final device, thus establishing their drilling as a critical step in developing modern miniaturized systems [3,4,8]. However, the properties that make glass to be valuable are optical transparency and electrical insulation, which present significant drilling challenges [4,9]. Its inherent brittleness makes it highly susceptible for cracking and chipping under mechanical stress, while its low thermal conductivity causes heat to accumulate locally during the drilling processing [3,6]. This concentrated heat leads to extensive thermal damage, including large heat-affected zones (HAZ), micro-fractures and recast layers, which reduce the material's structural integrity and optical properties [4]. To overcome these issues, laser drilling has developed as the main non-contact technique. Its ability to remove material without mechanical contact by using only focused light makes it perfect for precision drilling of brittle materials that would be damaged by traditional methods [5].

This work investigates the improvement of nanosecond laser drilling of micro holes for glass through the strategic application of an absorbing coolant. The use of laser itself provides precision and the thermal energy from nanosecond pulses that can still cause cracking and damage [3]. The work evaluates three methods that were used for the drilling process. These are dry conditions (without cooling), top-surface cooling and bottom-surface cooling. The study demonstrates that cooling to the top surface locally provides the most effective thermal control. This configuration provides optimal source-level cooling heat and resulting in a decrease in the heat-affected zone [9]. Also, it overturns the crack formation and propagation while minimizing the deposition of fragments and molten material. This makes the results in micro-holes with significantly improved geometric accuracy and surface integrity. These results have shown that using the laser parameters themselves and controlled cooling methods are important to obtain the high levels of precision, accuracy and quality holes that are required for the advanced laser drilling of micro holes in transparent materials.

II. MATERIALS AND METHODS

1. Materials:

a) Experiment setup

The experimental arrangement for nanosecond laser drilling of micro-holes in transparent media consisted of silica glass, a laser source, Galvano scanner, beam delivery optics, focusing lens, precision lifting system and a computer-controlled (Figure 1). Precise control of laser parameters and sample positioning was essential to ensure hole uniformity, minimize thermal effects, and maintain structural integrity. The fused silica glass sample used in this study had a diameter of 30 mm and a thickness of 15 mm.

b) Laser source

A pulsed nanosecond laser ($\lambda = 1.07$ μm, pulse duration = 10 ns, repetition rate = 10 kHz, focal length = 250 mm) was employed [2]. These parameters provided an optimal balance between pulse energy and absorption efficiency for silica glass.

979-8-3315-4784-4/26 $31.00 © 2026 IEEE

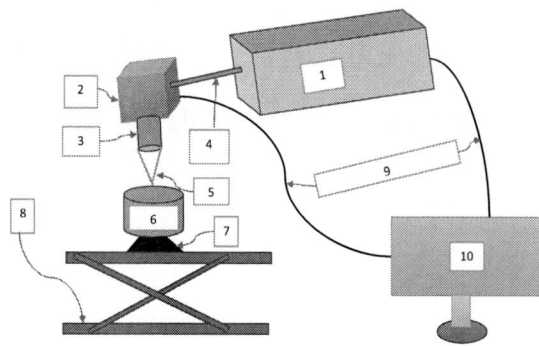

Fig. 1.   The diagram illustrates the laser drilling setup for a micro hole in glass; 1-Laser Source; 2- Galvo Scanner; 3- Focusing lens; 4-Laser beam; 5 -Focal length; 6-Glass; 7- Edge   8- Precision lifting system; 9- Connection cables; 10- Computer.

2. Methodologies:

An analytical comparison of three laser drilling methods for silica glass micro-holes was conducted. These are drilling without coolant (dry), drilling with cooling on top surface, and drilling with cooling on bottom [1].

i. Drilling without coolant (dry).

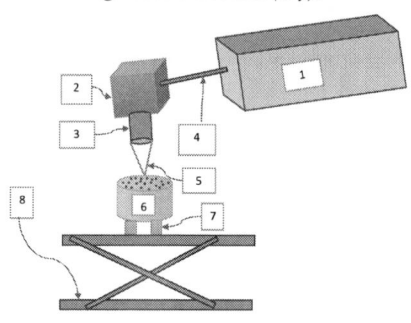

Fig. 2.   Laser drilling of micro-holes in glass without cooling; 1-Laser Source; 2- Galvo Scanner; 3- Focusing lens; 4-Laser beam; 5 -Focal length; 6-Glass; 7- Edge   8- Precision lifting system.

ii. Cooling on top surface

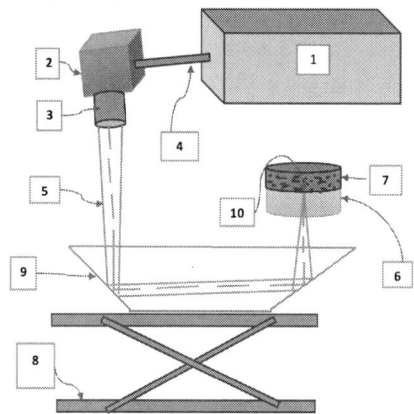

Fig. 3.   Laser drilling with cooling on top; 1-Laser Source; 2- Galvo Scanner; 3- Focusing lens; 4-Laser beam; 5 -Focal length; 6-Glass; 7-Cooling top   8- Precision lifting system; 9- Prism; 10-Beam Spot.

i) Cooling on bottom surface

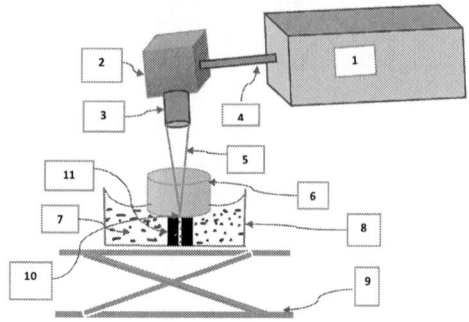

Fig. 4.   Laser drilling of micro-holes in glass with cooling on bottom; 1-Laser Source; 2- Galvo Scanner; 3- Focusing lens; 4-Laser beam; 5 -Focal length; 6-Glass; 7-solution; 8- Cooling container on bottom; 9- Precision lifting system; 9- Prism; 10-Beam Spot; 11-Edge.

III.   RESULTS AND DISCUSSION

a) Dry drilling.

Drilling without a cooling solution resulted in extensive thermal damage, poor hole geometry, and reduced dimensional accuracy, as illustrated in figure 6. Dry drilling caused overheating and plasma shielding, resulting in microcracks, rough edges, and an enlarged heat-affected zone. Consequently, it was significantly less efficient and produced lower quality results than methods using coolant.

b) Cooling on top surface

The results for nanosecond laser drilling with a cooling solution applied to the top surface are presented in figure 5. This configuration demonstrated clear advantages over the other methods. It effectively reduced thermal damage at the hole entrance, producing clean geometry with minimal melting and well-defined edges. Top-surface cooling also enhanced material removal efficiency and contributed to stable and consistent drilling performance.

c) Cooling on bottom surface

The results obtained with bottom-surface cooling are shown in Figure 7. The UV laser used for the drilling process in this method makes induced structural changes in the glass, resulting in the formation of color centers. Consequently, the laser beam could not be focused through the glass and was therefore not employed. Applying the cooling solution to the lower surface effectively dissipated heat from the exit side, resulting in fewer cracks and smoother edges. This improvement is attributed to the limited thermal conductivity of silica glass and the placement of the coolant opposite the region of laser energy deposition. Thus, bottom-surface cooling promotes indirect heat extraction through conduction, reducing backside damage and improving the overall drilling quality.

979-8-3315-4784-4/26 $31.00 © 2026 IEEE

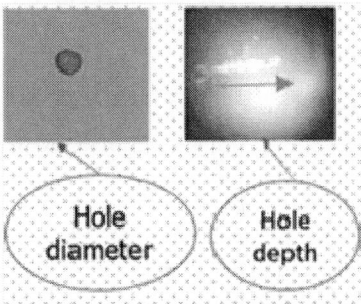

Fig. 5. Cooling on top

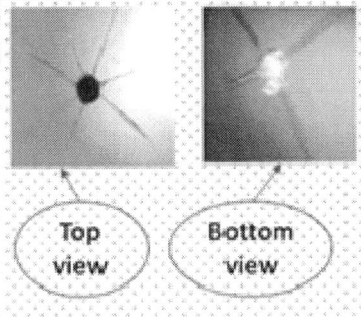

Fig. 6. Dry drilling

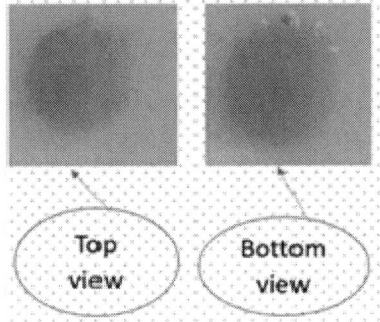

Fig. 7. Cooling on bottom

TABLE. I. FEATURES OF DRY DRILLING.

Feature	Dry drilling
Goal	Speed, Simplicity
Recast Layer	Thick layer
Heat affected zone	Large
Top Surface Spatter	High
Exit Quality	Poor
Complexity	Low
Application	Rapid prototyping, non-critical micro-holes, thick materials.

Dry drilling is characterized by its primary goals of speed and simplicity, resulting in a process with low complexity as shown in table 1 above. This method operates without any cooling mechanism, which leads to significant thermal effects on the workpiece. Consequently, it produces a thick recast layer and a large heat-affected zone (HAZ). Surface quality is also compromised, with high levels of top surface

spatter and poor exit hole quality. Given these characteristics, dry drilling is unsuitable for high-precision or final-product applications. Instead, it is best applied in scenarios where process speed outweighs quality concerns, such as in rapid prototyping, creating non-critical through-holes, or drilling through thick materials where other approaches would be harder to carry out.

TABLE. II. FEATURES OF TOP COOLING DRILLING.

Feature	Cooling on Top
Goal	Clean, maximize internal quality and reduce HAZ
Recast Layer	Minimum layer
Heat affected zone	Smallest
Top Surface Spatter	Very Low
Exit Quality	Very good
Complexity	High
Application	PCB, surface-sensitive components and thin sheets.

The cooling on top method prioritizes precision and internal quality, with the primary goals of producing a clean result, maximizing internal integrity, and reducing the heat-affected zone as shown in figure 5 and table 2 above. This method achieves superior material preservation by creating a minimum recast layer and the smallest heat affected zone. The surface quality is excellent and characterized by very low top surface spatter and very good exit hole integrity. However, this high level of quality comes at the cost of high complexity in the process setup as shown in figure 3. Given its precision and minimal thermal damage, this method is ideally suited for delicate and surface-sensitive applications, such as the manufacturing of printed circuit boards (PCBs), surface-sensitive components, and thin sheets.

TABLE. III. FEATURES OF BOTTOM COOLING DRILLING

Feature	Cooling on Bottom
Goal	Minimum internal quality and eliminate recast
Recast Layer	Moderate layer
Heat affected zone	Smaller
Top Surface Spatter	Low to Moderate
Exit Quality	Good
Complexity	Medium
Application	Aerospace turbine blades, fuel injectors

The primary objective of the bottom cooling approach is to preserve internal quality and ensure the complete absence of a recast layer as shown in table 3 above. This method offers a balanced compromise, effectively creating a smaller heat-affected zone (HAZ) and a moderate recast layer compared to other methods. It maintains respectable quality with good exit hole integrity and low to moderate top surface spatter. With a medium level of complexity, it provides a practical solution for demanding industrial components. This makes it particularly well-suited for critical applications where internal integrity and the absence of a recast layer are paramount, such as in aerospace turbine blades and fuel injector nozzles.

Based on the comparison of drilling methods described above the bar chart graphs has also shown below in figure 8.

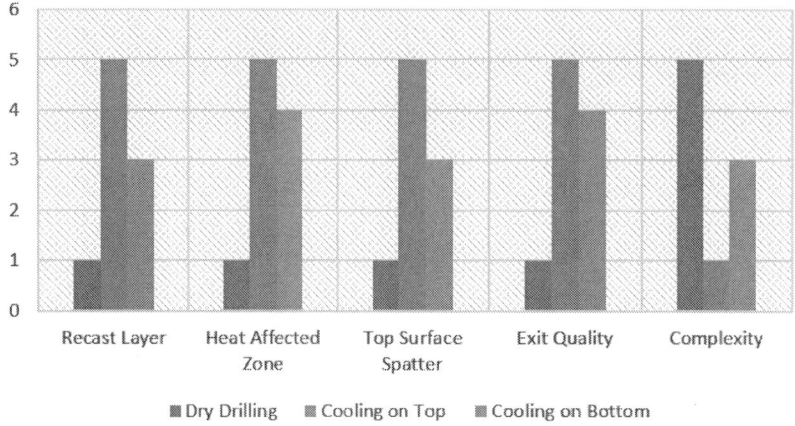

Fig. 8. Comparison of drilling methods based on features.

IV. CONCLUSION

This comparison reveals a direct trade-off between process efficiency and output quality among dry drilling, top cooling, and bottom cooling methods. Dry drilling of silica glass offers the simplest and fastest process but results in the poorest quality, with significant spatter, recast layer, and heat damage, making it suitable only for non-critical applications like rapid prototyping. In contrast, drilling of micro hole using top cooling method provides the highest quality by minimizing the recast layer, heat-affected zone, and spatter which made it ideal for sensitive components such as PCBs. The drilling of silica glass using the bottom cooling method offers a middle-ground solution with moderate quality and complexity. It is suitable for applications like aerospace components where internal quality is critical but some surface spatter is acceptable. However, further studies are required to optimize the laser parameters and drilling conditions to prevent sample cracking or structural failure.

ACKNOWLEDGMENT

The work was carried out with the full support of the department of Photonics, Saint Petersburg Electrotechnical University "LETI, Russia.

REFERENCES

[1] Zheng, H. Y., Zhou, W., & Huang, H. (2013). Femtosecond and nanosecond laser drilling of semiconductors and transparent materials

[2] Li, C., Wang, Q., & Wang, Z. (2018). A review of laser drilling of transparent materials. Optics & Laser Technology, 108, 439-451.

[3] Tan, B., & Venkatakrishnan, K. (2006). A femtosecond laser-induced microfluidic channel. Journal of Micromechanics and Microengineering, 16(8), 1490

[4] Ning, J., Zhang, Y., Zhang, K., & Wang, S. (2019). Experimental study on nanosecond laser drilling of sapphire. Journal of Materials Processing Technology, 273, 116243

[5] Zhou, W., Li, Z., & Li, D. (2017). Experimental investigation on nanosecond laser drilling of glass. The International Journal of Advanced Manufacturing Technology, 91(5-8), 1977-1985

[6] Chien, W. T., & Hou, S. C. (2007). Investigating the recast layer formed during the laser trepan drilling of Inconel 718 using the Taguchi method. The International Journal of Advanced Manufacturing Technology, 33(3-4), 308-316.

[7] Matsuo, S., Tabuchi, Y., & Ikeno, J. (2008). Water-assisted laser drilling of glass. Journal of Laser Micro/Nanoengineering, 3(1), 13-17.

[8] Ngoi, B. K., Venkatakrishnan, K., & Tan, B. (2001). Water-jet assisted laser drilling of transparent materials. Materials and Manufacturing Processes, 16(2), 205-212.

[9] Li, Z. Z., Wang, X. W., & Yuan, S. (2015). Experimental study on water-immersion laser drilling of silicon wafer. The International Journal of Advanced Manufacturing Technology, 78(1-4), 241-249

[10] Wang, X., Zheng, H., Chu, P. L., et al. (2012). Laser drilling of micro-holes in glass substrates by a long-pulse laser with bottom cooling. Optics and Lasers in Engineering, 50(2), 215-220.

[11] Zheng, H. Y., & Huang, H. (2007). Ultrasonic vibration-assisted laser drilling of glass. Journal of Micromechanics and Microengineering, 17(8), 1463.

[12] Wang, J., Wang, X., & Li, M. (2019). Comparative study of dry and water-immersion laser drilling of CVD diamond. Journal of Materials Processing Technology, 264, 223-231.

[13] Wang, X., Zheng, H., & Chu, P. L. (2010). Laser drilling of micro-holes in glass substrates with a long-pulse laser: a comparative study. Key Engineering Materials, 447-448, 772-776

[14] Wang, X., & Zheng, H. (2013). Laser drilling of micro-holes in brittle materials: mechanisms and quality improvement. In Laser Machining of Brittle Materials (pp. 67-98). Springer, London

[15] Wang, X., Zheng, H., & Lim, G. C. (2011). A comparative study of laser drilling in glass substrates with and without a layer of water. Proceedings of the Institution of Mechanical Engineers, Part B: Journal of Engineering Manufacture, 225(7), 1123-1130

Development and Investigation of a Reduced Graphene Oxide-Based Impedimetric Immunosensor for the Detection of FGF21 Protein

Elisavata I. Epifanova
Department of micro- and nanoelectronics
Saint-Petersburg Electrotechnical University ETU "LETI"
St. Petersburg, Russia
lizbon767@mail.ru

Andrey A. Ryabko
Department of micro- and nanoelectronics
Saint-Petersburg Electrotechnical University ETU "LETI"
St. Petersburg, Russia
a.a.ryabko93@yandex.ru

Alexey A. Kolobov
Institute of Power Electronics and Photonics
Saint-Petersburg Electrotechnical University ETU "LETI"
St. Petersburg, Russia
alexey.kolobov.spb@gmail.com

Ivan K. Khmelnitskiy
Department of micro- and nanoelectronics
Saint-Petersburg Electrotechnical University ETU "LETI"
St. Petersburg, Russia
khmelnitskiy@gmail.com

Yuri V. Cheburkin
Institute of Power Electronics and Photonics
Saint-Petersburg Electrotechnical University ETU "LETI"
St. Petersburg, Russia
yucheburkin@gmail.com

Nikita.O. Sitkov
Department of micro- and nanoelectronics
Saint-Petersburg Electrotechnical University ETU "LETI"
St. Petersburg, Russia
sitkov93@yandex.ru

Abstract— This study presents the design and fabrication of an impedimetric immunosensor based on reduced graphene oxide (rGO) for detecting fibroblast growth factor 21 (FGF21), a promising biomarker for metabolic disorders. Interdigitated nickel-chromium electrodes were fabricated on a glass substrate using photolithography. rGO films were synthesized by chemical reduction of graphene oxide using hydrazine or NaBH$_4$, with isopropanol providing optimal coating uniformity. Surface modification involved silanization with APTMS followed by immobilization of Protein A and anti-FGF21 antibodies. Impedance spectra were recorded using a PS-50 potentiostat, revealing a clear dependence of charge-transfer resistance on FGF21 concentration. The sensor exhibited minimal nonspecific response to chymotrypsinogen. The proposed platform demonstrates the potential of rGO-based microelectrode systems for label-free, low-cost, and selective detection of protein biomarkers in biomedical diagnostics.

Keywords— *reduced graphene oxide (rGO), biosensor, nanomaterials, FGF21, impedance spectroscopy, protein detection*

I. Introduction

Early diagnosis of metabolic disorders is critical for their effective prevention and treatment. The incidence of metabolic syndrome, type 2 diabetes, and nonalcoholic steatohepatitis is rapidly increasing worldwide [1]. These diseases pose a serious public health threat and require effective preventive measures to prevent their development into more severe pathologies, such as cardiovascular disease, stroke, and liver cirrhosis. One marker of such conditions is fibroblast growth factor 21 (FGF21), a stress-induced hormone of the liver and adipose tissue closely associated with energy homeostasis and inflammation [2]. It has attracted attention due to its ability to mimic the beneficial effects of diet and exercise, including weight loss, improved insulin sensitivity, and normalization of lipid profiles [3]. The biological role of FGF21 is to coordinate metabolic processes through interactions with key hormones and signaling pathways, such as insulin, leptin, adiponectin, and insulin-like growth factor 1 (IGF1). It acts as a systemic

regulator, participating in the body's adaptation to starvation, fatty acid oxidation, and appetite control. Due to its abnormally high levels, fibroblast growth factor 21 (FGF21) can serve as an indicator or biomarker of certain metabolic diseases, including diabetes and coronary heart disease (CHD) [4]. FGF21-based drugs are being studied as potential treatments for metabolic disorders, as they play a key role in regulating systemic glucose and lipid metabolism.

Therefore, the development of highly sensitive and specific biomarker analysis methods that enable the timely detection of abnormalities and monitoring of pathological process dynamics is particularly relevant. One promising area in this field is the use of electrochemical biosensors [5]. These devices provide high-speed and accurate analysis of biological samples, enabling the creation of compact and portable systems for real-time monitoring. Furthermore, they can directly detect target molecules without the use of complex labels, significantly simplifying the analysis procedure and reducing diagnostic costs. Impedimetric immunosensors [6], which measure changes in electrical resistance upon specific binding of biomolecules, are particularly noteworthy. The advantage of this approach is the elimination of the need for enzymatic or fluorescent labels, significantly simplifying the analysis procedure, reducing its time and reducing the overall cost of diagnostic procedures.

Specific binding of the target protein to the antibody results in a noticeable change in the electrical double layer parameters, which can be detected using EIS. Comb electrodes, due to their shape, provide high sensitivity and enable the generation of a stable and repeatable signal even at low concentrations of the target protein. Modifying electrodes with reduced graphene oxide significantly transforms the interface properties, directly affecting the mechanism for detecting antibody binding to the target protein. The high specific surface area of rGO facilitates dense immobilization of biomolecules, increasing detection sensitivity, while the structural homogeneity of the material minimizes noise, which is critical for the accurate measurement of electrochemical signals. In immunosensors, antibodies immobilized on the rGO surface ensure selective

---

The reported study was funded by Russian Science Foundation, project number 25-79-10055.

979-8-3315-4784-4/26 $31.00 © 2026 IEEE

antigen binding, while the high conductivity of the material enhances the impedance response, enabling accurate determination of biomarker concentrations [7]. Key changes are related to the electrochemical and structural properties of reduced graphene oxide. First, high electron conductivity and significant surface area reduce the baseline charge transfer resistance and increase the number of immobilized antibodies. This enhances sensor sensitivity, as even minor changes in electrical properties upon antigen binding become more pronounced. In addition, reduced graphene oxide modifies the structure of the electrical double layer due to the interaction of functional groups (carboxyl, hydroxyl) with electrolyte ions, which changes the capacitance of the electrical double layer [8]. As a result, the formation of the antibody-antigen complex causes more noticeable shifts in the capacitive component of the complex impedance [9]. Functional groups on the surface of reduced graphene oxide also improve the stability and immobilization efficiency of antibodies, ensuring their strong fixation and reducing non-specific adsorption. This increases the selectivity of the sensor, since the antibodies retain activity for binding the target antigen.

Although the basic operating principle of impedance sensors remains unchanged (recording changes in charge transfer resistance and bilayer capacitance during antibody/antigen interaction) modifying reduced graphene oxide shifts the basic interface parameters. Thus, a reduction in the initial impedance allows for the detection of more significant relative changes during binding, while increased sensitivity to variations in biolayer thickness and permittivity improves resolution even at low protein concentrations [10]. Therefore, the aim of this study is to develop and fabricate a prototype of an effective impedimetric immunosensor based on reduced graphene oxide for the detection of FGF21.

II. PREPARATION OF TARGET PROTEIN AND SPECIFIC ANTIBODIES

Preparation of the FGF21 protein and specific antibodies for subsequent analysis involves a series of steps aimed at obtaining highly pure recombinant protein, ensuring its correct refolding, and producing immunoreagents that ensure reproducible and selective detection of the target analyte. In the first step, FGF21 was expressed in the E. coli BL21 bacterial system by growing an overnight culture of the producer and then inducing it with isopropyl-β-D-1-thiogalactopyranoside at a reduced temperature, which stimulated the formation of recombinant rhFGF21 in the cells. Following induction, the culture was centrifuged and lysed in the presence of lysozyme and a protease inhibitor, supplemented by ultrasonic disintegration, which disrupted the cell walls and released inclusion bodies containing denatured protein. The FGF21-enriched precipitate was dissolved in a guanidine-containing buffer, which effectively solubilized protein aggregates, followed by metal affinity chromatography on a Ni-NTA sorbent. The presence of a His6 tag in the protein structure ensured selective binding of FGF21 molecules to nickel-containing Sepharose, while elution with high concentrations of imidazole allowed for the isolation of fractions containing predominantly the target protein.

The resulting FGF21 solution was dialyzed through a membrane with a 10 kDa cutoff, which removed low-molecular-weight impurities, salts, guanidine buffer, and residual imidazole. Dialysis was performed in a mixture of PBS with glycerol and L-arginine, which stabilized the protein and facilitated the correct restoration of its tertiary structure, including the formation of disulfide bonds. Refolding under mild buffer exchange conditions was critical, as the biological activity of FGF21 depends on its correct spatial configuration. Protein purity at each stage was assessed using SDS-PAAG (12%) denaturing polyacrylamide gel electrophoresis, which allowed visualization of protein fractions, monitoring for impurities, and comparing the sizes of protein components (Fig. 1). Following electrophoresis, the gel was stained with Coomassie, and Western blotting with protein transfer to a nitrocellulose membrane was used to confirm signal specificity and the presence of rhFGF21 in the purified sample.

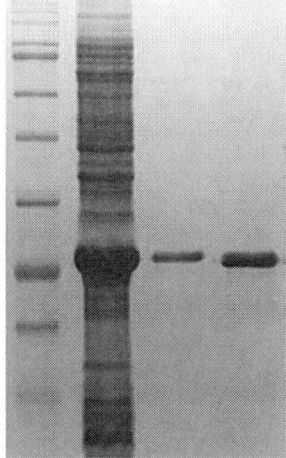

Fig. 1. The result of gel electrophoresis obtained after purification on Ni-NTA columns.

An integral part of the biosensor's development was the production and preparation of antibodies required for the immunodetection of FGF21. Rabbit polyclonal antibodies against rhFGF21 were generated by a standard cycle of immunization of animals with purified protein, followed by purification of the immunoglobulins on protein-A-Sepharose columns, which removed nonspecific serum proteins and produced a highly specific immunoreagent. Before use, the antibodies were blocked for nonspecific interactions on the membrane in 1×NET gelatin, minimizing background signals during immunoblotting. The membranes were then incubated with primary antibodies at the optimal working concentration, which ensured selective binding of FGF21 to the membrane surface. This was followed by a series of washes and a secondary incubation with antibodies conjugated to horseradish peroxidase. Final detection was performed using a chemiluminescent substrate, allowing visualization of even low protein concentrations.

III. XPS ANALYSIS OF THE STAGES OF FUNCTIONALIZATION OF THE BIOSENSOR STRUCTURE

Fabricating a reduced graphene oxide (rGO) layer for use in impedimetric biosensors requires a sequential step of dispersing the starting graphene oxide, selecting the optimal reducing agent and chemical reduction conditions, and applying the material to a substrate with controlled coating thickness and uniformity. Three approaches to GO reduction were used in this study: reduction in a water bath with hydrazine, chemical reduction in ethanol using NaBH4, and an improved reduction method in isopropyl alcohol, which ensures the best coating uniformity, adhesion, and stability. Regardless of the chosen method, the first step was preparing a GO dispersion, including ultrasonic treatment to break up aggregates and form a stable suspension of graphene sheets. Chemical reduction was then performed to remove oxygen-containing groups and partially restore the π-system of graphene, which increases conductivity and improves the electronically active properties of the finished material. A key criterion for the suitability of an rGO layer for subsequent application in biosensors is its surface morphology, which determines the active interface area and the potential for biomolecule immobilization. The most informative results were obtained by examining the samples using atomic force microscopy (AFM) and scanning electron microscopy (SEM). AFM analysis of the coatings obtained after GO reduction in a water bath demonstrated a pronounced surface relief with an irregularity amplitude of approximately 650–800 nm and a length of individual sheet fragments of 700–800 nm. SEM images obtained for samples reduced in a water bath (Fig. 2) reveal folded, non-uniformly oriented graphene fragments interspersed with nanopores 30–40 nm in diameter. These pores are formed during the reduction process by the removal of functional groups and partial destruction of the original oxide structure. At magnifications of ×10,000 and ×40,000, brighter areas associated with localized compaction and agglomeration of layers are visible, further confirming the presence of a folded structure. Despite the significant active surface area, such coatings proved less durable and required optimization of the formation method.

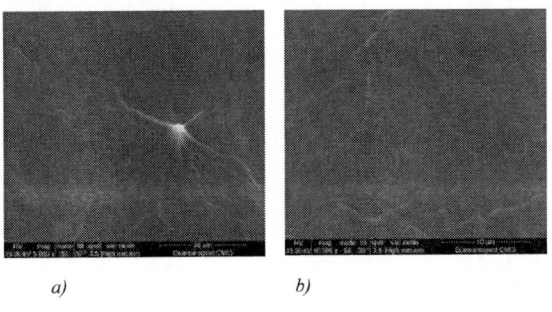

a)   *b)*

Fig. 2. SEM image of rGO obtained by the water bath reduction method: a – 10,000x zoom, b – 40,000x zoom

The chemical reduction of rGO in isopropyl alcohol proved to be the most promising method for implementing an impedimetric immunosensor. This method ensured the formation of a dense, uniform coating with high adhesion to the substrate. Due to the lower polarity and longer evaporation time of isopropanol, the graphene sheets are evenly distributed over the surface, reducing the likelihood of large aggregate formation. SEM images of these samples (Fig. 3) demonstrate a developed layered surface consisting of overlapping rGO sheets several atomic layers thick. High magnification reveals numerous nanoscale cracks and fine pores forming a complex network of channels. The presence of such a developed structure is an important advantage, as it provides improved electrical conductivity, increases the contact area with the solution of analyzed molecules, and facilitates subsequent surface functionalization with biorecognition elements.

To effectively immobilize antibodies on the electrode surface, the resulting rGO layer was silanized with 3-aminopropyl (trimethoxysilane) APTMS, creating a hydrophilic and chemically active surface. This allowed protein A to be immobilized on the surface, through which the antibodies were ligated.

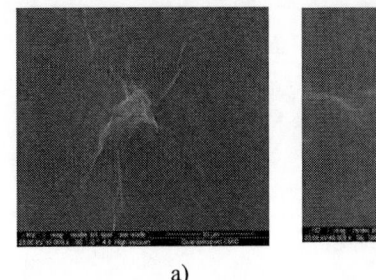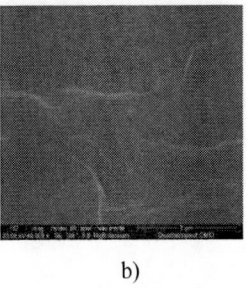

a)   b)

Fig. 3. SEM image of rGO obtained by reduction in isopropanol: a – 10,000x zoom, b – 40,000x zoom

Fourier transform infrared spectroscopy confirmed the effectiveness of the chemical modification, demonstrating the formation of a strong silane network, a decrease in the intensity of the O-H bands after annealing, and the appearance of distinct bands associated with the amino groups of APTMS. This modification ensures high adhesion and accessibility of reactive groups on the surface, which is a key requirement for creating a stable and sensitive biosensor.

IV. SENSOR PERFORMANCE TESTING

The procedure for measuring impedance characteristics as a function of the FGF21 target protein concentration was based on recording the response of a modified electrode surface containing immobilized antibodies to the sequential addition of protein solutions over a wide range of concentrations. Each measurement was conducted under strict control of the solution composition, frequency range, and stability of the electrode-solution interface, which allowed for the generation of reproducible hodographs and the assessment of the contribution of individual electrochemical processes to the overall impedance spectrum.

Before measurements, the accuracy of antibody immobilization was assessed, as the presence of serum protein impurities and nonspecific protein fragments can distort the impedance response. The use of antibodies isolated directly from serum resulted in significant variations in the low-frequency region and a lack of consistent concentration dependence. Impedance plots for FGF21 concentrations from 100 pg/mL to 10 ng/mL showed only slight changes in the low-frequency region, while the curves practically overlapped in the high-frequency region. This result was interpreted as a consequence of the presence of competing proteins occupying unoccupied surface areas and reducing the likelihood of forming specific antigen-antibody complexes.

Purification of antibodies on protein-A-Sepharose yielded pure IgG with high specificity, which was confirmed by SDS-PAGE. A subsequent series of measurements using purified IgG revealed a pronounced concentration dependence in the range from 1 pg/mL to 100 ng/mL. A set of impedance plots for various FGF21 concentrations (Figure 4) demonstrates an increasing diameter of the low-frequency semicircle with increasing analyte concentration, indicating an increase in charge transfer resistance. The presence of such a tendency corresponds to the classical mechanism of formation of a bioelectrochemical barrier on the electrode surface when an antibody binds to an antigen.

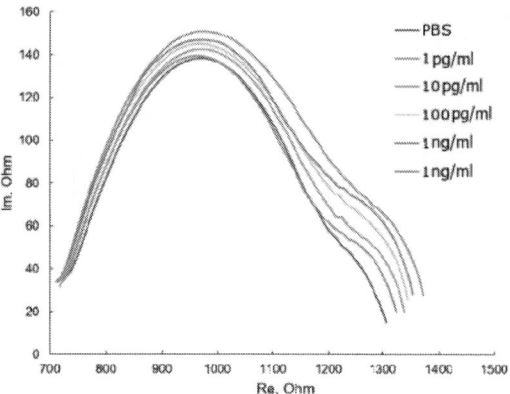

Fig. 4. Impedance plot for different concentrations of the target protein FGF21.

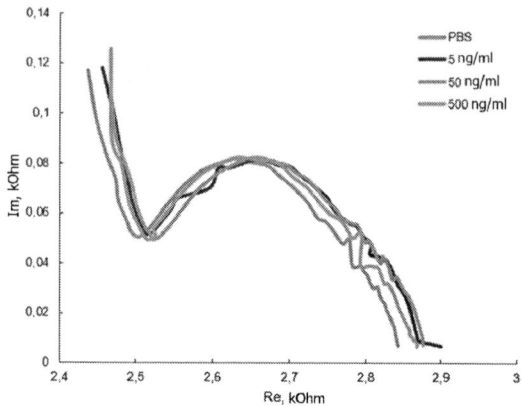

Fig. 6. Impedance hodograph of nonspecific chemotrypsinogen protein.

For quantitative analysis of the spectra, we used equivalent circuit simulations based on the extended Randles model. This simulation included the solution resistance, parameters describing the conductivity and capacitance of the reduced graphene oxide layer, and key elements of the biorecognition interface: charge transfer resistance and electrical double layer capacitance. Analysis revealed that R_3, which corresponds to the charge transfer resistance on the antibody layer, is the most informative parameter, increasing linearly with FGF21 concentration. Based on the simulation data, a dependence of charge transfer resistance on FGF21 concentration was constructed (Fig. 5). This calibration curve is linear over the studied concentration range and demonstrates the possibility of highly accurate quantitative determination of the analyte. This linearity results from gradual saturation of the surface with antibodies, reflecting a consistent increase in the density of the formed complexes and an increase in the barrier to electron transfer between the electrode and the solution.

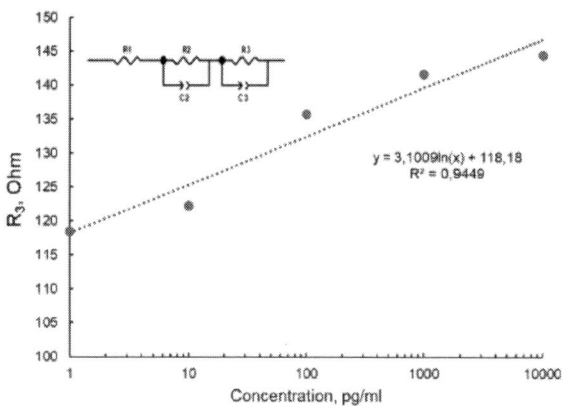

Fig. 5. Concentration dependence of the parameter R3.

To test the selectivity of the sensor platform, measurements were performed with the model nonspecific protein chymotrypsinogen. The resulting hodographs (Fig. 6) showed no signal dependence on concentration, indicating the absence of significant nonspecific binding and the correct operation of the blocking layer (BSA). This result confirms that the recorded impedance changes using purified antibodies are due to antigen-antibody interactions and not nonspecific processes.

V. CONCLUSIONS

The study demonstrated that the developed impedimetric immunosensor based on reduced graphene oxide is capable of detecting the FGF21 protein.

Analysis of impedance changes revealed that the system's response depends on the FGF21 concentration. The primary detection mechanism is an increase in charge transfer resistance across the biorecognition layer upon binding of FGF21 to immobilized antibodies. This change is most pronounced in the low-frequency region of the spectrum, as evidenced by a shift in the phase maximum and an increase in the impedance modulus. The high-frequency component of the signal associated with rGO remained stable, confirming the reliability of the underlying electrochemical platform.

The Randles model, which includes elements corresponding to electrolyte conductivity, rGO layer properties, and biorecognition interface parameters, accurately describes the system's behavior. Experimental data were successfully approximated using this model, confirming the correct interpretation of the electrochemical processes. Control experiments with a nonspecific protein (chymotrypsinogen) were performed to examine the sensor's selectivity. This demonstrated that nonspecific BSA sites are blocked and the antibodies are correctly immobilized. Thus, the developed method enables the creation of a stable, selective, and reproducible immunosensor. This method can be used not only for detecting FGF21 but also for analyzing other protein biomarkers.

ACKNOWLEDGMENT

The study was supported by a grant from the Russian Science Foundation No. 25-79-10055.

REFERENCES

[1] S. Bahijri *et al.*, "Fibroblast Growth Factor 21: A More Effective Biomarker Than Free Fatty Acids and Other Insulin Sensitivity Measures for Predicting Non-alcoholic Fatty Liver Disease in Saudi Arabian Type 2 Diabetes Patients," *Cureus*, Dec. 2023

[2] A. Kharitonenkov *et al.*, "FGF-21 as a novel metabolic regulator," *Journal of Clinical Investigation*, vol. 115, no. 6, pp. 1627–1635, Jun. 2005.

[3] I. Lakhani *et al.*, "Fibroblast growth factor 21 in cardio-metabolic disorders: a systematic review and meta-analysis," *Metabolism*, vol. 83, pp. 11–17, Jun. 2018

[4] Y. Guo, L. Li, and S. Hu, "Fibroblast growth factor 21 may be a strong biomarker for renal outcomes: a meta-analysis," *Renal Failure*, vol. 45, no. 1, Apr. 2023

[5] N. Sitkov, A. Ryabko, V. Moshnikov, A. Aleshin, D. Kaplun, and T. Zimina, "Hybrid Impedimetric Biosensors for Express Protein Markers Detection," *Micromachines*, vol. 15, no. 2, pp. 181–181, Jan. 2024

[6] N. O. Sitkov *et al.*, "Impedimetric Biosensor Coated with Zinc Oxide Nanorods Synthesized by a Modification of the Hydrothermal Method for Antibody Detection," *Chemosensors*, vol. 11, no. 1, pp. 66–66, Jan. 2023

[7] S. Mahari and S. Gandhi, "Electrochemical immunosensor for detection of Avian Salmonellosis based on electroactive reduced graphene oxide (rGO) modified electrode," Bioelectrochemistry, p. 108036, Dec. 2021

[8] S. Taniselass, M. K. M. Arshad, S. C. B. Gopinath, M. F. M. Fathil, C. Ibau, and P. Anbu, "Impedimetric cardiac biomarker determination in serum mediated by epoxy and hydroxyl of reduced graphene oxide on gold array microelectrodes," *Mikrochimica acta*, vol. 188, no. 8, p. 257, Autumn 2021

[9] F. Ambrosetti, B. Jiménez-García, J. Roel-Touris, and A. M. J. J. Bonvin, "Modeling Antibody-Antigen Complexes by Information-Driven Docking," Structure, vol. 28, no. 1, pp. 119-129.e2, Jan. 2020

[10] Y.-E. Shin *et al.*, "An ice-templated, pH-tunable self-assembly route to hierarchically porous graphene nanoscroll networks," *Nanoscale*, vol. 6, no. 16, pp. 9734–9741, 2014

Fabrication of Non-Lithographic Metal Masks for Metal-Assisted Chemical Etching of Silicon

Ekaterina Glushets

Peter the Great St. Petersburg Polytechnic University (SPbPU)
St. Petersburg, Russia
glushets.eyu@edu.spbstu.ru

Abstract — **A strategy for the fabrication of non-lithographic metal masks for metal-assisted chemical etching of silicon based on thermally induced dewetting of continuous nickel oxide films deposited by atomic layer deposition is proposed. Annealing induces dispersion with simultaneous reduction of nickel oxide to nickel and the formation of nickel islands. To enhance the catalytic activity, a selective silver metallization method is applied, which provides preferential growth of silver on the nickel islands and enables control over the amount of deposited catalyst. The resulting masks localize the etching process and make it possible to obtain porous silicon layers with a thickness of about 0.3 μm and tunable porosity; the approach is reproducible, compatible with large-area substrates, and promising for industrial upscaling.**

Keywords — *porous silicon; metal-assisted chemical etching; atomic layer deposition; thermally induced dewetting; selective silver metallization method; nickel.*

I. INTRODUCTION

Porous silicon formed as a layer on silicon wafers is an extremely promising material for microelectronics and related high-technology industries that employ planar systems to create microdevices [1]. Such applications include energy generation and storage devices (for example, solar and fuel cells, miniature high-capacity electrical accumulators, efficient hydrogen generators), photonics, vacuum electronics, and various types of sensors based on different principles (for example, detectors of various types of radiation or of the chemical composition of liquid media and the atmosphere), among many others [2]. The key obstacle to its industrial use remains the complexity and high cost of forming porous structures with varying degrees of regularity over large areas, which hinders industrial scaling of current technologies. In particular, the use of vacuum techniques and nanolithography makes material production economically inefficient for many applications.

The aim of this work is to develop a cost-effective and scalable lithography-free method for the formation of porous silicon. To achieve this goal, we propose an approach based on metal-assisted chemical etching (MACE) using masks obtained by chemical metal deposition from solution onto a selectively activated substrate surface

II. STATE OF THE ART IN THE FIELD

Modern methods for the formation of porous silicon include reactive ion etching, MACE, electrochemical etching, and laser ablation. Reactive ion etching, while providing high anisotropy and a low reflection coefficient [3], requires expensive equipment and complex lithographic masks. Electrochemical etching, despite the flexibility in controlling pore morphology, is unsuitable for creating thick graded layers in thin-film solar cells, since the porous layer itself begins to absorb a significant fraction of the light [4].

Laser ablation eliminates the need for photolithography but is characterized by difficulties in ensuring uniformity and by defect formation [5]. Against this background, the MACE method stands out for its simplicity, low cost, and ability to form highly uniform nanoporous structures [6]. In this method, etching of silicon in a solution of hydrofluoric acid (HF) and hydrogen peroxide (H_2O_2) is catalyzed by metal particles.

In MACE, the most common catalysts are gold (Au), silver (Ag), platinum (Pt), iron (Fe), and nickel (Ni). When noble metals (Ag, Au, Pt) are used, the metal particles remain on the silicon surface after treatment in the etching solution. The key advantage of nickel as a catalytic mask is its complete dissolution without redeposition onto silicon, which ensures a clean surface [7]; in addition, Ni reduces the process cost, which is important for scalable production. For silver, the etching process is self-limiting as the catalyst is consumed and silver hexafluorosilicate (Ag_2SiF_6) is formed, the process stops, which makes it possible to control the etching depth [8]; residual Ag is removed by nitric acid (HNO_3).

The conventional lithographic method for fabricating metal masks, although it offers advantages such as high pattern accuracy and reproducibility, has significant drawbacks: high cost, the need for complex equipment, the use of toxic materials, and limitations on the size of the processed wafers. The latter is especially critical for applications of porous silicon as a solar cell material, where large-area wafers are required: here the key factor is the processing cost, and the larger the area that can be covered in a single technological step, the more cost-effective the production. An alternative solution is the method of thermally induced dewetting, which consists in dispersing a film at a temperature significantly below its melting point, while it remains in the solid state. This technique makes it possible to create nanoscale-thickness masks without lithography [9] but requires a highly uniform metal film. For such thicknesses, only two methods can provide the required uniformity: electron-beam evaporation, which, although it ensures high coating quality, is characterized by a high cost and limitations on the treated area, and the more promising method of atomic layer deposition (ALD). The key advantages of ALD include the absence of limitations on substrate area and high thickness uniformity over the entire surface [10]. Although ALD of metals is still insufficiently developed, the technology makes it possible to deposit various metals via an intermediate oxide stage with subsequent reduction to the metallic state. Nickel is of particular interest as the most cost-effective material, which can also be efficiently transformed into an island-like metallic form through reduction from the oxide state.

III. PROBLEM STATEMENT AND METHOD OF ITS SOLUTION

The aim of this work is to develop a technology for the formation of porous silicon by metal-assisted chemical etching using metal masks fabricated by a combination of ALD, thermally induced dewetting, and chemical deposition from solution.

To achieve this goal, a concept was formulated that includes activation of the silicon substrate surface by depositing a continuous metal-containing film, thermochemical treatment to transform the continuous film into an island structure, control of island parameters in order to obtain a mask with specified properties, and implementation of MACE using the resulting mask. In this context, two key tasks are addressed. The first task is the fabrication of metal masks, which requires the creation of an activated surface—in this case, a nickel oxide (NiO) film—its subsequent thermochemical treatment to form nickel islands, and catalytic growth of these islands by chemical deposition methods. The second task involves using the obtained masks to carry out MACE and to form a nanoporous silicon structure.

IV. EXPERIMENTAL SECTION

The experimental work was carried out in accordance with the described methodology and included three consecutive stages: formation of the catalytic mask, its modification, and etching. To monitor the morphology and composition of the films at all stages, a combination of analytical methods was used: scanning electron microscopy was employed for direct surface visualization and structural analysis, and optical microscopy was used for rapid quality control of the samples throughout the process.

A. Formation of the catalytic mask

As initial substrates, p-type silicon wafers with a nickel oxide (NiO) film 1–4 nm thick, preliminarily deposited by ALD, were used. NiO films were deposited in a Picosun R-150 reactor at a working pressure of 1 kPa. The reaction chamber is equipped with two gas lines: the first supplies nickelocene, and the second supplies ozone. Nitrogen, used as a purge gas, is fed through the same lines. The nickelocene pulse time into the chamber was 3 s, the ozone pulse time was 6 s, and the purge time was 20 s. Nickel oxide deposition was carried out at a nickelocene evaporator temperature of 110 °C, a reactor temperature of 250 °C, and an ozone flow rate of 200 mL/min. The number of deposition cycles was 250.

Transformation of the continuous NiO film into an island-like metallic nickel structure by thermally induced dewetting was performed in a specialized laboratory setup, whose schematic is shown in Fig. 1. Prior to loading into the reactor, the samples were cleaned by boiling in acetone for 3 min. The process was carried out at a base pressure of 300 Pa with argon (50 mL/min) and ammonia (10 mL/min) flow. Dispersion of the 4-nm-thick NiO film was performed at a substrate temperature of 450 °C. The dwell time at the treatment temperature was 20 min.

Chemical silvering was carried out from a stock solution containing 5 g/L silver nitrate ($AgNO_3$) and 5 g/L potassium hydroxide (KOH), and a reducing solution containing 2.5 g/L inverted sugar. Before processing, the solutions were mixed in a 2:1 ratio, taking into account a 100-fold dilution of the reducer to slow down the process rate.

B. Metal-assisted chemical etching

Before etching, the samples were cleaned by sequential treatment in acetone and distilled water to remove organic contaminants. Etching was carried out for 30 min at room temperature in a solution containing 6.6 M hydrofluoric acid (HF) and 0.10 M hydrogen peroxide (H_2O_2) [11]. The high HF/H_2O_2 concentration ratio was required to ensure selective pore formation strictly beneath the catalytic islands. After completion of the process, the samples were thoroughly rinsed with a large amount of distilled water to completely remove residual etchant and reaction products.

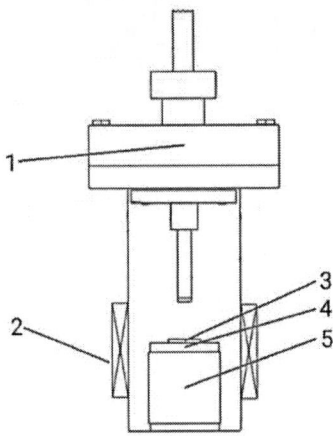

Fig. 1. Simplified schematic of the reactor unit: 1 – top flange with gas inlet; 2 – external furnace; 3 – sample; 4 – pedestal; 5 – copper shield

V. RESULTS

A. Formation of the island structure

The efficiency of the thermally induced dewetting process for the nickel oxide film was confirmed by SEM. As can be seen in Fig. 2, a clearly pronounced island structure was formed on the initially continuous film after thermal treatment. The appearance of such contrast indicates successful breakup of the film into individual nanoscale clusters, which is a necessary condition for its subsequent use as a mask.

B. Analysis of the kinetics of silver mask deposition

To control the deposition of the silver mask by chemical reduction, the concentration of the reducing agent (inverted sugar) was decreased by a factor of 100. This made it possible to slow down the process kinetics and to study in detail the film growth as a function of immersion time in the solution. Figure 3 shows the evolution of the morphology of the silver coating on silicon substrates with a preliminarily deposited nickel oxide film (2 nm thick). Optical microscopy revealed that, as the deposition time increases from 2 to 4 min, silver clusters grow and coalesce.

The key influence of the pre-formed nickel island structure on the silvering process is demonstrated by the SEM images in Fig. 4. At the same deposition time (4 min), a more uniform and fine-dispersed silver film is formed on the substrate with island-like nickel oxide (Fig. 4b) compared to deposition on bare silicon (Fig. 4a). This indicates a change in the nucleation and growth mechanism of silver nuclei, which predominantly form on the active parts of the nickel islands. The nanostructured silver film obtained in this way was used as a mask for metal-assisted chemical etching.

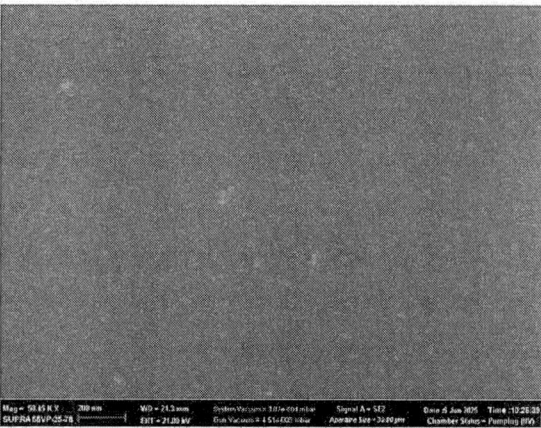

Fig. 2. SEM image of the sample after thermally induced dewetting

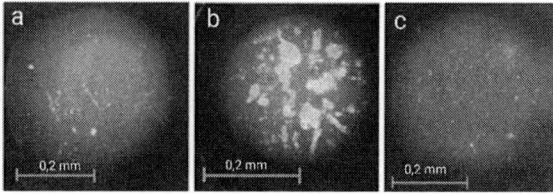

Fig. 3. Micrographs of the silver coating obtained by silvering with a diluted reducer for (a) 2 min, (b) 3 min, and (c) 4 min

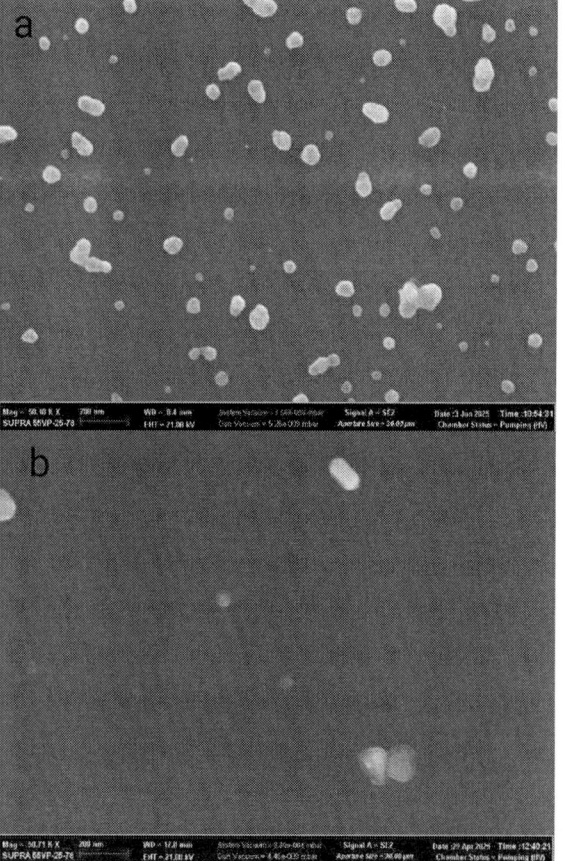

Fig. 4. SEM images of the silver coating obtained by silvering with a diluted reducer for 4 min on (a) bare silicon and (b) silicon with an island-like nickel oxide structure

C. Metal-assisted chemical etching

The result of 30-min MACE is shown in Fig. 5. In the cross-section, the emission characteristics differ from those of the substrate, which makes it easy to determine the thickness of the layer, while the surface clearly exhibits pores. In addition, defects caused by the pores are observed in the cross-section, and white nickel dots are visible at the pore bottoms. This is due to the fact that metals exhibit secondary-electron emission under the action of the electron beam and retain their structure.

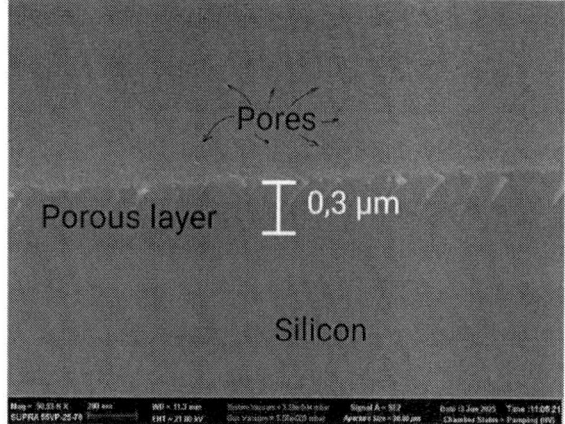

Fig. 5. SEM image of porous silicon obtained by MACE using island-like silver on nickel oxide

VI. CONCLUSION

The conducted studies have demonstrated the effectiveness of the proposed technology for the formation of porous silicon. This technology employs a combination of two metals: nickel, which serves as a "sketch" material for the pattern, and silver, subsequently deposited from solution onto the nickel-covered regions, which creates the actual nanoscale pattern. Silver exhibits high catalytic activity during the subsequent silicon etching. At the same time, the technology uses the expensive metal and its reagents very economically, owing to the fact that the target reaction proceeds only on the active part of the substrate surface. The process was successfully implemented in three stages. At the first stage, a well-defined nanoscale island structure was obtained from the initial continuous nickel oxide film by thermally induced dewetting, as confirmed by SEM analysis. This structure served as the basis for subsequent silver deposition.

The study of the kinetics of silver mask deposition showed that a reduction in the reducer concentration made it possible to control film growth. A key result is the demonstration of the influence of the nickel underlayer: on a surface with pre-formed nickel oxide islands, a more uniform and fine-dispersed silver film is formed compared to deposition on bare silicon. This indicates that nickel islands act as active nucleation centers, altering the mechanism of silver nucleus growth. The final stage—MACE using the obtained mask—resulted in the formation of a porous silicon layer 0.3 µm thick. The SEM image of the cross-section confirmed the presence of pores and residual nickel particles at their bottoms, which is a characteristic signature of MACE.

Thus, sequential implementation of thermally induced dewetting of the nickel oxide film, controlled silver deposition, and final etching makes it possible to obtain porous silicon structures. The successful outcome confirms the simplicity, controllability, and low cost of the proposed multistep technology.

REFERENCES

[1] G Otto M., Algasinger M., Branz H., Gesemann B. Black Silicon Photovoltaics // Advanced Optical Materials. — 2015. — Vol. 3 — №2. — P. 147.

[2] Föll H., Christophersen M., Carstensen J., Hasse G. Formation and application of porous silicon // Materials Science and Engineering: R. — 2002. — Vol. 39. — P. 93–141.

[3] Dussart R., Mellhaoui X., Tillocher T., Lefaucheux P., et al. Silicon columnar microstructures induced by an SF_6/O_2 plasma // Journal of Physics D: Applied Physics. — 2005. — Vol. 38—№ 18. — P. 3395–3402.

[4] Striemer C., Fauchet P. Dynamic etching of silicon for broadband antireflection applications // Applied Physics Letters. — 2002. — Vol. 81 — №16. — P. 2979–2982

[5] Her T, Wu C., Deliwala S., Mazur E. Microstructuring of silicon with femtosecond laser pulses // Applied Physics Letters. — 1998. — Vol. 73. — P. 1673–1675.

[6] Chen K., Pasanen T.P., Vähänissi V., Savin H. Effect of MACE parameters on electrical and optical properties of ALD passivated black silicon // IEEE Journal of Photovoltaics. — 2019. — Vol. 9 — № 4. — P. 2156–3381M. Young, The Technical Writer's Handbook. Mill Valley, CA: University Science, 1989.

[7] Yue, Z., Shen, H., Jiang, Y., Jin, J. Formation and mechanism of silicon nanostructures by Ni-assisted etching. // Journal of Materials Science: Materials in Electronics. — 2014. — № 25(3). — C. 1559–1563

[8] Volovlikova O.V., Dronov A.A., Gavrilov S.A., Sysa A.V. Method for forming silicon wires by metal-assisted etching using silver. RU Patent 2624839 C1. Patent holder: Federal State Autonomous Educational Institution of Higher Education "Siberian Federal University". No. 2016135859; filed 05.09.2016; published 05.07.2017, Bull. No. 19. — 8 p.

[9] Reiter G. Dewetting of Highly Elastic Thin Polymer Films // Physical Review Letters. — 2001. — Vol. 87, No. 18. — P. 186101-1 – 186101-4.

[10] George S. M., Ott A. W., Klaus J. W. Surface Chemistry for Atomic Layer Growth // The Journal of Physical Chemistry. — 1996. — Vol. 100 — №3. — C. 13121–13131.

[11] Matsumoto A. [et al.] Metal-Assisted Etching of n-Type and p-Type Silicon Using Patterned Platinum Films: Spatial Distribution of Mesoporous Layer 59 and Open Circuit Potential of Silicon // Journal of The Electrochemical Society. — 2023. — Vol. 170 —№ 5. — P. 052505-1 – 052505-9.

Direct Laser Writing As An Alternative To Existing Methods Of Silicon Microstructuring

Olga S. Gribovskaya
Photonics department
Saint-Petersburg Electrotechnical University "LETI"
Saint Petersburg, Russia
0009-0009-7765-2609

Abstract— **This article discusses various aspects of Direct Laser Writing (DLW) technology as a potential alternative to traditional silicon microstructuring techniques. Experimental results for silicon nitride are also presented and discussed.**

Keywords—**laser, silicon, microstructuring, direct laser writing**

I. INTRODUCTION

In recent years, the electronic industry has been one of the largest sectors of the economy. Increasing the consumer market requires increasing production capacity and speeds. At the moment, the most widely used semiconductor material in electronics is silicon. Its processing technologies, including lithography, chemical and mechanical processing, are expensive and inaccessible to small-scale industries. This article suggests using direct laser writing technology, in particular laser ablation.

II. OVERWIEV OF EXISTING SILICON MICROSTRUCTURING TECHNOLOGIES

A. Epitaxy

Epitaxy is a group of methods that make it possible to build up materials on the surface of a substrate to create various structures. In general, all these methods can be divided into two groups – chemical and physical epitaxy. In the first case, precipitation occurs due to a chemical reaction, for example, in the case of gas epitaxy between the precursor gas and the substrate material [1]. Such methods are slow and expensive due to the complexity of the design, and can also pose a risk to the environment and personnel due to the use of various reagents. In the case of physical epitaxy, precipitation occurs by spraying a solid and using various spraying methods [2]. The disadvantage of all types of epitaxies as a method of creating microstructures is the impossibility of creating complex geometries, as well as the difficulty of controlling the parameters of the grown layer.

B. Etching

For the manufacture of complex structures, methods related to the removal of material are more suitable. Such methods are easier to control in the process. To remove materials, you can use etching – "dry" or liquid [3]. Despite the availability, the disadvantage of such methods is the difficulty in obtaining clear lines in the geometry of the microstructure, which can be critical for small sizes.

C. Litography

The leader in the production of microstructures are various methods of lithography [4]. These technologies are multi-stage, where the sequential execution of various steps leads to the desired structure. Figure 1 shows the various lithography methods and their brief characteristics.

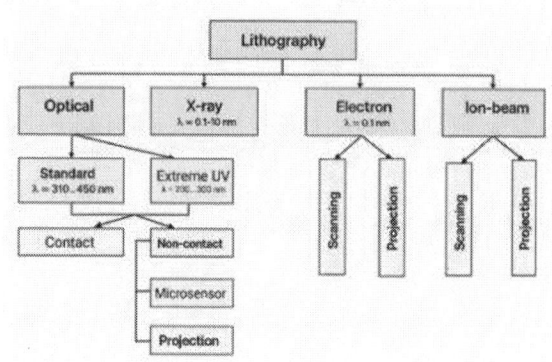

Fig. 1. Classification of litography processes

A variety of system designs make it possible to adapt to the demands of different fields of micro- and nanoelectronics, as well as to produce topologies of varying complexity, which is why this group of methods is a leader in global electronics. The disadvantages of lithography are the low availability of equipment, the complexity of preparing templates, and the need for large production areas.

III. THE NEED FOR SILICON MICROSTRUCTURE

The creation of volumetric structures of micro-dimensions is one of the fundamental tasks in modern electronics. For example, the initial polishing of silicon substrates can be performed using microstructuring technologies. Through the action of pulsed laser radiation, short-term heating of the material occurs, as a result of which micro-dimensional defects are eliminated without the disadvantages of methods with prolonged thermal exposure [5]. The need for silicon polishing is to minimize the number of defects on the surface, which in turn makes it possible to create integrated circuits with smaller element sizes and increases the mechanical strength of products [6, 7].

Microstructuring of semiconductor materials can also be used as a solution to the problem of energy losses in solar power plants. Thus, with the help of special volumetric structures, it is possible to significantly reduce the reflection of light from the surface of solar panels, which improves the ratio of the received solar energy to the generated electric energy due to an increase in the spectral range of absorption of matter [8]. Figure 2 shows the calculated and experimental spectral characteristics of a conventional solar cell structure and a modified one modeled after the wings of a butterfly P. aristolochiae, from which it can be concluded that the modified surface has a greater absorption of solar radiation energy.

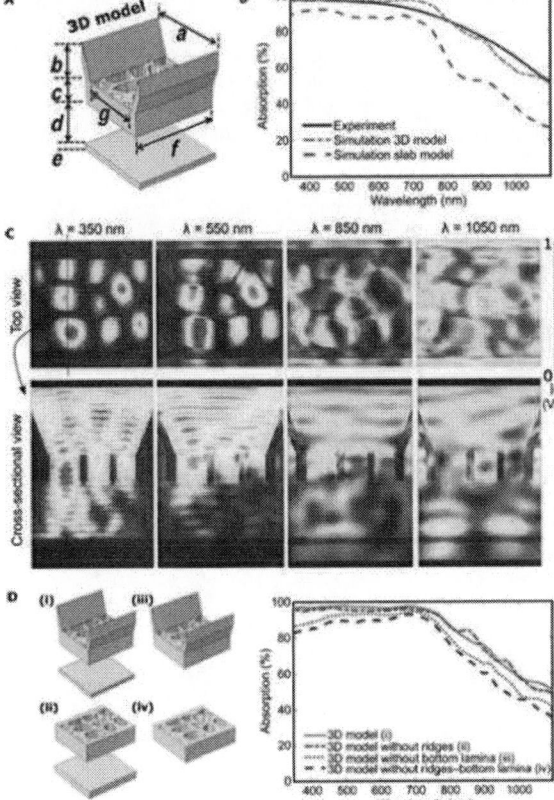

Fig. 2. 3D optical modeling of micro- and nanostructures of the P. aristolochiae butterfly by the FEM method. (A) A 3D model of a butterfly at scale and the corresponding parameters extracted from the images by the SEM method. The dimensions of this model are as follows: a = 2.0 microns, b = 1.2 microns, c = 0.8 microns, d = 1.2 microns, e = 0.22 microns, f = 5.0 microns, and g = 1.5 microns. (B) Comparison of experimental and simulated absorption spectra at normal incidence. The experimental data were measured using an integrating sphere in a matte black area (solid line). The dotted line corresponds to a 3D simulation model with a 59% air fill rate. For comparison, a slab model has been created without a template, with the compression of the original 3D model into a slab with the same bottom area and volume (long dotted line). (C) The normalized electric field strength distribution calculated for normal incidence and wavelengths of 350, 550, 850 and 1050 nm. (D) The influence of various structural components on the large-scale architecture of P. aristolochiae [8]

Microstructures can be used to create microprocessors, which are still used in harsh environments, as well as in military and industrial equipment. An important area of modern microelectronics is the manufacture of microelectromechanical systems (MEMS). Their applications include the creation of various medical devices and sensors, including implantable ones, the design of sensors for the oil and gas industry, modern automation systems for production processes (sensors for unmanned vehicles and robotic systems), as well as in the defense industry [9].

IV. DIRECT LASER WRITING

Direct laser writing (DLW) refers to a class of material processing methods that do not utilize masks and are based on the interaction between light and different materials. These methods find application in various electronic fields, such as solar energy, where they are used for microstructuring perovskite materials, manufacturing quantum dots, and creating LEDs and lasers. DLW can be divided into 6 subgroups [10]:

- Laser ablation

- Laser-induced crystallization
- Laser-induced ion migration
- Laser-induced phase segregation
- Laser-induced photoreaction
- Other laser-induced transitions.

The most common and user-friendly method is laser ablation. This involves the removal of melted material from a surface by using laser radiation. When exposed to laser light, different processes occur, such as evaporation, boiling, and localized explosions, which cause the substance to be ejected from the surface in different forms, including free molecules, atoms, and ions, depending on the power applied [11]. Furthermore, during laser ablation, plasma formation may occur, which can have a negative impact on subsequent processing steps [12].

A unique aspect of laser processing of silicon and other crystalline materials is that the ablation process in these materials cannot be fully explained by thermal mechanisms alone. In other words, while the transition to the liquid phase may not always occur depending on the laser irradiation parameters, simultaneous generation and accumulation of dislocations can occur. This can lead to the formation of microcracks and other defects. These processes, in turn, lead to the removal of material from the surface [13]. With their accumulation, they can cause more significant imperfections and even fractures. However, these effects are not always negative. With the appropriate laser parameters, they can be used to remove a specific amount of material that corresponds to a predetermined thickness. This material can then be used for layer-by-layer processing in 3D technologies. In addition, controlled extended effects can be utilized for cutting and marking operations on crystalline materials.

V. EXPERIMENTAL SETUP

The experiments were conducted using polished wafers made of silicon nitride Si_3N_4. A MicroSET system, using ytterbium pulsed fiber laser with a wavelength of 1.06 microns, an average power of 20 watts, and a beam diameter of 30 microns, was used as the experimental tool. The experiments were carried out under standard laboratory conditions. Before the experiment, the surface of the wafers was thoroughly cleaned using isopropyl alcohol and cotton swabs to remove any impurities.

VI. EFFECTS ON THE SURFACE

In order to investigate the effect of various laser irradiation conditions on the surface of silicon nitride, a series of experiments were conducted. These experiments included the production of a reference sample with a predefined roughness pattern. The roughness of each sample was measured using a Norgau Industrial NSRT-100 profilometer. Three measurements were taken for each sample, and an arithmetic average was calculated for each measurement. Figure 3 shows a control sample with a roughness of 0.05 microns.

979-8-3315-4784-4/26 $31.00 © 2026 IEEE

Fig. 3. Reference sample compared to an untreated surface

Due to the limitations in the accuracy of available measuring instruments, it is not possible to accurately measure the surface roughness of untreated material. Therefore, for the purposes of this study, a value of 0 microns has been assigned for this parameter.

The reference sample was prepared by creating diagonal lines on it at angles of 45 and 135 degrees, with a spacing between lines of 0.05 microns.

After laser treatment, all samples were cleaned using isopropyl alcohol, cotton swabs, and lint-free wipes.

Figure 4 presents photographs of the surface, acquired using different pulse durations of laser radiation.

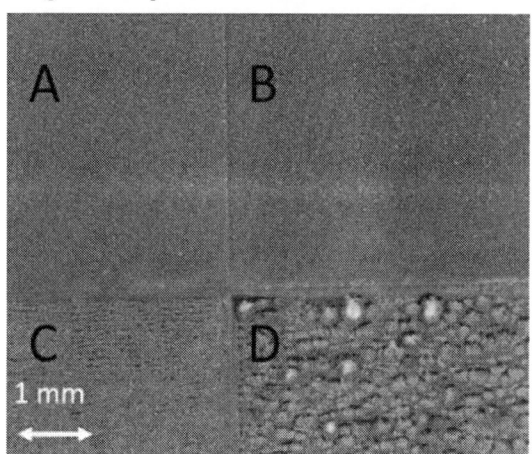

Fig. 4. Surface after treatment with different pulse duration, A) 2 ns, B) 4 ns, C) 8 ns, D) 14 ns

The images presented in this document illustrate the correlation between the pulse duration and the surface state. As the pulse duration increases, there is a gradual transformation of the surface, which becomes less homogeneous. The surface exhibits distinct changes in color, ranging from gray to light orange, yellow, and white.

These color variations can be attributed to the oxidation process of silicon when it interacts with atmospheric oxygen. This process results in the formation of silicon dioxide SiO_2, which can be seen as porous structures on the surface. These porous structures are formed through the deposition and recrystallization of molten silicon.

The roughness values for samples A, B, C, and D were 0.273, 0.840, 2.095, and 7.694 microns, respectively. These results are consistent with the findings from visual analysis.

Therefore, it can be concluded that when processing silicon, it would be beneficial to use processing modes with a lower pulse duration for laser radiation. Additionally, given the absence of silicon dioxide on the surface, these parameters can be utilized in developing an effective polishing and cleaning process for previously treated surfaces.

VII. LASER DRILLING

One of the crucial aspects in the microstructuring of silicon for electronic applications is the process of creating and drilling holes in the material. The use of laser radiation enables the creation of holes of various shapes and depths, as well as easy adjustment of the spacing between them.

Figures 6 and 7 show arrays of circular and square holes in silicon nitride wafers, each with a thickness of 100 microns.

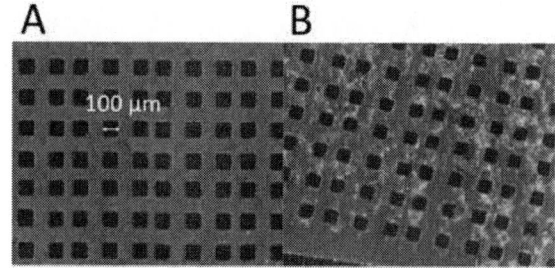

Fig. 5. Silicon wafer with square holes, A) front side, B) back side

The square holes have a size of 100 microns by 100 microns. There is visible rounding of the edges, which can be addressed by using a laser with a smaller beam diameter.

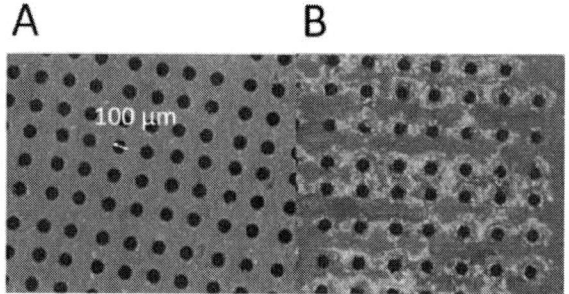

Fig. 6. Silicon wafer with round holes, A) front side, B) back side

The round holes have a diameter of 100 microns.

Figures 6 (B) and 7 (B) illustrate that silicon dioxide also accumulates on the underside of the treated plate, accompanied by the formation of cracks and other defects. These issues may be addressed through the development of a customized cleaning procedure. To thoroughly assess the viability of this approach, it is essential to conduct measurements of hole taper – the reduction in diameter as treatment progresses deeper.

VIII. 3D MICROSTRUCTURING

One of the most promising technologies in the field is 3D laser microstructuring, which allows for the fabrication of custom-made electronic components and microelectromechanical systems (MEMS) as well as modification of the properties of pre-existing systems.

This process involves the removal of material layer by layer from a surface following a trajectory corresponding to the layers of a 3D model of the object. It is crucial to carefully select the parameters during this process to prevent defects on the surface caused by overheating of the silicon material and to ensure that no defects occur within the depth of the material. On a figure 8 we can see different defects, including cotton-like debris on a surface, which consists of a molten silicon that crystalized on a surface of a sample.

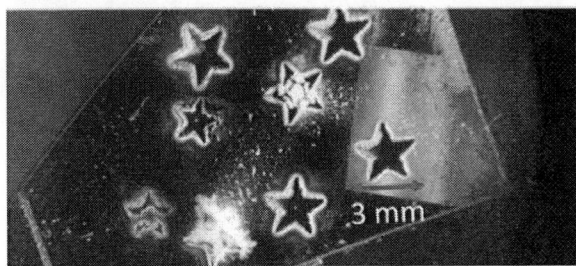

Fig. 7. Various defects of a silicon 3D microstructuring

To remove these defects, special "soft" cleaning modes can be used, with lower power and higher processing speed in order to avoid damage to the created structure.

Laser processing also makes it possible to cut out finished structures from a common array of material without using mechanical processing methods, which significantly reduces the manufacturing time of finished products. Figure 9 shows a sample of a 3D structure consisting of meanders, cut from a silicon wafer with a thickness of 100 microns.

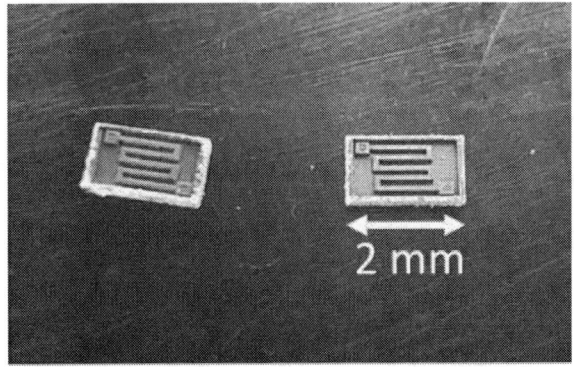

Fig. 8. Cut-out silicone structures

Therefore, it can be concluded that 3D laser microstructuring allows for the simultaneous substitution of two processes in the production of sophisticated electronic devices.

IX. CONCLUSIONS

The analysis of silicon microstructuring techniques has identified a need for a more cost-effective approach. The methods discussed here, which involve the use of laser technology, are part of a broader category of laser ablation techniques known as direct laser writing (DLW). The experimental results describe both existing processing techniques and technologies that can serve as a basis for the development of alternative methods for silicon surface treatment.

REFERENCES

[1] V. M. Anishchik, Ed., *Interaction of Radiation with Solids: Proc. 7th Int. Conf.*, Minsk, Belarus, Sep. 26–28, 2007. Minsk: BSU Publishing Center, 2007, pp. 53–55.

[2] M. N. Rumyantseva, E. A. Makeeva, and A. M. Gaskov, "Influence of microstructure of semiconductor sensor materials on oxygen chemisorption at their surface," *Russ. Chem. J.*, vol. 52, no. 2, pp. 122–129, 2008.

[3] V. Yu. Zheleznov, T. V. Malinsky, S. I. Mikolutskiy, V. E. Rogalin, S. A. Filin, Yu. V. Khomich, V. A. Yamshchikov, I. A. Kaplunov, and A. I. Ivanova, "Method for producing microstructures on semiconductor surface," U.S. Patent 2 756 777 C1, Mar. 10, 2021.

[4] R. H. Siddik, M. Gomez, E. Mendoza et al., "Bioinspired phase-separated disordered nanostructures for thin photovoltaic absorbers," *Sci. Adv.*, vol. 3, art. e1700232, 2017.

[5] V. Ya. Raspopov, *Micromechanical Devices.* Moscow: Mashinostroenie, 2007, 400 p.

[6] K. Petera, R. Kopeceka, P. Fatha, E. Buchera, and C. Zahedib, "Thin film silicon solar cells on upgraded metallurgical silicon substrates prepared by liquid phase epitaxy," *Sol. Energy Mater. Sol. Cells*, no. 74, pp. 219–223, 2002.

[7] V. A. Klyueva, "Review of silicon coating deposition methods," *Molodoi Uchenyi*, no. 10 (114), pp. 236–246, 2016.

[8] S. V. Ivanov and E. Yu. Karelin, "Fundamentals of silicon micromachining technology," in *Proc. Int. Symp. Reliability and Quality*, 2011, vol. 2, pp. 158–160.

[9] B. A. Lapshinov, *Lithographic Processes Technology: A Textbook.* Moscow: Moscow State Institute of Electronics and Mathematics, 2 Newton, 2011, 95 p.

[10] Y. Sheng, X. Wen, B. Jia, and Z. Gan, "Direct laser writing on halide perovskites: from mechanisms to applications," *Light: Adv. Manuf.*, vol. 4, p. 1, 2024, doi: 10.37188/lam.2024.004.

[11] N. A. Inogamov et al., "Laser ablation: physical concepts and applications (review)," *High Temp.*, vol. 58, no. 4, pp. 689–706, 2020.

[12] V. V. Pavlovich et al., "Laser ablation of monocrystalline silicon under pulsed-frequency fiber laser," *J. Sci. Tech. Inf. Technol. Mech. Opt.*, vol. 97, no. 3, pp. 426–434, 2015.

[13] S. I. Anisimov and B. S. Luk'yanchuk, "Selected problems in the theory of laser ablation," *Phys. Usp.*, vol. 45, no. 3, pp. 263–294, Mar. 2002.

Understanding and Predicting Signal Degradation in Printed Circuit Board Transmission Lines Using Neural Networks

Alexey V. Moskvin
Microprocessor System Group
JSC "NTC ELINS"
Zelenograd, Russia
moskvin@elins.ru

Ivan Yu. Gritskevich
Programmable Logic Integrated Circuit Group
JSC "NTC ELINS"
Zelenograd, Russia
gritskevich@elins.ru

Abstract — **Signal degradation in high-speed printed circuit boards (PCBs) is primarily caused by conductor and dielectric losses in transmission lines leading to amplitude reduction, dispersion, and intersymbol interference. This work presents an analytical and data-driven approach for modeling these effects in microstrip lines. Frequency-dependent RLGC parameters are derived to capture skin effect, dielectric loss, and group delay across 0,1-10 GHz. To accelerate analysis, a convolutional neural network (CNN) is trained on simulation data to predict insertion loss (IL) and amplitude-frequency characteristics based on geometric and material parameters. The hybrid analytical AI model achieves prediction accuracy within ±0,1 dB while reducing computation time by over two orders of magnitude compared to full-wave electromagnetic solvers. The results demonstrate the potential of neural networks for rapid signal-integrity evaluation and optimization of PCB transmission lines in high-speed digital systems up to 10 Gb/s.**

Keywords — insertion loss, skin effect, complex permittivity, propagation constant, eye diagram, intersymbol interference.

I. INTRODUCTION

The purpose of this article is to derive a mathematical model of a printed circuit board (PCB) transmission line of a given type and geometry based on analytical dependencies of its frequency-dependent linear parameters $R(\omega)$, $L(\omega)$, $G(\omega)$, $C(\omega)$ (hereinafter referred to as RLGC); to investigate its properties and to quantitatively explain the observed degradation of digital signals.

A significant portion of specialized publications [1, 3, 4] are presented in a mathematically dense style, which complicates the practical application of the results. The emphasis of the work is on a transparent physical interpretation of the phenomena, verified approximate formulas in a reproducible numerical path (using FFT/IFFT), with subsequent acceleration of the calculations due to the neural network model. In the course of the article, the nature of inter symbol interference (ISI) jitter [7] in matched long lines will be clarified).

II. HOMOGENEOUS MATCHED MICROSTRIP PCB TRANSMISSION LINE

A. PCB parameters

We consider a microstrip transmission line at the boundary of two media with one continuous reference plane specified by the geometric parameters shown in Fig. 1.

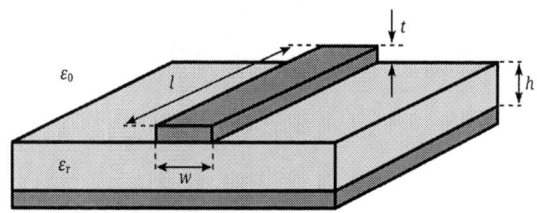

Fig. 1. Microstrip transmission line with a set of parameters: l – transmission line length, w – transmission line width, t – metallization layer thickness, h – distance to the reference plane, ε_r– relative permittivity of the dielectric, ε_0– permittivity of air.

In addition to the geometric parameters, we also have the data sheet for the printed circuit board dielectric — permittivity ε_{ref} and loss tangent $\tan_{\delta.ref}$ at a frequency of 1 GHz.

The following assumptions are made: the line is long, consistent, and geometrically constant in length; materials are homogeneous; reference plane is continuous; input signal is normalized to 1. The parameters used are as follows:

$\{l, w, h\} = \{50см\ 0.15мм, 18мкм, 0.127мм\}$,

$\varepsilon_{ref} = 4.2$, $\tan_{\delta.ref} = 0.0195$.

B. Finding a solution

The complex transmission coefficient characterizing the frequency properties of the line is defined in [2] and has the following form:

$H(\omega) \overset{\mathrm{def}}{=} e^{-l_{trace}\gamma(\omega)}$, [relative units of measurement]
where $l_{trace} = l$ – line length,
[m]; $\left[\dfrac{\text{relative units of measurement}}{\text{m}}\right]$;

$\gamma(\omega)$ – propagation constant,

ω – circular $\left[\dfrac{rad}{s}\right]$; frequency,

$\gamma(\omega) \overset{\mathrm{def}}{=} \sqrt{Z_{series}(\omega)\,Y_{shunt}(\omega)} =$
$= \sqrt{(R(\omega) + j\omega L(\omega))(G(\omega) + j\omega C(\omega))}$,

where $Z_{series}(\omega)$– series impedance of a transmission line, $[\Omega]$;

$Y_{shunt}(\omega)$ – parallel conductivity of a transmission line, [S]; $R(\omega)$, $L(\omega)$, $G(\omega)$, $C(\omega)$ – respectively, the frequency dependences of the active resistance $[\Omega]$, inductance [H] , active conductivity $[S]$, capacitance [F] of the line (hereinafter referred to as RLGC dependences or RLGC parameters).

Thus, the problem is reduced to constructing the RLGC frequency dependences, after which $H(\omega)$ is applied to the

979-8-3315-4784-4/26 $31.00 © 2026 IEEE

input signal spectrum, and the result is reconstructed in time using IFFT [8]. The RLGC dependencies are formed by referencing the analytical models to a reference point of 1 GHz.

III. TEST DIGITAL SIGNAL X(T)

To visualize the result of the signal passing along the transmission line, we feed it with a test digital signal, which is a pseudo-random bit sequence.

For illustration purposes, a pseudorandom binary sequence (PRBS) of $N_{bits} = 256$ with a probability of "1" equal to 0.5 is used. The signal is sampled so that there are $N_{samples} = 256$ per bit (convenient for Fast Fourier Transformation (FFT)), yielding $N_{samples} = 65536$. The bit rate is chosen to be $F_{bit} = 2.5$ GHz (Peripheral Component Interconnect Express (PCIe) level), and the IL is estimated at the first harmonic of $F_{bit}/2 = 1.25$ GHz. To simulate realistic waveforms, the original sequence is additionally smoothed with a first-order low-pass filter generating $f_g = 3 \times F_{bit}$, which is fed to the line input.

We perform the filtration: X_{f_k}

$$X_f(k) = X(k) \times A_{pre}(k \times \Delta\omega),$$

$x_f = IFFT(X_f)$ where x_f – a signal at the filter output; X_f – its spectrum.

Fig. 2 shows that the x_f signal now has a finite slope, and we are to use it further.

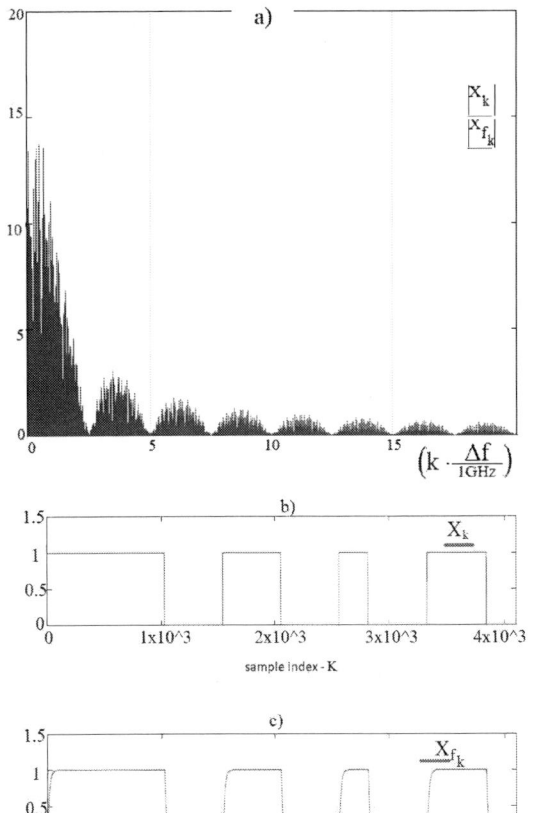

Fig.2. a) spectrum of the original signal (X) and filtered signal (X_f), b) oscillogram of the original signal (x), c) oscillogram of the filtered signal (x_f).

IV. SERIES IMPEDANCE OF THE LINE

$Z_{series}(\omega) = R(\omega) + j\omega L(\omega)$, and its frequency dependence is due to the so-called skin effect the essence of which is the displacement of the current density to the surfaces of the conductor from the center at high frequencies [2]. The skin depth δ is the penetration distance at which the electromagnetic wave is attenuated by a factor of e [1]:

$\delta = \frac{1}{\alpha}$, where $\alpha = \sqrt{\frac{\omega\mu\sigma}{2}}$ is the attenuation coefficient, and σ is the metal conductivity.

Assuming $\delta = t_{trace}$, we determine the cutoff frequency at which the skin effect begins to operate:

$f_\delta = \frac{1}{t_{trace}^2 \mu\sigma\pi} = 13.5 MHz$ for our example.

Inductive loops formed by the transmission line and the return path C_{air} is the capacitance of the transmission line in air (without a dielectric); the formulas for calculation can be found, for example, in [1].

Fig. 3 shows the resulting dependence $L_x(\omega)$

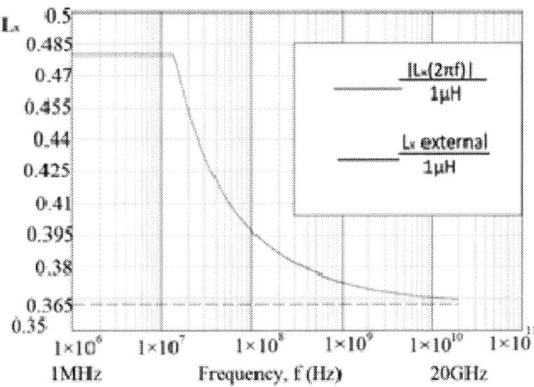

Fig. 3. Linear inductance as a function of frequency (the dashed line shows the inductance at infinite frequency).

A physically correct (causal) model of the line requires that the real and imaginary parts of $Z_{series}(\omega)$ obey the Kronig–Kramers relations [1], which is why roughness cannot be taken into account only through the growth of $R_x(\omega)$ without a consistent change in $\omega L_x(\omega)$. Methods for such an account are given in [1,3, 6] and are not considered here; however, the contribution of roughness to the series impedance is significant and should be verified by field solvers.

V. PARALLEL CONDUCTIVITY OF THE LINE

$Y_{shunt}(\omega) = G(\omega) + j\omega C(\omega)$; as for dielectric losses, they are caused by the polarization of the medium and are described by complex permittivity: $\varepsilon = \varepsilon' - j\varepsilon''$, where $\varepsilon' = \varepsilon_0 \varepsilon_r$ – direct permittivity; ε'' - dielectric losses expressing the loss tangent $\tan_\delta = \varepsilon'/\varepsilon''$; ε_r – relative permittivity.

For digital channels up to ~20 GHz, $\varepsilon'(\omega)$ is approximated by an "infinite pole model" with logarithmic roll-off; its parameters are chosen to pass through ε_{ref} at 1 GHz [1].

As is shown in [1], ε' for high-speed digital designs up to 20 GHz can be approximated with acceptable accuracy by the infinite pole model according to which the dependence $\varepsilon(\omega)$ is a logarithmic decay by the value $\Delta\varepsilon'$ starting from the frequency ω_1 and up to the frequency ω_2 to some value ε'_∞, and then independent of the frequency:

$$\varepsilon'(\omega) = \varepsilon'_\infty + \frac{\Delta\varepsilon'}{m_2 - m_1} \frac{\ln\left(\frac{\omega_2 + j\omega}{\omega_1 + j\omega}\right)}{\ln(10)} \xrightarrow{\omega_1 \ll \omega \ll \omega_2}$$

$$\varepsilon'_\infty + \frac{\Delta\varepsilon'}{m_2 - m_1} \frac{\ln\frac{\omega_2}{\omega}}{\ln 10},$$

where m_1, m_2 – exponents of ω_1, ω_2, respectively.

The parameters of the $\varepsilon'(\omega)$ model were calculated so that it passes through the reference point ε_{ref} @ 1 GHz. It is necessary to additionally clarify that the actual $\varepsilon'_{ref} = 2.96$ at a frequency of 1 GHz has a value, which is less than the rated $\varepsilon'_{ref} = 4.2$ due to the fact that for a microstrip line, part of the field passes through air.

The impulse response, whose characteristic asymmetric appearance confirms the model's compliance with the causality criterion, is shown in Fig. 4.

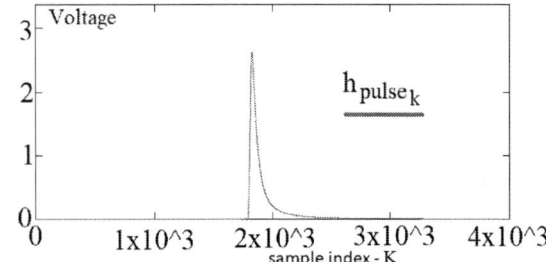

Fig. 4. Impulse response of a transmission line.

VI. EYE DIAGRAM

The original signal $x_f(k)$ at the end of the transmission line under consideration is of the asymmetric microstrip type. We multiply its spectrum by the discrete gain, then reconstruct it in the time domain:

$$Y = X_f \times H_d$$

$$y = IFFT(Y)$$

Fig.5 shows the spectra of the original and final signals superimposed, and Fig.6 shows the same signals in the time domain.

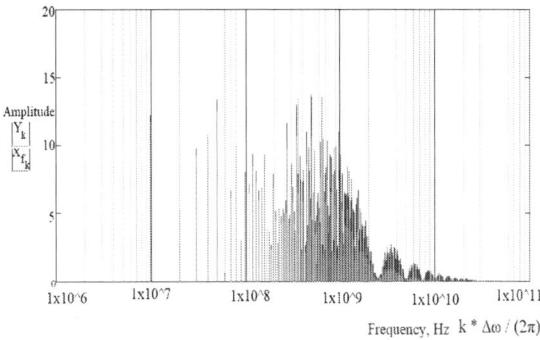

Fig. 5. Original ($X_f(k)$) and final ($Y_f(k)$) signals.

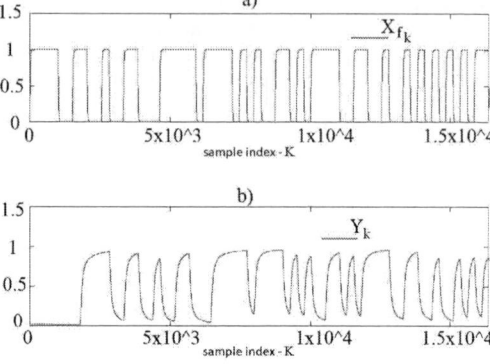

Fig. 6. Oscillograms of the original ($X_f(k)$) () and final ($Y_f(k)$) signals

The signal at the line output can be characterized as:

1) delayed by ~1800 samples (2.8 ns);

2) reduced in amplitude compared to the original, with high-frequency components suppressed more than low-frequency ones;

3) distorted in shape (dispersion);

4) the one that does not reach the "pure" 0 and 1 signals applied to the line input;

5) oscillated in DC component.

Considering characteristics 1) and 3), additionally construct the dependence of the group delay of the complex transmission coefficient of the line (Fig. 7):

$$\varphi(\omega) = Arg(H(\omega))$$

$$\tau(\omega) = d(\varphi(\omega))/d\omega$$

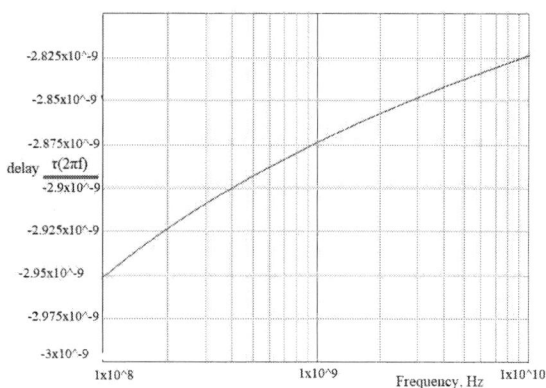

Fig. 7. Dependence of the line group delay on frequency.

Since the line's phase-frequency response $\varphi(\omega)$ becomes nonlinear at high speeds, the group delay ceases to be constant, and different signal components will propagate along the line at different speeds. This means that at the receiver, the signal will no longer be reassembled into its original form but will be distorted; i.e., the signal "disintegrates," resulting in signal dispersion.

Considering characteristic 2), we find that at the frequency of the 1st harmonic $H_x(\pi F_{bit}) = 0.62$ (-4.2dB); i.e., the IL for the microstrip under consideration is 38% of the original signal amplitude.

Considering characteristics 1) and 3), we note that at the line output, a drift of the constant component is observed due to which the current bit becomes a function of the channel state over the N previous symbols: a series of "1"s leads to charge accumulation, a series of "0"s leads to a discharge. Such channel memory corresponds to intersymbol interference and causes noticeable ISI jitter [5, 6, 7]. A quantitative assessment is performed using the eye diagram $y(k)y(k)y(k)$ (Fig. 9) based on the comparison with the diagram of the original signal $xf(k)x\_f(k)xf(k)$ (Fig. 8).

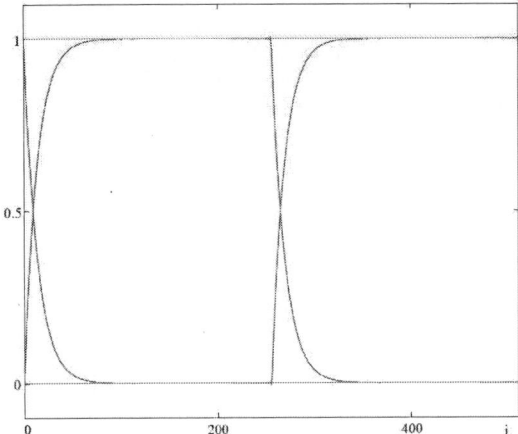

Fig. 8. Eye diagram of the signal at the beginning of the line

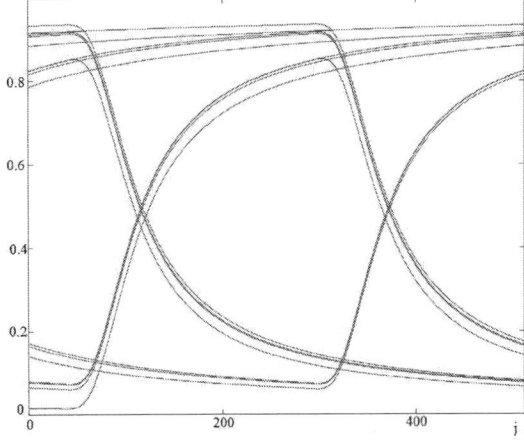

Fig. 9. Eye diagram of the signal at the end of the line.

The eye diagram represents 16 superimposed transitions at a scale of 2UI (Unit Interval, 1 UI = $1T_{bit}$) and helps us understand intersymbol interference, which is the effect of adjacent bits on each bit transmitted along the line. We can estimate the ISI jitter, which in this case is $12 \times T_s = 19$ ps.

VII. CONVOLUTIONAL NEURAL NETWORK FOR ACCELERATED INSERTION LOSS ESTIMATION

Parametric SI studies on field solvers are expensive, so a CNN-based surrogate for fast IL prediction was proposed [9,10], and machine learning methods were used [9].

In particular, the CNN was trained on simulation data, predicting the amplitude-frequency response and the IL value based on the input line parameters (w, t, h, l, ε_r, and loss tangent $\tan_\delta$). Similar approaches, using the models based on deep neural networks, are successfully applied in modern SI-modeling systems [10].

For training, a dataset, based on the above-mentioned parameters and the corresponding frequency dependencies (derived synthetically from an analytical model linked to a 2D solver), was formed. For each set of parameters, the transmission coefficient $H(f)$ and the IL value were calculated in the range of 0.1–10 GHz.

To quickly predict IL and key SI metrics, a CNN was trained; its accuracy and speed were compared with an analytical model and a field solver. Training was performed using the Adam optimizer and the MSE loss function. The training set size was 1000 parameter combinations; 70% of the data was used for training; 30% was used for validation. The neural network structure is shown in Fig. 10.

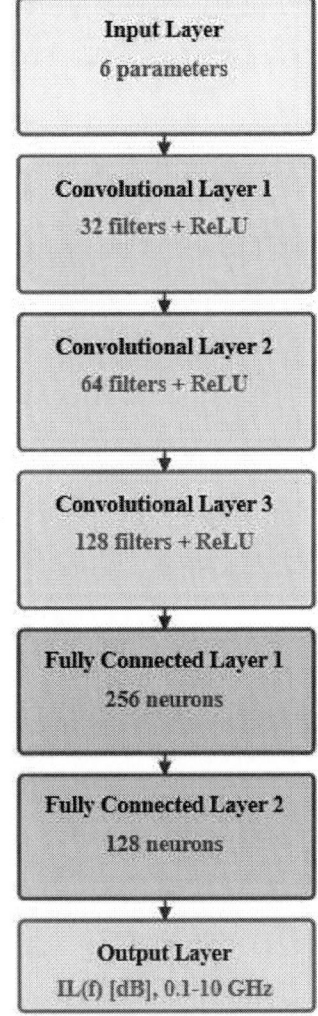

Fig. 10. Structure of neural network.

After training, the network provided an average absolute prediction error of no more than ±0.1 dB and a computation time of about 10 ms on the CPU, which is more than 100 times faster than a traditional field solver (see Table 1).

TABLE I. COMPARISON OF THE ACCURACY AND PERFORMANCE OF IL ESTIMATION METHODS

Modeling method	Average error, dB	Calculation time for one route	Data used	Note
Field model (3D solver)	0 (standard)	12 s	Geometry, materials	Full calculation of the field
Analytical RLGC model	±0.2	1.0 s	w, h, t, ε_r, $\tan_\delta$	Approximation, without roughness
CNN model (proposed)	±0.1	0.01 s	w, h, t, ε_r, $\tan_\delta$	Trained on 1000 samples
MLP model (without convolutions)	±0.3	0.02 s	Same parameters	Low accuracy at high f

Also, based on MathCAD, we generated (analytically) input data to train CNN in Python (PyTorch) and to further compare the CNN-predicted IL with the RLGC/EM reference response. The input data is represented in Table 2.

TABLE II. EXAMPLE OF INPUT DATA (A FRAGMENT OF THE TRAINING SAMPLE)

No	w, mm	h, mm	t, mm	ε_r	$\tan_\delta$	f, GHz	IL (dB/50 cm)
1	0.15	0.127	0.018	4.2	0.0195	1.0	−1.8
2	0.15	0.127	0.018	4.2	0.0195	5.0	−3.7
3	0.20	0.127	0.018	3.7	0.0150	5.0	−3.1
4	0.10	0.127	0.018	4.5	0.0250	10.0	−6.5
5	0.15	0.200	0.018	4.2	0.0195	10.0	−4.8

Fig. 11 confirms that the CNN surrogate closely matches the RLGC/EM reference across 0.1–10 GHz, with deviations within ±0.1 dB. The model generalizes well in the training range but degrades outside the learned frequency/ε_r domain.

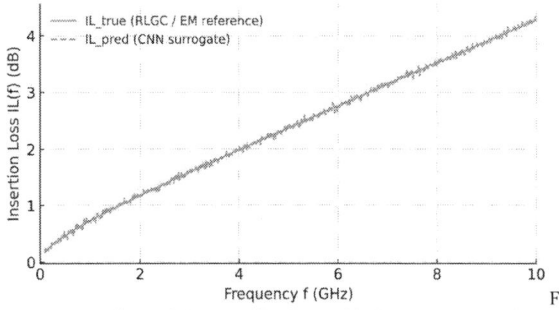

g. 11. Comparison of CNN-predicted IL with the RLGC/EM reference response over the 0.1–10 GHz band.

Robustness can be improved by enlarging the dataset to multilayer and differential structures, applying transfer learning with measured S-parameters, and adding PINN-based physical constraints.

VIII. CONCLUSION

In this paper, we developed a frequency-dependent RLGC model of a microstrip line that accounts for the skin effect, dielectric loss, and group delay dispersion in the 0.1–10 GHz range. It was shown that these mechanisms determine the amplitude roll off, front distortion, and ISI growth in high-speed channels. Based on analytical and field simulation data, we trained a CNN surrogate model for predicting IL(f) based on geometric and material science parameters, ensuring an accuracy of up to ±0.1 dB while accelerating calculations by more than two orders of magnitude relative to full-wave solvers. The resulting hybrid approach is suitable for operational signal integrity assessment and PCB routing optimization in digital systems with data rates up to 10 Gbps.

REFERENCES

[1] S. H. Hall and H. L. Neck, Advanced Signal Integrity for High-Speed Digital Designs. Hoboken, NJ: John Wiley & Sons, Inc., 2009.

[2] H. W. Johnson and M. Graham, High-speed Digital Design: A Handbook of Black Magic. Englewood Cliffs, NJ: Prentice Hall, 1993.

[3] B. Zhong, Lossy Transmission Line Modeling and Simulation Using Special Functions. Diss. Tucson, AZ: The University of Arizona, 2006.

[4] W. Maichen and B. Krsnik, A Practical Guide to Lossy Differential Lines. North Reading, MA: Teradyne, Inc.

[5] U. Tietze, Ch. Schenk, Halbleiter-Schaltungstechnik. Berlin, Heidelberg: Springer, 1985.

[6] V. Dmitriev-Zdorov, B. Simonovich, and I. Kochikov "A Causal Conductor Roughness Model and its Effect on Transmission Line Characteristics", Signal Integrity Journal, November 2018.

[7] Keysight Technologies Application Note. Choosing the ISI Filter Size for EZJIT Plus Arbitrary Data Jitter Analysis. Santa Rosa, CA: Keysight Technologies.

[8] DSPL-2.0. https://ru.dsplib.org/

[9] T. Lu, K. Wu, Z. Yang, and J. Sun, "High-speed channel modeling with deep neural network for signal integrity analysis," 2017 IEEE 26th Conference on Electrical Performance of Electronic Packaging and Systems, Vol. 2018-January, pp. 1–3, 2017, doi:10.1109/EPEPS.2017.8329733.

[10] J. Withöft, W. John, E. Ecik, R. Brüning, and J. Götze, "Machine Learning Methods for Elaborating the Feasible Region for Signal Integrity Analysis in Differential Pair PCB Structures," 2024 International Symposium on Electromagnetic Compatibility – EMC Europe, Brugge, Belgium, pp. 151–156, 2024, doi: 10.1109/EMCEurope59828.2024.10722241.

Features of Formation and New Areas of Application of Composite Structures Based on Aluminum Oxide

Ekaterina N. Muratova
Department of Micro and Nanoelectronics
Saint Petersburg Electrotechnical University "LETI"
Saint Petersburg, Russia
sokolovaeknik@yandex.ru

Abstract— The basis of a scientific and technological breakthrough at the nanoscale is the use of previously unknown properties and functionality of materials. New devices require the development of new materials with atomically smooth interfaces. Therefore, this paper examines modern metamaterials based on porous anodic alumina (PAA), and presents prospects for the development of micro- and nanoelectronics, including a wide range of innovative nanodevices and miniature systems using PAA. The features of anodizing aluminum foil are considered. Nanosized layers and membranes based on PAA with ordered nanoscale capillaries with a pore diameter from 20 nm to 200 nm and an aspect ratio of up to 500 are obtained. Study of structure and morphology using AFM, SEM and RBS techniques. The results of studying the optical properties (in the range of 200 nm - 16 μm) made it possible to propose a method for diagnosing a porous structure based on an analysis of optical transmission spectra. Various areas of application of structures based on PAA, including in popular new generation devices, have been demonstrated.

Keywords— *porous aluminum oxide; surface roughness; areas of application; membranes; nanoelectronics; biomedical application; optical transmission spectra; Rutherford backscattering*

I. INTRODUCTION

So far, several efficient techniques for fabricating nanostructures have been developed, with self-assembly-based methods being particularly significant. [1-9]. Self-organization methods involve various mechanisms for organizing nanoparticles into structures of a given shape and size. Electrochemical methods for forming porous materials are relatively simple to implement. This allows for the modification of material properties by varying synthesis parameters, facilitating the study of physical phenomena in nanostructures. The most studied materials obtained by the electrochemical etching method are porous silicon and porous anodic alumina (PAA) [9-15].

The most common model of PAA formation is based on the hypothesis about the mechanism of formation of the ordered honeycomb structure of the oxide layer. According to this hypothesis, a possible source of forces arising between neighboring cells is mechanical stress associated with the increase in volume during the formation of aluminum oxide. More detailed theoretical ideas about pore formation can be found in the works [2, 10-13].

This study primarily focuses on the development of nanoporous aluminum oxide. Membranes made from this material are highly sought after in fields such as nanotechnology and microbiology [12, 16-17], as well as nuclear physics [18-22], due to their distinctive characteristics (high aspect ratio, optical transparency, biological inertness), along with excellent mechanical strength, thermal stability, and chemical durability.

The difficulty in separating such porous anodic films from the substrate limits their potential use as free-standing membranes (masks, templates) with specific pore topologies and geometries. Therefore, developing mechanically robust PAA membranes with ordered through-pores—micro- and nanocapillaries in thin (~10 μm) aluminum foil—remains a relevant challenge.

The goal of this work is to conduct a comprehensive investigation of the processes involved in the controlled formation and self-assembly of nano- and microporous aluminum oxide membranes, as well as exploring their potential applications as nanoscale capillary matrices and masks for various micro- and nanoengineering purposes.

II. SYNTHESIS OF THE STUDIED MATERIAL

Since the primary goal of this work was to create uniform nanoporous membranes made of aluminum oxide, the main focus was on aluminum foil with a thickness of approximately 10 μm. Special attention was given to the initial surface preparation methods, the electrochemical anodization process itself, and the analysis of the structure of the resulting porous membranes.

To ensure reproducibility in membrane fabrication, it is essential to examine the microstructure of the original aluminum foil, as the grain size and crystallographic orientation influence the uniformity of the layer and the pore morphology. Throughout the study, particular focus was placed on the initial surface preparation of the ~10 μm thick aluminum foil and the electrochemical anodization process itself. The first stage involved preparing the foil through mechanical faceting, surface polishing, and recrystallization via thermal annealing. Figure 1 shows the results of the SEM (scanning electron microscopy) analysis of the original aluminum after the faceting process. Surface faceting significantly impacts the topological and geometric characteristics of pore formation and the pore shape.

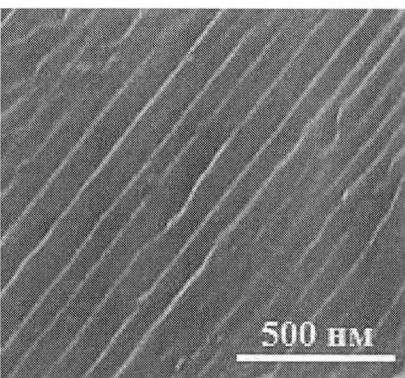

Fig. 1. SEM image of the original aluminum after faceting process

To fabricate PAA membranes, aqueous solutions of sulfuric and orthophosphoric acids with 10-15% glycerol were employed as electrolytes in this study. The detailed production process is described in references [23, 24], while Table 1 summarizes the key parameters, including U_A (anodizing voltage), t (duration), T (temperature), d_{pore} (pore diameter), and C_{pore} (pore concentration).

TABLE I   ALUMINUM ANODIZING PARAMETERS

electrolyte	H_2SO_4+15% glycerol	H_3PO_4	H_3PO_4+10...15% glycerol
U_A, V	20–27	80	120–150
t, min	3...5	20	10...17
T, ℃	5...7	-5...+5	0...7
d_{pore}, nm	20–30	80–100	180–220
C_{pore}, $1/\mu m^2$	350	50	20

To produce through membranes, additional steps involving chemical and electrochemical etching from the backside of the samples were implemented. The characteristics of the porous structures were examined using scanning electron microscopy, atomic force microscopy, and optical microscopy. Additionally, a technique based on analyzing optical transmission spectra across a broad wavelength range (from 200 nm to 15 μm) was employed.

III. RESULTS AND DISCUSSION

A comprehensive investigation of PAA demonstrated [15, 18, 23] that by adjusting the anodization parameters—such as electrolyte composition, voltage/current density, temperature, and duration—it is possible to control the structural characteristics of the porous layer. These include pore diameter (ranging from 20 to 200 nm, as shown in Fig. 2a and 2b, respectively), the spacing between pore centers (from 40 to 250 nm), and the thickness of the porous layer (from 200 nm to 30 μm).

a

b

Fig. 2. SEM images of PAA with different pore geometries: a) 20nm, b) 200 nm

The results indicated that by selecting the initial surface treatment method of the aluminum foil—such as faceting, polishing, or thermal annealing—it is possible to influence the pore shape, resulting in square, hexagonal, or round pores.

The results of the optical transmission spectra studies are presented in Fig. 3 [25].

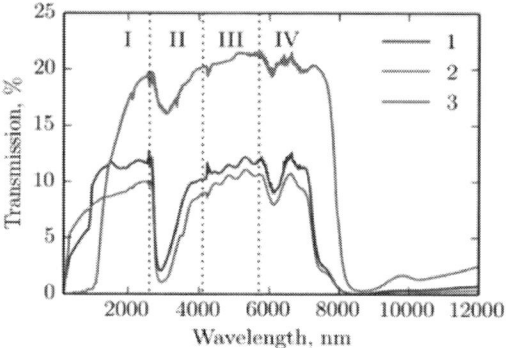

Fig. 3. Optical transmission spectrum for PAA-based membranes in a wide range of wavelengths (I–IV in the text)

The transmission spectrum includes regions that enable rapid characterization of the membrane: estimating the average pore size and size distribution (Section I), identifying the membrane material as aluminum oxide (Section II), determining membrane thickness (Section III), and detecting residual products (Sections I and IV).

IV. APPLICATION AREAS

This article introduces new applications of nanoporous membranes featuring a system of uniaxially aligned capillaries: as materials for reducing thermal radiation from the human body, nanoscale masks for directing ion beams, platforms for lab-on-a-chip growth, and filters for liposome extruders.

A. Shielding Coatings

The first application is related to the fact that nanoporous PAA membranes significantly reduce the transmission of IR radiation in the range from 8 to 14 μm [25]. This is particularly relevant because biological objects exhibit peak thermal radiation within this range, specifically around 10 μm. Fourier Transform Infrared (FTIR) spectroscopy reveals a notable reduction in radiation transmission for PAA membranes from 7.5 to 14 μm (Fig. 3, sections IV).

The results from FTIR spectroscopy were corroborated using a thermal imager (SDS INFRARED hotfind LXT). For instance, these membranes were employed to minimize the thermal radiation emanating from the human body. Fig. 4 illustrates thermal imaging results, indicating that a 15 μm-thick PAA membrane, mounted on a 5 mm-thick fluoroplastic ring, effectively shields thermal radiation. Specifically, the difference in recorded temperature with and without the membrane was at least 4–5°.

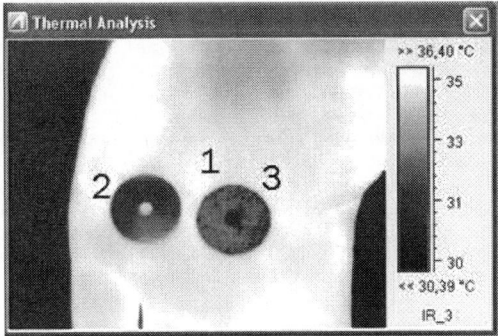

Fig. 4. Thermal imaging studies: 1 – surface of biological object, 2 – fluoroplastic membrane holder, 3 – membrane on holder

Therefore, although PAA membranes are optically transparent in the visible spectrum (Fig. 3, section I), they significantly attenuate radiation in the mid-IR range (λ = 8-13.5 μm). This property makes them suitable for use as radiation-absorbing filters for biological objects.

B. Templates for High-Energy Ion Beams

Furthermore, PAA membranes with through-nanocapillaries have been investigated as nanoscale templates for high-energy ion beams [18, 20, 23]. Specifically, their behavior when exposed to a helium ion beam with an energy of 1.5–2 MeV was studied. Figure 5 shows the Rutherford Backscattering (RBS) spectrum obtained using a mask-template with a pore diameter of approximately 20 nm. A "recording" medium-target, containing a heavier element (hafnium, Hf), was positioned behind the membrane.

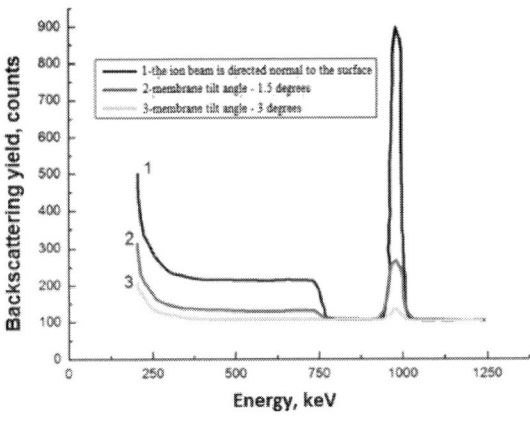

Fig. 5. The energy spectrum of backscattered He+ ions with an energy of 1200 keV. The black curve is the spectrum along the normal direction; the red curve is the spectrum when the deviation is 1.5 degrees from the normal; the blue curve is the spectrum when the deviation is 3 degrees

Even a small angular deviation (1.5...3°) of the membrane relative to the ion beam's trajectory significantly reduces the backscattered ion yield from the HfO2 target,

resulting in a lower recorded current. Thus, the membrane-templates developed and studied are capable of channeling a high-energy ion beam.

It has been found that the transmission coefficient of accelerated ions through the membrane can reach up to 60% [18, 23]. This allows for spatially controlled ion bombardment of the substrate with nanoscale precision. Consequently, PAA matrices, depending on their geometric properties, can enhance the capabilities of RBS measurements.

C. Biomedical Application

Porous ordered structures are of particular interest for filtering various media. The authors of this work consider the possibility of using PAA membranes as a growth platform for bacterial colonies, as well as a filter for homogenizing liposomes. In this case, it is important to select membranes with certain geometric parameters [2, 18].

To achieve the stated goals, microporous membranes obtained by both anodization through a photoresistive mask and microchannel plates were used in the work (Fig. 6).

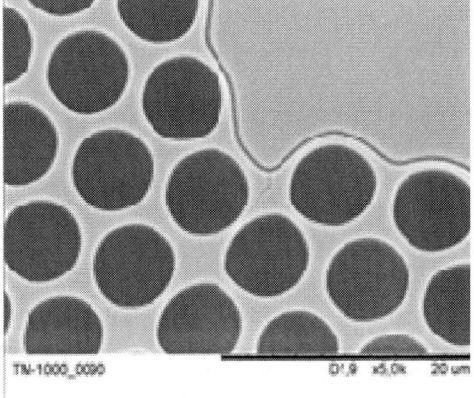

Fig. 6. SEM image of the edge of a microchannel plate

The method of layer-by-layer molecular deposition of Al_2O_3 was considered as a method for precision control of the pore diameter. As a result of the work, growth cells were formed, which represent a microporous structure (Fig. 7), the bottom of which is a nanoporous layer of aluminum oxide.

a

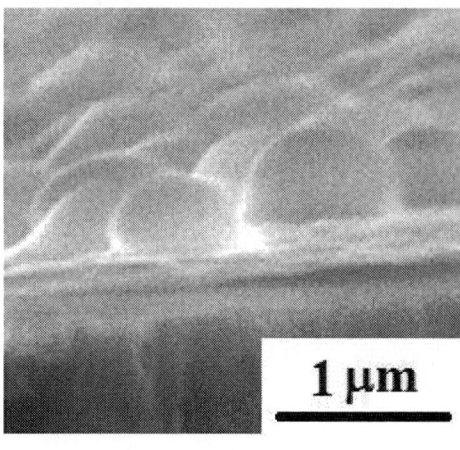

b

Fig. 7. SEM images of a PAA membrane: a) with a photolithographic pattern of micropores, b) with bacterial colonies

Through this porous bottom, the nutrient medium entered the bacteria located in the micropore due to capillary action. As a result, the bacterial colonies were able to survive.

In addition, a liposomal extruder was developed for homogenizing liposomes. The key element of such a device is a nanoporous membrane. The diameter of the membrane pores determines the size of the liposomes passing through it. The optimal pore size in this case is 200 nm in diameter. The process of electrochemical anodization under properly selected technological conditions makes it easy to achieve this.

V. CONCLUSIONS

This article details the self-organized fabrication of porous membranes composed of ordered, uniaxially oriented nano-capillaries (with pore diameters of 20 nm and larger), formed from aluminum oxide via the electrochemical anodization of micron-thick aluminum foil. The structure and optical properties of these ordered through-pore PAA membranes were investigated as a function of processing conditions.

A method for characterizing the porous structure based on the analysis of optical transmission spectra is proposed. The research results demonstrate that specific zones within the transmission spectrum enable rapid characterization of the membrane based on key structural and geometric parameters. Furthermore, it is shown that PAA membranes significantly attenuate IR radiation in the 8 to 14 μm range, a spectral region relevant to the IR emission of biological objects. Finally, their ability to channel a high-energy (1.5–2 MeV) helium ion beam is demonstrated.

The possibilities of using PAA membranes for biomedical purposes, namely as a growth platform for bacterial colonies and a filter for homogenization of liposomes, are considered.

REFERENCES

[1] Hamad L.B., et al. Analysis of static and dynamic characteristics of strain gradient shell structures made of porous nano-crystalline materials, *Adv. Mater. Res.* 2019, 8, pp. 179–196.

[2] Muratova E.N., Moshnikov V.A. The use of nanomaterials from electronics to medicine / International Conference on Advanced Materials and Nanotechnology for a Green and Sustainable future (ICAGS-2023). Maharaja's College, Ernakulam, Kerala, India, 5th to 6 th of December 2023, pp.7–9.

[3] Feng S., Ji W. Advanced nanoporous anodic alumina-based optical sensors for biomedical applications, *Front. Nanotechnol.* 2021, 3, pp. 678275.

[4] Domagalski J.T., et al. Recent Advances in Nanoporous Anodic Alumina: Principles, Engineering, and Applications, *Nanomaterials* 2021, 11, pp. 430.

[5] Spivak Y., et al. Improving the Conductivity of the PEDOT:PSS Layers in Photovoltaic Cells Based on Organometallic Halide Perovskites, *Materials*, 2022, 15 (3), pp. 990.

[6] Moshnikov V., et al. Controlled Crystallization of Hybrid Perovskite Films from Solution Using Prepared Crystal Centers, *Crystals.* 2024; 14(4), pp. 376.

[7] Nenashev G. V., et al. Effect of barium doping on the behavior of conductivity and impedance of organic-inorganic perovskite films" *Solid State Communications*, Vol. 388, 2024, pp.115554,

[8] Muratova E.N., Moshnikov V.A., Zhilenkov A.A. Development of Model Representations of Materials with Ordered Distribution of Vacancies, *Crystals* 2024, 14, pp.1095.

[9] Muratova E., et al. AFM for monitoring interphase interfaces of solar cells / European Materials Research Society. B. Advancing sustainable organic photovoltaics: from experiments and materials to applications and device models. Convention & Exhibition Centre of Strasbourg (France), from May 27 to 31, 2024, pp. 00370

[10] Ruiz-Clavijo A., Caballero-Calero O., Martín-González M. Revisiting anodic alumina templates: From fabrication to applications, *Nanoscale* 2021, 13, 2227–2265.

[11] Davoodi E., et al. Nano-porous anodic alumina: Fundamentals and applications in tissue engineering, *J. Mater. Sci. Mater. Med.* 2020, 31, 60.

[12] Alvarez-Carrizal R.P., et al. Manufacture of Al2O3/Ti composite by aluminum bonding reaction for their use as a biomaterial, *Adv. Mater. Res.* 2021, 10, pp. 331–341.

[13] Manzoor S., et al. Recent Progress of Fabrication, Characterization, and Applications of Anodic Aluminum Oxide (AAO) Membrane, *A Review. Comput. Model. Eng. Sci.* 2022, 135, pp. 1007–1052.

[14] Romero V., Benavente J. Electrochemical Characterization of Nanoporous Alumina-Based Membranes with Different Structure and Geometrical Parameters by Membrane Potential Analysis, *Micro* 2022, 2, pp.475–487.

[15] Lushpa N.V., et al. Using the method of SEM image processing to study the morphology of porous anodic alumina films, Chap. 13 in book "New Materials: Preparation, Properties and Applications in the Aspect of Nanotechnology" Ed. by A.G. Syrkov and K. L. Levine, Published by Nova Science Publishers 2020, pp. 125-132.

[16] Feng S., Ji, W. Advanced Nanoporous Anodic Alumina-Based Optical Sensors for Biomedical Applications, *Front. Nanotechnol.* 2021, 3, pp.678275.

[17] Muratova E., et al. Technology of formation and new areas of application of composite metastructures based on aluminum oxide / The 15th International Conference on Metamaterials, Photonic Crystals and Plasmonics. META 2025, MALAGA - SPAIN, JULY 22 – 25, 2025, pp. 1110.

[18] Shemukhin A.A., Muratova E.N. Investigation of transmission of 1.7-MeV He+ beams through porous alumina membranes, *Tech. Phys. Lett.* 2014, 40, pp.219–221.

[19] Pinilla S., et al. Highly ordered metal-coated alumina membranes: Synthesis and RBS characterization", *Surf. Coat. Technol.* 2019, 377, pp.124883.

[20] Luchinin V.V., et al. Formation of ordered nanoscale capillary membranes based on anodic alumina, *Jour. of Physics: Conference Series*, 2015. V. 586. pp. 012008.

[21] Zhu Z., Zhu D., Lu R. The experimental progress in studying of channeling of charged particles along nanostructure, *Proc. of SPIE, Bellingham.* WA, 2005. V. 5974. pp.13.

[22] Stolterfoht N. and Yamazaki Y. Guiding of charged particles through capillaries in insulating materials, Phys. Rep. 2016, 629, pp. 1–107.

[23] Muratova E.N., et al. The Influence of the Structural Parameters of Nanoporous Alumina Matrices on Optical Properties, *Metals* 2024, 14, pp.651.

[24] Muratova E., et al. Effect of Aluminum Foil Polishing on the Morphology of Membranes Based on Porous Anodic Aluminum Oxide / IEEE Conference of Russian Young Researchers in Electrical and Electronic Engineering (ElConRus), 2025, P.1017-1019

[25] Matyushkin L.B., Muratova E.N., Panov M.F. Determination of the alumina membrane geometrical parameters using its optical spectra, *Micro Nano Lett.* 2017, 12, pp.100–103.

Modeling and Analysis of the Design of Sensitive Element of a Capacitive Type Micromechanical Accelerometer

Paing Soe Thu
Institute of Nano and Microsystem Technology
National Research University of Electronic Technology "MIET"
Moscow, Russian Federation
paingsthu7@gmail.com

Viktor V. Kalugin
Institute of Nano and Microsystem Technology
National Research University of Electronic Technology "MIET"
Moscow, Russian Federation
viktor118@mail.ru

Elena S. Kochurina
Institute of Nano and Microsystem Technology
National Research University of Electronic Technology "MIET"
Moscow, Russian Federation
kochurinaes@gmail.com

Ye Min Hlaing
Institute of Nano and Microsystem Technology
National Research University of Electronic Technology "MIET"
Moscow, Russian Federation
yeminhlaing588@gmail.com

Satt Naing Moe
Institute of Nano and Microsystem Technology
National Research University of Electronic Technology "MIET"
Moscow, Russian Federation
sattnaingmoe59@gmail.com

Abstract— In this paper present the modeling and analysis of the design of a capacitive type micromechanical accelerometer (MMA) based on microelectromechanical system (MEMS) devices. The purpose in doing this paper is to study the design of silicon sensitive element (SE), results of mechanical deformation and mechanical stress of SE design with comb fingers at the linear accelerations (10g-50g). In addition, the effect of increasing the number of comb fingers to obtain optimal capacity values was analyzed. The torsions used in the sensitive elements and the resonant frequencies of the MMA sensitive elements were studied. Results of deformation in MMA design at the temperature ranges (from -45°C to +65°C) along the X-axis have investigated. The sensitivity of MMA can be achieved by selecting the design of parameters in silicon SE, gaps between electrodes, torsions, anchors and comb fingers. Results of the proposed design of MMA as a function of the applied acceleration ranges demonstrates the optimized sensitivity of the capacitive type MMA design.

Keywords—micromechanical accelerometer, sensitive element, acceleration range, temperature range, mechanical deformation, mechanical stress, comb fingers, capacitive changes

I. INTRODUCTION

Today, the global MEMS market is rapidly growing and dynamically changing. MEMS components and sensors based on them have great commercial attractiveness. Among these sensors, MMA is a sensor which widely used in various orientation, stabilization, guidance, and navigation systems. Although the primary purpose of MMA is to measure the acceleration, these sensors are also capable of measuring tilt, motion, position, impact, and vibration. MMA allows to measure the magnitude of change in speed of an object, measures the magnitude of this change in units of g (1g = 9.81 m/s2). Depending on the manufacturing technology and design, the accelerometer can measure along one, two or three axes. MMA are capable of measuring range in frequency including the static or dynamic accelerations while having special stability and optimized sensitivity along the working axes. In this paper, the results of mechanical deformation in

the design of MMA along the X-axis by acting the linear acceleration ranges from 10g to 50g were calculated to obtain the capacitive changes between comb fingers by using the ANSYS program based on finite element analysis (FEA). Capacitive changes between comb fingers in silicon SE design were analyzed based on the results of deformation.

Capacitive type accelerometers are very attracted to many consumers for the development of MEMS devices because of their small dimensions and weight, low power consumption and thermal sensitivity, static and dynamic response, high accuracy and sensitivity [1]. Recently, MEMS devices (i.e. accelerometers) based on capacitive sensing techniques have been significantly developed not only for a variety of industrial applications but also for many consumers electronic devices [2].

The integration of micromechanical elements and electronic components on a silicon substrate using the micro fabrication technology. MEMS contain extremely small mechanical elements and operational fabrication process. Modern MEMS-fabrication technologies are largely based on integrating electronic processing circuits for sensing devices. MEMS technologies and designs are developing and remain an important component of modern MEMS devices and equipment. Currently, MMA are widely used in various applications, such as automotive industries, aerospace and robotic systems (robot vacuum cleaner), cell phones, medical products and inertial navigation system due to their light weight, low power consumption, compactness and etc. [3].

II. DESIGN DESCRIPTION OF CAPACITIVE TYPE MMA

The design of MMA SE consists of a proof mass which attached with comb fingers (capacitive plates), elastic torsion bars, and anchors for a fixed support (Fig. 1. a and b). The proof mass is attached to the comb fingers, whose displacement causes capacitive changes between electrodes. Anchors are used to control the elastic torsion bars by the displacement of proof mass. This displacement of the design is converted into capacitive changes and electrical signal using capacitive, piezoresistive, and other sensors. During acceleration, the moving plate shifts and changes the capacitance of the capacitor, which is recorded by an

979-8-3315-4784-4/26 $31.00 © 2026 IEEE

electronic circuit. These accelerometers are compact and sensitive.

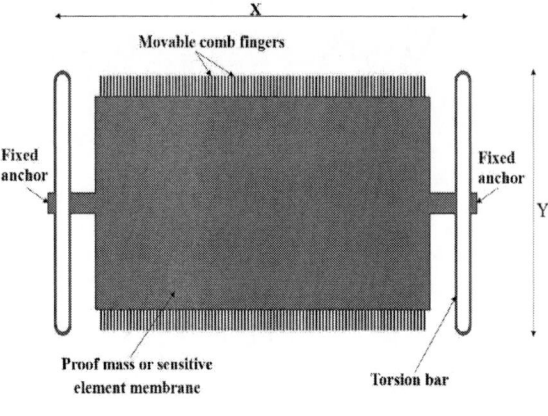

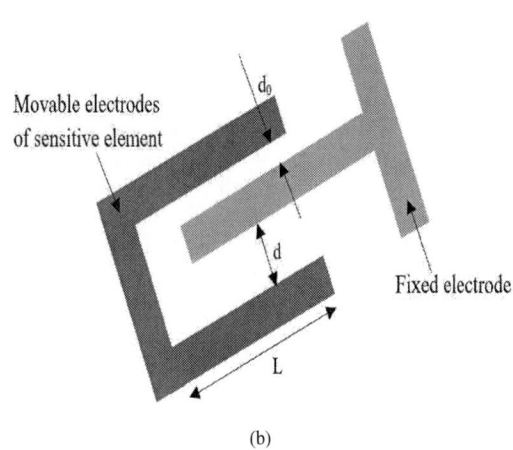

Fig. 1. (a) The design of a capacitive micromechanical accelerometer and (b) a schematic representation of a single pair of capacitance combs of the SE

where, d and d_0 - distances between movable and fixed electrodes before the displacement, L is the length of comb fingers.

The dimensions of SE design of MMA play an important role and they are in the micrometer range. The dimensions of silicon SE are 3090μm x 2100μm x 50μm, the comb finger (200μm x 10μm x 50μm), the torsion bar (1100μm x 400μm x 50μm) and the gap between the comb fingers is 10μm. Obviously, the larger the element size, the higher its sensitivity due to the increased number of comb fingers and proof mass. Silicon material possesses a unique combination of properties and is used to create MMA SE. Technological processing of silicon material allows for the creation of the required SE structure shapes and sizes [4].

III. BACKGROUND THEORY AND WORKING PRINCIPLE

Working principle of a capacitive type MMA is based on the change in capacitance caused by the displacement of a moving mass under acceleration. Typically, one set of finger-shaped electrodes is fixed to the base, while the second is connected via suspensions and can move freely along one or more axes. When the external acceleration is applied to the SE, the suspensions deform, and the structure of comb fingers move relative to the moving mass [4].

The operating principle of the accelerometer is based practically on measuring the inertial force developed by a proof mass as it moves with acceleration. A movable mass is attached by means of two torsion bars to the substrate, which remains fixed electrodes (capacitive plates). All fixed electrodes are connected together as well as the movable electrodes. A lot of connected capacitors in parallel are created by this way between the movable and fixed plates. For a pair of two capacitive plates can be expressed as:

$$C_1 = \frac{\varepsilon A}{d+x} = C_0 - \Delta C \text{ and } C_2 = \frac{\varepsilon A}{d-x} = C_0 + \Delta C, \qquad (1)$$

where A is the area of capacitive plates, d is the distance between the plates corresponding to a null acceleration of the movable mass, ε is the dielectric permittivity, and x is the displacement between the movable and fixed plates. It follows that [5]:

$$C_2 - C_1 = 2\Delta C = 2\varepsilon A \frac{x}{d^2 - x^2} \text{ or}$$

$$\Delta C x^2 + \varepsilon A x - \Delta C d^2 = 0 \qquad (2)$$

For $x \ll 1$ it results

$$x \approx d\Delta C/C_0 \qquad (3)$$

When designing a capacitive type MMA, there is considerable scope for varying the shape and size of the SE and suspensions. Finite element analysis (FEA) is used to model the sensitive element design. FEA is a numerical method that allows complex systems to be represented as a set of simple elements. The SE structure of capacitive type MMA was analyzed using computer modeling in ANSYS.

The following mathematical relations were used to generate and create the model in the simulation program of ANSYS. Stress is related to strain by the expression:

$$\{\sigma\} = [D] \cdot \{\varepsilon^{el}\} \qquad (4)$$

where $\{\sigma\} = \sigma_x \sigma_y \sigma_z \sigma_{xy} \sigma_{yz} \sigma_{xz}{}^T$ is the mechanical stress tensor; $[D]$ is the stiffness matrix, the inverse matrix; $\{\varepsilon^{el}\} = \{\varepsilon\} \cdot \{\varepsilon^{th}\}$ is the elastic strain vector; $\{\varepsilon\} = \varepsilon_x \varepsilon_y \varepsilon_z \varepsilon_{xy} \varepsilon_{yz} \varepsilon_{xz}{}^T$ is the total strain vector; $\{\varepsilon^{th}\}$ is the thermal strain vector.

In case of a volumetric model:

$$\{\varepsilon\} = \Delta T \; a_x^{se} \; a_y^{se} \; a_z^{se} \; 000 \; {}^T \qquad (5)$$

where $a_x^{se} \, a_y^{se} \, a_z^{se}$ are the linear coefficients of thermal expansion in the directions of x, y and z, respectively [6].

The main requirement of signal processing devices for capacitance changes is high sensitivity. The mechanical deformation results in silicon SE were calculated by acting the applied linear acceleration ranges along the working X-axis. Capacitive changes between the gaps of movable and fixed electrodes in SE design were calculated using the mechanical deformation results obtained during the analysis of MMA design.

IV. SIMULATION RESULTS IN MMA DESIGN

A. Mechanical Deformation Results

The design of silicon SE MMA was studied using the SOLIDWORKS and ANSYS programs. The objective of this paper was to study the effect of parameter changes on the sensitivity of design to consider the optimal dimension values to magnify the sensitivity of the proposed MMA SE design. When the linear acceleration ranges occur, the SE of MMA sensor deforms. These deformation changes were allowed to determine the capacitive changes by acting the

979-8-3315-4784-4/26 $31.00 © 2026 IEEE

applied acceleration and analyze the sensitivity of MMA sensor.

The deformation results of the MMA SE structure along the X-axis under an acceleration range of 50g were investigated. The results were also analyzed for the case of acceleration along the working X-axis. Illustrations of the simulation result in the proposed design are shown in Figure 2.

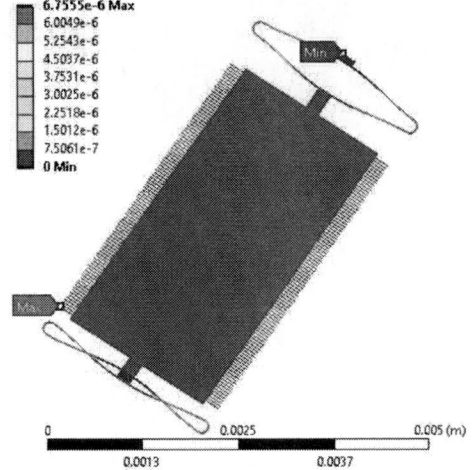

Fig. 2. Illustration of the simulation result of mechanical deformation in MMA design with 100 pairs of comb fingers under an acceleration of 50g along the X-axis

TABLE I. RESULTS OF MECHANICAL DEFORMATIONS IN MMA DESIGN ALONG THE WORKING AXIS OF X, CALCULATED IN ANSYS PROGRAM

Number of comb fingers	Applied Acceleration, g	Mechanical deformation along the X-axis, μm
100	10	1.3511
	20	2.7022
	30	4.0533
	40	5.4044
	50	6.7555

Measuring the deformation results along the working axis was allowed to study the change in suspension geometry under the applied acceleration and determine the sensitivity of capacitive type accelerometer. To obtain the optimized sensitivity of the MMA SE design, it is necessary to use the obtained deformation results by the displacement of proof mass.

B. Calculations of Mechanical Stress

Effect of mechanical stress plays a critical role in calculations, as it allows us to assess the tolerance of materials and predict their form under various loading conditions. Advantages of mechanical stress help people to create more reliable and safe designs, as well as optimize the application of existing and new material properties. The suspension movement during the working vibrations is relatively small and the mechanical stress is within the strength limits of material. The values of mechanical stress under acceleration were determined. Illustrations of mechanical stress simulations on a SE structure with 100 comb fingers using ANSYS software are shown in Figure 3.

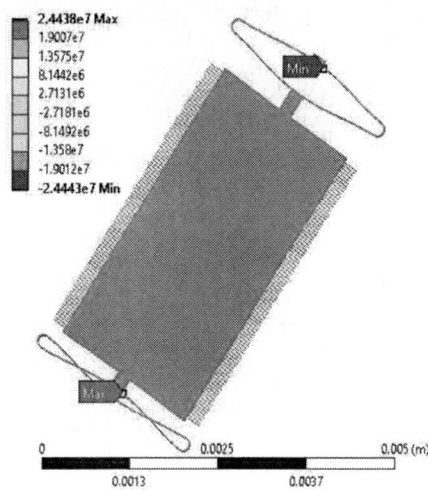

Fig. 3. Illustration of the simulation result of mechanical stress in the design of SE under the acceleration range of 50g along the X-axis

TABLE II. RESULTS OF MECHANICAL STRESSES IN MMA SE DESIGN ALONG THE WORKING X-AXIS, CALCULATED IN ANSYS PROGRAM

Number of electrodes	Applied Acceleration, g	Mechanical stress along the X-axis, MPa
100	10	4.8875
	20	9.7751
	30	14.663
	40	19.55
	50	24.438

C. Capacitive Changes between Comb Fingers of SE in MMA Design

The analysis of the silicon design of the capacitive type MMA SE includes a combination of electrostatic electrodes [7]. The capacitive changes happen due to several pairs of movable and fixed electrodes, which are connected in parallel on the SE MMA design. The sensitivity of the SE design is studied in terms of capacitive and displacement sensitivity, which are improved through optimal selection of design parameters [6]. The dependence of capacitive changes along the X-axis on the number of electrodes under the acceleration of 50g is shown in Fig.4. Thus, the total values of capacitance appear as a combination of the capacitances between the electrode.

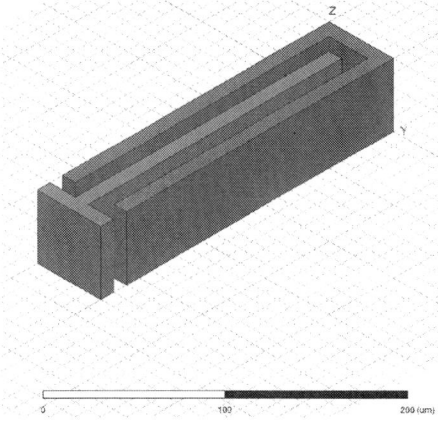

Fig. 4. Change in capacitance with one pair of fingers between moving and fixed electrodes under the acceleration of 50g

The values of capacitive changes for the gap of single pair of movable and fixed comb fingers were studied using ANSYS software. An electrical voltage is applied to the U-shaped movable electrode, and ground is applied to the T-shaped fixed electrode [7].

The results of capacitive changes between the movable and fixed electrodes of the MMA design along the working X-axis under the acceleration range (10g-50g) in ANSYS program are presented in Table 3.

TABLE III.   RESULTS OF CAPACITIVE CHANGES BETWEEN THE MOVABLE AND FIXED ELECTRODES OF SILICON SE DESIGN

Number of comb fingers	Acceleration range, g	Mechanical deformation along the X-axis, μm	Capacitive changes between comb fingers along the X-axis, pF
100	0	0	1.9274
	10	1.3511	1.961
	20	2.7022	2.0654
	30	4.0533	2.2659
	40	5.4044	2.6334
	50	6.7555	3.3622

The design of the capacitive MMA sensor with 100 pairs of combs is optimal in terms of sensitivity and weight and size parameters.

D. Study of Natural Frequencies of Oscillations of MMA Design under the Acceleration

One of the most parameters in this paper is the resonant frequency. It is necessary to apply different boundary conditions for each of operation modes. When the acceleration is applied to the design of SE, different vibration modes can be represented by comparing either with a moving free design on a vibrating or non-vibrating surface (i.e. proof mass, sensitive comb fingers, fixed anchor and torsions). Here, a torsion bar (torsion suspension) is an element of the accelerometer design that consists of an elastic crossbar on which a sensing element is suspended. The shape and size of the torsion bars determine the sensitivity of the device.

In MEMS accelerometers, the spring constant of the torsion spring depends on its shape and the material used. The spring constant in a given direction, and its ratio to the same spring constants in other directions, depend on its design.

By depending on the application fields, the sensitivity and operating resonance frequency of the accelerometer is different [8]. For example, an accelerometer used in navigation system should be more sensitive than smart phones. In another case, the required operating frequency of the accelerometer is low for the earth-quake detection, but it must be extremely high in missile intelligence [8]. The design o analyze is shown in Fig 1. This design is simple and easy to research and manufacture. For the modeling of MMA design, the following parameters were determined: the proof mass is fixed at the end of each of the supported torsion bars (springs) and anchors. The comb layers (movable and fixed capacitors), give a total thickness of 50μm and have a total length and width of 200μm and 10μm each. The length and the width of the torsion bars are 2400μm and 200μm respectively, including the central hole.

When an acceleration is applied to the accelerometer's sensitive axis, which coincides with the torsion bar axis.

This causes a change in the frequency and capacitance deposited on the deformable part of the torsion bar.

For the modal analysis of the capacitive type MMA design, the simulations were performed by using the ANSYS software.

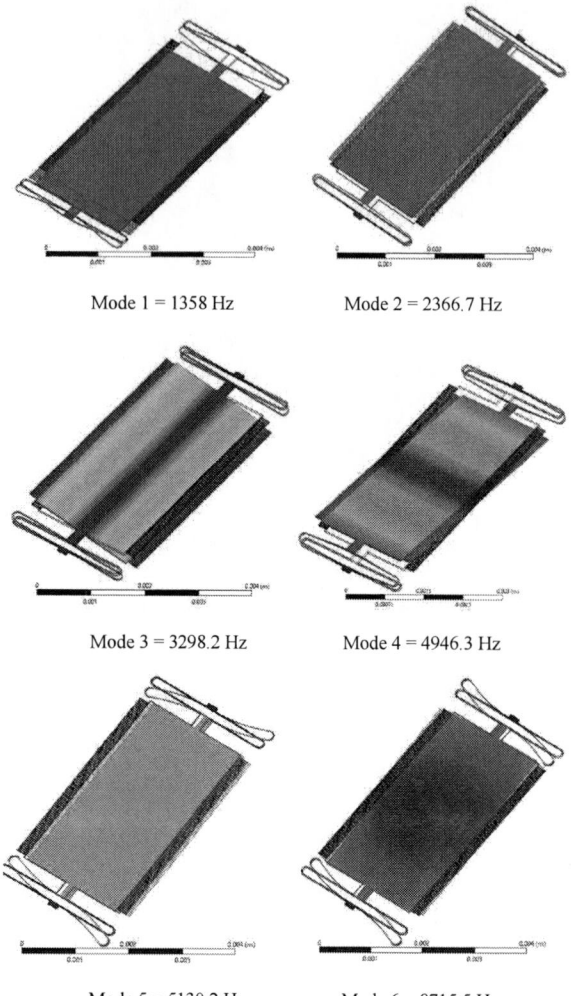

Fig.5. Modal analysis of frequency vibrations by the silicon SE design of the capacitive type MMA, calculated in the ANSYS program

Results of the modal analysis in the natural frequency and vibration modes shown in Fig 5. By the mechanical vibrating theory, the dimension of resonant frequency may consider its working frequency ranges from the elasticity system. As more apparent mechanical deformation of the film-beams corresponds with higher resonant frequency, the operational frequency will also expand measurement range [9]. The resonant frequency dimension is exactly related to the geometry size of the spring structure. As a result, all parameters and characteristics of the design are carried out the optimization before the manufacturing process.

According to the presented figures in modeling of the SE design, it can be seen different displacement forms depending on the resulting vibrating modes. It is clear that some vibration modes should be avoided since performance as an accelerometer will be adversely affected, due to gravity is applied only along one axis and the relative position of the sensor is affected only on a single plane.

Modal analysis of the vibrating modes allows us to determine the natural frequencies. The sensor's bandwidth decreases as the natural frequency of the design decreases.

E. Study of Temperature Effects on the Design of the SE

The most important of environmental factor affecting the performance of MEMS sensors, and therefore their reliability, is ambient temperature. Temperature changes effect the design of capacitive type MMA in various ways: through changes in material and thermal properties, thermomechanical deformation, and mechanical stress. The value of Young's modulus (E) for different temperatures can be expressed as follows:

$$E = K_{\varepsilon} \cdot T \cdot E_0 \qquad (6)$$

where K_{ε} is the coefficient of thermal expansion (for silicon at 25°C, it is $2.55 \cdot 10^{-6}$ K^{-1}) and E_0 is the Young's modulus of the silicon material (at a temperature of 0 K).

Three types of crystal surface orientation of silicon wafers are mainly used to create MEMS devices: (100), (110), and (111). The values of the elastic modulus (Young's modulus) of single-crystal silicon are $1.3 \cdot 10^{11}$, $1.69 \cdot 10^{11}$, and $1.87 \cdot 10^{11}$ Pa, respectively. The coefficient of thermal expansion of silicon is $(2.5\text{-}5.5) \cdot 10^{-6} \cdot$K^{-1} [10].

The results of deformation were studied under temperatures from -45°C to +65°C in the ANSYS environment, and illustrations of the simulation for a temperature of +65°C are shown in Fig. 6.

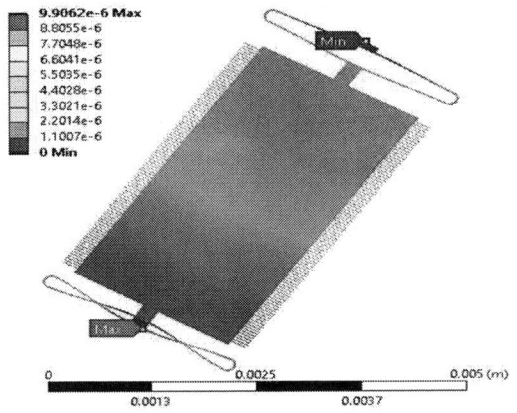

Fig. 6. Illustration of modeling result for thermal deformation of the SE design at the temperature range (+65°C)

Temperature stability is an important factor determining the accuracy and reliability of a micromechanical accelerometer.

TABLE IV. RESULTS OF THERMAL DEFORMATION IN MMA DESIGN ALONG THE AXIS OF X, CALCULATED IN ANSYS PROGRAM

Applied acceleration, g	Temperature range, °C	Thermal deformation along the X-axis, μm	Capacitive changes between comb fingers, pF
50	-45	6.7841	3.3834
	-25	7.3515	3.9323
	0	8.0611	5.0975
	+25	8.7707	7.6331
	+45	9.3385	13.573
	+65	9.9062	91.437

Temperature can significantly affect the accelerometer's sensitivity. The results of deformation and capacitance changes between the comb fingers of MMA design with 100 pairs of electrodes along the working X-axis are described in Table 4.

V. CONCLUSIONS

Modeling and structural analysis of the silicon SE MMA sensor along the working X-axis by acting the applied linear acceleration ranges from 10g to 50g in the simulation software of ANSYS to optimize the sensitivity of the design were studied in this paper. Based on the influence of linear acceleration ranges (10g-50g) along the working X-axis, the results of mechanical deformations and stresses of the capacitive type of MMA design were calculated. According the obtained maximum mechanical stress of the SE design along the working axis is 24.438 MPa, which does not exceed the acceptable value of mechanical stress for silicon material (440 MPa) [11]. The corresponding capacitive changes between the gaps of sensing comb fingers of the SE MMA design along the working X-axis is 3.3622pF by using the obtained values of mechanical deformation of 6.7555μm in the acceleration range of 50g. For the case where the design is used as a sensor for dynamic acceleration ranges, it results that the natural frequency is more 8kHz, which corresponds to the vibration mode of the sensor and its application. The influence of temperature effects in ranging from -45˚C to +65˚C on the displacement of the MMA SE structure with a thickness of 50μm was studied. The values of thermal deformation in MMA design were calculated based on the temperature influences. According to the analytical results of the deformation results in the design of silicon SE under the influence of acceleration range in 50g, the maximum deformation of the design along the X-axis is 9.9062μm and the effect of these temperature ranges from -45°C to +65°C on the displacement of a 50μm in thickness of MMA SE structure was studied. As a result of the impact of the thermal deformation results in the sensor's design with the 100 pairs of comb fingers under the influence of acceleration, the maximum value of the capacitive changes is 91.437pF in the temperature ranges from -45˚C to +65˚C.

ACKNOWLEDGEMENT

The authors of the study extend their special thanks to the Institute of Nano- and Microsystem Technology, Director of NMST Sergey P. Timoshenkov for his help on the explanation in this research project and this paper was supported by National Research University of Electronic Technology (MIET).

REFERENCES

[1] A. Piltan, M. Piltan, R. Ghodsi, "MEMS Technology in Automobile Industry: Trends and Applications," 1st International, 5th National Conference on Management of Technology. Tehran, Iran, May 2014, pp. 1-8.

[2] [2] R. Gholamzadeh, K. Jafari, and M. Gharooni, "Design, Simulation and Fabrication of a MEMS Accelerometer by using Sequential and Pulsed-Mode DRIE Processes," Journal of Micromechanics and Microengineering. Quchan, Iran, vol. 27, DOI: 10.1088/1361-6439/27/1/015022, pp. 1-10, November 2016.

[3] R. Kang, W. Xiaoli, Z. Mengqi, H. Chenyuan, L. Huafeng and T. Liang-Cheng. A MEMS Micro-g Capacitive Accelerometer Based on Through-Silicon-Wafer-Etching Process. Micromachines. DOI: 10.3390/mi10060380. 2019. C.1-14.

[4] MEMS Sensors and Actuators / Vishwas N. Bedekar and Khalid Hasan Tantawi // Microsystems and Nanosystems. 2017. P. 195-216. http://DOI 10.1007/978-3-319-32180-6_10.

[5] A CAPACITIVE ACCELEROMETER MODEL / Florin Constantinescu, Alexandru Gabriel Gheorghe, Miruna Nițescu // 2017. P.163-172.

[6] Micromechanical Accelerometers: Modeling of Structural Elements and Manufacturing / A. Boyko, A. Zavodyan, B. Simonov // ELECTRONICS: Science, Technology, Business. 2009. P. 100-103.

[7] Structural Design and Modeling of MEMS-based Single Axis Capacitive Accelerometer / Paing Soe Thu, V.V. Kalugin, E.S. Kochurina, Thu Ta // Proceedings of the 2024 Conference of Young Researchers in Electrical and Electronic Engineering (2024-EICon). 2024. ISBN 979-8-3503-6064-6. P. 559-564.

[8] Modal analysis of a structure used as a capacitive MEMS accelerometer sensor / G.S. Abarca-Jiménez, M.A. Reyes-Barranca, S. Mendoza-Acevedo, J.E. Munguía-Cervantes and M.A. Alemán-Arce // 11th International Conference on Electrical Engineering, Computing Science and Automatic Control (CCE). 2014. P. 422-425. DOI: 10.1109/ICEEE.2014.6978263.

[9] Resonant Frequency Analysis for Spring-mass Structure in High-g MEMS Accelerometer / Zhenya Geng, Yi Shen, Miao Zhang, Muhua Li, Jing Jin. P. 1-4.

[10] Analysis of Temperature Stability and Change of Resonant Frequency of a Capacitive MEMS Accelerometer / Xuan Luc Le, · Kihoon Kim, Sung-Hoon Choa // International Journal of Precision Engineering and Manufacturing. 2022. P. 347-359. https://doi.org/10.1007/s12541-021-00602-1

[11] Vavilov V.D., Timoshenkov S.P., Timoshenkov A.S.," Microsystem sensors of physical quantities," electronic ISBN 978-5-94836-498-8, pp.529-533.

Application of an Autoencoder for Improving the Accuracy of OTDR Measurements

Pavel Petrov
The Bonch-Bruevich Saint Petersburg State University of Telecommunications
St. Petersburg, Russia
mrfreej@outlook.com

Nikolay Kolybelnikov
The Bonch-Bruevich Saint Petersburg State University of Telecommunications
St. Petersburg, Russia
ya.nikolai-kolyb@yandex.ru

Abstract— We evaluate noise suppression for Optical Time-Domain Reflectometry (OTDR) reflectograms with the goal of reducing backscatter fluctuations while preserving diagnostically relevant event signatures. A residual neural-network autoencoder is trained on paired Measured/Clean signals generated by a physically consistent 50-km simulator with instrument effects (impulse-response convolution, saturation-induced dead zone, dynamic-range limiting, and ADC quantization) and four defect classes (splice, reflective connector, bend, crack). The autoencoder is compared with wavelet denoising and one-dimensional Wiener filtering under low-end and high-end OTDR profiles. For the low-end profile, the autoencoder reduces event-region RMSE from 1.128 to 0.619 dB (vs. 1.165/1.127 dB for wavelet/Wiener) and background RMSE from 0.061 to 0.019 dB; similar but smaller gains are observed for the high-end profile.

Keywords— *reflectogram denoising, Rayleigh backscatter noise, event signature preservation, wavelet thresholding, Wiener filtering, residual autoencoder.*

I. INTRODUCTION

The relevance of fiber-optic communication line diagnostics is driven by the increasing length of backbone and distribution networks, as well as by stricter requirements for the detection time and localization of degradations. One of the fundamental tools of operational monitoring remains Optical Time-Domain Reflectometry (OTDR), which enables the distribution of losses to be assessed from a reflectogram, local inhomogeneities to be identified, and reflective events to be registered. However, the informativeness of OTDR reflectograms is substantially limited by backscatter noise, the finite bandwidth of the receiver chain, and dynamic effects of the instrument. These limitations are particularly pronounced for mass-market devices with a low dynamic range and limited analog-to-digital converter (ADC) resolution, where increasing the number of averages only partially compensates for noise while simultaneously increasing the measurement duration.

The problem of noise suppression in reflectograms has a fundamental peculiarity: it is necessary not only to reduce fluctuations in the background region but also to preserve the local event structure used for defect interpretation. For non-reflective inhomogeneities (e.g., a splice or a bend), the key parameters are the step shape and amplitude. For reflective events (e.g., a connector or a crack), an additional reflection peak and a dead zone arise due to receiver saturation and finite recovery time. Under these conditions, aggressive smoothing can visually improve the background while simultaneously shifting edges and amplitudes, thereby reducing the accuracy of parametric defect estimation and complicating comparison with the physically expected signature [1].

Classical approaches to noise suppression in OTDR measurements include wavelet denoising and one-dimensional Wiener filtering. Their appeal is associated with simple tuning procedures and effective attenuation of stationary noise components. At the same time, in the presence of sharp edges and reflection peaks, these methods often exhibit limited performance in event regions [1].

In this study, we investigate noise suppression in OTDR reflectograms with the aim of improving the reliability of defect detection and parametric estimation without modifying the instrument hardware. To this end, we propose using a neural-network autoencoder, i.e., a model that encodes the input signal into a compact latent representation and subsequently reconstructs it [2, 3]. In contrast to linear filters, an autoencoder acts as a nonlinear regularizer capable of capturing the typical structure of OTDR signals: it suppresses noise fluctuations while preserving diagnostically critical elements (loss steps, local regions of increased attenuation, and reflection peaks) without introducing nonphysical artifacts. Thus, the proposed processing is considered a software-based means of increasing reflectogram informativeness, comparable in purpose to hardware improvements, yet achievable without changes to the measurement chain.

The objective of this work is to evaluate the effectiveness of an autoencoder for noise suppression in OTDR reflectograms while preserving diagnostically significant defect shapes, and to compare the obtained results with classical filtering methods.

To achieve this objective, the following tasks are addressed: development of a physically consistent model of reflectogram formation for two OTDR profiles; construction of a set of defect scenarios (splice, connector, bend, and crack) and an experimental protocol; and configuration and comparative evaluation of wavelet denoising, Wiener filtering, and a pre-trained autoencoder.

II. MATERIALS AND METHODS

Generation of synthetic reflectograms. First, an "ideal" Rayleigh backscatter profile is specified, which decays with distance according to the fiber attenuation. Then, defects of four classes are repeatedly placed along the link: a splice (a loss step of approximately 0.2 dB), a reflective connector (a step of approximately 0.3 dB and a reflection peak), a bend (a spatially extended region of increased loss of approximately 1 dB), and a crack (a short segment with a strong loss of approximately 3 dB with a reflective component). After superimposing the defects, the OTDR instrument response is taken into account: the profile is convolved with an equivalent impulse response (accounting for pulse duration and receiver bandwidth). For reflective events, an additional dead zone is modeled, which arises due

979-8-3315-4784-4/26 $31.00 © 2026 IEEE

to receiver saturation and subsequent recovery of the signal level.

Reference and measurement. For quantitative comparison, two signals are introduced. Clean corresponds to a "noise-free instrumental" reflectogram: it includes all deterministic instrument effects (convolution, saturation, and the dead zone) and all defects, while excluding noise and discretization distortions. Measured corresponds to the output of a real OTDR: noise components and discretization effects are added to Clean, averaging over a specified number of realizations is performed, after which dynamic-range limiting and ADC quantization are applied [4, 5]. The parameters of the two OTDR profiles (low-end and high-end), which determine the pulse duration, receiver bandwidth, ADC resolution, dynamic range, and the number of averages, are provided in Table I.

TABLE I.   OTDR System Parameters Used in the Simulations

Parameter	Symbol	Low-end OTDR	High-end OTDR
Probe pulse duration	τ_{imp}	80 ns	30 ns
Receiver bandwidth (−3 dB)	f_{rx}	60 MHz	200 MHz
Dynamic range	DR	26 dB	42 dB
Analog-to-digital converter (ADC) resolution	N_{ADC}	12 bits	14 bits
Number of averages	N_{avg}	16	64
Equivalent backscatter noise variance (at peak level)	σ_{noice}	0.8 dB	0.4 dB
Reflective-event dead zone	$L_{dz,ref}$	5 m	3 m
Recovery region after a reflective event	L_{rec}	20 m	10 m

Noise suppression methods. Processing is applied to Measured and aims to bring the result closer to Clean. The following filtering methods are compared.

Wavelet denoising. We denoise the reflectogram in the dB domain using DWT with the Symlet-8 wavelet (sym8) and $L = 7$ decomposition levels. Soft-thresholding is applied to all detail coefficients with the universal threshold $\lambda = \hat{\sigma}\sqrt{2\log N}$, where $\hat{\sigma} = \text{MAD}(d_1)/0.6745$ is estimated from the first-level detail coefficients. Symmetric signal extension is used at the boundaries.

Wiener filtering. We apply a 1D local Wiener filter with a sliding window of $W = 101$ samples (symmetric padding at the boundaries). The noise variance is fixed to $\sigma_n^2 = 0.8^2\,\text{dB}^2$ for the low-end profile and $\sigma_n^2 = 0.4^2\,\text{dB}^2$ for the high-end profile; we use $\varepsilon = 10^{-8}$ for numerical stability.

Autoencoder. We use a 1D residual convolutional autoencoder with a U-Net–like encoder–decoder. The encoder consists of four strided Conv1D blocks (kernel size 7, stride 2) with channel widths 16–32–64–128 and ReLU activations; the decoder mirrors the encoder with upsampling and skip connections. The last 1×1 Conv1D layer predicts the residual $\Delta x(z)$, and the output is $\hat{x}(z) = x(z) + \Delta x(z)$; the model has ≈ 0.52M parameters. Training is performed on synthetic Measured/Clean pairs by extracting windows of length 4096 samples with 50% overlap, yielding 120k windows per OTDR profile (80/10/10 train/val/test split); inputs are normalized to zero mean and unit variance per window. We minimize MSE between $\hat{x}(z)$ and Clean using Adam (learning rate 10^{-3}, batch size 64, weight decay 10^{-5}) for up to 40 epochs with early stopping (patience 5); random seed is fixed to 42 [2, 3].

Wavelet denoising and Wiener filtering are selected as baseline methods for comparison in the experiments for two reasons [1]. First, both approaches are de facto standards for one-dimensional noise suppression in measurement signals and are widely used for processing OTDR reflectograms due to their ease of implementation, reproducible parameter tuning, and low computational cost. Second, these methods represent two fundamentally different filtering classes: wavelet denoising implements sparsity-promoting thresholding in a multiscale representation (effectively suppressing local fluctuations), whereas the Wiener filter performs linear, statistically optimal adaptation (in the MSE sense under Gaussian assumptions) to the local noise variance. This combination provides a meaningful reference: if the autoencoder yields improvements relative to both methods, the gain can be interpreted as going beyond classical linear and multiscale regularization under comparable computational constraints and without resorting to instrument-specific heuristics.

Experimental setup. For each OTDR profile, a test set of reflectograms is generated that is not used for parameter tuning of the methods. For each realization, quality metrics are computed for Measured and for the three processed variants (Wavelet, Wiener, and AE). To evaluate errors in defect regions, an event mask is defined that includes windows around all inserted defects; a background region is identified separately, located away from defects and unaffected by dead zones of reflective events. In defect analysis, non-reflective events (splice and bend) are distinguished from reflective events (connector and crack), since for the latter the saturation and recovery region may constrain interpretation of the local loss "step" in the immediate vicinity of the reflection peak.

III. Results

The results obtained are conveniently interpreted through two complementary aspects. First, the overall proximity of the processed reflectogram to the reference is important and is characterized by the root mean square error (RMSE) over the entire trace, in the vicinity of defects, and in the background region. Second, it is necessary to understand which specific changes lead to the improvement, i.e., suppression of the structural backscatter noise, preservation of defect shapes, or a trade-off between these factors. In what follows, the comparison is performed using summary RMSE values (Fig. 1) and representative defect signatures (Figs. 2–3).

We begin with the overall RMSE trends. For the low-end OTDR, the autoencoder provides the largest error reduction in all three regions: the total RMSE decreases from 0.131 to 0.066 dB, the RMSE in defect regions decreases from 1.128 to 0.619 dB, and the background RMSE decreases from 0.061 to 0.019 dB. For wavelet denoising and Wiener filtering, the improvement is mainly observed in the background region: the background RMSE decreases to 0.021 and 0.031 dB, respectively. At the same time, no gain is observed in defect regions: for the wavelet method, RMSE_{ev} even slightly increases (1.165 dB versus 1.128 dB for Measured), whereas for the Wiener filter it remains essentially unchanged (1.127 dB). Thus, under a low initial signal-to-noise ratio (SNR), classical methods reliably "stabilize" the background but do not improve agreement with the reference shape in event neighborhoods, whereas the autoencoder yields a pronounced error reduction precisely where diagnostics is performed.

For the high-end OTDR, a similar but less pronounced pattern is observed. The total RMSE decreases from 0.096 to 0.059 dB when using the autoencoder, the RMSE in defect regions decreases from 0.933 to 0.571 dB, and the background RMSE decreases from 0.015 to 0.005 dB. Wavelet denoising and Wiener filtering again provide noticeable background suppression (down to 0.006 dB), but they almost do not change the error in defect regions: RMSE_{ev} equals 0.940 dB for the wavelet method and 0.933 dB for the Wiener filter, compared with 0.933 dB for Measured. This is consistent with the fact that, at a high initial SNR, the potential for improving event shapes is limited, and linear methods predominantly act as background "cleaning" without a substantial move toward the reference instrumental event signatures.

■Low-end ■High-end

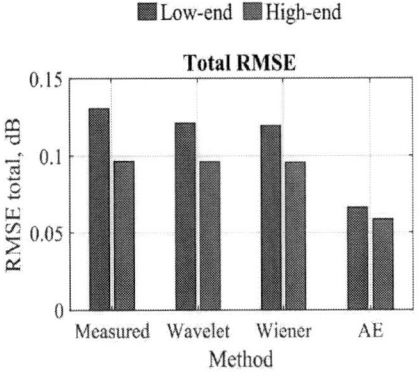

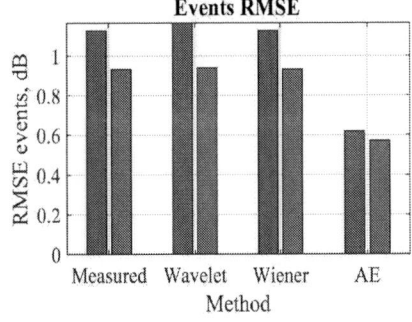

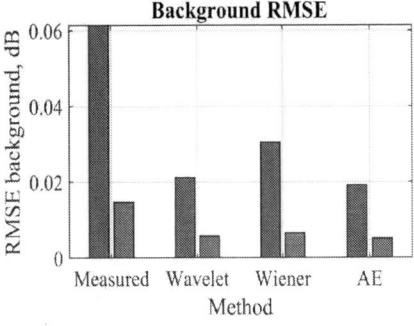

Fig. 1. Comparison of denoising performance in terms of RMSE for four processing methods (Measured, wavelet denoising, Wiener, AE) and two OTDR profiles: low-end (blue) and high-end (red) Turning to the analysis of defect signatures in Figs. 2–3, it is reasonable to treat them as an illustration of the mechanisms underlying the numerical trends. For the low-end OTDR, wavelet denoising exhibits the most aggressive noise suppression, which is consistent with the low background RMSE. However, in this case the distortion of the local shape of non-reflective defects is also the most pronounced: the step edge and level transitions become smoother and more stretched. Given that RMSE_{ev} for the wavelet method does not decrease, this behavior can be interpreted as a "transfer" of part of the informative event structure energy into components that are suppressed by the denoising procedure.

Wiener filtering, in contrast, provides a milder suppression of background fluctuations and leaves more residual oscillations, which is reflected in a higher background RMSE compared with wavelet denoising. At the same time, the transition shapes in defect regions are, on average, preserved better; therefore, RMSE_{ev} remains close to its initial value. This behavior appears expected: the Wiener filter adapts to the local variance and, in the vicinity of events, reduces the degree of smoothing to avoid blurring abrupt level changes. As a result, a compromise is achieved in which the background improves, but no pronounced convergence to the reference is obtained specifically in defect regions.

The autoencoder differs in that it simultaneously reduces both the background RMSE and the RMSE in defect regions. For the low-end OTDR, this manifests as the most substantial error reduction for events while maintaining a low level of residual fluctuations. For the high-end OTDR, an improvement in events is also observed, albeit to a smaller extent, which is consistent with the lower impact of noise in the original signal. Taken together, these observations support interpreting the autoencoder as a nonlinear method capable of suppressing coherent fluctuations without the event-region degradation of agreement with the reference instrumental shape that is typical of linear filtering.

A separate comment is required regarding the results of step-amplitude estimation. For non-reflective defects (splice and bend), estimation accuracy is strongly determined by how well a method preserves the edge and the pre-/post-event levels. In the low-end regime, a splice with the true magnitude of 0.2 dB is estimated as 0.147 dB in Measured; after wavelet processing, the value decreases to 0.070 dB, indicating the largest bias, consistent with oversmoothing of the shape. For Wiener filtering and the autoencoder, the estimates are 0.139 and 0.129 dB, respectively, i.e., they remain closer to the original level, although they still underestimate the true value. For a bend with the true loss of 1 dB, all methods yield similar values; however, the wavelet method again shows a larger deviation (0.922 dB) compared with Wiener filtering (0.984 dB) and the autoencoder (0.966 dB). In the high-end regime, the step-estimation errors for the splice and bend are small for all methods, which is consistent with the high initial reflectogram quality.

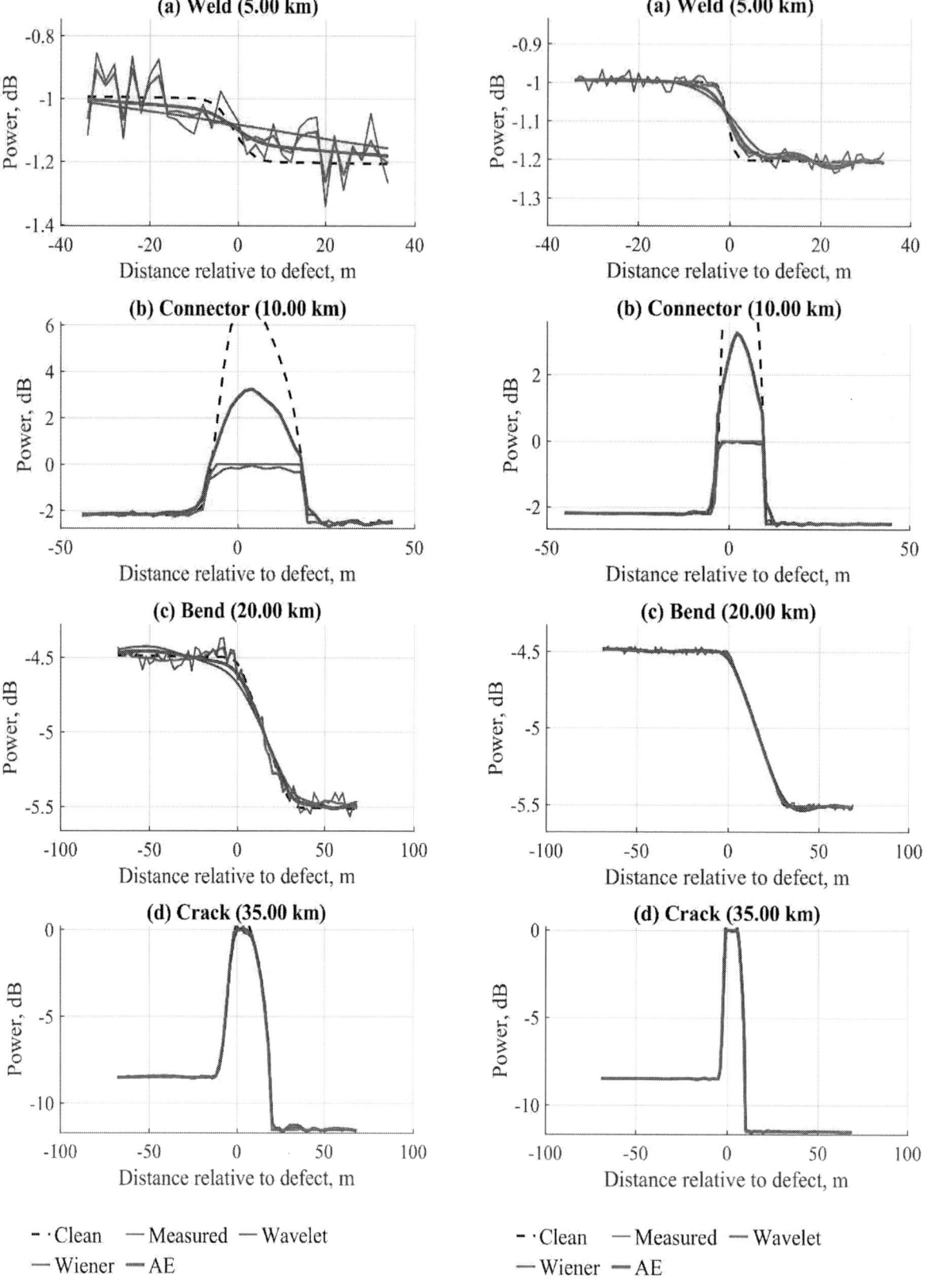

Fig. 2. Typical defect signatures for the low-end OTDR profile: (a) weld, (b) reflective connector, (c) bend, (d) crack

Fig. 3. Typical defect signatures for the high-end OTDR profile: (a) weld, (b) reflective connector, (c) bend , (d) crack

For reflective defects (connector and crack), the "step" measured in the immediate vicinity of the reflection peak is not a stable estimate of the insertion loss due to saturation and the dead zone. This is particularly evident in the low-end regime for the crack: for a true loss of 3 dB, the estimated "step" remains on the order of 0.16–0.20 dB for all methods. Therefore, under these conditions, interpreting reflective events based solely on the local step is limited by the instrument dynamics and requires a dedicated procedure that excludes the saturation/recovery region. Within the present comparison, for reflective defects it is more appropriate to focus on integral metrics in the event region and on the consistency of the signature with the reference, rather than on a direct comparison of the "step" with the true loss magnitude.

Overall, the numerical results (Fig. 1) and their consistency with the behavior of the methods on representative signatures (Figs. 2–3) indicate that the classical methods provide reliable background suppression but have limited effectiveness in defect regions, whereas the autoencoder yields a systematic error reduction specifically in event areas while maintaining a competitive level of noise suppression.

IV. DISCUSSION

The obtained results indicate differences in the mechanisms by which reflectogram quality is improved. Wavelet denoising and Wiener filtering, as classical regularization methods, primarily reduce the variance of fluctuations in the background region. This is natural for OTDR, where a substantial fraction of distortions is due to structured backscatter noise. However, the same suppression of high-frequency components inevitably affects edges and local level changes; therefore, background improvement does not guarantee error reduction in defect regions. The observed stability of $RMSE_{ev}$ for the Wiener filter and the increase of $RMSE_{ev}$ for the wavelet method in the low-end regime are consistent with the fact that smoothing partially blurs the instrumental shape of steps and localized losses.

The autoencoder exhibits a different behavior: the RMSE reduction in defect regions is accompanied by background suppression at a level not inferior to that of the classical methods. This can be interpreted as an effect of nonlinear approximation, in which the model trained on Measured/Clean pairs suppresses statistical fluctuations while preserving structural elements typical of physically plausible instrumental defect signatures. The particularly pronounced advantage in the low-end regime can be attributed to a combination of factors: lower ADC resolution, narrower receiver bandwidth, and fewer averages increase the contribution of noise and discretization distortions; therefore, the potential for improvement via "informed" regularization is higher. In the high-end regime, the initial SNR is substantially higher, and thus the gain in event regions is expectedly smaller.

Reflective defects require a separate interpretation. In the region of the reflection peak and the subsequent dead zone, the signal is governed by receiver saturation and recovery; under these conditions, no method should "recover" information that is physically absent from the measurement. Therefore, insertion-loss estimation based on the local step adjacent to the peak becomes unstable, especially in the low-end regime. In practice, this implies that for reflective events it is more appropriate to rely on integral metrics within the event region and on parameters that are not affected by saturation (e.g., peak characteristics and recovery length), or to estimate the levels before and after the event while excluding the dead-zone region.

The limitations of the proposed approach include sensitivity to mismatch between the instrument model and the actual measurement device, as well as to the ranges of defect parameters represented in the training data. In particular, if the receiver bandwidth, pulse duration, or saturation dynamics change substantially, adaptation or fine-tuning of the autoencoder may be required. A promising direction is to transfer the method to real OTDR traces by calibrating the instrument-response parameters and estimating reconstruction uncertainty, which would enable the autoencoder to be used as a component of a practical monitoring system without sacrificing the physical interpretability of the results.

V. CONCLUSION

In this work, we investigated noise suppression in OTDR reflectograms using an autoencoder and compared its performance with wavelet denoising and Wiener filtering under two instrument profiles representing mass-market (low-end) and professional (high-end) OTDR classes. The obtained results indicate that the classical methods provide robust reduction of residual fluctuations in the background region; however, their effectiveness in defect regions is limited, as background improvement is not accompanied by a comparable decrease in reconstruction error in event neighborhoods.

The considered autoencoder demonstrates a stable error reduction both in the background region and in defect regions. The most pronounced effect is observed in the low-end regime, where the contribution of coherent structured backscatter noise and discretization distortions is higher and, consequently, the practical value of software-based enhancement of reflectogram informativeness is greater. For the high-end regime, the gain persists but is expectedly smaller due to the higher initial signal-to-noise ratio.

REFERENCES

[1] X. Chen, H. Yu, J. Xu, and F. Gao, "An SNR Enhancement Method for Φ-OTDR Vibration Signals Based on the PCA-VSS-NLMS Algorithm," Sensors, vol. 24, Art. no. 4340, Jul. 2024, doi: 10.3390/s24134340.

[2] S. Chen and W. Guo, "Auto-Encoders in Deep Learning—A Review with New Perspectives," Mathematics, vol. 11, no. 8, Art. no. 1777, Apr. 2023, doi: 10.3390/math11081777.

[3] A. A. Neloy and M. Turgeon, "A comprehensive study of auto-encoders for anomaly detection: Efficiency and trade-offs," Machine Learning with Applications, vol. 17, Art. no. 100572, Jul. 2024, doi: 10.1016/j.mlwa.2024.100572.

[4] N. Lalam, P. S. Westbrook, J. Li, P. Lu, and M. P. Buric, "Phase-Sensitive Optical Time Domain Reflectometry With Rayleigh Enhanced Optical Fiber," IEEE Access, vol. 9, pp. 114428–114434, 2021, doi: 10.1109/ACCESS.2021.3105334.

[5] Y. Muanenda, "Recent Advances in Distributed Acoustic Sensing Based on Phase-Sensitive Optical Time Domain Reflectometry," Journal of Sensors, vol. 2018, pp. 1–16, May 2018, Art. ID 3897873, doi: 10.1155/2018/3897873.

Study of Sensing Elements with Various Resonator Designs on a Square Membrane for High-Precision MEMS Pressure Sensors

Phyo Win Tun
Department of Nano and Micro System Technology
National Research University of Electronic Technology, "MIET"
Moscow, Russia
kophyowinhtun0@gmail.com

Sergey P. Timoshenkov
Department of Nano and Micro System Technology
National Research University of Electronic Technology, "MIET"
Moscow, Russia
spt111@mail.ru

Boris M. Simonov
Department of Nano and Micro System Technology
National Research University of Electronic Technology, "MIET"
Moscow, Russia
serbosel@mail.ru

Abstract - **Resonant micromechanical pressure sensors exhibit high stability due to the conversion of pressure into a frequency output that is resistant to electronic drift. This study presents a comparative finite-element analysis of three resonator designs: a single beam (SR), a double-ended tuning fork (DETF), and a triple-ended tuning fork (TETF) integrated on an identical square silicon diaphragm. Two types of loads on the sensing element are considered: static pressure in the range 100 - 1000 kPa and uniform temperature variation from -40 to +85 °C. Based on static and modal analyses performed in the COMSOL Multiphysics environment, the deformations, mechanical stresses, frequency sensitivity of the SE to pressure changes, and the temperature coefficient of frequency are determined. It is shown that the TETF design provides the optimal combination of sensitivity, linearity, and thermal stability of the SE under identical boundary conditions. The results provide a basis for selecting the SE design at the development stage of high-precision resonant pressure sensors.**

Keywords - MEMS, resonant pressure sensor, tuning-fork resonator, sensitivity, thermal stability.

I. INTRODUCTION

The development of functional sensor modules with enhanced robustness is one of the priority tasks faced by designers of modern electronic equipment [1]. Resonant micromechanical pressure sensors represent one of the most promising classes of precision transducers, since they utilize a frequency output that is inherently protected from electronic drift and degradation of analog front-end circuits [2], [3]. Unlike capacitive or piezoresistive sensing structures, resonant sensors register the applied load through variation of the natural frequency of a mechanical resonator, which provides enhanced long-term stability and straightforward compatibility with digital signal processing.

In our earlier work [4], it was shown that the shape and parameters of the membrane strongly affect the deformation–frequency characteristic of the sensor for a given resonator architecture. However, the architectural topology of the resonator itself, placed on the membrane, is an independent design degree of freedom. Its impact on the sensitivity and thermal stability of the sensing element (SE) has not previously been systematically analyzed.

The choice between a single-beam structure (SR), a double ended tuning-fork structure (DETF), and a triple ended tuning-fork architecture (TETF) determines several key parameters: the transfer coefficient of membrane-induced stress into the resonator; the stability of the resonant frequency against temperature variations; the linearity of the frequency–pressure characteristic; the technological

feasibility; and the sensitivity to fabrication tolerances and process-induced variations. Double-ended tuning-fork resonators with differential response (DETF) are well described in the literature [5–8]. However, those works do not compare DETF resonators with triple-ended (TETF) structures under identical boundary and material conditions.

The aim of this work is to perform a consistent finite-element comparison of three architectures (SR, DETF, TETF) integrated on square silicon membranes, under static pressure and uniform temperature loading. The results are intended to provide a rational basis for selecting the SE structure prior to mask fabrication and process development stages for high-precision resonant MEMS pressure sensors.

II. THEORETICAL BACKGROUND AND OPERATING PRINCIPLE

A. Overall Sensor Design and Working Principle

The resonant principle of pressure-to-frequency transduction relies on the fact that membrane deformation induces mechanical stresses, which, in turn, cause a shift of the natural frequency of the resonator. Since the measured quantity is frequency rather than an analog electrical signal, this approach provides high immunity to parasitic electrical influences on sensor operation, particularly to zero drift and degradation of electronic components [2], [3].

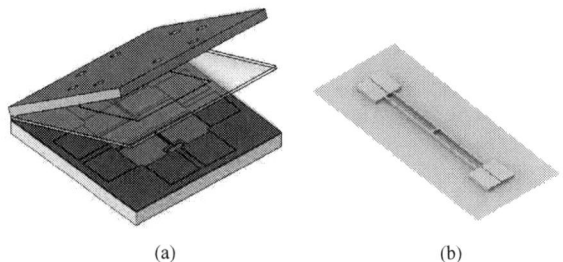

(a)   (b)

Fig. 1. Schematic of the sensing element of the resonant pressure sensor: (a) overall layout; (b) DETF beam resonator integrated on the silicon membrane.

Single-beam resonators represent the simplest form of resonant transducers, but they exhibit high sensitivity of their parameters to technological deviations and thermally induced stresses [2]. Double-ended tuning fork (DETF) resonators are used for partial compensation of temperature effects on the SE: one beam experiences tensile stress, while the other is under compression, which suppresses the contribution of uniform perturbations to the resulting resonant frequency [5].

In recent years, resonator structures with tripled or multi-beam symmetry (e.g., TETF) have appeared, allowing a

979-8-3315-4784-4/26 $31.00 © 2026 IEEE

reduction of mode asymmetries, a decrease in stress concentration, and improved utilization of axial stress for pressure sensing [7], [8]. Such structures can potentially provide higher sensitivity and thermal stability without complicating the readout principle.

In our previous work [4], it was shown that the choice of membrane shape affects the stress distribution and, consequently, sensor sensitivity. The present work extends and complements the results in [4] by focusing not on the membrane, but directly on the resonator architecture under fixed membrane conditions. Comparing SR, DETF and TETF allows us to isolate the contribution of the resonator topology, i.e. the design factor that determines the layout of structural layers prior to device fabrication.

TABLE I. DESIGN PARAMETERS OF RESONANT PRESSURE SENSOR

Parameter	Value	Unit
Overall sensor dimensions	10×10×1	mm³
Pressure sensitive membrane dimensions	5×5×0.3	mm³
Dimensions of resonator	1400×18×40	μm³
Gap between two beams (DETF and TETF)	40×18×40	μm³

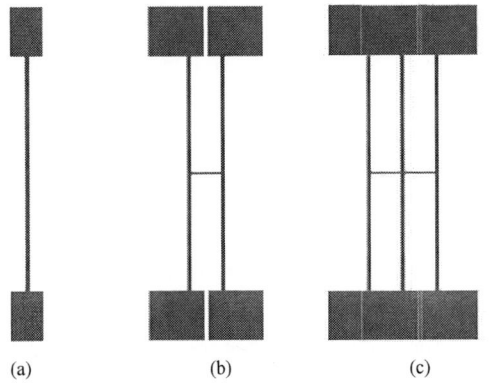

(a) (b) (c)

Fig. 2. Schematic diagram of the three resonator designs: (a) single beam resonator (SR); (b) double-ended tuning-fork resonator (DETF); (c) triple-ended tuning-fork resonator (TETF).

B. Sensitivity of Micromechanical Resonant Pressure Sensors

To interpret the simulation results, it is convenient to use simple analytical relations that describe the dependence of the resonant frequency on axial stress and temperature. For a clamped–clamped beam resonator of length L, cross-sectional area A, bending stiffness EI, and density ρ, the first natural frequency f_0 of the fundamental flexural mode can be approximated as [2]:

$$f_0 = \frac{\beta_1^2}{2\pi L^2}\sqrt{\frac{EI}{\rho A}} \qquad (1)$$

where f_0 is the first resonant frequency, $\beta_1 \approx 4.73$ is the modal constant for the fundamental mode of a clamped–clamped beam, E is Young's modulus, I is the second moment of area of the beam cross-section, and ρA is the mass per unit length.

When an axial stress σ is induced in the resonator, the effective stiffness increases, and the resonant frequency is shifted. In a first approximation, this effect can be written in the form [5]:

$$f(\sigma) = f_0\sqrt{1 + \frac{\sigma}{\sigma_c}} \qquad (2)$$

where $f(\sigma)$ is the resonant frequency under axial loading, σ is the average axial stress in the resonator, and σ_c is a characteristic stress that depends on the resonator geometry and boundary conditions and defines the scale at which axial loading becomes comparable to bending stiffness.

For relatively small stresses, $\sigma \ll \sigma_c$, (2) can be linearized as:

$$\frac{\Delta f}{f_0} = \frac{1}{2}\frac{\sigma}{\sigma_c} \qquad (3)$$

where $\Delta f = f(\sigma) - f_0$ is the stress-induced frequency shift. Relations (2), (3) show that the relative frequency change is directly proportional to the axial stress transmitted from the membrane and inversely proportional to σ_c. Thus, not only the membrane deformation but also the efficiency of stress transfer – determined by the resonator architecture, symmetry, and anchoring – governs the sensitivity and linearity of the frequency–pressure characteristic. In particular, multi-beam architectures (DETF, TETF) can provide higher effective axial stress in the working beams and more favorable distribution of this stress compared to a single-beam SR structure.

For silicon MEMS resonators, the temperature dependence of the frequency is mainly governed by the temperature dependence of Young's modulus $E(T)$ and by thermal expansion with coefficient α [3]. For small temperature variations, the temperature coefficient of frequency (TCF) can be expressed as:

$$TCF = \frac{1}{f_0}\frac{df}{dT} = \frac{1}{2E}\frac{dE}{dT} + \alpha \qquad (4)$$

where df/dT is the temperature derivative of the resonant frequency, dE/dT is the temperature derivative of Young's modulus, and α is the coefficient of linear thermal expansion. The first term in (4) is usually negative for silicon, while the second term is positive, so the net TCF is the result of their partial compensation. Resonator architectures with higher symmetry and out-of-phase strain perception (DETF, and especially TETF) reduce the influence of uniform thermoelastic changes described by (4), whereas a single-beam structure (SR) lacks such internal compensation and therefore exhibits the largest thermal drift.

III. MODELING RESULTS AND DISCUSSION OF THE SENSING ELEMENTS OF A RESONANT PRESSURE SENSOR

To ensure comparability of results, the resonators are compared using a fixed membrane platform for the sensing element and identical boundary conditions. As the main structural element of the SE, we use a square silicon membrane with dimensions 5 × 5 mm and a thickness of about 300 μm, formed in the device layer of an SOI wafer.

Three resonator structures are considered: a single-beam structure (SR), placed in the membrane region with maximum stress gradient; a double-ended tuning-fork resonator (DETF), implementing differential perception of tension/compression in two beams; a triple-ended tuning-fork resonator (TETF), providing higher symmetry and more uniform stress distribution due to the third longitudinal beam connected through common cross-bars.

All three resonators have identical thickness, position on the membrane, and the same anchor configuration, which excludes the influence of any factors other than the resonator architecture itself.

The material properties used in the simulations correspond to two typical materials for integrated resonant

MEMS structures: single-crystal silicon, silicon dioxide. The elastic and density parameters used in the modeling are taken from reference data [2], [8].

The membrane edges are assumed to be rigidly clamped, approximating the conditions of a hermetic Si glass bond. The resonator anchors are assumed to be monolithically connected to the membrane. Table 2 summarizes the properties of the materials used: single-crystal silicon, silicon dioxide.

TABLE II.   MATERIAL PROPERTIES USED IN THE MODELING

Material	Density ρ (kg/m³)	Young's modulus E (GPa)	Poisson's ratio v	CTE α (×10⁻⁶ 1/°C)	Thermal conductivity k (W/m·K)
Single-crystal silicon (Si)	2330	169	0.28	2.6	148
Silicon dioxide (SiO₂)	2200	73	0.17	0.55	1.4

When pressure is applied, the membrane deformation corresponds to the classical bending behavior of a thin plate: the deflection increases monotonically with pressure. However, the critical factor is not the absolute deflection, but how the membrane stress state is transferred into the resonator region. The modelling results for deformation and stresses in SR, DETF, and TETF structures on square diaphragms under pressure of 1000 kPa are shown in Fig. 3. Static membrane deflection and maximum stresses at pressures of 100–1000 kPa for SR, DETF, and TETF are summarized in Table 3.

For the single-beam structure (SR), a pronounced stress concentration is observed near the anchors, which potentially increases sensitivity to technological deviations and long-term relaxation. For the DETF, the stress is redistributed between two parallel beams, reducing local concentration and providing symmetric tension/compression states. The TETF architecture exhibits the most uniform stress field, owing to its higher geometric symmetry and the participation of the third beam in load transfer.

This more uniform stress distribution leads to improved linearity and stability of the frequency response to pressure.

p(10)=1E6 Pa   Surface: Displacement magnitude (m)

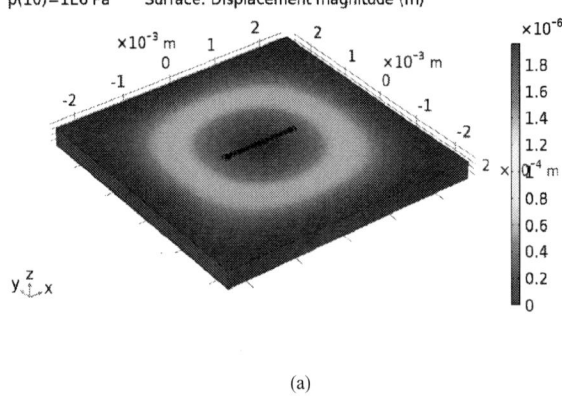

(a)

p(10)=1E6 Pa   Surface: von Mises stress (N/m²)

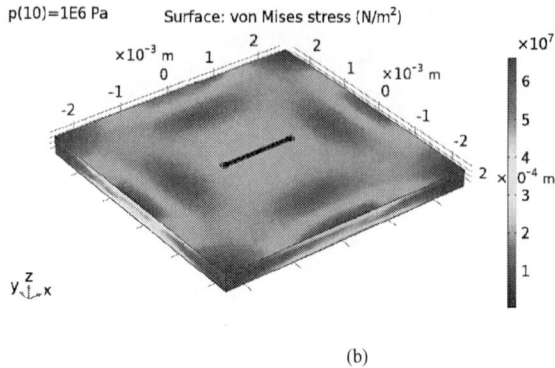

(b)

Fig. 3. Simulation results of deformation in square membrane (a) and mechanical stress distribution in square membrane (b) under an applied pressure of 1000 kPa.

TABLE III.   SIMULATION RESULTS OF MEMBRANE DEFLECTION AND MAXIMUM STRESSES DISTRIBUTION UNDER AN APPLIED PRESSURE 100-1000 KPA

Pressure (kPa)	Membrane deflection on SR (µm)	Deflection DETF (µm)	Deflection TETF (µm)	Max stress SR (MPa)	Max stress DETF (MPa)	Max stress TETF (MPa)
100	0.1845	0.1777	0.1612	42.78	38.33	36.5
300	0.5274	0.4951	0.4736	125.22	116.2	108.0
500	0.8625	0.8122	0.7877	210.08	195.4	182.3
700	1.221	1.162	1.101	295.54	274.4	258.1
1000	1.757	1.659	1.562	420.22	390.8	365.8

The natural (resonant) frequencies computed under pre-stress conditions vary monotonically with pressure, in agreement with the physical nature of membrane-induced axial stresses. The single-beam resonator (SR) on the membrane provides the highest natural frequency due to greater stiffness; however, its relative sensitivity $\Delta f/f_0$ is significantly lower than that of SEs with multi-beam resonator architectures.

Eigenfrequency=7.1788E5 Hz Surface: Displacement magnitude (µm)

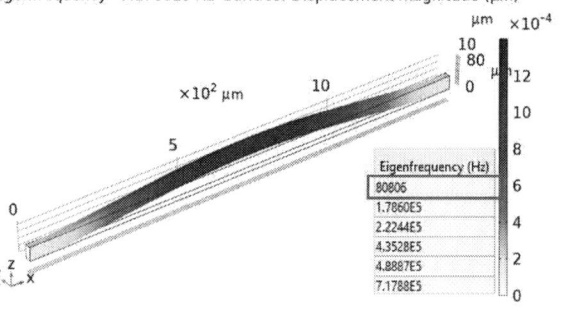

(a)

Eigenfrequency=80309 Hz Surface: Displacement magnitude (µm)

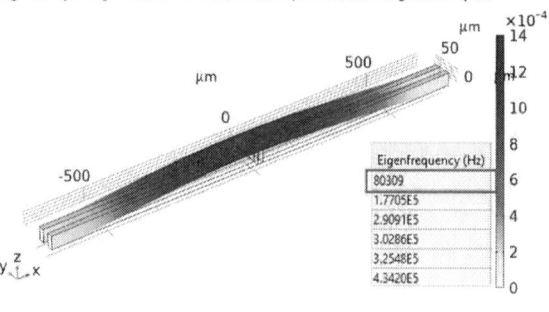

(b)

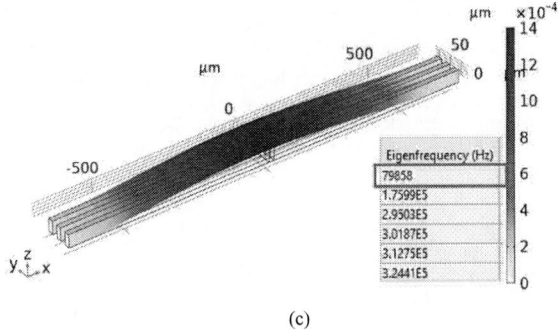

(c)

Fig. 4. Simulation results of resonant frequency under pre-stressed conditions of three different types of resonators: (a) single beam resonator, (b) DETF resonator and (c) TETF resonator.

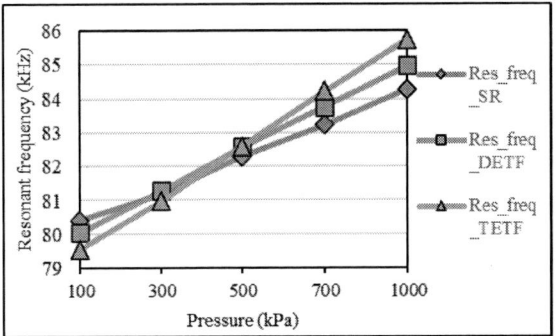

Fig. 5. Resonant frequency versus applied pressure of three different types of resonators.

The dependences of the first-mode resonant frequency on pressure for the three resonator architectures are shown in Fig. 5. For all structures, the frequency increases almost linearly with pressure in the range 100 -1000 kPa. For the SR resonator, the fundamental frequency increases from about 80.806 kHz at 100 kPa to 84.266 kHz at 1000 kPa. For the DETF and TETF structures, the corresponding ranges are approximately 80.055 - 84.969 kHz and 79.858 - 85.728 kHz, respectively. Thus, although the absolute frequency levels of the three designs are close, the relative frequency shift with pressure is significantly larger for the TETF architecture. This results in the highest pressure sensitivity for the TETF resonator, whereas the DETF structure exhibits intermediate sensitivity and the SR resonator provides the lowest sensitivity. The sensitivities df/dp and linearity coefficients R^2 in Table 4 were obtained by linear regression of the simulated resonant frequencies over the pressure range 100 -1000 kPa.

TABLE IV. FREQUENCY SENSITIVITY AND LINEARITY OF DIFFERENT TYPES OF RESONATOR ON A SQUARE MEMBRANE

Resonator type	Resonant frequency f_0 (kHz)	Sensitivity df/dp (Hz/kPa)	Linearity coefficient R^2
SR	80.806	3.84	0.992
DETF	80.055	5.46	0.996
TETF	79.858	6.52	0.998

Under uniform temperature variation, the frequency decreases due to two factors: the reduction of silicon's Young's modulus with heating and thermal elongation of the resonator. For the SR, the temperature influence is most pronounced because its architecture has no internal compensation mechanisms. In the DETF resonator, the temperature contribution is partially compensated due to

antipolar signals from the two beams. The TETF resonator demonstrates the smallest absolute temperature coefficient of frequency (TCF) among the three architectures, indicating an enhanced compensation effect due to geometric symmetry and distributed stiffness. The average TCF values and $\Delta f/f_0$ at $\Delta T=125\ °C$ in Table 5 were extracted from linear fits of the simulated resonant frequencies over the temperature range -40 to $+85\ °C$.

TABLE V. TEMPERATURE COEFFICIENTS OF FREQUENCY (TCF)

Resonator type	Average TCF (ppm / °C)	$\Delta f/f_0$ at $\Delta T = 125\ °C$ (%)	Drift characteristic
SR	−30	−0.38	pronounced negative drift
DETF	−15	−0.19	partial compensation
TETF	−7	−0.09	minimal drift, high thermal stability

When both pressure and temperature vary, the single-beam resonator exhibits a noticeable "cross-sensitivity" of pressure response to temperature, which degrades the predictability of the sensor output. The DETF shows partial suppression of such cross-influence. The TETF maintains an almost invariant slope of $f(p)$ over a wide temperature range, demonstrating high robustness to thermo-induced errors under operating conditions.

The comparative characteristics of the three architectures, summarized in Tables 3 and 4, clearly indicate the advantage of using the three-beam TETF resonator structure in the sensing element of high-precision resonant pressure sensors. The gains in sensitivity, linearity, and thermal stability are achieved at the cost of only moderate complexity increase in technological implementation.

IV. CONCLUSION

A comparative finite-element study of three resonator architectures: single-beam, double-ended tuning fork, and triple-ended tuning fork integrated on a common square silicon membrane of a pressure sensor sensing element has been carried out.

The COMSOL Multiphysics modeling was performed under identical material properties, boundary conditions, and two types of loading of the SE: static pressure and uniform temperature change.

The obtained results demonstrate the following:

- Resonator symmetry is a key factor determining the efficiency of stress transfer from the membrane and the linearity of the frequency response.

- The triple-ended tuning fork (TETF) resonator within the SE provides the highest sensitivity to pressure while maintaining high linearity of the transfer characteristic.

- The TETF resonator exhibits the smallest thermal drift of the resonant frequency due to more complete internal compensation of thermoselastic effects.

- The superiority of the TETF resonator over the other architectures according to the combined criteria is achieved without altering the basic SE design, which makes the choice of resonator architecture a decisive step prior to photomask design.

979-8-3315-4784-4/26 $31.00 © 2026 IEEE

Thus, the triple-ended resonator structure is the most rational candidate for fabrication and experimental verification of sensing elements in high-precision resonant MEMS pressure sensors.

REFERENCES

[1] L. S. Chelyshev, S. P. Timoshenkov, N. I. Plis, and V. O. Orlov, "Multifunctional sensor module with enhanced protection and satellite navigation," Journal, News of Higher Educational Institutions. Electronics, vol. 29, no. 2, pp. 194–202, 2024.

[2] W. P. Eaton, J. H. Smith, "Micromachined pressure sensors: Review and recent developments," Smart mater. Struct, vol. 6, pp. 530–539, 1997.

[3] S. P. Beeby, G. Ensel, M. Kraft, MEMS Mechanical Sensors. Artech House, pp. 97-117, 2008.

[4] Phyo Win Tun, S. P. Timoshenkov, B. M. Simonov, "Study of the Sensitive Element of a Resonant Pressure Sensor with Membranes of Various Shapes," Proc. IEEE, 2023 Seminar on Microelectronics, Dielectrics and Plasmas (MDP), no. 8, pp. 807-810, 2023.

[5] C. Xiang, Y. Lu, P. Yan, J. Chen, D. Chen, "A resonant pressure microsensor with temperature compensation based on differential outputs," Micromachines, vol. 11, pp. 1-14, 2020.

[6] Y. Lu et al., "A resonant pressure microsensor with a stress isolation layer," IEEE Sensors Journal, vol. 19, no. 18, pp. 7875-7883, 2019.

[7] L. Zhu et al., "A high-quality resonant pressure microsensor with through-silicon-via electrical interconnections," Transducers 2017, Kaohsiung, Taiwan, pp. 82-85.

[8] K. Seshan, Handbook of Thin Film Deposition. 2nd ed., Noyes Publications, 2002.

Modeling of a SiC trench gate MOSFET

Mikhail A. Potapov
Department of Micro- and Nanoelectronics
St. Petersburg state electrotechnical University "LETI"
St. Petersburg, Russia
mapotapov@stud.etu.ru

Abstract— The technological process flow of trench gate structures based on SiC is discussed in this article. Trench gated structures show lower specific resistance which is significant in power application. Although silicon carbide is a suitable material, the difference in dielectric constants affects gate oxide reliability. Angular ion implantation and thick trench oxide are suggested to lower the electric field in the bottom oxide region in order to improve the structure's breakdown capability. The influence of the trench depth on device characteristics is studied using TCAD and original trench-MOSFET model. Optimal structure parameters were calculated using physical models with specific-on-resistance of 80 mOhm and breakdown voltage 1500 volts. The proposed device model exhibited low resistance and a reasonable threshold voltage of approximately 3.5 volts.

Keywords—Silicon Carbide (SiC); Trench MOSFET; TCAD; Breakdown Voltage; Specific On-Resistance; Gate Oxide Reliability

I. Introduction

Silicon carbide (SiC) has emerged as the leading wide-bandgap semiconductor for next-generation power electronics, offering a critical electric field an order of magnitude higher than silicon, superior thermal conductivity, and the ability to form a native oxide (SiO_2) [1]. These properties enable power devices with significantly reduced on-state losses and higher operating temperatures. Among various topologies, the trench-gated MOSFET is particularly attractive as it provides a higher channel density and eliminates the JFET effect present in planar structures, leading to lower specific on-resistance ($R_{DS,on}$) [2].

Despite these advantages, the commercialization of SiC trench MOSFETs has been hindered by a critical reliability concern: the extremely high electric field that develops at the bottom corners of the trench during the off-state. The dielectric constant of SiC (ϵ_r, SiC $\approx$ 9.7) is significantly higher than that of it's native oxide (ϵ_r, $SiO_2 \approx 3.9$). This mismatch, combined with the sharp trench geometry, causes severe electric field crowding in the gate oxide, often exceeding its long-term reliability limit of ~4 MV/cm [3]. This can lead to time-dependent dielectric breakdown (TDDB) before the intrinsic avalanche breakdown of the SiC drift region is reached, thus underutilizing the material's high-voltage potential [4].

This work focuses on a holistic technological and device-level co-design to engineer the electric field distribution and enhance oxide reliability. We propose a structure that combines three key features: an optimized angled ion implantation for creating uniform p-type shielding regions, a locally thickened gate oxide (TBOX) layer at the trench bottom, and a systematic analysis of how trench depth influences both on-state and off-state device characteristics. Through 2D numerical TCAD simulations, we demonstrate a structure that successfully meets the target specifications for a 1500 V power switch.

II. Technology Simulation Methodology

The design methodology is based on a comprehensive TCAD simulation flow that replicates a feasible fabrication process for 4H-SiC devices. The initial epitaxial structure consists of a highly doped n+ substrate, lightly doped n− drift layer, a p-well layer, and a heavily doped n+ source. All simulations were performed using a 2D finite-element solver with physical models calibrated for 4H-SiC, including field-dependent mobility, impact ionization, and Shockley-Read-Hall statistics

A. Structure Formation and Process Flow

The simulation begins with a standard epitaxial wafer stack: n+ 4H–SiC substrate: doping $N_D = 1 \times 10^{19}$ cm^{-3}, thickness $\approx$ 4 μm; n− drift layer: $N_D = 1.5 \times 10^{15}$ cm^{-3}, thickness = 13 μm, optimized to support 1500 V blocking with a 20% safety margin.

The drift layer thickness was determined from the critical field expression (1) for 4H–SiC ($E_{crit} \approx 3$ MV/cm)

$$Wd = (2\epsilon_{SiC}V_{BR}/qN_D)^{(1/2)} \qquad (1)$$

which aligns closely with the adopted 13 μm design, ensuring full depletion at avalanche.

The key process steps modeled are:

Trench Etching. The trench depth is a primary variable in our study, with values ranging from 1.2 μm to 1.6 μm. Gate Dielectric Formation - a thermal oxide is grown, but with a critical modification: the oxide thickness at the trench bottom is intentionally increased to be approximately ten times thicker than on the sidewalls (TBOX concept). Polysilicon Gate and Metallization - standard process steps for gate and contact formation are assumed. Fig. 1 represents the structure material and geometric scheme

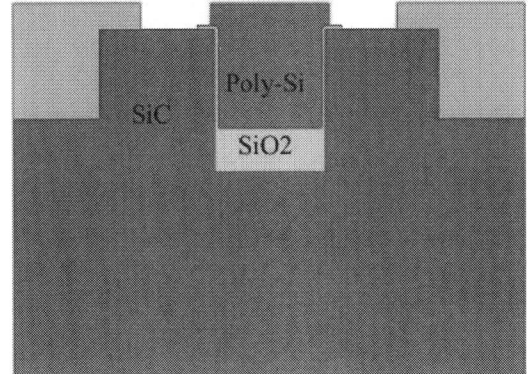

Fig. 1. Schematic representation of TMOSFET

B. Angular Ion Implantation

A major issue in trench processing is the non-uniform doping profile on vertical sidewalls achieved by conventional ion implantation. This non-uniformity can create weak spots in the electric field profile. To overcome this, implant using an optimized tilt angle of approximately 7°-10° was proposed. This angled implant ensures a more homogeneous lateral distribution of the p-type dopant. This uniformity is critical for creating a smooth and predictable electric field contour, effectively pushing the peak field away from the dielectric and into the SiC bulk [5].

Ion Implantation: Two critical implants are modeled:

p-well formation: Al^+ at 320 keV, dose $7 \times 10^{12} cm^{-2}$, tilt angle = 11°; n^+ source: N^+ at 20 keV, dose $3 \times 10^{14} cm^{-2}$, tilt angle = 7°.

A post-implantation anneal at 1650°C for 30 min in argon ambient is simulated to activate dopants and repair crystal damage. A non-uniform gate oxide is implemented: 50 nm thermal SiO_2 on the trench sidewalls (for efficient channel inversion); 500 nm thick bottom oxide (TBOX) at the trench base, approximated as a composite of thermal oxide and PECVD-deposited layer. This structure is inspired by reliability-enhancing architectures in modern SiC trench patents. The final optimized device structure is presented in Fig. 2.

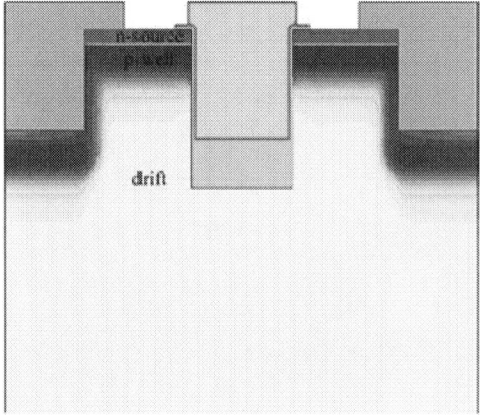

Fig. 2. Optimized device structure

III. DEVICE OPTIMIZATION AND ELECTRIC FIELD ENGINEERING

The core of this work is the management of the electric field to simultaneously achieve high breakdown voltage and reliable oxide operation.

A. Thick Bottom Oxide (TBOX)

The primary innovation is the implementation of the TBOX. In a standard trench MOSFET with uniform oxide thickness, the electric field lines from the drain converge directly onto the sharp corner of the thin gate oxide at the trench bottom, creating a field hotspot. By locally increasing the oxide thickness in this region, we effectively create a field-plate effect. The thicker oxide acts as a buffer, significantly reducing the electric field intensity in the dielectric at this critical location. This design allows the device to leverage the full avalanche breakdown capability of the SiC material without being limited by oxide failure [6].

B. Influence of Trench Depth on Device Characteristics

The trench depth is a crucial parameter that affects both on-state and off-state performance, and its optimization is central to our study.

A deeper trench increases the effective channel length along the vertical sidewalls and, more importantly, provides more space for the current to spread laterally downward the drift region. This significantly reduces the JFET-like constriction resistance in the drift layer.

The trench depth has a minor effect on threshold voltage, since the threshold is primarily governed by the channel doping and gate oxide thickness on the sidewalls. In all simulated cases, threshold voltages remained stable at 3.5 V, which is compatible with standard gate drivers.

Although the threshold voltage remains practically unchanged with trench depth (as the inversion channel forms exclusively along the upper sidewalls), the transconductance (gm) exhibits a mild dependence on trench geometry. A deeper trench increases the total vertical channel length, thereby enlarging the effective channel area available for inversion. This results in a slight increase in drain current at a given overdrive voltage ($V_{GS} - V_{Th}$), which manifests as a rise in the maximum transconductance. However, this effect saturates beyond a depth of ~1.4 μm, as the channel formation remains confined to the top 0.5 μm of the trench due to the shielding by the p-well and the thick bottom oxide (TBOX). Consequently, while trench deepening improves on-resistance through better current spreading in the drift region, its impact on the gate-controlled transfer characteristics is secondary-primarily enhancing current drive without altering the threshold or subthreshold slope. This behavior aligns with observations in trench MOSFETs where gate control is decoupled from deep trench regions by design [7].

IV. SIMULATION RESULTS

The simulation results are summarized as follows.

The device achieves a drain-source breakdown voltage of 1500 V. The peak field in the thick bottom oxide is maintained safely below the 3 MV/cm reliability threshold, confirming that breakdown is bulk-limited, not oxide-limited. This validates the TBOX field-plate concept, which decouples the oxide reliability from the drift-layer avalanche capability.

The specific on-state resistance is lower than 80 mΩ·cm². This low value is a direct result of the trench architecture, the optimized trench depth of 1.5 μm for current spreading, and the effective channel formation on the vertical sidewalls. The results are illustrated on the Fig. 3 The low JFET contribution is a direct consequence of the trench depth, which enables efficient lateral current spreading before the carriers enter the drift region. This is in agreement with the analytical model, where the drift resistance is expressed as (2)

$$R_D = \rho_D \cdot [\ln((p - W_T)/W_T) + t/W_T] \qquad (2)$$

with p=3.5 μm (half-cell pitch), W_T =1.0 μm (trench width), and t=13 μm (drift thickness). The simulation result deviates by <5% from this analytical estimate, confirming the accuracy of the TCAD model.

The transfer characteristics yield a threshold voltage of 3.5 V, which is a practical value for standard 5 V or 15 V gate drivers.

A comparative analysis with a reference structure (without TBOX and with standard normal-incidence implantation) shows a significant improvement in both breakdown voltage (from ~600 V to 1500 V), validating the effectiveness of our proposed technological solutions.

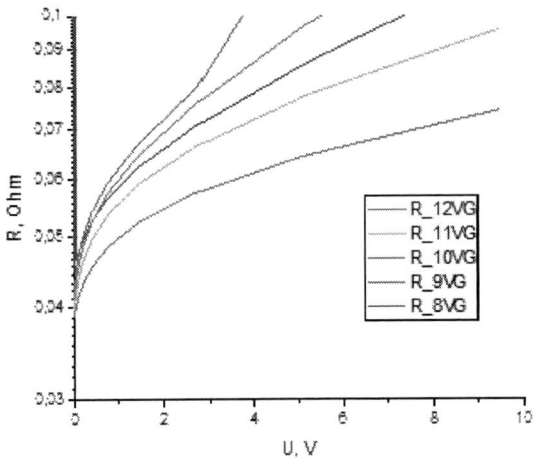

Fig. 3. RDS,on

V. CONLUSION

In this work, a 1500 V 4H-SiC trench MOSFET structure was successfully designed and simulated. That directly addresses the key reliability challenge of electric field crowding in the gate oxide. The co-optimization of a thick bottom oxide (TBOX), and a system-level analysis of trench depth has proven highly effective. An optimal trench depth of 1.5 μm was identified as the best trade-off between low on-resistance and manageable gate charge.

The simulated device meets all target specifications: a 1500 V breakdown voltage, an 80 mΩ·cm² specific on-resistance, and a 3.5 V threshold voltage. The electric field analysis confirms that the peak field is safely contained within the SiC semiconductor, ensuring long-term oxide reliability. This work demonstrates a viable pathway for the development of robust, high-performance SiC trench MOSFETs for next-generation power conversion systems in electric vehicles, renewable energy, and industrial automation.

ACKNOWLEDGMENT

This work was conducted with the help of Institute of Power Electronics and Photonics based in St. Petersburg state electrotechnical University "LETI", we express our gratitude to Sergey A. Shevchenko for all the efforts and facilities provided.

REFERENCES

[1] Peters, D., Friedrichs, P., Schorner, R., Mitlehner, H., Weis, B., & Stephani, D. (1999, May). Electrical performance of triple implanted vertical silicon carbide MOSFETs with low on-resistance. In 11th International Symposium on Power Semiconductor Devices and ICs. ISPSD'99 Proceedings (Cat. No. 99CH36312) (pp. 103-106). IEEE.

[2] M. Chaturvedi, C. Li, and A. K. Agarwal, "Comparison of commercial planar and trench SiC MOSFETs by electrical characterization of performance-degrading near-interface traps," IEEE Trans. Electron Devices, vol. 69, no. 11, pp. 6225-6230, Nov. 2022.

[3] T. Kimoto and H. Watanabe, "Defect engineering in SiC technology for high-voltage power devices," Appl. Phys. Express, vol. 13, no. 12, p. 120101, Dec. 2020.

[4] T. Yang, Y. Wang, and R. Yue, "SiC trench MOSFET with reduced switching loss and increased short-circuit capability," IEEE Trans. Electron Devices, vol. 67, no. 9, pp. 3685-3690, Sep. 2020.

[5] Seok, O., Ha, M. W., Kang, I. H., Kim, H. W., Kim, D. Y., & Bahng, W. (2018). Effects of trench profile and self-aligned ion implantation on electrical characteristics of 1.2 kV 4H-SiC trench MOSFETs using bottom protection p-well. Japanese Journal of Applied Physics, 57(6S1), 06HC07.

[6] Agarwal, A., Han, K., & Baliga, B. J. (2018, October). Analysis of 1.2 kV 4H-SiC trench-gate MOSFETs with thick trench bottom oxide. In 2018 IEEE 6th Workshop on Wide Bandgap Power Devices and Applications (WiPDA) (pp. 125-129). IEEE.

[7] Alatise, O., Parker-Allotey, N. A., Jennings, M., Mawby, P., Kennedy, I., & Petkos, G. (2011). Modeling the impact of the trench depth on the gate-drain capacitance in power MOSFETs. IEEE electron device letters, 32(9), 1269-1271.

Investigation of the Formation of an Impedance Biosensor Response for a Structure Based on Zinc Oxide Nanorods and Peptide Recognition Elements

Alina S. Printseva
Department of micro- and nanoelectronics
Saint-Petersburg Electrotechnical University ETU "LETI"
St. Petersburg, Russia
alya.printseva@inbox.ru

Andrey A. Ryabko
Department of micro- and nanoelectronics
Saint-Petersburg Electrotechnical University ETU "LETI"
St. Petersburg, Russia
a.a.ryabko93@yandex.ru

Alexey A. Kolobov
Institute of Power Electronics and Photonics
Saint-Petersburg Electrotechnical University ETU "LETI"
St. Petersburg, Russia
alexey.kolobov.spb@gmail.com

Valentina V. Trushlyakova
Department of micro- and nanoelectronics
Saint-Petersburg Electrotechnical University ETU "LETI"
St. Petersburg, Russia
vvtrushliakova@mail.ru

Nikita O. Sitkov
Department of micro- and nanoelectronics
Saint-Petersburg Electrotechnical University ETU "LETI"
St. Petersburg, Russia
sitkov93@yandex.ru

Abstract— Impedimetric biosensors are currently becoming one of the popular tools for diagnosing various diseases. An important task is to increase the sensitivity of their electrical response, which can be achieved by modifying the electrodes with nanostructured materials. In this article, we investigated the coatings of ridged electrodes with zinc oxide nanorods at different stages of sensor operation. The study is devoted to the development of impedance measuring biosensors based on zinc oxide nanorods designed to detect lactoferrin, a marker protein associated with various pathological conditions. As part of this work, an (XPS) X-ray photoelectron spectroscopy analysis of all stages of biosensor functionalization was performed, and the possibilities of using an impedance technique for specific lactoferrin recognition were studied.

Keywords— *zinc oxide, biosensor, nanomaterials, nanorods, impedance spectroscopy, XPS*

I. INTRODUCTION

The development of highly sensitive and selective biosensor platforms is a key area of modern biomedical analytics [1]. The rapid expansion of miniaturized diagnostic methods and the transition to personalized point-of-care systems are driving a steady demand for sensor systems capable of detecting biomarkers without complex sample preparation, using small sample volumes, and delivering highly reliable responses [2]. Among the various classes of biosensors, impedimetric devices occupy a special place, as they enable the detection of subtle changes in the electrical properties of the working electrode surface that arise during specific biomolecular interactions.

One key area of impedimetric biosensor development is increasing their sensitivity by modifying the working electrodes with nanostructured materials, which significantly increase the active surface area and amplify the recorded signal. One promising material for creating such systems is zinc oxide (ZnO) nanorods [3], which are a direct-gap n-type semiconductor with high charge carrier mobility, a developed surface, and controlled morphology. The crystal chemical properties of ZnO enable the formation of dense arrays of

The reported study was funded by Russian Science Foundation, project number 25-79-10055.

one-dimensional nanostructures sensitive to changes in the electrode/solution interface [4, 5]. The significant number of hydroxyl groups on the surface of the nanorods enables efficient silanization and subsequent immobilization of biorecognition molecules. This is particularly important when working with peptide aptamers, which, due to their small size, high stability, low cost, and ease of modification, are considered an alternative to antibodies. Peptides ensure dense packing on the surface of nanostructures, minimize steric hindrance during binding of large proteins, and contribute to increased analytical sensitivity of impedimetric sensors [6]. The use of ZnO nanorods in combination with peptide biorecognition structures opens up opportunities for the creation of compact, stable, and highly specific platforms for early biomedical diagnostics. Non-faradaic impedance spectroscopy is particularly sensitive to changes in the thickness, polarizability, and permittivity of the interfacial layer upon binding of large-molecular proteins to ligands [7].

The significant demand for sensor technologies is driven by the need to analyze key biomarkers of inflammatory and infectious processes. One such marker is lactoferrin, an iron-binding glycoprotein of the transferrin family, which plays a critical role in innate immunity and the regulation of iron homeostasis. Since ZnO nanorod arrays have high resistivity and form a pronounced capacitive interface [8, 9], the impedance change upon lactoferrin binding can be significantly more pronounced than on flat metal electrodes.

The aim of this study is to analyze the various stages of biosensor modification and create a prototype that provides accurate impedance measurements in laboratory conditions while maintaining the required sensitivity.

II. TECHNOLOGY FOR THE FORMATION OF BIOSENSOR STRUCTURES

The impedance response of biosensor structures designed for protein detection is determined by a combination of physicochemical processes occurring on the surface of modified electrodes. The creation of a nanostructured substrate based on ZnO nanorods, their subsequent functionalization with silane agents, the introduction of a cross-linker, and the immobilization of a peptide form a multicomponent system in which each structural layer contributes to the electrical behavior of the biosensor.

First, a nanostructured ZnO array is formed, which significantly increases the specific surface area and creates a dielectric, high-resistance interface sensitive to changes in the thickness and electrical properties of the surface layer. The use of low-temperature hydrothermal synthesis after deposition of a seed layer allowed us to obtain nanorods with a diameter of less than 50 nm and interrod spacing of more than 100 nm, which is optimal for subsequent biofunctionalization. The resulting surface is a dense array of predominantly vertically oriented nanorods (Fig. 1), significantly increasing the specific surface area of the electrode. This structure facilitates the efficient interaction of silane reagent solutions and biomolecules with the active surface, and also expands the dynamic range of impedance measurements. From the perspective of the electrochemistry of the impedance interface, ZnO nanorods act as a capacitive layer, in which the parameters of the electrical double layer, in particular the effective capacitance and charge transfer resistance, are determined not only by the morphology of the nanostructures but also by subsequent chemical modifications.

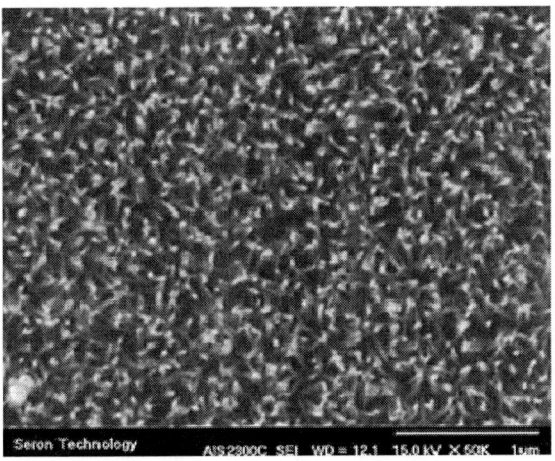

Fig. 1. SEM image of ZnO nanorod coating on silicon substrate

After removing organic residues from the ZnO surface by annealing, silanization is performed using (3-aminopropyl)trimethoxysilane (APTMS). Silane molecules form covalent bonds with the surface hydroxyl groups of ZnO, forming a monolayer with terminal amino groups. This layer plays a key role in creating the biorecognition interface, as the amino groups serve as the initial reaction sites for the introduction of the cross-linker. The next step is surface activation with the hybrid reagent MBS (maleimidobenzoyl-N-hydroxysuccinimide ester). The MBS molecule connects two functional fragments: an N-hydroxysuccinimide residue, which has a high affinity for the amino groups of APTMS, and a maleimide group, which is designed to form strong bonds with the sulfhydryl fragments of the peptide. After formation of the APTMS-MBS conjugate, the LETI-13 peptide [10], specific for lactoferrin, is immobilized. The peptide is introduced into PBS as a solution and then applied to the sensor surface, where it is covalently fixed via the reactive maleimide group of the cross-linker.

III. XPS ANALYSIS OF THE STAGES OF FUNCTIONALIZATION OF THE BIOSENSOR STRUCTURE

X-ray photoelectron spectroscopy (XPS) is one of the most informative methods for studying surface chemical states [11] and a suitable tool for analyzing the stepwise modification of nanostructured materials in the biosensor system studied in this study. Since the formation of the sensor's impedance response is directly related to changes in the chemical composition and thickness of the surface layers, sequential XPS monitoring not only confirms the success of functionalization but also sheds light on the mechanisms of formation of the multilayer structure, including zinc oxide, APTMS, the MBS cross-linker, the LETI-13 peptide, and the target protein lactoferrin.

A comprehensive analysis of the surface modification of ZnO nanorods using X-ray photoelectron spectroscopy revealed changes in the intensity and shape of the spectra of the Zn, O, and N core levels, indirectly indicating sequential successful modification and immobilization. However, the most illustrative information is provided by analysis of the Si2p spectrum (Fig. 2).

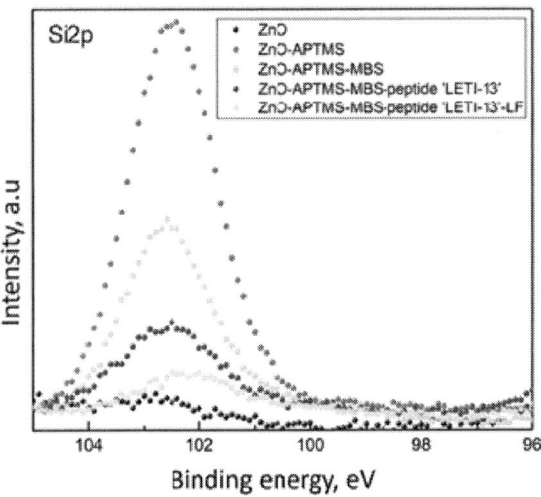

Fig. 2. XPS spectrum of the Si2p core level of nanostructured ZnO coatings with stepwise functionalization and conjugation.

Silicon atoms are present exclusively in aminopropyltriethoxysilane (APTMS), and the binding process occurs sequentially with the APTMS-functionalized ZnO surface. As a result, some areas of the ZnO surface may remain unfunctionalized. Before silanization, the surface of the nanostructured ZnO layer is free of silicon atoms, indicating complete coverage of the silicon substrate with the nanostructured ZnO layer, eliminating voids at the bases of the nanorods. Silanization leads to the maximum concentration of silicon-containing compounds, particularly O-Si-O bonds. Further surface treatment is accompanied by a gradual decrease in the Si2p peak, confirming the sequential attachment of various functional groups to the APTMS amino groups, such as the MBS cross-linker, the LETI-13 peptide, and the lactoferrin biomarker.

The detection of iron after lactoferrin binding is a key confirmation of successful conjugation. Since lactoferrin contains Fe^{3+} ions within its domains, the appearance of a characteristic Fe2p doublet in the spectrum directly indicates its surface binding. Figure 3 shows the Fe2p XPS spectrum of a functionalized ZnO-APTMS-MBS-peptide-LETI-13-lactoferrin sample. The Fe2p line intensity (1–1.5 at%)

979-8-3315-4784-4/26 $31.00 © 2026 IEEE

corresponds to the amount of protein adsorbed within the method's information depth (5–10 nm) and confirms the specific binding of lactoferrin to the peptide biorecognizer. The peak position is consistent with the valence state of iron in the protein structure, ruling out its random adsorption from solutions.

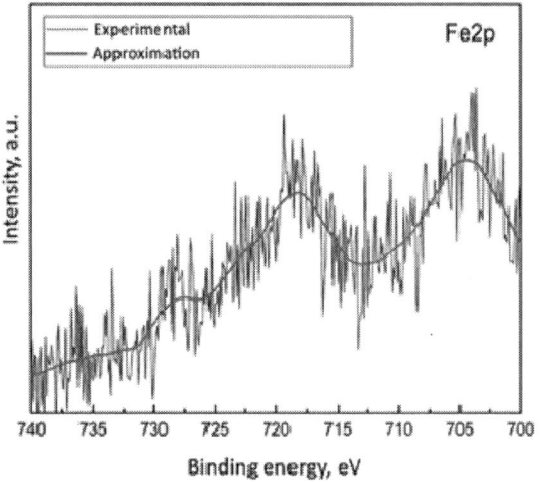

Fig. 3. XPS spectrum of the Fe2p core level of the ZnO-APTMS-MBS-peptide-LETI-13-lactoferrin sample.

As a result, silanization forms an amino-grouped platform, the MBS cross-linker creates an oriented base for strong covalent binding, the LETI-13 peptide is anchored to the electrode surface, and lactoferrin specifically interacts with the peptide motif, which is directly confirmed by the appearance of the Fe2p line. Sequential changes in the shapes and intensities of the Zn2p, O1s, N1s, Si2p, and Fe2p spectra demonstrate the success of multilayer functionalization and provide a detailed understanding of the chemical mechanisms underlying the impedance response of the developed sensor system.

IV. STUDY OF IMPEDANCE RESPONSE OF BIOSENSOR STRUCTURE

A PS-50 potentiostat-galvanostat (SmartStat, Russia) was used to conduct impedance spectroscopy. The frequency range was from 1 Hz to 0.5 MHz, and the amplitude of the alternating signal was 10 mV. To bind the biomarker protein, a PBS solution was applied to the electrode area, forming a drop. The solution was then left to incubate for 20 minutes. After this, the drop was rinsed in PBS. PBS was then applied to the biosensor surface again to measure the impedance spectrum. Figure 4 shows a Nyquist plot reflecting the impedance response of a typical lactoferrin biosensor structure in the concentration range from 20 ng/mL to 2000 ng/mL.

The Nyquist plot shows that biomarker binding alters the spectrum shape, which is associated with changes in the electrical double layer parameters. As biomarker concentration increases, the inflection point of the graph shifts. The impedance phase is complex (Fig. 5(a)), with biomarker binding to the sensor structure leading to significant changes in the impedance phase spectrum in dual-frequency ranges.

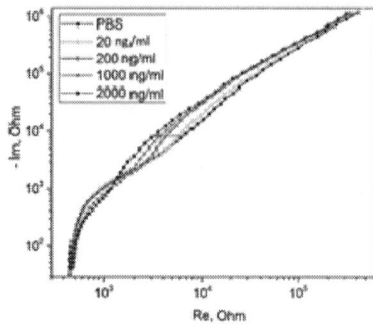

a)

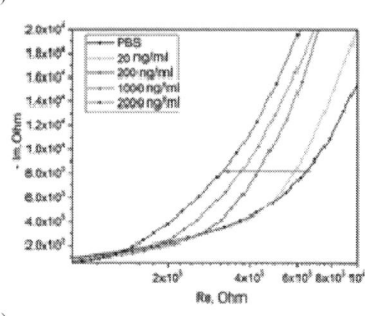

b)

Fig. 4. Nyquist plot for the lactoferrin biosensor with variation in the concentration of the biomarker in the solution from 2 to 2000 ng/ml: a) in a double logarithmic scale; b) in a semi-logarithmic scale

Fig. 5(b) shows the frequency range where the concentration dependence of the impedance modulus on biomarker concentration is observed.

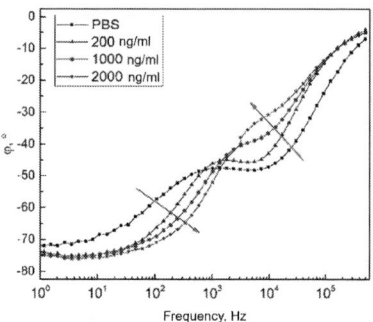

a)

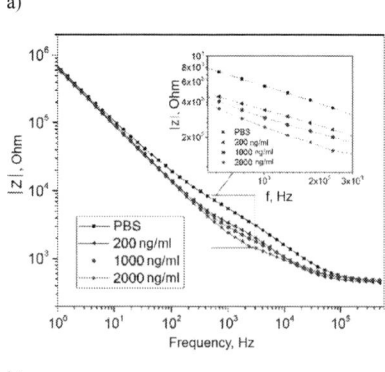

b)

Fig. 5. Phase (a) and modulus (b) spectra of the impedance of the lactoferrin biosensor with variations in the concentration of the biomarker in the solution from 2 to 2000 ng/ml

When analyzing experimental impedance data from a biosensor based on electrodes coated with ZnO nanorods, the

979-8-3315-4784-4/26 $31.00 © 2026 IEEE

resistance and capacitance are influenced by both the parameters of the nanostructured layer itself and potential changes in these parameters due to the influence of electrical double layer charges on the Debye screening length in the ZnO surface layer.

The impedimetric response of the biosensor is complex and is associated with both changes in the electrical double layer (EDL) parameters due to target biomarker binding and potential changes in the parameters of the nanostructured ZnO layer itself induced by changes in the EDL. The binding specificity of the developed lactoferrin biosensor was studied using myoglobin solutions, which also contain iron ions. As can be seen from the Nyquist plot (Figure 6), no changes in the spectra are observed when the samples are incubated in the myoglobin protein solution. It is worth noting that the characteristic shape of the spectra in the form of a Nyquist diagram differs from the spectra obtained upon binding to the target protein, which indicates the absence of binding and the formation of a characteristic double electrical layer due to the lack of binding of myoglobin molecules to the LETI-13 peptide.

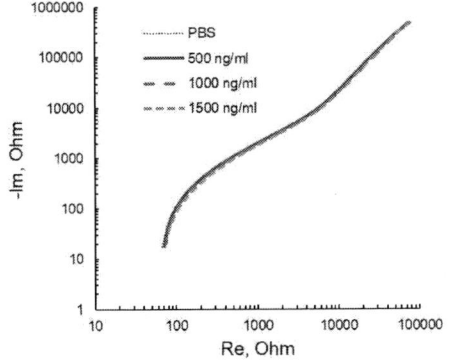

Fig. 6. Nyquist plots of impedance spectra during incubation of lactoferrin biosensors in myoglobin solutions

Thus, the developed structure demonstrates a specific response to the target protein, lactoferrin, in the range from 20 to 2000 ng/ml.

V. Conclusions

The study confirmed the potential of the developed approach for creating highly sensitive protein marker diagnostic platforms. A sequential study of the surface functionalization stages using X-ray photoelectron spectroscopy revealed consistent changes in the chemical composition and thickness of the formed layers, enabling the entire transformation of the initial nanostructured material into a multilayered bioselective interface to be traced. The observed dynamics of the XPS spectral signals reflected the gradual formation of a functional coating, beginning with silanization of the APTMS surface, through the formation of an active intermediate MBS layer, to the attachment of the LETI-13 peptide and the specific binding of lactoferrin. Of particular significance was the detection of the Fe2p peak after incubation with the biomarker, directly confirming the effectiveness of conjugation and demonstrating the formation of a selective layer capable of specific interaction with the target protein. Impedance spectroscopy results demonstrated that the developed biosensor structure exhibits pronounced sensitivity to lactoferrin binding, demonstrating characteristic changes in the Nyquist diagram shape and the frequency dependences of the impedance modulus and phase. The dependence of the electrical response on the analyte concentration revealed a specific detection range of 20 to 2000 ng/mL, corresponding to significant lactoferrin concentrations in clinical samples. The absence of spectral changes upon exposure to myoglobin confirmed the selectivity of the LETI-13 peptide recognition element and the absence of nonspecific adsorption of proteins containing iron ions, which is an important parameter for the reliability of the developed system.

Thus, the developed approach to modifying electrodes with ZnO nanorods and peptide ligands creates a highly effective biosensor interface in which structural, chemical, and electrical characteristics are interrelated and enhance the analytical sensitivity of the system. A comprehensive analysis, including morphological, spectroscopic, and electrochemical methods, confirmed the feasibility of the chosen biosensor architecture and its potential for further development into portable point-of-care diagnostic systems. The results provide a scientific basis for optimizing functionalization parameters, scaling the technology, and expanding the range of biomarkers detectable using similar impedimetric platforms.

Acknowledgment

The study was supported by a grant from the Russian Science Foundation No. 25-79-10055.

References

[1] F. Malvano, R. Pilloton, and D. Albanese, "Label-free impedimetric biosensors for the control of food safety – a review," vol. 100, no. 4, pp. 468–491, Sep. 2019

[2] S. N. Prabhu, C. P. Gooneratne, K.-A. Hoang, and S. C. Mukhopadhyay, "IoT-Associated Impedimetric Biosensing for Point-of-Care Monitoring of Kidney Health," *IEEE Sensors Journal*, vol. 21, no. 13, pp. 14320–14329, Jul. 2020

[3] J. Eveness, L. Cao, J. Kiely, and R. Luxton, "Equivalent circuit model of a non-faradaic impedimetric ZnO nano-crystal biosensor," *Journal of Electroanalytical Chemistry*, vol. 906, p. 116003, Feb. 2022

[4] A. A. Ryabko, A. I. Maximov, V. N. Verbitskii, V. S. Levitskii, V. A. Moshnikov, and E. I. Terukov, "Two-Stage Synthesis of Structured Microsystems Based on Zinc-Oxide Nanorods by Ultrasonic Spray Pyrolysis and the Low-Temperature Hydrothermal Method," *Semiconductors (Woodbury, N.Y., Print)*, vol. 54, no. 11, pp. 1496–1502, Nov. 2020

[5] H. Beitollahi, S. Tajik, F. G. Nejad, and M. Safaei, "Recent advances in ZnO nanostructure-based electrochemical sensors and biosensors," *Journal of Materials Chemistry B*, vol. 8, no. 27, pp. 5826–5844, Jan. 2020, doi: 10.1039/d0tb00569j

[6] M. Tertis, O. Hosu, B. Feier, A. Cernat, A. Florea, and C. Cristea, "Electrochemical Peptide-Based Sensors for Foodborne Pathogens Detection," *Molecules*, vol. 26, no. 11, p. 3200, May 2021

[7] H. Adam, S. C. B. Gopinath, T. Adam, M. A. Fakhri, E. T. Salim, and S. Subramaniam, "Exploring faradaic and non-faradaic electrochemical impedance spectroscopy approaches in Parkinson's disease diagnosis," *Heliyon*, vol. 10, no. 5, pp. e27433–e27433, Mar. 2024

[8] N. O. Sitkov *et al.*, "Impedimetric Biosensor Coated with Zinc Oxide Nanorods Synthesized by a Modification of the Hydrothermal Method for Antibody Detection," *Chemosensors*, vol. 11, no. 1, pp. 66–66, Jan. 2023

[9] H. S. Magar, R. Y. A. Hassan, and A. Mulchandani, "Electrochemical Impedance Spectroscopy (EIS): Principles, Construction, and Biosensing Applications," *Sensors (Basel, Switzerland)*, vol. 21, no. 19, p. 6578, Oct. 2021

[10] T. M. Zimina *et al.*, "Design of Peptide Ligand for Lactoferrin and Study of Its Binding Specificity," *Chemosensors*, vol. 11, no. 3, pp. 162–162, Feb. 2023

[11] T. J. Frankcombe and Y. Liu, "Interpretation of Oxygen 1s X-ray Photoelectron Spectroscopy of ZnO," *Chemistry of Materials*, vol. 35, no. 14, pp. 5468–5474, Jul. 2023

AFM Investigation of Surface Morphological Changes in Microplastics under Photocatalytic Decomposition

Dmitry G. Radaykin
Department of Electronics
Saint Petersburg Electrotechnical University "LETI"
Saint Petersburg, Russia
dima19980219@gmail.com

Abstract—Environmental contamination by persistent micro- and nanoplastics represents a global crisis requiring advanced remediation strategies. Photocatalysis is a promising approach for degrading these polymers, necessitating precise methods to evaluate catalytic efficiency. This work reviews the application of Atomic Force Microscopy (AFM) as a multifaceted tool for characterizing plastic degradation at micro- and nanoscales. AFM provides quantitative data on key morphological and mechanical parameters—such as surface roughness, Young's modulus, average particle size, and adhesion—that serve as direct indicators of polymer breakdown. To standardize the assessment of catalytic performance, we propose a novel quantitative methodology centered on an Integral Degradation Index (IDI). The IDI integrates normalized changes in the aforementioned parameters into a single metric, weighted by their relative significance to the degradation process. This approach enables a comparative and systematic evaluation of different catalytic systems, facilitating the rational design of high-efficiency materials for environmental cleanup. The reviewed studies and the proposed framework establish a foundation for developing a standardized, high-precision protocol to advance photocatalytic solutions for plastic pollution.

Keywords—Microplastics, Nanoplastics, Photocatalysis, Polymer Degradation, Atomic Force Microscopy (AFM), Catalytic Systems, Integral Degradation Index (IDI), Surface Characterization, Environmental Remediation

I. INTRODUCTION

Environmental contamination by microplastics is one of the most extensive and alarming environmental problems of our time, having reached the level of a global crisis. Microplastics have now been detected in all ecosystems: in water, soil, air, and living organisms. It is a borderless problem, affecting even the planet's pristine remote areas. During its life cycle, microplastic does not degrade into safe components but rather breaks down into smaller fragments, ultimately reaching the nanoscale range [1]. After 75 years of exponential plastic production, the environmental density of nanoplastics is expected to exceed that of microplastics [2].

Nanoplastics (<1 μm) are becoming the predominant form of plastic pollution, though their detection and characterization are still hindered by the constraints of conventional techniques [3]. Meanwhile, biodegradable plastics (such as polyhydroxyalkanoates) may fragment in the environment into persistent nanoparticles, calling their environmental safety into question.

Photocatalysis stands out as a promising approach for addressing micro- and nanoplastic pollution [4]. Semiconductor materials can be used to accelerate the degradation of polymer compounds into harmless substances. Developing high-efficiency catalysts is a crucial objective for contemporary science and society [5-7]. By applying principles of nanoengineering and nanoarchitectonics, materials with the requisite properties for effectively breaking down persistent polymers can be designed [8-11].

The successful design of catalytic systems requires the development of methods for detecting and characterizing plastics in the micro- and nano-ranges. Establishing a precise methodology for the quantitative analysis of the polymer degradation process will allow for a proper assessment of catalytic activity, thereby facilitating the development of highly effective solutions for environmental cleanup from this type of pollution [12-13].

Atomic force microscopy (AFM) stands out as a promising tool for evaluating the degradation of plastics across micro and nano scales [14]. It enables surface investigation with nanoscale resolution, providing both qualitative and quantitative analysis of changes in surface morphology [15-17]. Such quantitative data can establish a baseline for subsequent assessment of catalytic system performance.

Key parameters for this quantitative evaluation may include average particle size, surface roughness, Young's modulus, and adhesion (the latter being potentially less critical). Analyzing the evolution of these parameters offers a pathway to a comprehensive evaluation of system efficacy.

The present work provides an overview of the current state of analytical techniques for polymer degradation assessment via AFM. The findings from the reviewed studies may serve as a foundation for developing a novel, high-precision methodology to evaluate catalytic systems designed for micro- and nanoplastic breakdown.

II. ATOMIC FORCE MICROSCOPY (AFM) AS A CHARACTERIZATION TECHNIQUE FOR MICRO- AND NANOPLASTIC DEGRADATION PROCESSES

Atomic force microscopy (AFM) represents a unique tool capable of providing a complete profile of a material's surface properties. It holds particular promise for studying the degradation processes of polymer materials, specifically micro- and nanoplastics.

The aging or degradation of plastic manifests through several clear signs, including increased surface roughness, altered elastic modulus, reduced average particle size, and changes in adhesion.

Research [18] emphasized studying microplastic degradation under UV exposure over a 14-day period. AFM analysis enabled the evaluation of polymer aging through the following markers:

979-8-3315-4784-4/26 $31.00 © 2026 IEEE

- Reduction in Young's modulus (a change of approximately 50% was reported), indicating surface softening due to polymer chain breakage and the formation of surface defects.

- Nanoscale topographic evaluation and comparison, which helps track the dynamics of the degradation process.

- Variation in surface roughness, serving as a key quantitative indicator of degradation intensity.

- Nanoparticle detection capability, allowing the instrument to monitor particle fragmentation, specifically the decrease in average size, which further characterizes the process.

The study compared the dynamics of plastic degradation in natural environments versus laboratory conditions using UV radiation. The results indicated that natural samples exhibited higher roughness (4-45 nm) compared to laboratory samples (0.1-0.2 nm).

Research [3] discusses two advanced nanospectroscopy techniques: AFM-IR and O-PTIR. AFM-IR (Atomic Force Microscopy based Infrared Spectroscopy) combines AFM with the chemical specificity of IR spectroscopy. O-PTIR (Optical Photothermal Infrared Spectroscopy) is a non-contact method with a resolution of about 500 nm, enabling the simultaneous acquisition of both IR and Raman spectra. AFM-IR allows obtaining IR spectra with nanoscale resolution, which enables tracking of:

- The disappearance of characteristic peaks of the original polymer (e.g., the C=O bond at 1735 cm^{-1} for polyesters);

- The emergence of new functional groups resulting from oxidation (carbonyls, hydroxyls, carboxylic acids);

- Chemical heterogeneity on the particle surface, indicating localized degradation sites;

- The formation of nano-sized degradation products.

The investigation presented in [19] complements the review topic on methods for studying microplastic degradation by examining them as objects for heavy metal adsorption—a crucial aspect for assessing the environmental impact of this type of pollution. The study involved UV degradation of PET microparticles. AFM analysis revealed a decrease in Young's modulus from 3.154 GPa to 0.91 GPa, a consequence of plastic softening. Additionally, the samples were examined using IR spectroscopy. The results showed a decrease in the intensity of peaks in the regions of 968, 848, and 894 cm^{-1} (ethylene glycol fragment), indicating polymer chain scission. Concurrently, signs of increased crystallinity emerged (peak at 1717 cm^{-1}). Regarding adsorption, it is noted that different polymers behave differently; for instance, rubber adsorbs copper more actively than PET particles.

The review highlights the pivotal role of Atomic Force Microscopy (AFM) in the multifaceted investigation of microplastic degradation. This unique instrument integrates nanoscale visualization with functionalities for both mechanical and chemical characterization. The data it provides are crucial for conducting a quantitative evaluation of the degradation process of polymeric compounds.

III. CREATING A UNIFIED QUANTITATIVE METHODOLOGY FOR CATALYTIC SYSTEM ASSESSMENT

To enable a more precise evaluation of the catalytic process, we propose developing a unified integral metric—the Integral Degradation Index (IDI). The IDI would integrate dimensionless values of heterogeneous physical parameters, including surface roughness, average particle size, Young's modulus, and adhesion. The index is normalized to a range from 0 to 1 (or 0% to 100%), where 0 corresponds to system ineffectiveness (comparable to a control sample) and 1 represents maximal degradation induced by the catalytic system across all parameters.

We will outline the formulas that allow for the normalization of the measured parameters. The assessment of parameter behavior was based on the results from the studies reviewed above:

- Young's modulus (U) - (a decrease in this parameter is expected during polymer degradation due to the observed softening process) (1):

$$N_U = \frac{U_{control} - U_{sample}}{U_{control} - U_{min}} \quad (1)$$

- average particle size (D) - (this parameter is expected to decrease during degradation) (2):

$$N_D = \frac{D_{control} - D_{sample}}{D_{control} - D_{min}} \quad (2)$$

- surface roughness (R) - (an increase in this parameter is anticipated as degradation progresses) (3):

$$N_R = \frac{R_{sample} - R_{control}}{R_{max} - R_{control}} \quad (3)$$

- adhesion (A) - (this parameter is expected to increase during degradation due to the formation of polar groups) (4):

$$N_A = \frac{A_{sample} - A_{control}}{A_{max} - A_{control}} \quad (4)$$

where "Control" is the value for the initial sample; "Sample" is the value for the plastic after treatment; "Max/Min" is the maximum or minimum observed value in the series of experiments.

The development of such an assessment methodology for catalytic systems requires evaluating the relative significance (weight) of each parameter in relation to the overall polymer degradation efficiency. For example, variations in Young's modulus and particle size are direct evidence of the degradation process, implying their quantitative contribution to the overall index should be more substantial than that of parameters such as roughness and adhesion. Determining appropriate weighting coefficients is therefore a critical step towards creating a standardized evaluation protocol.

Future work will involve determining the weighting coefficients for the IDI via correlation analysis. This analysis will relate the dimensionless changes in morpho-mechanical parameters to established integral degradation indicators (e.g., mass loss, carbonyl index) derived from calibration experiments. This statistical approach is intended to provide an objective and adaptable framework for the methodology's application across different polymer-catalyst systems.

We propose an approximate calculation method for the weighting coefficients.

- Young's modulus (weight coefficient K_{weight_U} = 0.3-0.4) – this parameter directly indicates the loss of the polymer's structural integrity, making it a crucial indicator;

- Particle size (weight coefficient K_{weight_D} = 0.3-0.4) – this is a direct indicator of fragmentation, which is also a crucial parameter for assessing the degradation process;

- Roughness (weight coefficient K_{weight_R} = 0.2) – an indicator of the degree of surface erosion; it can be considered a secondary characteristic for now;

- Adhesion (weight coefficient K_{weight_A} = 0.1-0.2) – an indirect chemical indicator, which may be of lesser importance in this study.

The total of all weighting coefficients should be 1: $K_{weight\_U} + K_{weight\_D} + K_{weight\_R} + K_{weight\_A} = 0{,}35 + 0{,}35 + 0{,}2 + 0{,}1 = 1$.

Using the dimensionless parameters and weighting coefficients determined above, the Integral Degradation Index (IDI) can be calculated as follows: $IDI = N_U * K_{weight\_U} + N_D * K_{weight\_D} + N_R * K_{weight\_R} + N_A * K_{weight\_A}$ (5).

Let's consider an example of calculating the efficiency of polymer degradation (table 1).

TABLE I.   AN EXAMPLE OF CALCULATING THE ACTIVITY OF A CATALYST

Parameter	Values of the original sample	The value after exposure to the catalyst	A dimensionless quantity
Young's modulus (U), GPa	3,154	0,91	$N_U = \frac{3{,}154 - 0{,}91}{3{,}154 - 0{,}5} = 0{,}845$
average particle size (D), nm	1000	750	$N_D = \frac{1000 - 750}{1000 - 300} = 0{,}36$
roughness (R), nm	10	20	$N_R = \frac{20 - 10}{40 - 10} = 0{,}33$
adhesion (A), nN	2,3	5,4	$N_A = \frac{5{,}4 - 2{,}3}{7 - 2{,}3} = 0{,}66$

$IDI = 0{,}845 * 0{,}35 + 0{,}36 * 0{,}35 + 0{,}33 * 0{,}2 + 0{,}66 * 0{,}1 = 0{,}554$ or 55,4 %. For this system, the calculated efficiency (IDI) was 55.4%. Applying the same methodology to another catalyst yields a result of, for example, 53.2%. In general, a higher IDI value corresponds to a system with greater effectiveness in degrading complex polymeric compounds.

Furthermore, the proposed methodology allows for visual representation of the system's activity (fig. 1). Prior to the final calculation of the integrated index (IDI), the dimensionless parameters can be plotted on a four-quadrant radar chart, creating a visual degradation profile (see Figure 1 for an example). This graphical analysis enables additional insights:

- The total area enclosed by the profile correlates with the overall degradation efficiency.

- The profile's shape reveals the dominant mode or pathway of degradation.

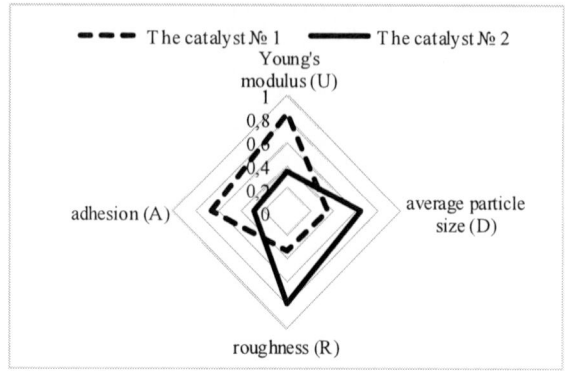

Fig. 1.   Catalyst Efficiency diagram

Analysis of this diagram allows for several conclusions:

- The catalyst № 2 (IDI = 53.2%) is more effective for the complete breakdown of plastic, as it induces deeper fragmentation without causing significant softening (structural integrity is largely retained). In terms of mechanism, this suggests deep photocatalytic oxidation of polymer chains, leading to fragmentation and chemical surface modification.

- The catalyst № 1 (IDI = 55.4%) is less effective, as it causes mainly surface degradation and does not lead to significant fragmentation. The degradation mechanism is limited to surface erosion.

The presented efficiency assessment is only approximate and in a nascent stage. Creating a more accurate evaluation requires refinement based on experimental data. Future work involves conducting a series of experiments with data interpretation to refine this assessment methodology.

IV. CONCLUSION

The advancements and outcomes discussed in this body of work provide a crucial foundation for studying the photocatalytic oxidation and breakdown of microplastics with semiconductor materials. The interpretation of these results and the establishment of a standardized framework for evaluating catalytic system performance in decomposing complex polymers create significant opportunities for designing next-generation, high-efficiency catalysts.

REFERENCES

[1] A. Singh; H. Sigloch, P. Laux, A. Luch, "Micro/nanoplastics: an emerging environmental concern for the future decade," Front. Nanosci. Nanotechnol. vol. 6, 2020.

[2] S. Agarwal "Biodegradable Polymers: Present Opportunities and Challenges in Providing a Microplastic-Free Environment," Macromol. Chem. Phys. vol. 221, no. 6, 2000017, 2020.

[3] S. Belontz, J. Brahney, C. Caplan, E. Dillon "Combining Submicron Spectroscopy Techniques (AFM-IR and O-PTIR) To Detect and Quantify Microplastics and Nanoplastics in Snow from a Utah Ski Resort," Environ. Sci. Technol., vol. 59, pp. 13362−13373, 2025.

[4] D. Radaykin "The problem of microplastics and methods of its elimination," Science of the present and the future, vol. 1, pp. 192-195, 2023.

[5] E. Maraeva, D. Radaykin et al. "Sorption analysis of composites based on zinc oxide for catalysis and medical materials science," Chimica Techno Acta, vol.9, no. 4, 20229422, 2022.

[6] D. Radaykin, A. Bobkov, V. Moshnikov "Plasmonics for photocatalysis. Achievements and perspectives," Nanoparticles, nanosystems and their application. Promising photoactive systems for solar energy, Saint Petersburg, Saint Petersburg Electrotechnical University "LETI" Publ., pp. 15-51, 2024.

[7] D. Radaykin, V. Moshnikov "Structures for photocatalysis based on ZnO with Ag nanoparticles," Condensed Matter and Interphases, vol. 27, no. 2, pp. 293-301, 2025.

[8] A. Bobkov, I. Kononova, V. Moshnikov "Materials Science of Micro- and Nanosystems. Hierarchical Structures, " Saint Petersburg, Saint Petersburg Electrotechnical University "LETI" Publ., 204 p., 2017.

[9] A. Karmanov, I. Pronin, N. Yakushova, V. Kondratiev et al. "Method for producing a photocatalyst based on ZnO/Cu2O-CuO heterostructure with enhanced photocatalytic activity," RU2794093C1, 2023.

[10] N. Yakushova, "Physical and technological features of the formation of semiconductor nanomaterials based on zinc oxide for photocatalytic applications," Candidate dissertation, National Research University "MPEI", 2023.

[11] I. Pronin, I. Averin, B. Donkova, D. Dimitrov et al. "Relationship between the photocatalytic and photoluminescence properties of zinc oxide doped with copper and manganese," Semiconductors, vol. 48, no. 7, pp. 842-847, 2014.

[12] D. Kozodaev (Ed.) "From Nanotechnology to Nanoarchitectonics," Ufa: Aeterna, 125 p., 2025.

[13] I. Plugin, A. Karmanov, I. Pronin, V. Sysoev et al. "Method for determining the average size of subnanopores in metal oxide nanomaterials with a hierarchical structure," RU2848413C1.

[14] D. Kozodaev, V. Moshnikov, E. Muratova, A. Solomorov et al. "Joint Center for Scanning Probe Microscopy of LLC 'NOVA SPB' - SPbGETU 'LETI' in the field of radio electronics - a new type center for solving priority tasks," Innovations, vol. 297, no. 1, pp. 10-18, 2024.

[15] V. Moshnikov, Yu. Spivak "Atomic Force Microscopy for Nanotechnology and Diagnostics", Saint Petersburg, Saint Petersburg Electrotechnical University "LETI" Publ., 80 p., 2009.

[16] A. Aglikov, M. Zhukov, T. Aliev, D. Kozodaev et al. "New metrics for describing atomic force microscopy data of nanostructured surfaces through topological data analysis," Applied Surface Science, vol. 670, p. 160640, 2024.

[17] V. Moshnikov, Yu. Spivak, P. Alekseev, N. Permiakov "Atomic Force Microscopy for the Study of Nanostructured Materials and Device Structures," Saint Petersburg, Saint Petersburg Electrotechnical University "LETI" Publ., 144 p.,2014.

[18] M. Galluzzi, M. Lancia, C. Zheng, V. Re et al. "Atomic Force Microscopy (AFM) nanomechanical characterization of micro- and nanoplastics to support environmental investigations in groundwater," Emerging Contaminants, vol. 11, 100478, 2025.

[19] K. Diego-Perez "Optimizing analytical methods for studying microplastics and nanoplastics at the microscopic level," Dissertation, 2024.

Nanostructuring by Local Anodic Oxidation Using Atomic Force Microscopy (AFM) for the Formation of Mask Structures

Matvey Repkin
St. Petersburg State Electrotechnical University "LETI"
Center of Scanning Microscopy (V. A. Moshnikov School)
Saint Petersburg, Russia
matveirepkin123@mail.ru

Abstract— **In this work, the principles of local anodic oxidation of the sample surface are studied using a scanning probe microscope. The study focuses on controlled oxidation, specifically using the probe as a local electrode to induce electrochemical reactions on the sample surface. By applying a bias voltage between the conducting probe and the sample in the presence of an electrolyte (often a thin layer of adsorbed water), we have successfully formed oxide nanostructures with precise control over their size and placement.**

Keywords—oxidation, probe, AFM

I. Introduction

Significant attention is paid to the development of nanotechnology and nanomaterials at ETU "LETI" [1]. This work was carried out at the AFM Center of ETU "LETI" as part of the leading research and educational school's plan "Atomic-molecular design and nanoarchitectonics" [2].

Among new materials, developments in fractal, especially fractal-percolation structures, and quasicrystalline structures are relevant. Theoretical aspects of fractals are discussed in [3]. As technological approaches, AFM is used for surface modification and nanostructuring [4]. The main methods of nanolithography based on scanning probe microscopy are reviewed in [5].

The AFM Center has successfully applied methods of force nanolithography to create one-dimensional quasicrystalline aperiodic structures. It is shown in [8] that the duality with respect to the Fourier transform between the distribution of non-trivial zeros of the Riemann zeta function along the critical line, on one hand, and the distribution of logarithms of primes and prime powers, on the other, can be used as a theoretical basis for creating new diffractive optical elements. In particular, an aperiodic diffraction grating was fabricated, with its slits ordered according to the distribution of non-trivial zeros of the Riemann zeta function. Atomic force lithography was used for nanoscale patterning. The resulting diffraction pattern is characterized by the presence of discrete diffraction maxima at the logarithms of primes and their powers, which is a direct experimental illustration of the Hilbert–Pólya hypothesis [6][7][8][9].

Methods of optical manipulation are also being developed [10].

The prospects of applying AFM local anodic oxidation for creating two-dimensional patterns were demonstrated in [11].

II. Aim

The aim of this work was to create two-dimensional aperiodic structures of the Penrose tiling type using Local Anodic Oxidation (LAO).

III. Fundumental Principles

Local Anodic Oxidation (LAO) is fundamentally an electrochemical process. The formation of an oxide nanostructure involves the transport of ionic species through an existing or growing oxide layer under the influence of a strong, localized electric field. The growth kinetics are often described by models adapted from conventional anodization, with specific modifications for the nanoscale, point-source nature of the AFM probe.

IV. Mechanism and Sequential Stages

Meniscus Formation: Under ambient conditions, a thin layer of adsorbed water exists on both the tip and sample surface. When the tip is brought into contact or near-contact, a nanoscale water bridge (meniscus) forms via capillary condensation. Its size and stability are controlled by relative humidity, surface hydrophilicity, and tip geometry.

Electrochemical Reactions: Upon application of a positive bias to the sample (or negative to the tip), the water in the meniscus electrolyzes.

Ion Transport and Oxide Growth: The applied field drives the oxidizing anions (O^{2-}/OH^-) towards the sample and the metal cations (e.g., Si^{4+} for silicon) towards the tip/oxide interface. The oxide grows at either the metal/oxide interface (cation movement) or the oxide/electrolyte interface (anion movement), depending on the material. For silicon, evidence suggests the dominant transport species are O^{2-}/OH^- anions migrating inward, leading to oxide growth at the Si/SiO_2 interface. This ion drift through the existing oxide is the rate-limiting step, governed by the high-field mechanism.

V. Influence of Tip Geometry and Material

The tip is not just a passive electrode; its properties critically define the LAO outcome.

Radius of Curvature: A smaller tip radius (R) produces a higher local electric field for the same voltage, enabling finer feature size and potentially lower threshold voltages. However, very sharp tips may wear faster.

Aspect Ratio: High-aspect-ratio probes (like carbon nanofibers or nanotubes) can access recessed features and may reduce lateral spreading of the electric field, improving pattern fidelity in complex topographies.

Tip Conductivity and Chemical Stability: The tip material must be conductive and electrochemically inert under the process conditions to avoid parasitic reactions, contamination, or rapid wear. Common coatings include Pt, PtIr, Au, and TiN. Diamond-coated or carbon-based tips offer exceptional wear resistance.

VI. Resolution Limits and Proximity Effects

The ultimate resolution of LAO is not solely determined by the tip radius. Several factors cause the fabricated feature to be larger than the tip-sample contact area.

Lateral Field Spread: The electric field extends laterally from the tip apex, causing oxidation in a region broader than the physical contact.

Meniscus Size: The water bridge has a finite lateral dimension, defining the electrochemical reaction zone. Controlling humidity is key to minimizing the meniscus.

Diffusion of Species: Ions can diffuse slightly within the water meniscus before reacting.

Stress Effects: The volume expansion during silicon oxidation (~2.2x) induces mechanical stress, which can affect the shape and even the crystalline structure of the surrounding material.

Empirically, line widths achievable with standard probes on silicon are typically in the range of 20-100 nm, while specialized ultra-sharp probes under optimal conditions can achieve sub-10 nm features.

VII. Material-Specific Considerations

While silicon is the most studied material, LAO is applicable to various others.

Metals (Ti, Al, Nb, etc.): Form their respective anodic oxides (TiO_2, Al_2O_3, Nb_2O_5). The kinetics and oxide properties differ.

Semiconductors (GaAs, InP, Ge): Oxidation mechanisms are more complex due to multi-component systems and possible selective oxidation.

2D Materials (Graphene, MoS_2): LAO can selectively oxidize the top layer or substrate beneath, enabling patterning and bandgap engineering.

The growth rate, final oxide stoichiometry, and etch resistance are highly material-dependent, requiring parameter optimization for each system.

VIII. Dynamic Mode LAO

While much early work used contact mode, LAO in amplitude modulation dynamic (tapping) mode offers advantages. Operating in the attractive force regime minimizes shear forces and tip/sample wear. The intermittent contact modulates the water meniscus formation and the electric field. Key parameters expand to include free oscillation amplitude and setpoint ratio, which control the effective tip-sample interaction time per oscillation cycle, adding another layer of control over the oxidation kinetics.

IX. Results

Using raster lithography in the nanolithography mode of the NTEGRA Prima AFM, images were obtained (Fig.1).

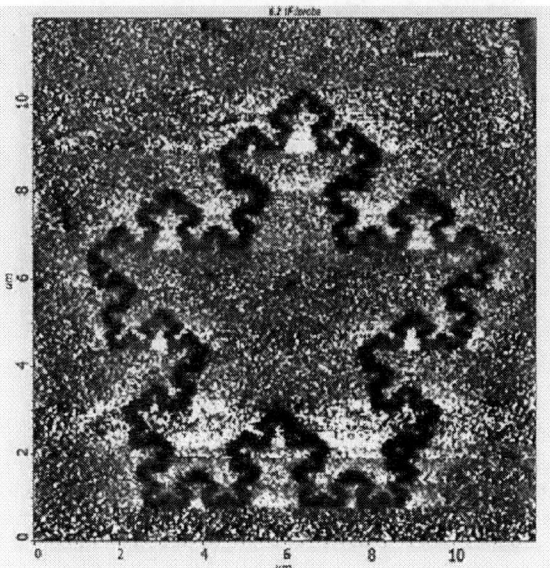

Fig. 1.   Complex mask nanostructure (IProbe signal)

The parameters used to obtain this oxide structure are given (Table 1).

TABLE I   LITHOGRAPHY PARAMETERS OF FIG.1

Voltage	Lithography Speed	Humidity	SetPoint
9 V	0.5 µm/sec	70%	0.35 nA

These images clearly show precise and contrast-rich fractal structures. It can be concluded that the nanostructure was formed with correctly selected parameters.

Furthermore, Figures 2 and 3 show images of a Penrose tiling, which is a non-periodic tiling of the plane.

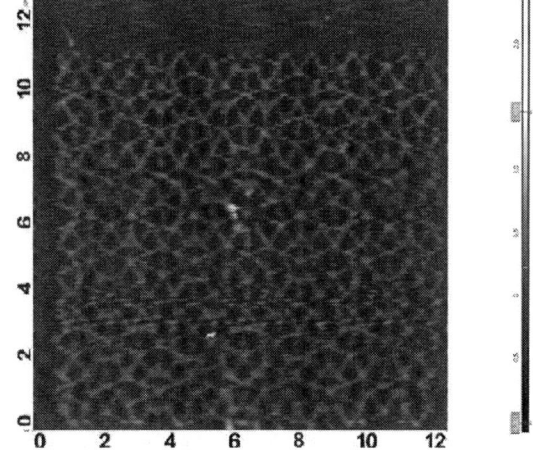

Fig. 2.   Penrose tiling (Height signal)

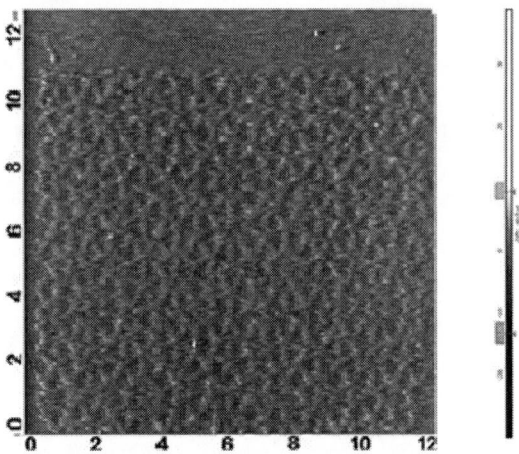

Fig. 3. Penrose tiling (DFL signal)

Selecting optimal parameters allowed for obtaining a high-quality nanostructure. Although classical lithography is based on periodic structures, the use of aperiodic Penrose patterns opens new possibilities for creating functional nanostructures with unique physical properties. It also shifts nanotechnology from the realm of creating simple periodic structures to the domain of designing complex functional systems.

X. PROSPECTS

Subsequent deposition of zinc oxide on such a tiling allows for "materializing" its unique geometric potential into specific physical properties—optical, electrical, and mechanical. This creates a foundation for developing a new generation of nanophotonic, sensory, and energy devices operating on principles inaccessible to traditional periodic nanostructures.

The developed methodologies, which are of practical interest, are being implemented at Active Photonics LLC in the development of modern equipment [12].

XI. CONCLUSION

In this work, the successful application of Local Anodic Oxidation (LAO) using Atomic Force Microscopy (AFM) for the fabrication of complex, mask-compatible nanostructures has been demonstrated. The investigation encompassed both the theoretical underpinnings of the LAO process—including the electrochemical mechanism, the critical role of the water meniscus, ion transport under a high electric field, and the influence of probe geometry—and its practical implementation.

Experimental results confirm the feasibility of creating high-precision two-dimensional aperiodic structures, specifically a Penrose tiling pattern, on a silicon substrate. The quality of the obtained nanostructures, evidenced by clear and contrast-rich images in Height, IProbe, and DFL signals, was achieved through the meticulous optimization of key process parameters: applied voltage, lithography speed, ambient humidity, and SetPoint. The use of raster lithography mode proved effective for transferring complex geometric patterns.

This study underscores the significant potential of AFM-based LAO as a versatile and accessible tool for direct-write nanolithography in research settings. It enables the prototyping of functional nanostructures with unique geometries that are challenging or impossible to achieve with conventional periodic patterning. The formation of such aperiodic mask structures, like the Penrose tiling, paves the way for developing novel optical, photonic, and sensory devices by leveraging their distinct physical properties. The methodologies developed are of direct practical interest for implementation in advanced equipment development.

REFERENCES

[1] Nanotechnology. Physics. Processes. Diagnostics. Instruments / ed. by V. V. Luchinin, Yu. M. Tairov. – Moscow : Fizmatlit, 2006. – 552 p.

[2] Kozodaev, D. A. The United Center of Scanning Probe Microscopy of NOVA SPB LLC - ETU "LETI" in the Field of Radioelectronics - A New Type Center for Solving Priority Tasks / D. A. Kozodaev, V. A. Moshnikov, E. N. Muratova, A. V. Solomonov, M. A. Trusov // Innovatsii. – 2024. – No. 1 (297). – P. 10–18.

[3] Moshnikov, V. A. Sol-Gel Technology of Micro- and Nanocomposites / V. A. Moshnikov, Yu. M. Tairov, T. V. Khamova, O. A. Shilova. – Saint Petersburg : Lan, 2013. – 304 p.

[4] Moshnikov, V. A. Atomic Force Microscopy for Nanotechnology and Diagnostics: Textbook / V. A. Moshnikov, Yu. M. Spivak. – Saint Petersburg : Publishing House of ETU "LETI", 2009. – 80 p.

[5] Moshnikov, V. A. Methods of Scanning Probe Microscopy in Micro- and Nanoelectronics / V. A. Moshnikov, A. A. Fedotov, A. I. Rumyantseva. – Saint Petersburg : Publishing House of ETU "LETI", 2003. – 84 p.

[6] Madison, A. E. Aperiodic Diffraction Grating Based on the Connection Between Prime Numbers and Zeros of the Riemann Zeta Function / A. E. Madison, D. A. Kozodaev, A. N. Kazankov, P. A. Madison, V. A. Moshnikov // Zhurnal Tekhnicheskoi Fiziki. – 2024. – Vol. 94, No. 4. – P. 658–663.

[7] Madison, A. E. Aperiodic Diffraction Grating Based on the Distribution of Zeros of the Riemann ζ-Function / A. E. Madison, P. A. Madison, D. A. Kozodaev, A. N. Kazankov, V. A. Moshnikov // Neva Photonics-2023: All-Russian Scientific Conference with International Participation: Collection of Scientific Papers. – Saint Petersburg, 2023. – P. 32.

[8] Madison, A. E. Aperiodic Diffraction Grating Based on the Distribution of Zeros of the Riemann ζ-Function / A. E. Madison, P. A. Madison, D. A. Kozodaev, A. N. Kazankov, V. A. Moshnikov // HOLOEXPO 2023: Abstracts of the 20th International Conference on Holography and Applied Optical Technologies. – Saint Petersburg, 2023. – P. 69–73.

[9] Madison, A. E. The Concept of Unit Cells in the Theory of Quasicrystals / A. E. Madison, P. A. Madison, V. A. Moshnikov // Zhurnal Tekhnicheskoi Fiziki. – 2024. – Vol. 94, No. 4. – P. 561–574.

[10] Near-Field Optical Microscope Probe : Pat. 2731164 Russian Federation : IPC G02B21/00 / Belorus A. O., Pastukhov A. I., Krasnoborodko S. Yu., Kozodaev D. A., Moshnikov V. A. ; applicant and patent holder A. O. Belorus [et al.]. – No. 2020110147 ; filed 11.03.2020 ; publ. 31.08.2020. – 13 p.

[11] Maksimov, A. I. Formation of titanium oxide semiconductor structures by the local anodic oxidation / A. I. Maksimov, V. A. Moshnikov, N. S. Pshchelko, A. V. Startseva, G. Suchanec // Smart Nanocomposites. – 2014. – Vol. 5, No. 1. – P. 19–28.

[12] Talkach, N. M. Modern Probe-Optical Industrial and Scientific Equipment for the Study of Amorphous and Microcrystalline Semiconductor Materials / N. M. Talkach, A. I. Kazankov, D. A. Kozodaev, E. V. Kuznetsov, M. A. Trusov // Amorfnye i mikrokristallicheskie poluprovodniki: Collection of Papers of the International Conference. – Moscow : National Research University "MIET", 2025. – P. 65–66.

Study of the Influence of DX-Centers on the Characteristics of a Pseudomorphic High-Electron-Mobility Transistor (pHEMT)

Alexander V. Sapozhnikov
Department of Physical Electronics and Technology
St. Petersburg Electrotechnical University «LETI»
Saint-Petersburg, Russia
slicer305@gmail.com

Roman S. Kryukov
Department of Electronics
St. Petersburg Electrotechnical University «LETI»
Saint-Petersburg, Russia
r.kryukov@svrost.ru

Vadim V. Perepelovsky
Department of Physical Electronics and Technology
St. Petersburg Electrotechnical University «LETI»
Saint-Petersburg, Russia
vvperepelovsky@mail.ru

Anatoly L. Dudin
Chief Technologist
JSC «Svetlana-Rost»
Saint-Petersburg, Russia
a.dudin@svrost.ru

Artyom I. Baranov
Laboratory of Renewable Energy Sources
St. Petersburg Academic University
Saint-Petersburg, Russia
baranov_art@spbau.ru

Abstract— This work presents a detailed analysis of the influence of deep donor traps – DX-centers – on the characteristics of a pseudomorphic high-electron-mobility transistor (pHEMT). Particular attention is paid to a systematic investigation of the output and transfer current-voltage characteristics (I-V curves) while varying the empirical coefficient Eb and the molar fraction of aluminum in the donor AlGaAs layer. It is shown that changes in these parameters cause modulation of the DX-center concentration, which significantly affects the formation of localized states in the bandgap. Numerical simulation is based on solving the fundamental system of semiconductor electronics equations (Poisson's, continuity, and transport equations) using a two-dimensional hydrodynamic model that accounts for quantum wells. The obtained results show good agreement with experimental I-V curves, confirming the correctness of the employed model. Based on the calculated dependencies, conclusions are drawn regarding the influence of the DX-center concentration on key pHEMT parameters – in particular, the saturation current and the cutoff voltage. The interpretation of the calculated I-V characteristics provides a methodology for the quantitative determination of DX-center concentration in real pHEMTs. The obtained concentration-dependent correlations allow for the optimization of epitaxial growth and transistor geometry to minimize the degrading influence of DX-centers on operating parameters, thereby increasing production yield.

Keywords— *pHEMT, high-electron-mobility transistor, GaAs/AlGaAs/InGaAs, TCAD, DX-centers, deep donor traps.*

I. INTRODUCTION

The presence of DX-centers in Si-doped $Al_xGa_{1-x}As$ layers grown by molecular beam epitaxy (MBE) [1] is one of the key challenges in the design and production of devices based on GaAs/InGaAs/AlGaAs heterostructures. Despite years of research, the microscopic nature of DX-centers and the mechanisms of their impact on device characteristics remain subjects of active study [2, 3]. It is established that DX-centers are impurity defects capable of existing in three charge states: singly positive (d^+), neutral (d^o), both associated with a shallow donor state, and singly negative (DX), corresponding to the deep state of the DX-center. Transitions between these states are determined by the position of the Fermi level in the bandgap, significantly affecting carrier distribution and the electrical characteristics of the structure.

Minimizing the influence of DX-centers is an important task in the development of pHEMT structures and can be implemented both at the technological level (doping control, selection of AlGaAs composition) and through physical modeling. The latter allows for a quantitative assessment of the influence of trap parameters and growth conditions on device characteristics. The aim of this study is the development and application of quantitative methods for modeling the influence of DX-centers on the I-V characteristics of pHEMTs while varying the composition of the donor layer and the energy levels of the traps. To achieve this goal, a two-dimensional hydrodynamic model, adapted to the specifics of quantum heterostructures, is used.

II. DEVICE STRUCTURE AND OPERATION

Numerical methods are widely used in the device-technology modeling of semiconductor elements, allowing the solution of the basic system of semiconductor electronics equations: Poisson's equation, as well as the continuity and transport equations for charge carriers. In this work, for submicron structures, a hydrodynamic model implemented in the TCAD environment and described in detail in [4,5] is used.

The electrophysical parameters of the pHEMT were investigated in the same TCAD software suite, which supports multi-criteria analysis and possesses broad capabilities for modeling semiconductor devices. The structural parameters of the device, including functional heterolayers, are presented in Table 1. During the calculations, the main focus was on the correct accounting of fundamental physical processes, as reflected in [4].

An adaptive computational mesh comprising approximately 5.6×10^4 nodes was generated for discretizing the transistor's region under study. In the region of the InGaAs channel, the mesh density was locally increased to ensure more accurate calculation of carrier distribution.

This work includes a detailed investigation of the influence of deep donor traps (DX-centers) on the characteristics of a GaAs pHEMT. A generalized theoretical model for interpreting DX-centers [6], as well as

979-8-3315-4784-4/26 $31.00 © 2026 IEEE

experimental data obtained by deep-level transient spectroscopy (DLTS) [7] and I-V measurements of real structures numerically modeled in this study, were used as the basis. A comparison of experimental results with calculated data on DX-center concentration and their influence on pHEMT characteristics is presented in this article.

The applied model for calculating the DX-center concentration in the GaAs pHEMT is based on the methodology outlined in [6, 8, 9]. According to DLTS measurements performed for pHEMT structures manufactured at JSC "Svetlana-Rost", with an Al content = 0.22 in the donor AlGaAs layer, the DX-center concentration can reach 1.2×10^{17} cm^{-3}. A key parameter of the universal DX-center model is the empirical coefficient Eb, which, within experimental error, can be considered a universal constant determining the binding energies of DX-centers in ternary and quaternary alloys based on the zinc-blende structure [6]. During numerical simulation, it was found that for the system under study, the optimal value of Eb is 0.14 eV. Its variation range was 0.1–0.2 eV with a step of 0.01 eV at a fixed Al fraction of 0.22. For each value of Eb, transfer and output I-V curves were calculated. The influence of varying the Al fraction in the donor AlGaAs layer within the range of 0.15–0.3 with a step of 0.01 at a fixed coefficient Eb = 0.14 eV was also investigated. For each value of the Al molar fraction, output and transfer I-V curves were computed. The properties and parameters of the alloys and materials used in this work are reflected in [10].

The calculated DX-center concentrations in this study for the variation of the Al molar fraction are: 1.98×10^{16} cm^{-3} - 1.16×10^{17} cm^{-3} in the range from 0.15 to 0.22 and 1.7×10^{17} cm^{-3} - 1.01×10^{18} cm^{-3} in the range from 0.23 to 0.3. The DX-center concentrations for the variation of the coefficient Eb are: 2.95×10^{16} cm^{-3} - 1.16×10^{17} cm^{-3} in the range from 0.1 eV to 0.14 eV and 1.88×10^{17} cm^{-3} to 1.08×10^{18} cm^{-3} in the range from 0.15 eV to 0.2 eV.

TABLE I.   GaAs pHEMT 0.5 switch type structure data

Layer Name	Layer thickness, Å	Doping type, concentration, см$^{-3}$
Passivation Si$_3$N$_4$	200	-
Contact layer GaAs	500	n, $3.3 \cdot 10^{18}$
Cap layer GaAs	200	n, $2.0 \cdot 10^{17}$
Etch-stop AlGaAs	25	n, $4.1 \cdot 10^{18}$
Etch-stop GaAs	50	n, $1.0 \cdot 10^{17}$
Barrier layer Al$_{0.22}$Ga$_{0.78}$As	175	n, $1.0 \cdot 10^{17}$
Donor layer Al$_{0.22}$Ga$_{0.78}$As	125	n, $3,4 \cdot 10^{18}$
Spacer Al$_{0.22}$Ga$_{0.78}$As	20	-
Channel In$_{0.22}$Ga$_{0.78}$As	120	-
Buffer layer GaAs	8000	-

III. Results and Discussion

This work presents an analysis of the impact of DX-centers on the electrophysical parameters of an AlGaAs/InGaAs/GaAs pHEMT. A comparative study was performed on the output and transfer I-V curves calculated for the numerical pHEMT model and experimental pHEMT structures manufactured at JSC "Svetlana-Rost" with identical topological parameters. The standard numerical model under investigation assumes an aluminum fraction in

the AlGaAs layer of 0.22 and an empirical coefficient Eb of 0.14 eV, which corresponds to a DX-center concentration of 1.16×10^{17} cm^{-3}, consistent with experimental measurements.

Experimental verification of the output characteristics (see Fig. 1 a) was carried out by varying the drain-source voltage from 0 V to 5 V at zero gate bias. Simultaneously, the study of transfer characteristics (see Fig. 1 b) was conducted at a fixed drain-source voltage of 5 V and by varying the gate potential from -2 V to 0.5 V to accurately determine the cutoff voltage. The results of the analysis of the output and transfer I-V curves demonstrate a saturation current of 28.62 mA, which falls within the experimentally recorded range of 27.40 mA to 31.55 mA, and a cutoff voltage of -1.3 V, consistent with experimental values ranging from -1.5 V to -1.19 V. The obtained correlation between theoretical and experimental data confirms the adequacy of the applied model.

The output (see Fig. 2 a) and transfer (see Fig. 2 b) I-V curves were investigated while varying the empirical coefficient Eb. In the Eb range from 0.1 eV to 0.13 eV, a slight decrease in the saturation current from 29.48 mA to 28.96 mA is observed, corresponding to a deviation from the value of the standard model under investigation of 3% to 1.21%, caused by the low concentration of DX-centers. With a further increase in Eb to 0.17 eV, the saturation current decreases to 27.35 mA with a deviation of up to 4.43%, demonstrating an enhanced influence of DX-centers. The most pronounced parameter degradation is observed in the Eb range from 0.17 eV to 0.2 eV, where the saturation current drops to 19.40 mA with a maximum deviation of 32.23%, which agrees with the sharp increase in DX-center concentration. The study of transfer characteristics revealed the stability of the cutoff voltage at -1.3 V in the Eb range from 0.1 eV to 0.14 eV, coinciding with the standard value of the model under investigation. However, for Eb $\geqslant$ 0.15 eV, a shift in the cutoff voltage to -1.38 V with a deviation of up to 5.03% is observed.

The output (see Fig. 3 a) and transfer (see Fig. 3 b) I-V curves, obtained by systematically changing the molar fraction of aluminum in the ternary AlGaAs solution in the range from 0.15 to 0.3 with a step of 0.01 at a fixed value of the empirical coefficient Eb = 0.14 eV, were investigated. In the Al molar fraction range from 0.15 to 0.21, slight fluctuations in the saturation current within the range of 28.71 mA to 28.86 mA are observed, with the maximum current at an Al fraction of 0.18 and the minimum at an Al fraction of 0.16. The deviation from the value of the standard model in this region does not exceed 0.83%, indicating weak sensitivity of the parameter to composition at low aluminum concentrations, particularly due to the low concentration of DX-centers. When the Al fraction increases to 0.3, the saturation current decreases to 24.07 mA with a deviation of 15.91% from the value of the standard model, demonstrating a pronounced nonlinear dependence. Analysis of the transfer characteristics revealed the stability of the cutoff voltage at -1.3 V within the Al fraction range from 0.19 to 0.22, corresponding to the standard value of the model under investigation. In the Al fraction range from 0.15 to 0.18, the cutoff voltage varies from -1.27 V to -1.29 V with a deviation of up to 2.03%, whereas for Al fractions from 0.23 to 0.3, its shift to -1.37 V with a maximum deviation of 4.43% is observed.

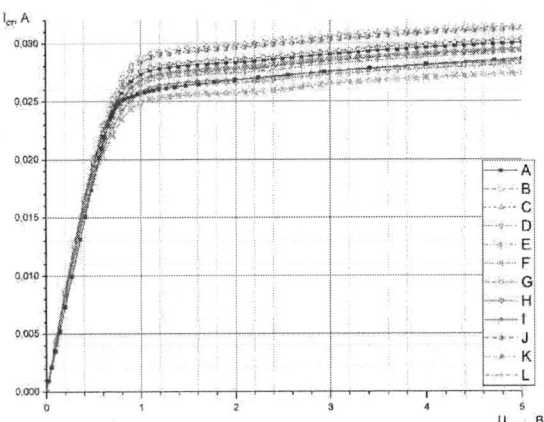

Fig. 1. A comparison of the output characteristics between simulated and fabricated pHEMTs. The simulation (curve A) incorporates a numerical model that includes DX centers, B-S curves present the corresponding experimental data.

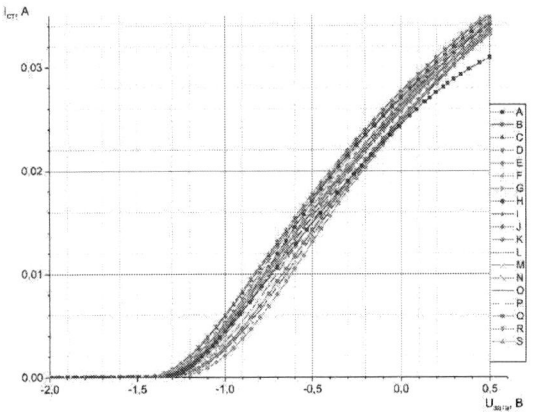

Fig. 2. A comparison of the transfer characteristics between simulated and fabricated pHEMTs. The simulation (curve A) incorporates a numerical model that includes DX centers, B-S curves present the corresponding experimental data.

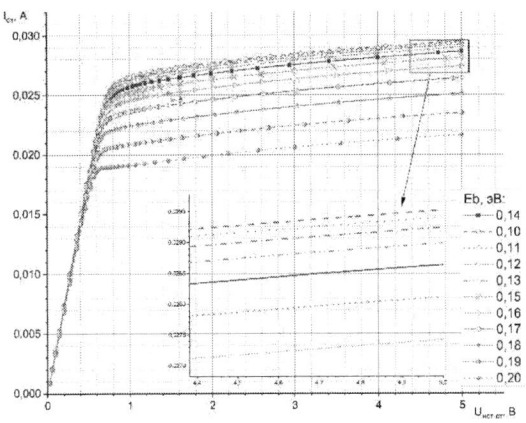

Fig. 3. Simulated output characteristics of the pHEMT, demonstrating the effect of varying the empirical coefficient Eb between 0.1 eV and 0.2 eV across a range of DX-center concentrations. These simulations were performed with a fixed aluminum molar fraction (x = 0.22) in the $Al_xGa_{1-x}As$ donor layer.

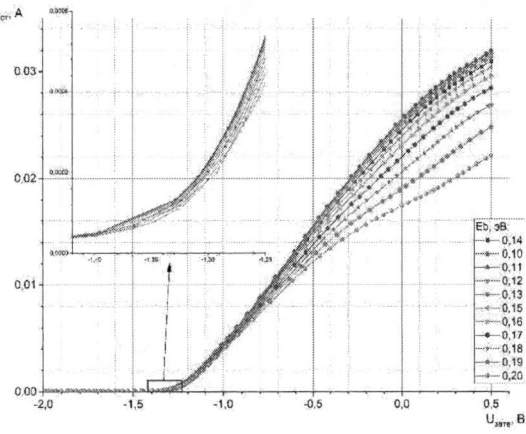

Fig. 4. Simulated transfer characteristics of the pHEMT, demonstrating the effect of varying the empirical coefficient Eb between 0.1 eV and 0.2 eV across a range of DX-center concentrations. These simulations were performed with a fixed aluminum molar fraction (x = 0.22) in the $Al_xGa_{1-x}As$ donor layer.

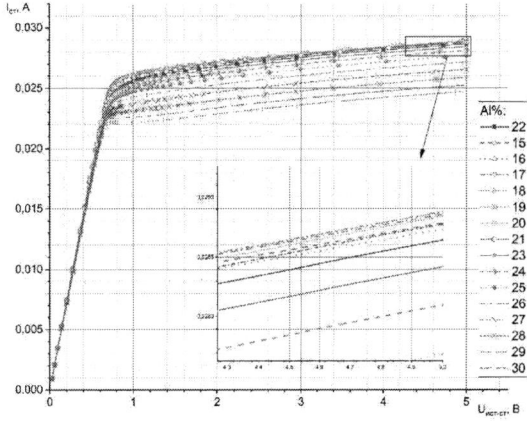

Fig. 5. Simulated output characteristics of the pHEMT model, showing the influence of varying the aluminum composition (x = 0.15 to 0.3) in the $Al_xGa_{1-x}As$ donor layer across different DX-center concentrations. For these simulations, the coefficient Eb was maintained at a constant value of 0.14 eV.

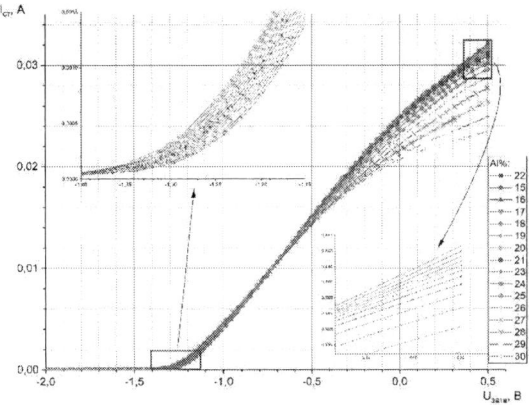

Fig. 6. Simulated transfer characteristics of the pHEMT model, showing the influence of varying the aluminum composition (x = 0.15 to 0.3) in the $Al_xGa_{1-x}As$ donor layer across different DX-center concentrations. For these simulations, the coefficient Eb was maintained at a constant value of 0.14 eV.

979-8-3315-4784-4/26 $31.00 © 2026 IEEE

IV. CONCLUSION

In the course of this study, a numerical model accounting for the influence of deep donor traps (DX-centers) on the characteristics of a GaAs/AlGaAs-based pHEMT was developed and applied.

Using real pHEMT structures as an example, the results of numerical simulation were compared with experimental data, which confirmed the adequacy of the proposed calculation methodology. The DX-center concentration, determined by the empirical coefficient Eb and the molar fraction of aluminum in the donor AlGaAs layer, plays a key role in the formation of localized states and, consequently, in the device operation. Varying these parameters makes it possible to influence key pHEMT characteristics, such as output and transfer current-voltage dependencies, and significantly affects the saturation current and cutoff voltage.

The obtained results open prospects for more precise adjustment of growth recipes and transistor design parameters to minimize the negative impact of DX-centers on the operating characteristics of pHEMTs. In the future, it is planned to expand the model to account for other types of deep centers to assess device reliability and longevity.

The interpretation of the calculated I-V characteristics provides a methodology for the quantitative determination of DX-center concentration in real pHEMTs. The obtained concentration-dependent correlations allow for the optimization of epitaxial growth and transistor geometry to minimize the degrading influence of DX-centers on operating parameters, thereby increasing production yield.

REFERENCES

[1] Mooney, P. M. DX centers in III-V alloys: recent developments. Deep Centers in Semiconductors. 2024, pp. 643-665.

[2] Chadi, D. J., Chang, K. J. Theory of the atomic and electronic structure of DX centers in GaAs and Al x Ga 1− x As alloys. Physical review letters. 1988, vol. 61, no. 7, pp. 873. doi: https://doi.org/10.1103/PhysRevLett.61.873

[3] Chadi, D. J., & Zhang, S. B. Atomic structure of DX centers: Theory. Journal of electronic materials. 1991, vol. 20, pp. 55-58. doi: https://doi.org/10.1007/BF02651965

[4] Sapozhnikov A.V., Pushnitsa I.S., Dudin A.L., Perepelovskiy V.V. Numerical Analysis of AlGaAs/InGaAs/GaAs pHEMT. Journal of the Russian Universities. Radioelectronics. 2025;28(3):116-128. (In Russ.) https://doi.org/10.32603/1993-8985-2025-28-3-116-128

[5] SentaurusTM Device User Guide, Ver. T-2022. 03, Synopsys TCAD Sentaurus, San Jose, CA, USA, 2022. 1446 p.Karpov S. Y. A Universal Model for DX - Center Binding Energy in Cubic III-V Compounds. physica status solidi (b). 2021

[6] Mari, Ruaz Hussain. DLTS characterisation of defects in III-V compound semiconductors grown by MBE. Diss. University of Nottingham, 2011.

[7] Karpov S. Y. A Universal Model for DX - Center Binding Energy in Cubic III-V Compounds. physica status solidi (b). 2021, vol. 258, no. 5, pp. 2000596. doi: https://doi.org/10.1002/pssb.202000596

[8] Tachikawa M., Mizuta M., Kukimoto H., Minomura S. A simple calculation of the DX center concentration based on an L-donor model. Japanese journal of applied physics. 1985, vol. 24, no. 10A, pp. L821. doi: 10.1143/JJAP.24.L821D

[9] Gordon L., Lyons J. L., Janotti A., Van de Walle C. G. Hybrid functional calculations of DX centers in AlN and GaN. Physical Review B. 2014, vol. 89, no. 8, pp. 085204. doi: https://doi.org/10.1103/PhysRevB.89.085204 '

[10] Adachi S. Properties of semiconductor alloys: group-IV, III-V and II-VI semiconductors. John Wiley & Sons. 2009

Controlled Hot-Injection synthesis of Lead Halide Perovskite Quantum Dots for Enhanced Photoluminescence and Sensor Application

Sohail Amir
Department of Photonics
Saint Petersburg Electrotechnical University "LETI"
St.-Petersburg, Russia
amir.sohail.ieeeuot@gmail.com

Abstract—This work presents a systematic investigation of hot-injection synthesis of lead halide perovskite quantum dots (QDs) using both hybrid organic-inorganic and full inorganic compositions, including MAPbBr3, FAPbBr3, and CsPbBr3. The synthesis was performed through a controlled bottom-up route in which precursor ratios and reaction temperatures were precisely tuned to regulate nucleation and crystal growth.

Photoluminescence (PL) spectroscopy revealed that emission wavelength, intensity, and bandwidth strongly depend on nanocrystal size and halide composition. Smaller QDs exhibited pronounced quantum confinement behavior, with MAPbBr3, and CsPbBr3 showing sharp green emission, while bromide based systems showed broader green spectra. The comparative results demonstrate that hot injection method enables reliable control over structural and optical properties across different compositions. These findings highlight the potential of perovskite QDs for high performances optoelectronic and sensor applications and provide new insights into structure property relationships governing their photoluminescence behavior.

Keywords—metal halide perovskite, quantum dots, photoluminescence, synthesis

I. INTRODUCTION

Lead halide perovskite quantum dots (QDs) have gained significant attention in recent years because of their exceptional optical characteristics, including high photoluminescence efficiency, narrow emission bandwidth, and strong size–dependent quantum confinement effects [1], [2]. These properties make them attractive materials for a variety of modern optoelectronic technologies such as light emitting devices, optical sensors, biomedical imaging, and high performance photonic systems [3]. Compared to traditional semiconductor nanocrystals, perovskite QDs can be synthesized through low temperature chemical routes, which reduces cost and enables scalable processing for device applications [4], [5].

Among the different fabrication strategies, the hot-injection remains one of the most effective techniques for producing highly uniform nanocrystals with controllable size and chemical composition [6]. The rapid temperature driven nucleation allows precise adjustment of crystal growth, making it possible to tune emission wavelength and crystallinity by simply modifying reaction conductions. Despite intensive research efforts, controlling structural homogeneity and understanding how synthesis parameters influence optical behavior remain active scientific challenges. These challenges becomes more pronounced when varying A-site cations (such as MA+, FA+ and Cs+) or

halide anions (Br-, I-), both of which significantly impact stability and optical performance [7],

Furthermore, establishing clear relationships between precursor ratios, reaction temperature, and resulting photoluminescence properties is crucial for applications requiring stable and high intensity emission, particularly in sensing and imaging technologies. while pervious works have investigated individual compositions, comprehensive comparisons between hybrid organic-inorganics system (MAPbl3, MAPbBr3, FAPbBr3) and fully inorganic CsPbBr3 nanocrystals under controlled hot-injection conductions remain limited in the current literature [8].

In this work, we address these gaps by performing a systematic hot-injection synthesis of MAPbBr3, FAPbBr3, and CsPbBr3 QDs. By carefully tuning precursor ratios and injections temperatures, we investigate how synthesis parameters affect particle size, emission wavelength and overall photoluminescence response. The findings provide new insight into structure property relationship of perovskite QDs and highlight their suitability for next generation optoelectronic sensor-based applications.

II. METHODOLOGY

A. Synthesis of MAPbBr₃, FAPbBr₃ and CsPbBr₃ Quantum Dots

The three perovskite quantum dots (MAPbBr₃, FAPbBr₃ and CsPbBr₃) were synthesized using the Hot Injection technique, following the reaction scheme shown in Fig.1.

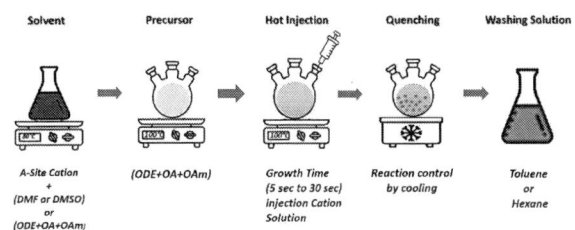

Fig. 1. Workflow diagram of hot injection synthesis process

B. Solvent perparation

In a separate vial, , 0.2 mmol of Cesium Bromide (CsBr) or Formamidinium Bromide (FABr) was dissolved in 1 mL of Dimethylformamide (DMF) or Dimethyl Sufoxide (DMSO) heated at 80-100°C

For Methylammonium Bromide (MABr) solution, 0.2 mmol of MABr was added in 1 mL Octadecene (ODE),

Oleic Acid (OA) Ligand and oleyamine (OAm) heated at 80-100ºC.

1. Precursor Solution Formation

10 mL of Octadecene (ODE), 1 mL Oleic Acid (OA), 1 mL of Oleyamine (OAm) and 0.2 mmol of Lead Bromide ($PbBr_2$) were mixed in a three-necked round-bottom flask. The solution was heated to 100-110ºC under continuous stirring until the $PbBr_2$ completely dissolved.

2. Hot Injection

A-site precursor solution (MA^+, FA^+ or Cs^+) was rapidly injected into the $PbBr_2$ solution at 100-110ºC using a preheated syringe. Growth time was controlled between 5-30 seconds depending on the desired crystal size.

3. Quenching

After required growth period, the reaction was immediately cooled using an ice bath to prevent further crystal growth. Color change was observed, confirming.

4. Purification

The product solution was diluted using toluene or hexane, following by centrifugation at 4000 rpm for 10 min. The precipitated quantum dots were collected, redispersed in fresh solvent, and 0.2 ml TOP was added to enhance surface passivation. Washing was repeated two times to remove residual ligands and unreached precursors.

III. RESULTS AND DISCUSSION

Fig. 2 shows the photoluminescence (PL) emission spectra of $CsPbBr_3$, $MAPbBr_3$ and $FAPbBr_3$ quantum dots, recorded under the same experimental conditions. All three samples exhibit emission within the green region, verifying the successful synthesis of bromide based perovskite nanocrystals.

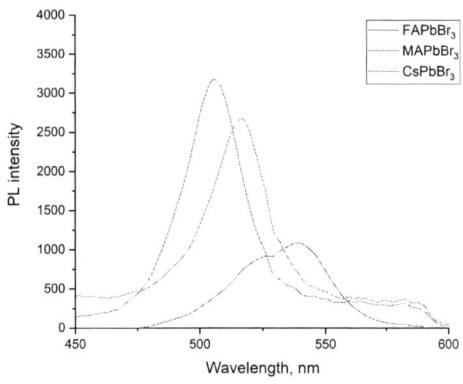

Fig. 2. Photoluminescence spectra of $MAPbBr_3$,$CsPbBr_3$ and $FAPbBr_3$

Among the prepared materials, $CsPbBr_3$ displays the strongest PL response, with a narrow emission centered around 506 nm ($\approx$2.44 eV). The high peak intensity indicates improved crystal quality and fewer non-radiative pathways. The $MAPbBr_3$ sample emits at 517 nm ($\approx$2.39 eV) with moderate intensity, suggesting a comparatively higher defect density or slightly lower radiative efficiency. In contrast, $FAPbBr_3$ shows a broader and red-shifted peak at 537 nm ($\approx$2.30 eV), which may be related to larger crystal domains, surface traps, or enhanced defect-assisted recombination.

The shift in emission wavelength from Cs → MA → FA confirms that optical bandgap can be tuned by changing the A-site cation, where FA-based QDs exhibit the lowest bandgap and Cs-based modification due to lattice expansion and compositional variation.

Overall, the findings indicate that $CsPbBr_3$ quantum dots possess superior optical performance, making them promising candidates for high-brightness optoelectronic applications such as LEDs. Meanwhile, the lower bandgap and broader emission of $FAPbBr_3$ may be advantageous for devices like photodetectors and sensing platforms.

A. Application of Perovskite Quantum Dots (PQDs)

Fig. 3 demonstrates the major applications of perovskite quantum dots (PQDs) in optoelectronic devices. The center shows the typical ABX_3 perovskite crystal structure, where A is a monovalent cation (MA^+/FA^+/Cs^+), B is a divalent metal (commonly Pb^{2+}), and X denotes halide ions (Cl^-, Br^-,I^-). PQDs possess remarkable optical characteristics including high photoluminescence quantum yield, narrow emission bandwidth, defect tolerance and band gap tunability, which make them suitable for advanced sensing and energy applications.

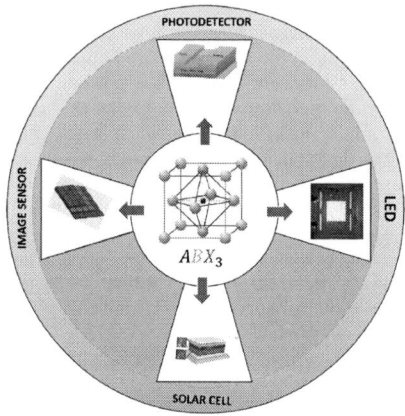

Fig. 3. Applications of (PQDs) in photodetectors, LEDs, solar cells, and image sensors.

The surrounding sections highlight four major PQD device applications:

a) Photodetectors

PQDs act as efficient light-absorbing layers with high carrier mobility and responsivity. They have been used in fast response and low-dark-current photodetectors, enabling high detection sensitivity for optical communication and low-light sensing [9].

b) Light-emitting diodes (LEDs)

Due to size and composition dependent emission, PQDs provide tunable, high purity light output. Perovskite based LEDs show strong brightness and are widely explored for displays and solid-state lighting applications [10].

c) Solar cells

PQDs enhance photon absorption and charge extraction extraction in photovoltaic structure. Their adjustable bandgap is useful for designing tandem and high-efficiency solar cells, making them attractive for next-generation PV technologies [11].

d) Image sensors

PQDs integrated in CMOS architectures improve color sensitivity, spectral selectivity and low-light image capture.

Their strong optical response enables high-quality image sensing for camera modules and biomedical imaging systems [12].

IV. CONCLUSION

In this work, MAPbBr$_3$, CsPbBr$_3$ and FAPbBr$_3$ perovskite quantum dots were successfully synthesized using the hot injection method. The research demonstrated that the ratio of Oleic Acid (OA) and Oleylamine (OAm) plays a critical role in crystal growth. As the concentration of the precursors was increased (e.g., from 0.2 mmol upward), the corresponding volumes of the ligands were also adjusted. Specifically, the Oleic Acid (OA) and Oleylamine (OAm) volumes were increased to 1 mL each. This increase in OA and OAm facilitated improved crystal growth and resulted in better-quality perovskite nanocrystals. It was also observed that increasing the volume of these ligands further enhanced surface passivation, leading to batter stability and crystallinity.

Reaction time was found to be another crucial parameter. By controlling the reaction time, smaller-sized quantum dots with better morphology were obtained, as evidenced in (Fig. 2). Shorter reaction time favored nucleation dominated growth, preventing large crystal formation and producing more monodisperse nanoparticles.

Photoluminescence (PL) results showed high emission intensity and improved quantum yield, particularly in CsPbBr$_3$ QDs, indicating strong radiative recombination and minimal trap states. The PL spectra (Fig. 3) confirm that the synthesized perovskite quantum dots exhibit excellent optical properties.

Based on the obtained PL data, these synthesized quantum dots show strong potential for optoelectronic applications, especially in LEDs, imaging sensors, photodetectors, and solar cells. Overall, controlled ligand ratio and reaction time enable the fabrication of high quality perovskite QDs suitable for future device integration.

REFERENCES

[1] H. Li et al., "Synthesis of Size-Adjustable CsPbBr3 Perovskite Quantum Dots for Potential Photoelectric Catalysis Applications," Materials, vol. 17, no. 7, p. 1607, Apr. 2024, doi: 10.3390/ma17071607.

[2] P. D. Badillo and A. E. Degterev, "Perspectives on Perovskite Solar Cells Under the Glass of Characterization and Model-based Research," in 2023 XXVI International Conference on Soft Computing and Measurements (SCM), May 2023, pp. 277–280. doi: 10.1109/SCM58628.2023.10159103.

[3] A. E. Degterev, M. M. Romanovich, I. I. Mikhailov, I. A. Lamkin, and S. A. Tarasov, "Ways to Slow Down the Degradation and Enhance the Stability of Perovskite Solar Cells," in 2021 IEEE Conference of Russian Young Researchers in Electrical and Electronic Engineering (ElConRus), Jan. 2021, pp. 1301–1304. doi: 10.1109/ElConRus51938.2021.9396607.

[4] M. Degtereva et al., "Influence of the Spectral Composition of Illuminating Light Sources on Biometric and Phytochemical Characteristics of Ocimum basilicum L.," Photonics, vol. 10, no. 12, Dec. 2023, doi: 10.3390/photonics10121369.

[5] K. Vighnesh, S. Wang, H. Liu, and A. L. Rogach, "Hot-Injection Synthesis Protocol for Green-Emitting Cesium Lead Bromide Perovskite Nanocrystals," ACS Nano, vol. 16, no. 12, pp. 19618–19625, Dec. 2022, doi: 10.1021/acsnano.2c11689.

[6] L.-C. Chen, K.-L. Lee, C.-Y. Huang, J.-C. Lin, and Z.-L. Tseng, "Preparation and Characteristics of MAPbBr3 Perovskite Quantum Dots on NiOx Film and Application for High Transparent Solar Cells," Micromachines, vol. 9, no. 5, p. 205, Apr. 2018, doi: 10.3390/mi9050205.

[7] A. S. Tarasov, A. E. Degterev, M. M. Romanovich, M. D. Pavlova, I. I. Mikhailov, and I. A. Lamkin, "Optical Properties of Bromine-Doped Perovskite Films," in 2023 Seminar on Fields, Waves, Photonics and Electro-optics: Theory and Practical Applications (FWPE), Nov. 2023, pp. 147–150. doi: 10.1109/FWPE60445.2023.10368505.

[8] H. Yu et al., "Green Light-Emitting Devices Based on Perovskite CsPbBr3 Quantum Dots," Front. Chem., vol. 6, p. 381, Aug. 2018, doi: 10.3389/fchem.2018.00381.

[9] L. Dou et al., "Solution-processed hybrid perovskite photodetectors with high detectivity," Nat. Commun., vol. 5, no. 1, p. 5404, Nov. 2014, doi: 10.1038/ncomms6404.

[10] X.-K. Liu et al., "Metal halide perovskites for light-emitting diodes," Nat. Mater., vol. 20, no. 1, pp. 10–21, Jan. 2021, doi: 10.1038/s41563-020-0784-7.

[11] A. E. Degterev et al., "Modeling of photoelectric characteristics and development of structures for solar cells based on CsPbI3 and CsPbBr3," J. Opt. Technol., vol. 91, no. 8, pp. 521–526, Aug. 2024, doi: 10.1364/JOT.91.000521.

[12] P. Lu et al., "Metal halide perovskite nanocrystals and their applications in optoelectronic devices," InfoMat, vol. 1, no. 4, pp. 430–459, Dec. 2019, doi: 10.1002/inf2.12031.

Design and Analysis of a Stacked Silicon Wafer-Based Technology for Multilayer Multi-Chip Modules

Ilya A. Solovyov
Institute of Nano and Microsystems Technology
National Research University of Electronic Technology
Moscow, Russia
ilya.a.solovyov@gmail.com

Andrey Yu. Titov
Institute of Nano and Microsystems Technology
National Research University of Electronic Technology
Moscow, Russia
kisrpi@mail.ru

Denis V. Vertyanov
Institute of Nano and Microsystems Technology
National Research University of Electronic Technology
Moscow, Russia
vdv.vertyanov@gmail.com

Abstract— **This paper investigates modern technological approaches to the design and fabrication of multichip modules (MCMs) with an emphasis on silicon-based multilayer integration. Key processes involved in the formation of silicon interposers are analyzed, with particular attention paid to manufacturing yield and structural robustness in multilayer silicon assemblies. A method for fabricating multilayer multichip modules (mMCM) based on wafer-level batch processing is proposed, where the upper wafer is intended for mounting integrated circuit dies and the lower wafer serves as a base for interconnection elements. Finite element analysis (FEA) was employed to evaluate the thermomechanical stresses induced during the wafer bonding process. The results demonstrate that the proposed approach improves manufacturing yield, reliability, and form factor of mMCMs, indicating its suitability for advanced multichip module fabrication.**

Keywords—multi-chip modules, three-dimensional integration, thermomechanical stress, interposer, finite element analysis

I. Introduction

Three-dimensional integration is one of the most rapidly developing directions of microelectronics. It makes it possible to significantly increase the functional density of devices while reducing their volume and power consumption. By reducing interconnection lengths and integrating heterogeneous components in a small volume, such solutions provide higher bandwidth and reduced latency compared to traditional assembly methods. One of the key aspects of 3D integration is ensuring reliable inter-tier interconnection between the layers of multilevel multichip modules (mMCM) [1].

At present, 2.5D and 3D integration technologies are used to fabricate mMCMs. In 2.5D integration, several dies are mounted on an intermediate substrate, which can be based on silicon or silicon–germanium or quartz; organic substrates based on polyimide or bismaleimide-triazine (BT) resins are also widely used. Silicon interposers have a number of advantages over other types of substrates. First, they provide a higher wiring density compared to polymer-based substrates, since they are manufactured using semiconductor fabrication processes. Second, they offer a complete match of the coefficient of linear thermal expansion (CTE) to that of the silicon die of the integrated circuit, which increases the reliability of the final product. Third, they are cheaper than the fabrication of quartz interposers. Such solutions are used

The work was carried out with the financial support of the Ministry of Education and Science as part of the state project FSMR-2025-0005.

by the largest die manufacturers, such as Huawei, TSMC, Qualcomm, Samsung, and Intel [2].

II. Methods Of Three-Dimensional Integration

There are four main classes of approaches to 3D integration (fig. 1): those employing redistribution layers (RDL, 3D-WLP, InFO), die stacking (including the use of DAF films) with wire bonding, TSV-based solutions within the die structure, and approaches employing interposers. The key elements of the latter approach are through-silicon vias (TSVs), which connect the two sides of the substrate. For inter-tier assembly, the use of solder balls or copper pillars (Cu-pillar) and wire bonding is widely applied. Recently, the use of copper wires coated with a thin palladium layer has begun to gain popularity [3–5].

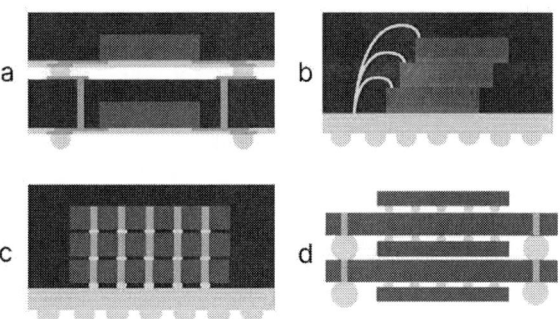

Fig. 1. Approaches to 3D integration (a – 3D-FOWLP, b – die stacking, c – TSV, d – interposer-based)

There are also solutions that propose the use of interposers in combination with silicon frames performing the functions of inter-tier interconnection. Such an approach has a number of advantages. First, due to the close placement of dies at different tiers, the interconnection lengths are significantly reduced, which increases the data transmission speed and reduces the power consumption of the module. Second, wafer-level integration makes it possible to use standard silicon materials and processing methods (TSV, lithography, laser processing), which simplifies the process flow and increases the reliability of the interconnection [6].

The multilevel architecture provides design flexibility: different types of dies (logic, memory, sensor elements) and passive components can be combined at different tiers. This creates the possibility of integrating heterogeneous technologies (RF, optoelectronics, MEMS, ADC/DAC, etc.) within a single module with minimal interface losses [7].

979-8-3315-4784-4/26 $31.00 © 2026 IEEE

However, this approach also has disadvantages. The complexity of the technological process of fabrication and alignment of the interposer with the frame leads to an increase in cost and a reduction in yield. The fragility of the frame significantly degrades the yield at the stage of multichip module assembly. The objective of this paper is to investigate a batch technology for the fabrication of mMCMs based on intermediate substrates and frames in order to increase the reliability and performance of the devices.

III. DESCRIPTION OF THE PROPOSED TECHNOLOGY

The proposed technology is based on batch fabrication of mMCMs by sequential stacking of two silicon wafers. The first wafer performs the functions of a substrate and contains contact pads and interconnection elements formed using batch microfabrication methods. This wafer provides sites for mounting integrated circuit dies. The second wafer serves as a base for forming silicon frames and includes a system of through holes intended for die placement and for providing inter-tier interconnection. The overall fabrication process of assemblies based on the proposed technology is shown in Fig. 2.

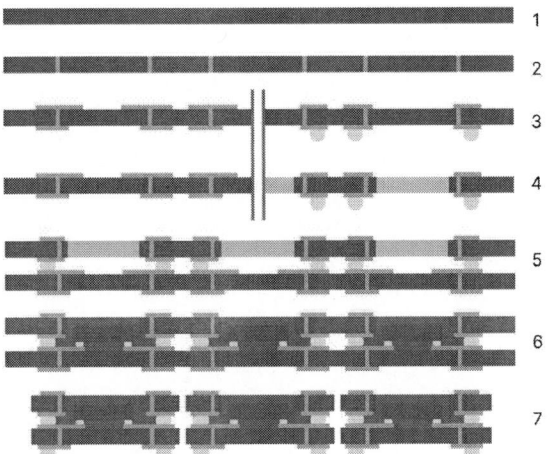

Fig. 2. Sequence of fabrication of multichip module tiers (1 – wafer preparation, 2 – TSV formation, 3 – topology formation, application of solder balls onto the wafer with holes for die mounting, 4 – formation of holes in the wafer, 5 – wafer assembly, 6 – die mounting and encapsulation, 7 – application of solder balls)

At the first stage, through-silicon vias (TSVs) are formed in silicon. For this purpose, deep reactive ion etching (DRIE) methods are used or, in some cases, laser drilling, which provides high speed and accuracy when processing thick substrates. It should be noted that laser drilling results in the formation of local melting zones on the wafer surface; therefore, to ensure the required planarity and quality of subsequent technological operations, chemical-mechanical polishing is recommended. The diameter and pitch of the holes are selected based on the requirements for interconnection density and the mechanical strength of the wafer [3,8].

After hole formation, silicon surface oxidation is performed to provide electrical insulation of the future metallized structures. A thin metallized layer with a thickness of approximately 1 μm is deposited on both sides of the wafer; this layer serves as a seed layer for subsequent electroplating of copper or other conductive materials to the required thickness (typically in the range of 5–20 μm), depending on the required design rules.

For deposition of the initial layer, it is preferable to use atomic layer deposition (ALD) methods, which provide uniform coating over the entire surface and inside the holes, or magnetron sputtering using specialized fixtures. As an adhesion sublayer, films of titanium, chromium, or titanium nitride, tantalum nitride with a thickness of 30–100 nm are typically used; these layers improve adhesion between the silicon substrate and the metallized layer [9].

Subsequently, photolithography and etching of the conductive pattern are performed on the wafer surface. After completion of topology formation, dielectric coatings are applied to the wafer. These can include polymer dielectrics, such as polyimide, as well as standard SiO_2 coatings. At the same stage, windows for die mounting are formed, as well as inter-tier interconnection regions that provide electrical and mechanical coupling between the layers during subsequent wafer alignment [10].

For wafers performing the functions of silicon frames, through holes are formed in a similar manner using DRIE or laser ablation, providing alignment with the underlying layers and placement of dies inside the formed openings. After completion of wafer fabrication, the wafers are aligned and bonded. For this purpose, solder ball terminals are formed on the wafer with holes using batch methods, for example, through a stencil (Fig. 3), or by laser-assisted attachment of terminals [11]. The wafer with holes is placed on top of the wafer on which the dies are to be mounted, and the solder balls are reflowed. Then, die mounting is performed at the wafer level. After placement, all dies undergo an encapsulation process or filling with a high-flow compound – underfill.

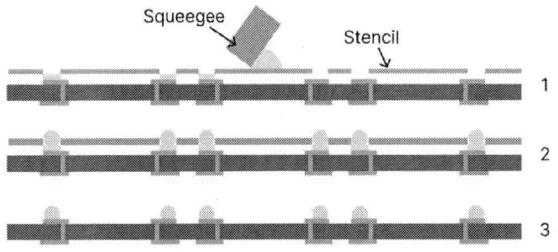

Fig. 3. Application of solder balls through a stencil (1 – flux application, 2 – ball distribution, 3 – reflow)

To improve the mass-and-size characteristics of the multichip module and to optimize the overall design, it is proposed to implement wafer-level formation of recesses for die mounting (Fig. 4). These recesses provide partial or complete embedding of the die into the substrate thickness, which contributes to a reduction in the overall assembly height and an increase in the mechanical stability of the structure.

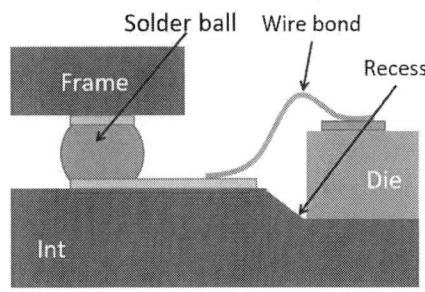

Fig. 4. Schematic of recess formation for die mounting

The formation of recesses can be carried out by anisotropic silicon etching, which makes it possible to obtain strictly controlled geometric parameters of the cavities with smooth sidewalls, or by laser ablation, which provides high flexibility in processing individual areas of the wafer. The choice of a specific method is determined by the requirements for the depth and shape of the recess, the type of die used, and the specifics of subsequent wire bonding operations.

A key feature of the proposed technology is batch wafer processing at all stages of the technological cycle prior to their separation into individual modules. The use of batch technology makes it possible to significantly increase productivity and process repeatability by unifying operations and reducing the number of individual assembly steps.

In contrast to individual assembly, where each assembly tier is fabricated separately, the proposed method significantly reduces labor intensity and lowers production cost. At the same time, improved reliability and quality control are achieved, since all key technological operations are performed at the wafer level rather than at the level of an individual die or assembly.

IV. SIMULATION OF YHE STRUCTURE

To assess the reliability of the proposed structure, thermomechanical stress simulation arising during assembly was performed using FEA. The temperature of 215°C was selected as it corresponds to the onset of solidification of SAC305 solder during cooling after reflow, which is the critical stage for the formation of residual thermomechanical stresses in the assembled structure. The wafer thickness was chosen as 460 μm, as standard for 100-mm wafers. The simulation was performed in COMSOL Multiphysics. The specified material parameters are presented in Table 1.

TABLE I. PARAMETERS DEFINED FOR THE SIMULATION

Material	Silicon (215℃)	SAC305 (215℃)
Young's modulus, GPa	140	45
Poisson's ratio	0.28	0.33
Density, kg/m³	2325	8420
Thermal conductivity, W/(m·K)	73.7	58
Coefficient of thermal expansion, 1/K	$3.2 \cdot 10^{-6}$	$24 \cdot 10^{-6}$
Heat capacity at constant pressure, J/(kg·K)	840	240

The main parameter varied during the simulation process was the effective bridge width, defined as twice the frame width, since two module tiers are located adjacent to each other in the wafer structure. The simulation result for an effective width of 4 mm is shown in Fig. 5. The maximum stress arising in the structure of the upper wafer does not exceed 40 MPa and is caused by the presence of contact with solder ball interconnects.

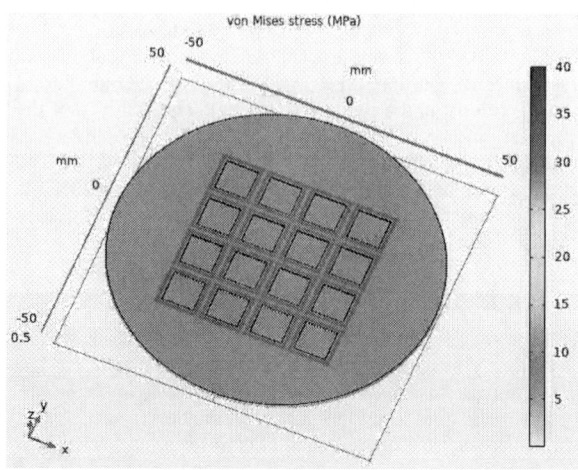

Fig. 5. Result of wafer structure simulation

A set of simulations was carried out, where the effective width ranged from 3 to 7 mm (Fig. 6). As shown in Fig. 6, the maximum stress decreases almost linearly with an increase in the effective bridge width. Measurements of the maximum stress were performed in a cross-section at the center of the bridge.

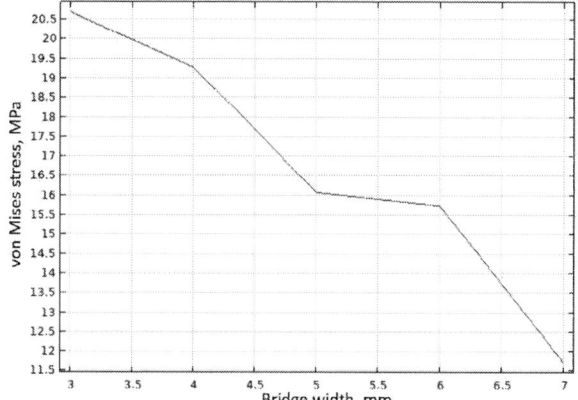

Fig. 6. Result of simulation with varying parameters

At a frame width of 3 mm, the maximum stress value is 20.69 MPa, which is acceptable for mMCM. However, thermocompression bonding – used, for instance, with copper pillars instead of solder balls – poses a higher risk of cracking due to the increased mechanical stress. Thus, for silicon elements it is considered that stresses in silicon should not exceed 50-80 MPa, however, it should be taken into account that the wafer contains TSV holes, which introduce additional nonuniformities into its structure, and excessive pressure may lead to wafer cracking [12, 13].

V. FABRICATION OF A DEMONSTRATION SAMPLE

To confirm the technological feasibility of implementing the proposed technology, a demonstration sample of wafer tiers with holes and an interposer with a mounted die was fabricated. Fig. 7 shows the fabricated sample containing an array of 16 identical cells, each of which is intended for die placement and formation of inter-tier interconnection. A silicon wafer with a diameter of 100 mm and a thickness of 460 μm was used as the substrate. Metallization and dielectric layers were deposited on its surface.

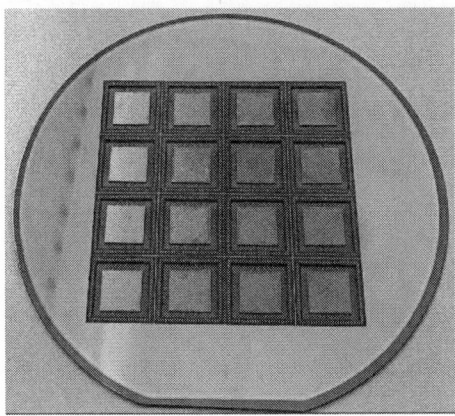

Fig. 7. Demonstration sample

The interconnection topology and windows for die installation were formed by laser micromachining, which made it possible to eliminate the use of photomasks and simplified the process of fabricating the demonstration sample. At the same time, the use of laser technologies led to an increase in the duration of individual operations, however, it ensured high flexibility and accuracy during prototype fabrication.

The demonstration sample illustrates the design and technological features of the proposed technology. The obtained result confirmed the possibility of batch formation of multichip modules at the wafer level, and also demonstrated the compatibility of the technology with existing microelectronic fabrication methods. The results of the performed work demonstrate the technological feasibility and prospects of the proposed approach for further development of multilevel MCMs.

VI. CONCLUSION

This paper presents an analysis of approaches to the design and fabrication of multichip modules, and also proposes an original method for fabricating multilevel multichip modules based on batch technological operations at the wafer level from the formation of functional layers to die mounting and singulation into individual modules. Based on the results of the performed work, the following recommendations aimed at improving the reliability and manufacturability of multichip modules were formulated:

- When designing wafer structures with an effective bridge width of less than 4 mm, it is recommended to perform additional analysis of mechanical stress distribution, since local increases in deformation may occur in this region during tier bonding. The gap between adjacent cells on the wafer can be selected arbitrarily within permissible technological limitations, based on the required packing density.

- When using die assembly technology based on wire bonding, it is advisable to apply batch formation of recesses in the substrate structure to improve mass-and-size characteristics. For this purpose, it is recommended to use anisotropic silicon etching or laser ablation methods.

- When using laser ablation for window formation, it is recommended to preliminarily coat the wafer surface with a temporary protective layer, for example, photoresist, which makes it possible to prevent metal contamination by silicon dust.

A limitation of the proposed technology is its sensitivity to defects in individual cells. Defects present on both wafers are cumulative; as a result, a defect in a cell on either wafer leads to failure of the entire cell. Therefore, electrical and structural testing stages should be implemented both prior to wafer bonding and after bonding, before die assembly.

Thus, the proposed technology can serve as a basis for further development of compact, high-performance, and reliable next-generation multichip modules.

ACKNOWLEDGMENT

The work was carried out with the financial support of the Ministry of Education and Science as part of the state project FSMR-2025-0005.

REFERENCES

[1] S. Zhang, Z. Li, H. Zhou, R. Li, S. Wang, K. W. Paik, and P. He, "Challenges and recent prospects of 3D heterogeneous integration," *e-Prime - Advances in Electrical Engineering, Electronics and Energy*, vol. 2, Art. no. 100052, 2022.

[2] F. Sheikh, R. Nagisetty, T. Karnik, and D. Kehlet, "2.5D and 3D heterogeneous integration: Emerging applications," *IEEE Solid-State Circuits Mag.*, vol. 13, no. 4, pp. 77-87, 2021.

[3] D. H. Cho, S. M. Seo, J. B. Kim, S. H. Rajendran, and J. P. Jung, "A review on the fabrication and reliability of three-dimensional integration technologies for microelectronic packaging: Through-silicon via and solder bumping processes," *Metals*, vol. 11, no. 10, Art. no. 1664, 2021.

[4] M. D. Kochergin, I. A. Solovyov, S. A. Batin, D. V. Vertyanov, S. P. Timoshenkov, and N. E. Korobova, "Problems of microsystems reliability design with redistribution layers in wafer-level packaging," in *Proc. 25th Int. Conf. Young Professionals in Electron Devices and Materials (EDM)*, 2024, pp. 220-224.

[5] Y. C. Huang, Y. X. Lin, C. K. Hsiung, T. H. Hung, and K. N. Chen, "Cu-based thermocompression bonding and Cu/dielectric hybrid bonding for three-dimensional integrated circuits applications," *Nanomaterials*, vol. 13, no. 17, Art. no. 2490, 2023.

[6] M. M. Burakov, D. V. Vertyanov, A. N. Boyko, and A. V. Sosnovsky, "Investigation of TSV metallization for MEMS packaging technology," in *Proc. IEEE Conf. Russian Young Researchers in Electrical and Electronic Engineering (EIConRus)*, 2018.

[7] N. Singh *et al.*, "Challenges and opportunities in engineering next-generation 3D microelectronic devices: Improved performance and higher integration density," *Nanoscale Adv.*, vol. 6, no. 24, pp. 6044–6060, 2024.

[8] F. C. Rufino, A. Alaferdov, L. C. J. Espindola, A. R. Silva, F. I. D. L. Leite, and J. A. Diniz, "Study of laser drilling and chemical polishing as an alternative TSV fabrication route," in *Proc. 39th Symp. Microelectronics Technology and Devices (SBMicro)*, vol. 1, 2025, pp. 1-4.

[9] H. Ma, Q. Wang, L. Tan, Y. Hu, Y. Zheng, and J. Cai, "Electroplating Cu on ALD TiN for small-size TSV," in *Proc. 25th Int. Conf. Electronic Packaging Technology (ICEPT)*, 2024, pp. 1-6.

[10] P. Nimbalkar, P. Bhaskar, M. Kathaperumal, M. Swaminathan, and R. R. Tummala, "A review of polymer dielectrics for redistribution layers in interposers and package substrates," *Polymers*, vol. 15, no. 19, Art. no. 3895, 2023.

[11] M. Z. Ding, L. C. Wai, S. Zhang, and V. S. Rao, "Evaluation of laser solder ball jetting for solder ball attachment process," in *Proc. IEEE 14th Electronics Packaging Technology Conf. (EPTC)*, 2012, pp. 23-29.

[12] R. F. Cook, "Strength and sharp contact fracture of silicon," *J. Mater. Sci.*, vol. 41, no. 3, pp. 841-872, 2006.

[13] S. Y. Jun, J. H. Bang, M. S. Kim, D. G. Han, T. Y. Lee, and S. Yoo, "Thermocompression bonding of Cu/SnAg pillar bumps with electroless palladium immersion gold (EPIG) surface finish," *Materials*, vol. 16, no. 4, Art. no. 173, 2023.

979-8-3315-4784-4/26 $31.00 © 2026 IEEE

Thermoelectric Numerical Simulation of a Memristor-based Microchip

Maxim V. Sozonov
School of Natural Sciences
University of Tyumen
Tyumen, Russia
ORCID: 0000-0003-1232-0389

Abstract — **Simulating a microchip with many memristors requires significant computational resources. This can be reduced by using the functional dependence of the memristive layer conductivity on the flowing current based on the I–V characteristic data, rather than directly modeling the filament or simulating the vacancy diffusion process. The result of this approach to simulation is a qualitative correspondence of the temperature field to experimental data and a reduction in calculation time. This approach is suitable for modeling large multilayer memristor-based microchips.**

Keywords — *memristor, simulation, heat, microchip, temperature*

I. INTRODUCTION

The memristor was postulated by L. Chua in 1971 [1] and first physically realized by HP Labs in 2008 [2]. It has emerged as a fundamental circuit element with significant potential for revolutionizing data storage and neuromorphic computing.

This is possible due to the memristive effect, which means that the resistance (and therefore the logical state) depends on the current flowing through the device. Its ability to combine memory and logic functions within a single, nanoscale, energy-efficient, and non-volatile device makes it attractive for next-generation electronics [3].

Despite these advantages, the widespread adoption of memristors is hindered by challenges in mass production and the absence of a unified physical model. The dominant switching mechanism in memristive structures is generally attributed to the formation and rupture of conductive filaments (CFs), driven by the drift of oxygen vacancies [3]. Most studies focus on single-device modeling with full nanoscale simulation of coupled equations for vacancy drift, electric potential, and heat transfer [4, 5]. Scaling such models to simulate microchips with thousands or millions of memristors requires substantial computational resources. Therefore, the development of a simplified approach to modeling the switching process is necessary and relevant.

Another critical aspect of memristor modeling is heat management. Joule heating can significantly impact device stability, switching characteristics, and operational lifetime [6, 19]. Predicting heat distribution across a microchip with multiple memristors is essential during the design stage. Experimental temperature measurements at the nanoscale present certain challenges. For a multilayer microchip with multiple memristors, it is also necessary to understand the heat distribution inside the structure, where direct temperature measurements are impossible. The use of computer-aided engineering (CAE) modeling for such problems is therefore highly relevant. Combined with a simplified approach to resistive switching modeling, this could accelerate the adoption of memristors in practical applications.

This work presents a practical electro-thermal model for a complete memristor-diode crossbar microchip. The primary novelty lies in its simplified approach to modeling the resistive switching mechanism. Simulations were performed under various operational conditions and provide critical insights into heat management and architectural optimization.

II. METHODS

A. Model Concept and Simplification

Instead of explicitly modeling the formation and rupture of CFs and the associated vacancy dynamics, the memristor state is defined directly through the resistivity ρ_e of its memristive layer. This resistivity is derived from experimentally measured current–voltage (I–V) characteristics of the real device [7], assigning specific values for the High-Resistance State (HRS or OFF) and the Low-Resistance State (LRS or ON). This approach allows simulation of the full microchip structure without solving oxygen-vacancy drift equations or embedding CF structures in each memristor.

B. Governing Equations

The governing heat equation used in the simulation is the standard transient energy balance with internal Joule heating and convective boundary conditions:

$$c_p\rho(\partial t/\partial T) = \lambda\nabla^2 T + \rho_e j^2 \qquad (1)$$

$$q_{surface} = \alpha(T_0 - T_{surface}) \qquad (2)$$

where c_p is specific heat capacity (J/(kg·K)), ρ is density (kg/m³), λ is thermal conductivity (W/(m·K)), ρ_e is electrical resistivity (Ω·m), j is current density (A/m²), α is the convective heat-transfer coefficient (W/(m²·K)), T_0 is ambient temperature, and $T_{surface}$ is the surface temperature.

Radiative heat transfer was neglected because it contributed negligibly to the overall heat balance while increasing solver time. Thus, convective cooling is the dominant external heat-transfer mechanism in the present model.

C. Geometry and Materials

The model geometry is based on an experimental 8-cell memristor–diode crossbar microchip [7]. Each cell comprises two memristors and a Zener diode, arranged in a shared-bitline topology. The topology of a cell is shown in Fig. 1. The chip includes a 0.5 mm silicon substrate (a glass substrate was also considered in one case) with a 100 nm SiO₂ layer. The memristive layer is a 30 nm Ti$_{0.93}$Al$_{0.07}$O$_x$ film. The chip is mounted on a 1 mm thick textolite board, with electrical connections modeled as small copper cylinders. Material properties were taken from [7] and the ANSYS material library.

D. Numerical Implementation

The model was implemented in ANSYS v19.2 using the Finite Element Method (FEM). The computational workflow is two-stage: (1) the steady-state electric problem is solved in the Electric module to obtain volumetric Joule heating distribution; (2) this Joule heating distribution is imported as a fixed internal heat source into the Transient Thermal module, where the time-dependent heat-transfer problem is solved. The time-step resolution corresponds to the pulse sequences used in the simulation cases.

Several simulation cases were investigated, corresponding to different operating modes of the chip, as summarized in Table 1. Case 1 represents a pulsed "reading" operation mode [7]: the signal is first fed to one cell, then to the other, and then to both at once. Cases 2-4 represent constant signal operating modes (current flow duration of 3 minutes). Cases 5 and 6 present high-stress operating modes. For Case 4, the influence of the substrate material on the temperature field was investigated. For cases 1 and 2, the non-active cells had an HRS+HRS memristor configuration. In case 3, non-active cells had an HRS+LRS memristor configuration.

TABLE I.   SIMULATION (OPERATING) CASES

Case	Voltage, V	Active Cells	Memristor Configuration in Active Cell
1	±1.5	2	LRS+HRS
2	±1.5	4	LRS+HRS
3	±1.5	4	LRS+HRS
4	±1.5	8	LRS+HRS
5	±4	8	LRS+HRS
6	±4	8	LRS+LRS

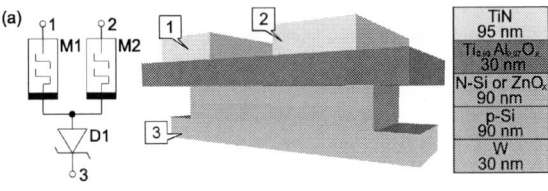

Fig. 1. Topology of a cell [7].

III. RESULTS

A. Temperature Dynamics

For the pulsed mode (Case 1), the average temperature in the active region showed a rapid response to the applied signal (Fig. 2). A residual-heating effect was observed: subsequent pulses started from a higher baseline temperature, indicating that heat does not fully dissipate between short pulses. For constant-signal operating modes (e.g., Case 2, Fig. 3), the chip temperature gradually increased, reaching a steady-state regime after approximately three minutes.

For Cases 1-4, maximum temperatures remain below ~40 °C (Case 1 peak ~32.15 °C), whereas extreme cases with ± 4 V applied to many cells (Case 6) produce peak temperatures up to ~109.48 °C. This highlights the need for active cooling in high-density, high-power applications.

Furthermore, the choice of substrate material proved important: replacing the silicon substrate with glass resulted in a maximum-temperature increase of 7.6 °C and an average-temperature increase of 3.9 °C due to the lower thermal conductivity of glass.

B. Spatial Temperature Distributions

Temperature fields of the active crossbar region reveal the spatial pattern of heating (Fig. 4). In simulated snapshots for Case 6, temperature maxima are located at memristors in the ON state; heating is concentrated within the cell area, while neighboring cells show moderate warming, plausibly due to parasitic currents through shared electrodes.

The model revealed significant asymmetry in the temperature field, with one specific memristor consistently reaching the highest temperature. This is attributed to the chip's position on the board, which creates a preferential heat-dissipation path toward the nearest board edge. This asymmetry could lead to non-uniform aging and reduced device reliability.

The configuration of inactive cells also influenced the thermal profile. A chip fully populated with data (many memristors in LRS state) operates at a higher temperature than an "empty" chip (all HRS) during read operations.

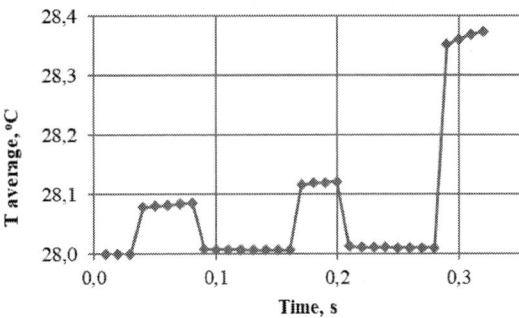

Fig. 2. Average temperature for Case 1 (comma is used as the decimal separator).

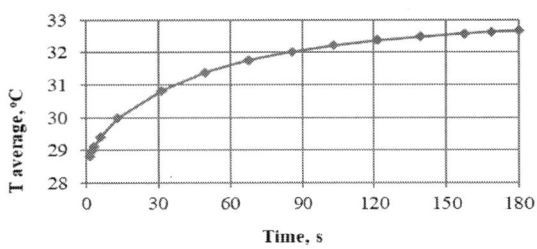

Fig. 3. Average temperature for Case 2.

979-8-3315-4784-4/26 $31.00 © 2026 IEEE

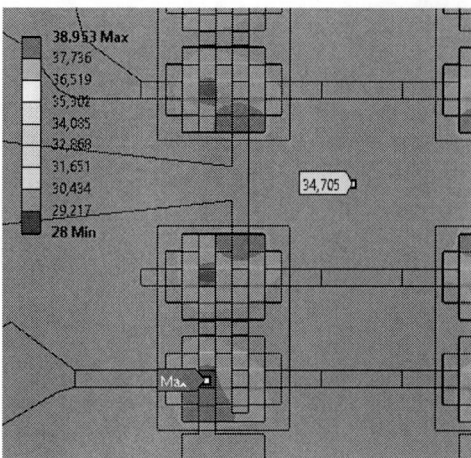

Fig. 4. Temperature field (values in °C) for Case 6.

IV. DISCUSSION

A. Advantages of the Model

The simplified resistivity-based modeling approach enables large-scale 3D simulation of the entire microchip without explicitly resolving CF dynamics. This approach captures chip-level heating trends, the influence of cell configuration and chip geometry, and allows rapid exploration of thermal-management strategies during design. The simulations reproduce expected behaviors: ON-state cells exhibit higher local heating; residual heating accumulates across pulses; substrate material and chip placement influence hotspot locations and magnitudes. The model also identifies thermal crosstalk due to parasitic currents and the need for cooling under extreme operating modes. These strengths make the model practical for early design decisions and for studying array-level thermal coupling.

B. Spatial-Heating Mismatch With Nanoscale Experiments

A key limitation of the simplified model is its inability to resolve nanoscale filament heating. Numerous high-resolution thermometry studies show that Joule heating in resistive switching is highly localized within the filament core (tens to hundreds of nanometres), resulting in sharp hotspots rather than the distributed heating predicted by a homogeneous-resistivity model.

For example, Nandi et al. [8] used in-situ scanning thermal microscopy (SThM) with ~100 nm resolution to map volatile switching in NbO$_x$ memristors and reported a nanoscale hotspot (Fig. 5) at the filament location. The authors emphasize that SThM reveals heating highly confined to the filament rather than distributed across the top electrode area. Similarly, Roldan et al. [9] used SThM for thermal mapping and also localized the heating to a small point.

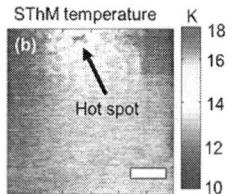

Fig. 5. Temperature field of memristor from Nandi et al. [8].

Because the presented model treats each memristor as a homogeneous resistor (uniform ρ_e across the memristive layer), the volumetric Joule heat is distributed over the entire volume in the simulation. This smears what is, in reality, a small high-power-density filament into a larger, lower-power-density region, producing a bulk-heating pattern.

Two major consequences follow:

- Underestimation of peak local temperature: a filament confined to a <100 nm cross-section with the same dissipated power reaches a much higher peak temperature than the cell-averaged model predicts. This is critical for switching mechanisms, local chemistry, and reliability.

- Overestimation of heated volume and thermal coupling: the model overpredicts lateral heat spreading and therefore overestimates thermal influence on neighboring cells. This may mislead design choices regarding spacing, thermal isolation, and cooling strategies.

C. Ways to Improve the Model

To bridge the gap between chip-scale simplified modeling and nanoscale thermal behavior, the following approaches may be pursued:

1) Effective filament submodel:

Introduce an effective filament heat source with calibrated geometry (radius, height) and contact resistance parameters, using data from SThM or thermal-resistance extraction. This preserves computational efficiency while restoring localized-heating physics.

2) Non-linear material properties:

Include temperature-dependent properties of matereials and non-uniform spatial resistance distribution. This approach focuses on physically accurate material behavior even without explicit filament geometry.

Implementing these improvements (both or choosing one) should bring simulated peak temperatures and spatial profiles into closer agreement with nanoscale thermometry while preserving chip-level predictive capability.

V. CONCLUSION

A computationally efficient 3D thermophysical model of a memristor–diode crossbar microchip has been presented. The model's key feature is the simplified representation of memristor states via layer resistivity derived from experimental I–V data. The model predicts chip-level temperature dynamics and spatial fields for multiple realistic operating cases and demonstrates how the substrate, chip architecture, and cell configuration influence the thermal regime. However, because the model treats memristors as spatially homogeneous resistive regions, it produces distributed heating across the device volume, whereas experimental studies show filament-localized hotspots. General recommendations are provided to address this discrepancy in future research.

ACKNOWLEDGMENT

Author thanks Center for Academic Writing "Impulse" at University of Tyumen for assistance in preparing the thesis.

REFERENCES

[1] L. Chua, "Memristor – the missing circuit," IEEE Trans. Circuit Theory, vol. CT-18, no. 5, pp. 507-519.

[2] D. B. Strukov, G. S. Snider, D. R. Stewart, R. S. Williams, "The missing memristor found," Nature, vol. 453, pp. 80-83.

[3] L. G. Alekseeva, A. S. Ivanov, V. V. Luchinin, A. A. Petrov, T. Tikeu, T. Nabatame, "Memristor — the new nanoscale element of multilevel neuromorphic logic," Biotechnosphere, no. 3-4, pp. 45-46.

[4] X. Gao, D. Mamaluy, P. R. Mickel, M. Marinella, "Three-dimensional fully-coupled electrical and thermal transport model of dynamic switching in oxide memristors," ECS Transactions, vol. 69, no. 5, pp. 183-193.

[5] J. P. Strachan, D. B. Strukov, J. Borghetti, J. J. Yang, G. Medeiros-Ribeiro, S. R. Williams, "The switching location of a bipolar memristor: chemical, thermal and structural mapping," Nanotechnology, vol. 22, no. 25.

[6] D. G. Pahinkar, P. Basnet, M. P. West, B. Zivasatienraj, A. Weidenbach, A. W. Doolittle, E. Vogel, S. Graham, "Experimental and computational analysis of thermal environment in the operation of HfO2 memristors," AIP Advances, vol. 10, no. 3.

[7] A. Pisarev, A. Busygin, A. Bobylev, A. Gubin, S. Udovichenko, "Fabrication technology and electrophysical properties of a composite memristor-diode crossbar used as a basis for hardware implementation of a biomorphic neuroprocessor," Microelectronic Engineering, vol. 236.

[8] S. K. Nandi, E. Puyoo, S. K. Nath, D. Albertini, N. Baboux, S. K. Das, T. Ratcliff, R. G. Elliman, "High Spatial Resolution Thermal Mapping of Volatile Switching in NbOx-Based Memristor Using In Situ Scanning Thermal Microscopy," ACS Appl. Mater. Interfaces, vol. 14, no. 25, pp. 29025–29031.

[9] J. B. Roldan, A. Cantudo, D. Maldonado, C. Aguilera-Pedregosa, E. Moreno, T. Swoboda, F. Jimenez-Molinos, Y. Yuan, K. Zhu, M. Lanza, M. M. Rojo, "Thermal Compact Modeling and Resistive Switching Analysis in Titanium Oxide-Based Memristors," ACS Appl. Electron. Mater., vol. 6, no. 2, pp. 1424–1433.

A Low-Power 12-bit SAR ADC with a Switching Energy Minimization Algorithm in 180-nm CMOS

Anastasiia Tsepilova
Institute of Integrated Electronics
National Research University of
Electronic Technology – MIET
Moscow, Russia
tsepilovanastya@yandex.ru

Vladimir Losev
Institute of Integrated Electronics
National Research University of
Electronic Technology – MIET
Moscow, Russia
dsd@miee.ru

Aleksandr Timoshenko
Institute of Integrated Electronics
National Research University of
Electronic Technology – MIET
Moscow, Russia
timoshneko@org.miet.ru

Abstract—This paper presents a low-power 12-bit Successive Approximation Register Analog-to-Digital Converter (SAR ADC) implemented in 180nm technology. The design employs a charge-redistribution capacitor DAC as its core. The conversion consists of two phases: sampling and bit cycling. During sampling, the input signal is stored on the capacitor array. The conversion phase begins by simultaneously switching segments of the DAC to establish an initial condition. A custom switching algorithm then sequentially toggles these capacitor segments based on the comparator's decisions. This method minimizes the average energy required for each bit decision by reducing the number of large capacitor switches per cycle. Measurement results demonstrate the efficiency of this approach, achieving a SNDR of 75 dB while consuming only 109 μA. This represents a significant improvement over a conventional SAR ADC baseline, which achieved 72 dB SNDR at 769 μA consumption, confirming the superior power efficiency and linearity of the proposed switching scheme.

Keywords—*SAR ADC, low power, energy-efficient switching, split-capacitor DAC*

I. INTRODUCTION

The proliferation of portable, implantable, and Internet of Things (IoT) devices creates a persistent demand for analog-to-digital converters with ultra-low power consumption and small chip area [1]. Among various architectures, successive approximation register ADCs are a predominant choice for medium-speed applications due to their power-efficient, bit-by-bit conversion principle which allows major analog blocks to be idle between cycles [2].

However, the core power bottleneck in a conventional SAR ADC is the capacitive DAC responsible for charge redistribution. In a binary-weighted array, each bit decision requires switching large MSB capacitors, incurring an energy cost proportional to the square of the voltage and the total capacitance [3]. This issue becomes critical for high-resolution designs (e.g., 12-bit and above), where the capacitor spread necessary for linearity leads to a large die area, increased parasitic capacitance, and consequently, higher power dissipation.

Thus, optimizing the DAC architecture and its switching sequence is crucial for minimizing energy per conversion without degrading key linearity parameters such as integral and differential nonlinearity (INL, DNL) and signal-to-noise and distortion ratio (SNDR).

II. ENERGY CONSUMPTION OF CLASSICAL CAPACITIVE DAC AND OPTIMIZATION METHODS

A. Energy Model of a Classical Binary DAC

The energy expended per switching event in an N-bit binary-weighted capacitor array depends on the initial and final states of the switched capacitors. For the common case where the bottom plate of a capacitor switches between the reference voltage V_{ref} and ground (GND), the average energy drawn from the reference supply per conversion cycle is approximated by [3]:

$$E_{avg} \approx 0.5 C_{total} V^2_{ref}(2^N-1) \tag{1}$$

where C_{total} is the total array capacitance. Equation (1) shows an exponential growth of energy with resolution. For a 12-bit ADC with a unit capacitance $C_{unit} = 20$ fF, this leads to significant energy dissipation, predominantly from switching the large MSB capacitors.

B. Overview of Switching Energy Reduction Methods

Several approaches to mitigate this problem are documented in the literature:

- Split-Capacitor Array: The array is divided into MSB and LSB segments connected via an attenuation capacitor (C_b) [4]. This drastically reduces the LSB segment's capacitance and its switching energy. However, effectiveness highly depends on the precise ratio $C_{MSB}:C_b:C_{LSB}$.

- Monotonic and Energy-Saving Switching: These algorithms minimize the number of switch toggles by starting the comparison from the sampled voltage and switching only necessary capacitors for subsequent steps [5].

- Low-Voltage Operation: Reducing V_{ref} and V_{dd} directly lowers energy but constrains dynamic range and complicates comparator design.

The solution presented in this work synthesizes and optimizes the first two approaches: a segmented capacitor array is controlled by a modified energy-saving switching algorithm specifically adapted for segmented operation.

III. PROPOSED ADC ARCHITECTURE AND SWITCHING ALGORITHM

A. Overall ADC Structure

The designed 12-bit SAR ADC (conceptual block diagram is shown in Fig. 1) comprises the following key blocks:

- Segmented Capacitive DAC: Consists of an MSB segment (6 bits), an LSB segment (6 bits), and a bridge capacitor C_b. The chosen capacitance ratio (C_{MSB}:C_b:C_{LSB}=1:1:1/32) provides the necessary attenuation while minimizing total area.

- Dynamic Latch Comparator: Features zero static power consumption, which is critical for overall efficiency.

- SAR Control and Switch Driver Logic: A digital block implementing a finite-state machine that generates clock signals for the comparator and controls the capacitor-switching algorithm.

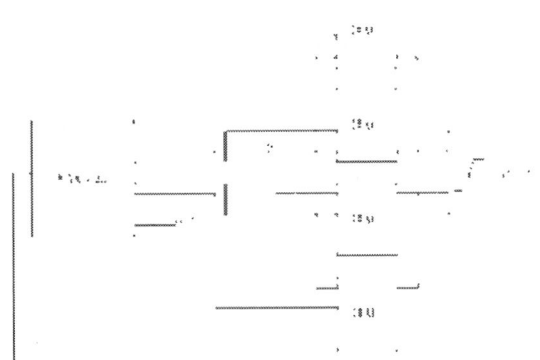

Fig. 1. ADC architecture

B. Detailed Switching Algorithm Description

The conversion process is divided into two phases: the sampling phase (SP) (Fig. 2 (a)) and the bit-cycling phase (BCP) (Fig. 2 (b, c)).

- Sampling Phase: All bottom plates of the capacitor array are connected to the input signal V_{in}, while the top plates are connected to the common-mode voltage V_{cm}. A charge proportional to V_{in} is stored on the array.

- Bit-Cycling Phase: Consists of 12 clock cycles:

  o Initial Condition (Cycle 1): The bottom plates of all capacitors in the MSB segment are switched to V_{ref} and all capacitors in the LSB segment are switched to GND. This creates a voltage at the comparator input equal to $-V_{in}+V_{ref}/2$. The first comparator decision (D_{11}) determines the sign of the input signal relative to the mid-scale.

  o Successive Approximation (Cycles 2-12): The algorithm follows logic aimed at minimizing movement of the large MSB capacitors.

- If the current cycle evaluates a bit belonging to the MSB segment, only the corresponding capacitor within this segment is switched. The state of the LSB segment remains unchanged.

- After all MSB segment bits are determined (6 bits), further approximation is conducted exclusively within the LSB segment. Capacitors inside the LSB segment are switched while the MSB segment remains fixed in its final state.

The key advantage of the algorithm is that after the first few cycles resolve the MSBs, energy-intensive switching of large capacitors ceases entirely. The entire fine-tuning process (LSBs) occurs within the low-power LSB segment, which has a substantially smaller total capacitance. This directly reduces the average switching energy, as confirmed by subsequent calculations and measurements.

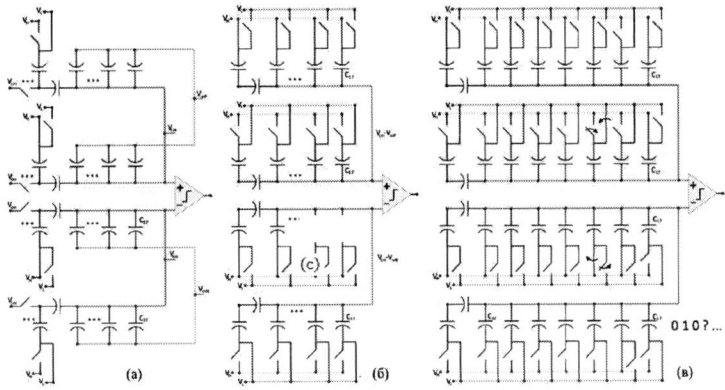

Fig. 2. DAC switching algorithm

IV. COMPARISON WITH THE CLASSIC SAR ADC

The key merits and advantages of the proposed architecture are most clearly demonstrated through a direct comparison with a conventional SAR ADC. To provide an objective benchmark, a standard 12-bit SAR ADC was designed, fabricated, and characterized using the same 180 nm CMOS process and the same 5.0 V supply voltage. This reference design employed a conventional binary-weighted capacitive DAC without segmentation and a standard successive approximation algorithm.

The fundamental distinction lies in the core DAC architecture. The conventional ADC utilizes a single array of twelve binary-weighted capacitors, where the total capacitance $C_{total} = 4095C_{unit}$ (with $C_{unit} = 20$ fF) directly dictates both its energy consumption and silicon area. In the proposed architecture, this monolithic structure is partitioned into two segments – Most Significant Bits (MSB) and Least

Significant Bits (LSB) – interconnected via a bridge capacitor, C_b. This segmentation leads to a fundamental reduction in the total array capacitance by more than a factor of eight, as the effective contribution of the LSB segment is attenuated.

The switching algorithm has also been substantially modified. In the conventional scheme, each approximation step, particularly during the early cycles, involves toggling the largest MSB capacitors, which constitutes the primary source of high-energy expenditure. Conversely, the proposed algorithm fixes the state of the entire MSB segment after the first six high-order bits are resolved. All subsequent fine-tuning operations are performed exclusively within the low-power LSB segment, which possesses orders-of-magnitude lower intrinsic capacitance. This fundamental shift in operational logic directly enables a radical reduction in the average energy consumed per conversion cycle.

These structural improvements are directly reflected in the hardware layout. The die area occupied by the conventional ADC core measures 1200x700 um². The optimized architecture, benefiting from the compactness of the segmented DAC, occupies only 900x700 um². Consequently, a 25% reduction in core area has been achieved. This is a significant result, particularly for cost-sensitive applications where every square millimeter of silicon directly impacts the final device cost.

V. SIMULATION AND MEASUREMENT RESULTS

A. Implementation and Measurement Setup

The ADC was designed and fabricated using a standard 180-nm CMOS process with one poly and four metal layers. The supply voltage for both analog and digital parts was 5.0 V. The sampling frequency was 1 MHz. Characterization was performed using a measurement system based on a high-precision signal generator and a spectrum analyzer.

B. Electrical Performance

Key measurement results are summarized in Table I.

TABLE I. PERFORMANCE COMPARISON: PROPOSED VS. CONVENTIONAL ADC

	Table Column Head		
	Conventional ADC	**Proposed ADC**	**Improvement**
Tecnology	180-nm CMOS	180-nm CMOS	-
Resolution	12-bit	12-bit	-
SNDR @1MHz	72 dB	75 dB	+3 dB
ENOB	11.7 bits	12.2 bits	+0.5 bits
Supply Current	769 uA	109 uA	-85.8% (7x)
Power Consumption	3.845 mW	0,545 mW	-85.8% (7x)
FoM	4.15 pJ/conv.-step	0.59 pJ/conv.-step	-85.8% (7x)
Area (core)	0.84 mm²	0.63 mm²	-25%
Architecture	Conventional	Segmented DAC with modified switching algorithm	

C. Analysis of Results

Linearity: The 3 dB improvement in SNDR (from 72 dB to 75 dB) and the corresponding increase in effective number of bits (ENOB) indicate enhanced linearity. This results from the reduced impact of process mismatch in the segmented structure compared to a full 12-bit binary ladder, positively affecting INL/DNL.

Energy Efficiency: The sevenfold reduction in supply current (from 769 µA to 109 µA) directly validates the efficiency of the proposed switching algorithm. The primary contributor to the power reduction is the decreased DAC switching energy. The achieved FoM of 0.59 fJ/conversion-step is competitive for this technology node and resolution [6].

Spectral Analysis: The FFT plot of the ADC output for a near-Nyquist input sine wave (Fig. 3) shows that the noise levels correspond to the reported SNDR, with the dominant noise being close to thermal noise.

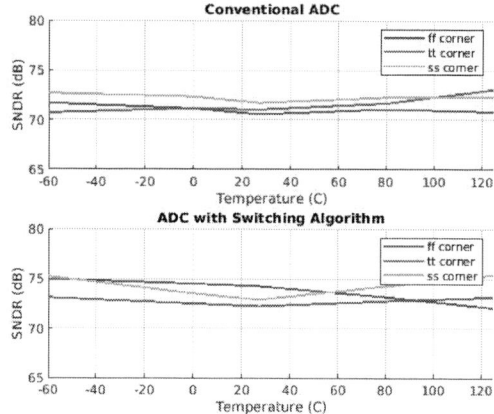

Fig. 3. SNDR Comparison in PVT analysis

D. Comparison with State-of-the-Art

A brief comparison with other recently published 12-bit SAR ADCs is provided in Table II. The proposed design demonstrates an excellent balance between energy efficiency (FoM) and linearity (SNDR) in the 180-nm node.

TABLE II. COMPARISON OF SWITCHING SCHEMES

	[7], 2010	[8], 2018	[9], 2020	This work
Tecnology (nm)	130	180	65	180
Architecture	SAR (mon.)	SAR (Split)	SAR (Sep.)	SAR (Segm.)
SNDR (dB)	57.0*	65.2*	61.35*	75.0
Power Consumption (mW)	0.826	0.62	0.071*	0.545
FoM (pJ/conv.-step)	29*	42.8*	3.71*	0.59
Area (mm²)	0.052	0.1225	-	0.63
Notes	10 bit, 50 MHz	11 bit, 10 MHz	10 bit, 20 MHz	Without calibration, V_{dd}=5V

a. Note: SINAD, ENOB and FoM values are directly dependent on the sampling frequency and bandwidth. To accurately compare data with different f_s, the Walden FoM was used. * – recalculated or taken from the article. ** – FoMS (SNDR FoM).

VI. CONCLUSION

This paper presented the design and experimental verification of an ultra-low-power 12-bit SAR ADC implemented in 180-nm CMOS technology. The core contribution is the proposal and successful implementation of a modified switching algorithm for a segmented capacitive DAC.

The proposed algorithm, which fixes the state of the MSB segment after determining the high-order bits and conducts all fine adjustments within the low-capacitance LSB segment, proved highly effective. Measurement results confirm the main thesis of the research: a simultaneous and significant improvement in both power consumption (7x reduction) and linearity (3 dB SNDR improvement) is achieved compared to a conventional baseline architecture.

The obtained FoM of 0.59 fJ/conversion-step confirms the competitiveness of the solution, despite the use of a mature 180-nm technology node, making it attractive for cost-sensitive, power-constrained applications such as IoT sensors, wearable devices, and battery-powered systems. Future work may focus on adapting the algorithm for sub-1V operation and exploring calibration techniques to achieve 14-bit accuracy.

ACKNOWLEDGMENT

The designed circuit is being manufactured with funding from the Russian Ministry of Education and Science as part of the federal project "Training Personnel and Developing a Scientific Foundation for the Electronics Industry" under the state assignment for research and development "Development of a Methodology for Prototyping Electronic Components in Domestic Microelectronic Manufacturing Using the MPW Service."

REFERENCES

[1] B. Murmann, "Energy limits in A/D converters," in Proc. IEEE Faible Tension Faible Consommation (FTFC), 2013, pp. 1-4.

[2] R. H. Walden, "Analog-to-digital converter survey and analysis," IEEE J. Sel. Areas Commun., vol. 17, no. 4, pp. 539-550, Apr. 1999.

[3] B. P. Ginsburg and A. P. Chandrakasan, "500-MS/s 5-bit ADC in 65-nm CMOS with split capacitor array DAC," IEEE J. Solid-State Circuits, vol. 42, no. 4, pp. 739-747, Apr. 2007.

[4] D. Lobankov, S. Yamaliev, E. Atkin, D. Normanov and A. Cherbov, "A 14-Bit 150 kS/s Hybrid Adc for Matrix Applications," 2025 IEEE 26th International Conference of Young Professionals in Electron Devices and Materials (EDM), Altai, Russian Federation, 2025, pp. 710-713.

[5] V. Hariprasath, J. Guerber, S.-H. Lee, and U.-K. Moon, "Merged capacitor switching based SAR ADC with highest switching energy-efficiency," Electron. Lett., vol. 46, no. 9, pp. 620-621, Apr. 2010.

[6] M. van Elzakker, E. van Tuijl, P. Geraedts, D. Schinkel, E. Klumperink, and B. Nauta, "A 10-bit charge-redistribution ADC consuming 1.9 μW at 1 MS/s," IEEE J. Solid-State Circuits, vol. 45, no. 5, pp. 1007-1015, May 2010.

[7] C.-C. Liu, S.-J. Chang, G.-Y. Huang, and Y.-Z. Lin, "A 10-bit 50-MS/s SAR ADC with a monotonic capacitor switching procedure," IEEE J. Solid-State Circuits, vol. 45, no. 4, pp. 731-740, Apr. 2010

[8] W. Tung and S.-C. Huang, "An Energy-Efficient 11-bit 10-MSps SAR ADC with Monotonic Switching Split Capacitor Array," in Proc. of IEEE Conf., 2018.

[9] A. Gusev, D. Osipov, and S. Paul, "An Energy Efficient SAR ADC Architecture with DAC Separation," in Proc. of IEEE Conf., 2020, pp. 119-121.

[10] M. M. Pilipko and M. E. Manokhin, "Design of a Low-Power 12-bit SAR ADC," 2019 IEEE Conference of Russian Young Researchers in Electrical and Electronic Engineering (EIConRus), Saint Petersburg and Moscow, Russia, 2019, pp. 129-131.

[11] M. M. Pilipko and D. V. Morozov, "An Algorithm for the Search of a Low Capacitor Count DAC Switching Scheme for SAR ADCs," in IEEE Transactions on Computer-Aided Design of Integrated Circuits and Systems, vol. 39, no. 12, pp. 5309-5313, Dec. 2020.

Quantum Dots in Nanoelectronics: From Fundamental Principles to Revolutionary Applications

Ivan Y. Yaroshenko
Institute of Computer Science and Cybersecurity
Peter the Great Saint-Petersburg Polytechnic University
Saint-Petersburg, Russia
yaroshenko.iyu@edu.spbstu.ru

Sergey R. Galimov
Institute of Computer Science and Cybersecurity
Peter the Great Saint-Petersburg Polytechnic University
Saint-Petersburg, Russia
galimov.sr@edu.spbstu.ru

Alexander V. Petruhanov
Institute of Computer Science and Cybersecurity
Peter the Great Saint-Petersburg Polytechnic University
Saint-Petersburg, Russia
petruhanov.av@edu.spbstu.ru

Dmitry M. Kuznetsov
Institute of Computer Science and Cybersecurity
Peter the Great Saint-Petersburg Polytechnic University
Saint-Petersburg, Russia
kuznetsov2.dm@edu.spbstu.ru

Abstract — **QDs (quantum dot) are nanoscale semiconductor structures in which the quantum confinement effect gives rise to discrete energy levels and the phenomenon of Coulomb blockade. This article examines the fundamental physical principles of quantum dots as well as their applications in single-electron transistors, non-volatile memory, and spin qubits for quantum computers. It is shown that, once the technological challenges of fabrication and cryogenic operating requirements are overcome, quantum dots have the potential to become the foundation for ultra-low-power, scalable post-CMOS electronics of the next generation.**

Keywords — *quantum dots, quantum confinement, Coulomb blockade, single-electron transistors, spin qubits, nanoelectronics, quantum computing, semiconductor nanostructures*

I. Introduction

The relentless trend toward miniaturization in the electronics industry, historically guided by Moore's law, is approaching fundamental physical limits. As device feature sizes shrink to the atomic scale, quantum-mechanical effects and manufacturing challenges pose serious obstacles to further scaling of conventional complementary metal-oxide-semiconductor (CMOS) technology. This impasse has spurred the search for alternative paradigms and materials capable of sustaining technological progress. Among the most promising candidates are quantum dots (QDs) — nanoscale semiconductor crystals typically ranging in diameter from 2 to 10 nanometers.

Quantum dots are often referred to as "artificial atoms" because their electronic properties are determined not only by their chemical composition but can also be finely tuned by varying their physical size and shape. This tunability arises from the quantum confinement effect, which endows QDs with a discrete energy spectrum in sharp contrast to the continuous bands found in bulk semiconductors.

The purpose of this article is to trace the evolution of quantum dots from a fundamental physical phenomenon to a breakthrough technological platform. We will examine the core principles governing their behavior, explore their most impactful applications in nanoelectronics, address the practical challenges hindering their commercialization, and outline prospective trajectories for this rapidly evolving field.

II. Fundamental Physical Principles of QD

A. Quantum Confinement and Discrete Energy Levels

In bulk semiconductors, electrons and holes (charge carriers) can move freely in all three spatial dimensions, resulting in the formation of continuous valence and conduction bands separated by a fixed bandgap. However, when the dimensions of a semiconductor crystal become comparable to or smaller than the natural exciton Bohr radius, the motion of charge carriers becomes spatially confined. This is the essence of quantum confinement [1-3].

In a quantum dot, which confines carriers in all three spatial dimensions, this confinement leads to quantization of the energy levels. Instead of a continuous band in which an electron can possess any energy (as in a bulk crystal), in a confined system its energy can only take certain discrete values—similar to the electron states in an atom. The energy of these states can be approximately described using the "particle-in-a-box" model:

$$E_n \propto \frac{\hbar^2 \cdot n^2}{2 \cdot m \cdot L^2} \qquad (1)$$

where:

E_n - energy of the nth quantum state, $\hbar$ - reduced Planck's constant, m - effective mass of the charge carrier, L - size (diameter) of the quantum dot, n - principal quantum number.

This relation reveals a critically important feature: the bandgap of a quantum dot is inversely proportional to its size. A smaller dot has a larger bandgap, while a larger dot has a smaller one. This size dependence enables precise tuning of the optical and electronic properties of quantum dots simply by controlling their physical dimensions during synthesis—for instance, allowing emission of specific wavelengths of light from the same base material [4-6].

B. Coulomb Blockade Effect

A direct consequence of the nanoscale size of quantum dots is their extremely small intrinsic capacitance (C). The energy required to add a single electron to a neutral quantum dot is determined by the charging energy:

$$E_c = \frac{e^2}{2C} \qquad (2)$$

where:

E_n — energy of the n'th quantum state, e — the elementary charge. For a typical quantum dot, this energy amounts to several millielectronvolts. C — intrinsic capacitance

At sufficiently low temperatures, when the thermal energy $k_B \cdot T$ (where k_B is the Boltzmann constant and T is temperature) is less than E_n Coulomb repulsion prevents sequential tunneling of electrons onto the dot. This phenomenon, known as Coulomb blockade, enables precise control over the number of electrons occupying the dot—one electron at a time. The condition for observing Coulomb blockade is:

$$\frac{e^2}{2 \cdot c} > k_B \cdot T \tag{3}$$

This effect constitutes the fundamental operating principle of one of the most iconic quantum-dot-based devices: the single-electron transistor [7-9].

III. APPLICATIONS IN ADVANCED NANOELECTRONIC DEVICES

The unique properties of quantum dots (QDs) have enabled the development of a new class of nanoelectronic devices with unprecedented capabilities.

A. Single-Electron Transistors (SETs)

A single-electron transistor is a three-terminal device (source, drain, and gate) in which a quantum dot serves as a central "island" connected to the source and drain via tunnel junctions. The gate electrode is used to electrostatically control the potential of the dot.

Operating principle: The Coulomb blockade effect suppresses current flow from source to drain at low bias voltages. By applying a gate voltage, the energy levels of the quantum dot can be shifted relative to the Fermi levels of the source and drain, lifting the blockade and allowing a single electron to tunnel through the island. This results in conductivity oscillations as a function of gate voltage, known as Coulomb oscillations [10-13].

Significance: SETs are the most sensitive electrometers available, capable of detecting fractions of a single electron charge. Their primary appeal lies in ultra-low power consumption, as they operate by manipulating individual electrons.

B. Quantum-Dot-Based Memory

Quantum dots are ideal candidates for replacing the floating gate in non-volatile memory cells. In conventional flash memory, the floating gate is a continuous conducting layer that stores charge. Replacing it with a dense array of quantum dots offers significant advantages:

Enhanced reliability: If a defect develops in the tunnel oxide, only the charge stored in the nearest quantum dots is lost, rather than the entire charge of a continuous floating gate, thereby preventing catastrophic failure.

Higher write/erase speeds and lower operating voltages: The discrete nature of the charge-storage nodes enables more efficient tunneling.

Improved endurance: Distributed charge storage reduces stress on the tunnel oxide.

C. Qubits for Quantum Computing

Perhaps the most revolutionary application of quantum dots is in quantum information processing. A quantum dot can confine a single electron whose spin state can serve as a quantum bit (qubit).

Spin qubits: The two spin states of the electron—spin-up ($|\uparrow\rangle$) and spin-down ($|\downarrow\rangle$)—can represent the $|0\rangle$ and $|1\rangle$ states of a qubit. Long electron-spin coherence times in materials such as silicon make them robust carriers of quantum information.

Control and readout: Qubit states can be manipulated using microwave pulses (for electron spin resonance) or fast electrical pulses (via spin–orbit coupling). Qubit-state readout relies on spin-to-charge conversion, typically implemented by using a single-electron transistor as a highly sensitive charge detector.

Scalability: The major advantage of quantum-dot-based qubits is their potential for scaling using mature semiconductor fabrication technologies, enabling the integration of millions of qubits on a single chip—a formidable challenge for most other qubit platforms.

IV. CHALLENGES AND FUTURE PROSPECTS

Despite their tremendous promise, the path toward widespread commercial adoption of quantum-dot-based nanoelectronics is fraught with difficulties.

A. Fabrication Challenges and Scalability

Practical implementation of devices such as large-scale SET arrays and multi-qubit systems demands atomic-level precision in the placement, size, and composition of quantum dots. Top-down fabrication approaches (e.g., electron-beam lithography) offer high precision but are slow and costly. Bottom-up methods (e.g., colloidal synthesis) can produce high-quality quantum dots, yet they face enormous difficulties in precise positioning and integration with existing silicon technology. Achieving uniform, defect-free arrays comprising millions of quantum dots remains a major obstacle.

B. Integration and Operational Requirements

Many quantum-dot devices, particularly SETs and spin qubits, require cryogenic temperatures (often below 1 K) for reliable operation, because thermal energy must be suppressed to preserve quantum coherence and to observe effects such as Coulomb blockade. Integrating the necessary cryogenic control electronics and readout circuitry represents a formidable engineering challenge [14-15].

C. Outlook for the Future

Future research efforts are focused on overcoming these barriers:

- **New materials**: Exploration of 2D materials (e.g., transition-metal dichalcogenides) and germanium to create quantum dots with more favorable properties, such as stronger spin–orbit coupling for faster electrical control.

- **Hybrid systems**: Integration of quantum dots with other quantum platforms, such as superconducting circuits, to combine the strengths of both approaches.

- **Advanced fabrication**: Development of directed self-assembly and other scalable nanofabrication techniques to produce large-scale ordered quantum-dot arrays.

- **Cryo-CMOS electronics**: Design of classical control circuitry capable of operating at cryogenic temperatures alongside quantum-dot processors.

With continued progress in these directions, quantum dots hold the potential to become a foundational technology for ultra-low-power classical electronics and large-scale fault-tolerant quantum computing in the coming decades.

ACKNOWLEDGMENT

The authors express their sincere gratitude to Peter the Great St. Petersburg Polytechnic University (SPbPU) for providing the opportunity to participate in the international conference and to present a review paper on the current state of research in the field of quantum dots.

REFERENCES

[1] Kastner M. A. Artificial atoms // Physics Today. – 1993. – Vol. 46, № 1. – P. 24–31.

[2] Bimberg D., Grundmann M., Ledentsov N. N. Quantum Dot Heterostructures. – Wiley, 1999. – 328 p.

[3] Loss D., DiVincenzo D. P. Quantum computation with quantum dots // Phys. Rev. A. – 1998. – Vol. 57, № 1. – P. 120–126.

[4] Likharev K. K. Single-electron devices and their applications // Proc. IEEE. – 1999. – Vol. 87, № 4. – P. 606–632.

[5] Novak, D., Kozhubaev, Y., Potekhin, V., Cheng, H., Ershov, R. (2025). Asymmetric Object Recognition Process for Miners' Safety Based on Improved YOLOv10 Technology. Symmetry, 17(9), Article 1435. https://doi.org/10.3390/sym17091435

[6] Muratbakeev, E., Novak, D., Kozhubaev, Y. (2025). Investigation of Industrial Bearing Fault Diagnosis Based on 1D-Cnn-Lstm. , 1059-1067. https://doi.org/10.1109/ICIEAM65163.2025.11028381

[7] Hanson R. et al. Spins in few-electron quantum dots // Rev. Mod. Phys. – 2007. – Vol. 79, № 4. – P. 1217–1265.

[8] Zwanenburg F. A. et al. Silicon quantum electronics // Rev. Mod. Phys. – 2013. – Vol. 85, № 3. – P. 961–1019.

[9] Petta J. R. et al. Coherent manipulation of coupled electron spins in semiconductor quantum dots // Science. – 2005. – Vol. 309, № 5744. – P. 2180–2184.

[10] Novak, D., Kozhubaev, Y., Kang, H., Cheng, H., Ershov, R. (2025). Intelligent System Study for Asymmetric Positioning of Personnel, Transport, and Equipment Monitoring in Coal Mines. Symmetry, 17(5), Article 755. https://doi.org/10.3390/sym17050755

[11] Muratbakeev, E., Kozhubaev, Y., Novak, D., Kuzmenko, E., Yiming, Y. (2025). Research of Control Systems and Predictive Diagnostics of Electric Motors. Symmetry, 17(5), Article 751. https://doi.org/10.3390/sym17050751

[12] Kozhubaev, Y., Ovchinnikova, E., Krotova, S., Il'in, A., Sabbgan, A. (2023). Controlling a combined polymer fuel cell and directional battery system maintaining efficiency at the optimum point. E3S Web of Conferences, 389, Article 02014. https://doi.org/10.1051/e3sconf/202338902014

[13] Veldhorst M. et al. A two-qubit logic gate in silicon // Nature. – 2015. – Vol. 526, № 7573. – P. 410–414.

[14] Yoneda J. et al. A quantum-dot spin qubit with coherence limited by charge noise // Nature Nanotechnology. – 2023. – Vol. 18, P. 331–337.

[15] Quantum Engineering in Semiconductor Nanostructures // Edited by A. A. Gorbatsevich and S. A. Tarasenko. – Moscow: Fizmatlit, 2022. – 512 p.

A Mini Review: Perovskite Solar Cells

Al Walo Walo
Saint Petersburg Electrotechnical University "LETI"
St. Petersburg, Russia
alwalowalo@gmail.com

Abstract — **This mini review analyzes the advancements and challenges associated with perovskite solar cells (PSCs) with potential to be the viable solution addressing the growing global energy demand. With energy conversion efficiencies attaining 27%, PSCs demonstrate considerable potential in comparison to conventional photovoltaic technologies. Central topics include the influence of processing techniques on film quality, strategies for enhancing stability, and the environmental implications of lead toxicity. Furthermore, the review emphasizes innovative materials such as graphene and investigates the integration of perovskite and silicon cells in tandem configurations to enhance efficiency and lower costs.**

Keywords—perovskite, stability, efficiency, degradation, perovskite/Si tandem solar cell.

I. INTRODUCTION

Energy is among the most crucial requirements for our world, both at present and for the foreseeable future. As traditional sectors evolve and new industries emerge, demand for electricity is growing at a faster rate than the overall demand for energy. This trend is expected to be particularly pronounced in the field of artificial intelligence, where electricity consumption is projected to rise substantially [1], [2].

Historically, there has been an ongoing pursuit of renewable energy, such as SunPower, to meet these needs. Currently, silicon-based photovoltaic cells comprise 90% of the global market due to their high efficiency (27.3%), long lifespan (exceeding 25 years), and relative affordability [3], [4].

II. FUNDAMENTAL PROPERTIES

Perovskite was first utilized by Tsutomu Miyasaka in 2009 and initially exhibited low efficiency. However, it quickly demonstrated rapid advancements in energy efficiency over a short period. Perovskites have unique optical and electrical properties, which allows them to be used in LEDs that provide high efficiency and brightness, as well as in agriphotovoltaics, where they are integrated into solar panels, allowing for the simultaneous generation of electricity and the maintenance of crop growth in the same area [4], [5].

Organic-inorganic perovskites offer several advantages, including tunable bandgaps, excellent charge transfer properties, ease of processing, low production costs [6], high absorption coefficients, extended diffusion lengths, excellent carrier mobilities, low exciton binding energies (below 10 meV) [7], and a unique electronic structure. For the previous decade, perovskite solar cells achieved remarkable advancements in performance, with structural innovations leading to superior PCE (power conversion efficiency) of approximately 27% [8]. This level of efficiency is equivalent to that of well-established PV technologies, like multicrystalline silicon (Si), CIGS (copper-indium-gallium diselenide), and CdTe (cadmium telluride) cells. [7].

Table 1 presents some distinctive characteristics of perovskite solar cells and other types, clearly demonstrating the superiority of perovskite solar cells over other varieties.

TABLE I   KEY PROPERTIES OF DIFFERENT SOLAR CELL TYPES [9].

Parameters	CdTe	CIGS	c-Si	Perovskite
Cost of source materials	Low	Moderate	Low	Low
Cost of processed materials	Low	High	High	Low
Manufacturing cost	Moderate	Moderate	High	Low
Energy return time	Moderate	High	High	Low
Levelized cost of electricity	Moderate	High	High	Low
Conversion efficiency	Moderate	Average	High	High

III. COMPOSITION NAD ARCHITECTURE FOR PEROVSKITE SOLAR CELLS

Perovskite solar cells (PSCs) composed from multiple layers, including a TCO (Transparent conductive oxide) layer that functions like a forward electrode. This layer is often made of FTO (fluorine-doped tin oxide) or ITO (indium tin oxide). Perovskite layer, absorbing light and represented by the chemical formula ABX_3, is positioned between the N-type ETL (electron transfer layer) and the P-type HTL (hole transfer layer). The metal electrodes typically are made from gold (Au) or silver (Ag) [6], [10].

Perovskite materials consist of a mixed organic or inorganic halide structure, typically represented by the formula ABX_3, the crystal lattice structure is visualized in Fig. 1. In this configuration, "A" includes organic components, that is methylammonium (MA) or formamidinium (FA), or inorganic elements like rubidium (Ru), cesium (Cs), or francium (Fr). The "B" component comprises a metal element—such as lead (Pb), tin (Sn), germanium (Ge), silicon (Si), or antimony (Sb)—with A being larger in size than B, while "X" represents an anionic halogen element (e.g., Cl-, Br-, I-) [10].

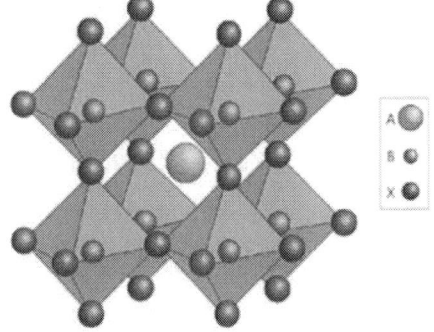

Fig. 1.   Crystal architecture of ABX_3 perovskite [11].

Charge-selective layers are essential to the performance of perovskite solar cells (PSCs). The ETL (electron transfer layer) and HTL (hole transfer layer) facilitate unidirectional movement of charge carriers. Specifically, the ETL enhances electron transfer while simultaneously blocking holes, whereas the HTL performs the complementary function. As a result, these charge-selective layers play a significant function in determining the efficiency of charge transportation and extraction processes, as well as the rate of charge recombination [12].

Typically, the electron transfer layer (ETL) comprises inorganic materials such as TiO2, ZnO, SnO2, CdS, Zn2SO4, Fe2O3, and In2S3, as well as organic components like C60 and C70 fullerenes, poly(vinylpyrrolidone) (PVP), and phenyl-C61-butyric acid methyl ester (PCBM). In contrast, the hole transfer layer (HTL) may consist of inorganic materials, including Cu2O, CuO, CuI, CuSCN, and CuS, or organic components such as Spiro-OMeTAD, PEDOT , and PTAA [10]. The choice of materials for the electron transfer layer (ETL) and hole transfer layer (HTL) depends on the energy band alignment compatibility within the device architecture and the solvents used during processing [11].

Figure 2 illustrates the energy level alignment of various materials commonly used in perovskite solar cells (PSCs). The diagram classifies the materials into electron extraction layers (EELs), organometal perovskites, small molecular hole extraction layers (HELs), polymer HELs, inorganic HELs, and metal electrodes. The vertical axis represents energy levels in electron volts (eV) relative to the vacuum energy level (Evac = 0), highlighting the conduction band minima, valence band maxima, and work functions of these materials. Proper energy level alignment is critical for ensuring efficient charge transfer, extraction, and minimizing recombination losses in PSC devices [13].

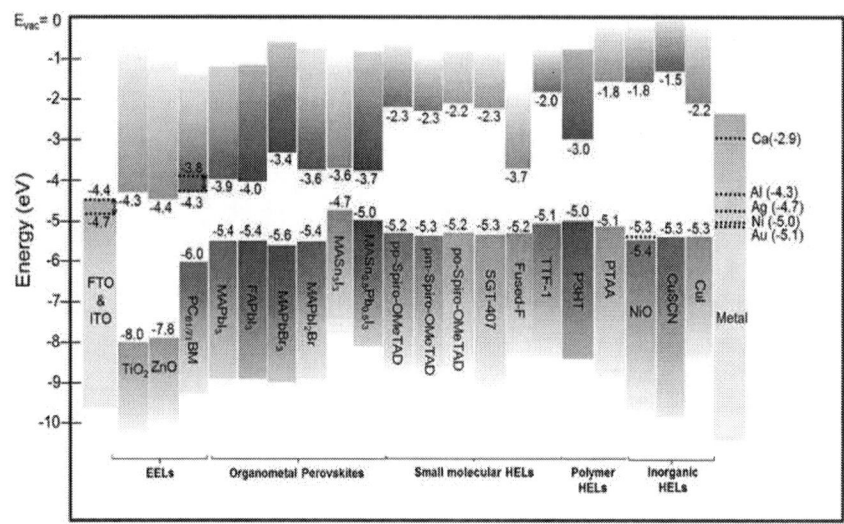

Fig. 2. Prominent materials utilized in the fabrication of perovskite solar cell (PSC) devices [13].

The formation of high-quality perovskite films is essential for the development of efficient devices, which is influenced by factors such as precursor modifications, thermal annealing [12], [14]. At the laboratory scale, the spin-coating method is commonly utilized for depositing various layers of perovskite solar cells (PSCs), as it enables high efficiency. However, to support large-scale production, researchers are exploring alternative deposition methods. Numerous manufacturing techniques for perovskite-based solar cells have been studied and reported. This includes techniques such as drop casting, spray coating, ultrasonic spray coating, slot-die coating, electrodeposition, chemical vapor deposition (CVD), thermal vapor deposition, vacuum deposition, screen printing, and inkjet printing, each adapted to suit specific device architectures [15].

The mesoporous and planar configurations are the most widely investigated structures in current perovskite solar cell (PSC) research. The planar configuration is further categorized into two types: the conventional planar structure (n-i-p) and the inverted planar structure (p-i-n). Figure 3 provides a schematic representation of these three structures, along with the direction of incoming light, highlighting the necessity of maintaining transparency in certain layers to ensure effective light transmission to the perovskite absorber for optimal performance. Each structure possesses unique characteristics and advantages. The mesoporous structure has demonstrated higher efficiency compared to planar designs, primarily due to the inclusion of an additional porous metal oxide layer, which serves as an efficient insulator to prevent recombination between electrons and holes. Planar structures, by contrast, are simpler and more adaptable, making them easier to modify and optimize for various performance requirements. Among planar designs, the inverted planar structure offers specific benefits over its conventional counterpart, including a simplified fabrication process, reduced hysteresis effects, and compatibility with low-temperature processing, making it particularly well-suited for applications involving flexible substrates [16].

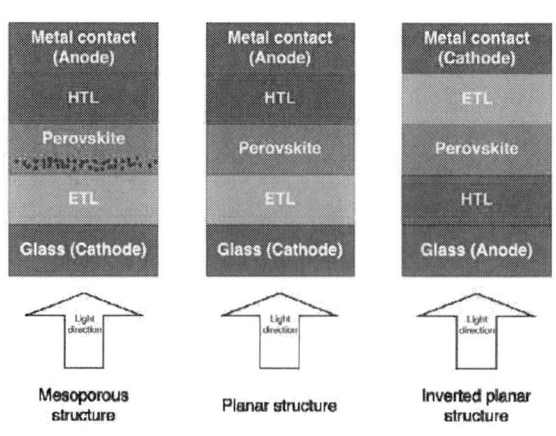

Fig. 3. Perovskite solar cell structures

979-8-3315-4784-4/26 $31.00 © 2026 IEEE 80

IV. CHALLENGES AND INNOVATIONS

Solar cells face several challenges, particularly the instability of perovskite solar cells, which experience performance degradation over time due to exposure to moisture and heat [17], [18], [19]. Moisture during the processing phase can adversely affect the morphology and stability of perovskites, necessitating treatment through simple drying under a nitrogen stream for several hours [19].

In study [18], utilized various analytical methods, such as UV-visible spectroscopy and X-ray analysis, demonstrated that thermal stress and light exposure can lead to significant deterioration of perovskite films, even in environments free from oxygen and moisture. The stability of complex lead halides follows the order: $CsPbBr_3 > CsPbI_3 > FAPbBr_3 > FAPbI_3 > MAPbI_3 > MAPbBr_3$.

Another significant challenge in advancing perovskite solar cells is addressing the toxicity of lead (Pb) content and the environmental concerns surrounding its potential widespread use in the future. The risk of lead contamination from operating a 1 GW photovoltaic (PV) manufacturing facility is comparatively lower than that from other lead-emitting sources, such as fossil fuels, mining operations, battery technologies, and various electronic devices. Furthermore, recent findings indicate that perovskite solar cells can be produced using recycled lead sourced from automobile batteries. This approach presents an opportunity to mitigate lead contamination in the environment by facilitating the reuse of lead in the perovskite PV industry [6], [20].

Graphene has recently gained attention as a promising material for perovskite solar cell (PSC) applications, owing to its exceptional optical, electrical, and mechanical properties. Its unique characteristics make graphene suitable for various roles within PSCs, including serving as a conductive electrode, a charge carrier transfer material, and a stabilizing agent. The use of graphene can enhance the efficiency and stability of PSCs by improving charge transfer and facilitating better electrical conductivity, which are critical factors in optimizing solar cell performance. Additionally, graphene's mechanical strength and flexibility may contribute to the durability and overall longevity of PSCs, thereby addressing some of the stability challenges associated with traditional perovskite materials [21].

Current research in the field of solar cells focuses on enhancing stability and efficiency. One approach to addressing stability issues is the incorporation of a layer, such as copper iodide (CuI), between the hole transfer layer (HTL) and the perovskite layer when the HTL consists of nickel oxide (NiO) [22]. Furthermore, novel materials are being explored for use as electron transfer layers (ETLs), such as γ-Ga_2O_3 nanocrystals. Devices based on Ga_2O_3 have demonstrated minimal hysteresis and have outperformed traditional TiO_2-based devices [23]. Additionally, new hole transfer materials (HTMs), such as TOP-HTM-α1 and TOP-HTM-α2, which do not require dopants or additives, unlike Spiro-OMeTAD, are being investigated due to their cost-effectiveness. Perovskite solar cells utilizing these new HTMs exhibit superior stability compared to those based on doped Spiro-OMeTAD, both in dark conditions and under 1 sun illumination [24], [25].

To enhance efficiency, the integration of perovskite and silicon solar cells in tandem configurations is employed, which has the potential to increase efficiencies beyond 30% while simultaneously lowering the cost per kilowatt. A remarkable power conversion efficiency (PCE) of 34.85% has been achieved with a monolithic two-terminal wide-bandgap perovskite/silicon tandem solar cell, thanks to the optimal integration of the spectral responses from both the top and bottom cells [4].

V. CONCLUSIONS

The outlook for perovskite solar cells is extremely promising as developments focus on improving their stability, efficiency and scalability for commercial applications. Innovative approaches, such as tandem silicon-cell configurations, can raise power conversion efficiency above 30 percent while solving problems such as moisture sensitivity, thermal degradation, and lead toxicity. New materials, including graphene and new electron and hole transfer layers, offer opportunities for improved device performance and durability. Recycling lead from existing sources and exploring alternative deposition methods are increasing the sustainability and practicality of perovskite production. These developments make perovskite technology a major player in renewable energy.

REFERENCES

[1] P.D. Badillo and A. E. Degterev, 'Perspectives on Perovskite Solar Cells Under the Glass of Characterization and Model-based Research', in 2023 XXVI International Conference on Soft Computing and Measurements (SCM), IEEE, May 2023, pp. 277–280. doi: 10.1109/SCM58628.2023.10159103.

[2] World Energy Outlook 2024 – Analysis - IEA. Accessed: Dec. 04, 2024. [Online]. Available: https://www.iea.org/reports/world-energy-outlook-2024

[3] S. K. Sharma and K. Ali, Eds., Solar Cells. Cham: Springer International Publishing, 2020. doi: 10.1007/978-3-030-36354-3.

[4] 'Best Research-Cell Efficiency Chart | Photovoltaic Research | NREL'. Accessed: Oct. 29, 2024. [Online]. Available: https://www.nrel.gov/pv/cell-efficiency.html

[5] A. Kojima, K. Teshima, Y. Shirai, and T. Miyasaka, 'Organometal halide perovskites as visible-light sensitizers for photovoltaic cells.', J. Am. Chem. Soc., vol. 131, no. 17, pp. 6050–6051, May 2009, doi: 10.1021/ja809598r.

[6] N. Elumalai, M. Mahmud, D. Wang, and A. Uddin, 'Perovskite Solar Cells: Progress and Advancements', Energies, vol. 9, no. 11, p. 861, Oct. 2016, doi: 10.3390/en9110861.

[7] K. Choi et al., 'A Short Review on Interface Engineering of Perovskite Solar Cells: A Self-Assembled Monolayer and Its Roles', Sol. RRL, vol. 4, no. 2, p. 1900251, Feb. 2020, doi: 10.1002/solr.201900251.

[8] M. A. Green et al., 'Solar Cell Efficiency Tables (Version 66)', Prog. Photovoltaics Res. Appl., vol. 33, no. 7, pp. 795–810, Jul. 2025, doi: 10.1002/pip.3919.

[9] S. K. Sahoo, B. Manoharan, and N. Sivakumar, 'Introduction: Why Perovskite and Perovskite Solar Cells?', in Perovskite Photovoltaics, vol. 12, no. 8, Elsevier, 2018, pp. 1–24. doi: 10.1016/B978-0-12-812915-9.00001-0.

[10] N. R. Pochont and Y. R. Sekhar, 'Numerical Simulation of Nitrogen-Doped Titanium Dioxide as an Inorganic Hole Transport Layer in Mixed Halide Perovskite Structures Using SCAPS 1-D', Inorganics, vol. 11, no. 1, p. 3, Dec. 2022, doi: 10.3390/inorganics11010003.

[11] Q. Tai, K. C. Tang, and F. Yan, 'Recent progress of inorganic perovskite solar cells', 2019. doi: 10.1039/c9ee01479a.

[12] A. S. Tarasov, A. E. Degterev, M. M. Romanovich, M. D. Pavlova, I. I. Mikhailov, and I. A. Lamkin, 'Optical Properties of Bromine-Doped Perovskite Films', in 2023 Seminar on Fields, Waves, Photonics and Electro-optics: Theory and Practical Applications (FWPE), IEEE, Nov. 2023, pp. 147–150. doi: 10.1109/FWPE60445.2023.10368505.

[13] C.-C. Chueh, C.-Z. Li, and A. K.-Y. Jen, 'Recent progress and perspective in solution-processed Interfacial materials for efficient and stable polymer and organometal perovskite solar cells', Energy Environ. Sci., vol. 8, no. 4, pp. 1160–1189, 2015, doi: 10.1039/C4EE03824J.

[14] T.-B. Song et al., 'Perovskite solar cells: film formation and properties', J. Mater. Chem. A, vol. 3, no. 17, pp. 9032–9050, 2015, doi: 10.1039/C4TA05246C.

[15] P. Kajal, K. Ghosh, and S. Powar, 'Manufacturing Techniques of Perovskite Solar Cells', 2018, pp. 341–364. doi: 10.1007/978-981-10-7206-2_16.

[16] P. D. Badillo and A. E. Degterev, 'Transport Layers as a Factor in Perovskite Solar Cells Improvement', in 2024 Conference of Young Researchers in Electrical and Electronic Engineering (ElCon), IEEE, Jan. 2024, pp. 698–701. doi: 10.1109/ElCon61730.2024.10468489.

[17] A. E. Degterev, M. M. Romanovich, I. I. Mikhailov, I. A. Lamkin, and S. A. Tarasov, 'Ways to Slow Down the Degradation and Enhance the Stability of Perovskite Solar Cells', in 2021 IEEE Conference of Russian Young Researchers in Electrical and Electronic Engineering (ElConRus), IEEE, Jan. 2021, pp. 1301–1304. doi: 10.1109/ElConRus51938.2021.9396607.

[18] A. F. Akbulatov et al., 'Light or Heat: What Is Killing Lead Halide Perovskites under Solar Cell Operation Conditions?', J. Phys. Chem. Lett., vol. 11, no. 1, pp. 333–339, Jan. 2020, doi: 10.1021/acs.jpclett.9b03308.

[19] M. L. Petrus et al., 'Capturing the Sun: A Review of the Challenges and Perspectives of Perovskite Solar Cells', Adv. Energy Mater., vol. 7, no. 16, p. 1700264, Aug. 2017, doi: 10.1002/aenm.201700264.

[20] World Energy Outlook 2025 – Analysis - IEA'. Accessed: Dec. 15, 2025. [Online]. Available: https://www.iea.org/reports/world-energy-outlook-2025

[21] A. A. Abrikosov and L. M. Falicov, 'Fundamentals of the Theory of Metals', 1990. doi: 10.1063/1.2810530.

[22] D. Saranin et al., 'Copper iodide interlayer for improved charge extraction and stability of inverted perovskite solar cells', Materials (Basel)., vol. 12, no. 9, 2019, doi: 10.3390/ma12091406.

[23] K.-H. Hu et al., 'γ-Ga$_2$O$_3$ Nanocrystals Electron-Transporting Layer for High-Performance Perovskite Solar Cells', Sol. RRL, vol. 3, no. 9, p. 1900201, Sep. 2019, doi: 10.1002/solr.201900201.

[24] H. Nishimura, I. Okada, T. Tanabe, T. Nakamura, R. Murdey, and A. Wakamiya, 'Additive-free, Cost-Effective Hole-Transporting Materials for Perovskite Solar Cells Based on Vinyl Triarylamines', ACS Appl. Mater. Interfaces, vol. 12, no. 29, pp. 32994–33003, Jul. 2020, doi: 10.1021/acsami.0c06055.

[25] S. H. Lee, S. Bin Lim, J. Y. Kim, S. Lee, S. Y. Oh, and G. M. Kim, 'An Alternative to Chlorobenzene as a Hole Transport Materials Solvent for High-Performance Perovskite Solar Cells', Crystals, vol. 13, no. 12, p. 1667, Dec. 2023, doi: 10.3390/cryst13121667.

A Comparative Study of ZnO and TiO₂ as ETL Performance in Cs₂SnI₆-Based Perovskite Solar Cells Using SCAPS-1D

Al Walo Walo
Saint Petersburg Electrotechnical University "LETI"
St. Petersburg, Russia
alwalowalo@gmail.com

Abstract — **A comprehensive simulation study was conducted to evaluate the influence of the electron transport layer (ETL) on perovskite solar cell performance. This investigation systematically examined the impact of key ETL physical parameters, including layer thickness, doping concentration, bulk defect density, and the density of interfacial traps at the ETL/perovskite junction.**

Keywords—Perovskite, Electron transport layer, Hole transport layer, TiO₂, ZnO, Cs₂SnI₆, Efficiency, SCAPS-1D, Simulation.

I. INTRODUCTION

Energy constitutes a foundational pillar of contemporary society and serves as a primary catalyst for technological innovation, national progress, and urban expansion. The advancement of industrialization and rising prosperity, especially within emerging economies, has precipitated a sharp increase in global energy requirements. This demand assumes heightened significance in the era of artificial intelligence (AI), wherein energy is indispensable for fuelling data-intensive computational processes and sustaining the operational capacity of complex systems. The efficacy and performance of AI technologies are profoundly contingent upon both the accessibility and caliber of energy supplies, given that sophisticated algorithms and large-scale data analytics necessitate substantial computational power correlated with considerable energy consumption [1].

The provision of energy has historically relied predominantly on fossil fuels. However, their protracted utilization results in the exhaustion of finite resources and the release of detrimental greenhouse gases, which intensifies anthropogenic climate change. In response, the pursuit of renewable and ecologically sustainable energy alternatives has ascended to a position of critical global importance. Within this spectrum of alternatives, solar energy—particularly through photovoltaic (PV) technology—presents a preeminent solution. PV systems operate by converting solar radiation directly into electricity via semiconductor materials, thereby supplying a clean and non-polluting source of power. This characteristic renders it highly suitable for broad societal application as well as for sustaining energy-intensive technological paradigms, including artificial intelligence [2].

In recent years, solar cell technology has attracted substantial research interest, driven largely by initiatives to lower production expenses and enhance sunlight-to-electricity conversion efficiencies beyond contemporary benchmarks. Within this pursuit, a novel and promising category of solar cells—perovskite solar cells—has gained prominence, deriving its name from the distinctive crystal structure of its constituent light-absorbing material. Perovskites are a class of compounds characterized by the general chemical formula ABX₃. These materials exhibit remarkable electrical, structural, optical, and ferroelectric properties, which collectively render them highly suitable for advanced photovoltaic applications. The deployment of organometal halide perovskites as photoactive materials in solar photovoltaics has catalyzed a period of unprecedented progress in power conversion efficiency (PCE). Following their initial implementation in a sensitized solar cell architecture in 2009, which featured self-organized nanocrystalline perovskite particles, the certified PCE of perovskite-based devices has undergone a rapid ascent. Over the subsequent decade, efficiencies have escalated from an initial 3.8% to a contemporary benchmark exceeding 27% [3], [4].

Perovskite solar cells (PSCs) have been realized through diverse device architectures. Among these, the planar heterojunction structure has garnered significant attention for its demonstrated capacity to achieve high power conversion efficiencies. A conventional planar PSC is constructed as a layered structure, featuring a perovskite absorber layer interposed between an n-type electron transport layer (ETL) and a p-type hole transport layer (HTL). The systematic optimization of device architecture and performance necessitates a fundamental comprehension of the operational mechanisms governing these distinct functional layers. Consequently, a detailed analysis of the specific role and influence of each constituent layer is paramount [5].

Within the widely adopted n–i–p planar configuration, the ETL is situated at the illuminated (front) side of the device, where its properties critically govern overall efficiency. The analysis is conducted utilizing a one-dimensional numerical simulation framework. Specifically, the Solar Cell Capacitance Simulator (SCAPS-1D) software serves as an indispensable tool for modeling, understanding, and optimizing device performance. SCAPS-1D is particularly instrumental in systematically evaluating the attributes and configurations of PSC layers, thereby facilitating the enhancement of their efficacy and operational reliability. This investigation, therefore, focuses explicitly on the impact of the electron transport layer on the photovoltaic parameters of planar PSCs [6].

II. METHODOLOGY

The following materials were selected for the structural configuration of the solar cell in this study: Cs₂SnI₆ was employed as the photoactive absorber layer, CuSCN served as the hole transport layer (HTL), and zinc oxide (ZnO) as well as titanium dioxide (TiO₂) were investigated as the electron transport layer (ETL). Device performance was simulated under standard test conditions (STC), utilizing the AM 1.5 G solar spectrum with an incident power density of 1000 W/m². Key photovoltaic parameters—such as open-circuit voltage (V_{OC}), short-circuit current density (J_{SC}) and power conversion efficiency (PCE), — were calculated for the modeled device at an operational temperature of 300 K.

In solar cell modeling, a basic set of physical properties for each layer is used, which are relatively constant for the material and do not vary significantly between different references, although they sometimes differ slightly due to differences in measurement methods or crystal structure. In this study, we based our calculations on typical values for these fundamental parameters (such as layer thickness, band gap, electron affinity, permittivity, and effective state density in the conduction and valence bands, as well as charge mobility and impurity levels) as reported in references [8] and [9]. Similarly, the defect characteristics in the materials (such as defect density and capture cross sections for electrons and holes) were determined based on the same sources, ensuring consistency with the studied structure of the solar cell system based on TiO_2, ZnO, Cs_2SnI_6, and CuSCN layers.

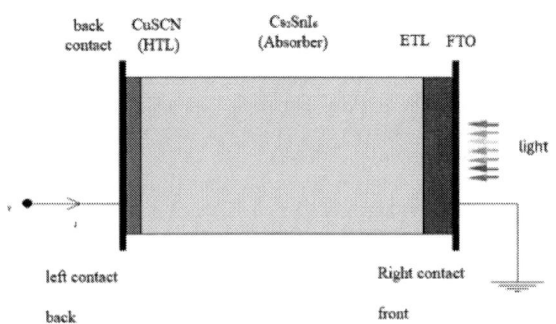

Fig. 1.   Structure of solar cell

Figure 1 presents a structural diagram of a typical device. The device is fabricated in a conventional stack, starting with a front-coated glass substrate of fluorine-doped tin oxide (FTO) front contact. Subsequent layers—comprising the electron transport layer (ETL), the absorber (perovskite), the hole transport layer (HTL), and a metal back contact metal back contact — are deposited atop this foundation. Photons from incident sunlight penetrate the glass substrate and are absorbed within their active layer, generating carriers that will later be will later be subsequently collected at electrodes, thereby producing electrical power [8].

III. RESULTS

A. *Effect of ETL thickness on solar cell*

Electron transport layer (ETL) thickness as a critical parameter influencing the overall performance of perovskite solar cells (PSCs). Impact the ETL thickness on key photovoltaic parameters was investigated by simulating a range between 20 nm to 200 nm, with results presented in Figure 2. Notably, the fill factor (FF) and open-circuit voltage (V_{OC}) exhibited minimal variation across this thickness range. Conversely, a marginal yet consistent decrease for both short-circuit current density (J_{SC}) and power conversion efficiency (PCE) was observed with increasing ETL thickness.

The stability of the fill factor can be attributed to its dependence on the intrinsic quality of the active layer and interfacial properties. Within the specific ETL/Cs_2SnI_6/CuSCN architecture, variations in ETL thickness appear to have a negligible effect on the primary factors governing FF, suggesting that interfacial quality is preserved irrespective of layer thickness.

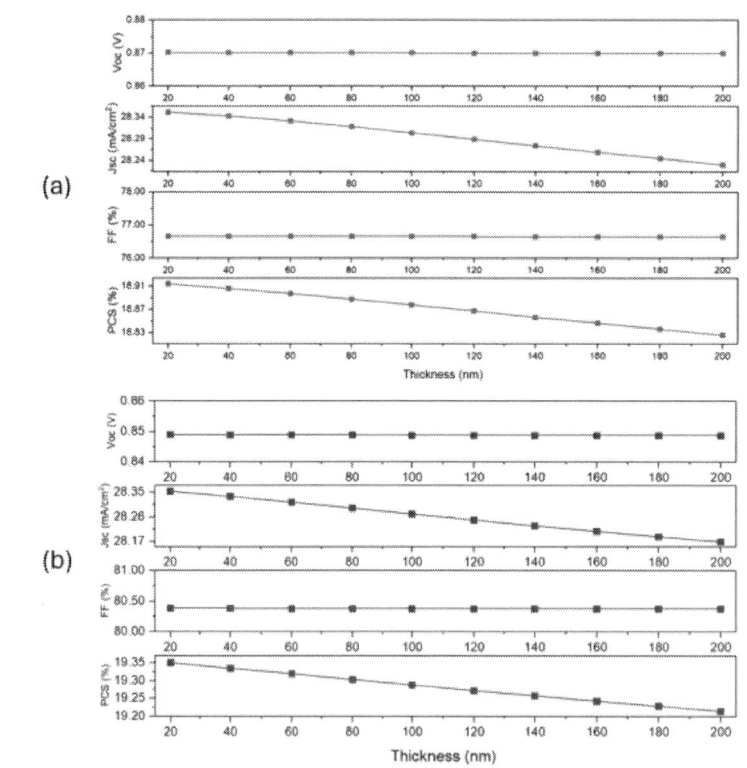

Fig. 2.   The effect of (a) TiO_2 and (b) ZnO thickness on solar cell performance

Similarly, the invariance of V_{OC} can be explained by its fundamental dependence on the alignment of the energy level

at material interfaces and the absorber's bandgap (E_g). As these properties are inherent to the selected semiconductor

979-8-3315-4784-4/26 $31.00 © 2026 IEEE

materials and given the effective band alignment between the ETL and Cs_2SnI_6, V_{OC} remains largely unaffected by charge transport layer thickness by charge transport layer thickness. Observed decline in J_{SC} with increasing ETL thickness can be ascribed to two principal mechanisms. First, parasitic optical absorption within the ETL increases with thickness, thereby reducing the photon flux reaching the perovskite absorber layer. This directly diminishes the generation rate of photocarriers. Second, a thicker ETL imposes a longer transport path for the optogenerated electrons from absorber to front electrode. This extended pathway increases the potential for surface and bulk recombination losses, further decreasing the collected current [9].

B. *Effect of ETL doping concentration on performance of solar cell*

The influence of electron transport layer (ETL) doping concentration on photovoltaic parameters was simulated across a range of 1×10^{15} cm^{-3} to 1×10^{20} cm^{-3}, with the ETL thickness held constant at 50 nm. The results are presented in Figure 3. The data indicate that varying the ETL doping density differentially affects device performance metrics. Specifically, both the open-circuit voltage (V_{OC}) and fill factor (FF) exhibit a moderate but consistent increase with higher doping concentrations. These enhancements can be attributed to improved electrical conductivity within the ETL, which promotes more favorable quasi-Fermi level splitting at the critical charge-selective interfaces and reduces the overall series resistance of the device. The introduction of shallow donor states through doping optimizes the energy band alignment between functional layers, thereby facilitating more efficient charge extraction and suppressing interfacial recombination losses [10].

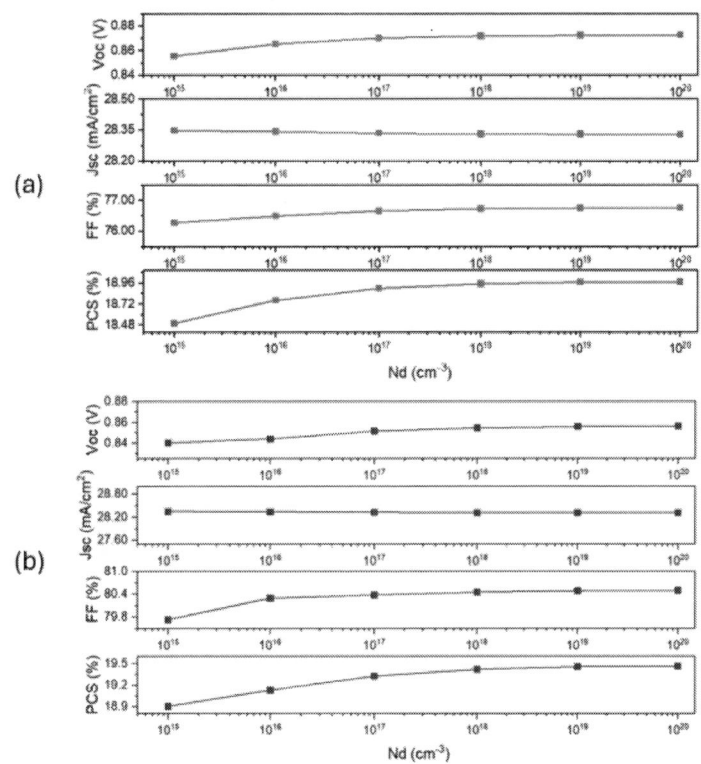

Fig. 3. The effect of (a) TiO_2 and (b) ZnO doping concentration on performance of solar cell

The short-circuit current density (J_{SC}) demonstrates minimal variation with ETL doping concentration, as light absorption is dominated by the Cs_2SnI_6 absorber and is therefore unaffected by changes in the transport layer's activation. The resultant increase in power conversion efficiency (PCE) is thus driven by the concurrent improvements in open-circuit voltage (V_{OC}) and fill factor (FF). This indicates that tuning the ETL doping density is a viable strategy for boosting solar cell performance by optimizing electrical properties while preserving the intrinsic optical characteristics of the device.

C. *Effect of ETL defects on the performance of solar cell*

The influence of defect density within the electron transport layer (ETL) on perovskite solar cell performance was investigated. Simulations were conducted with ETL defect concentrations ranging from 1×10^{13} cm^{-3} to 1×10^{20} cm^{-3}, while maintaining a constant ETL thickness of 50 nm.

As presented in Figure 4, the results indicate that as the ETL defect density increases, the open-circuit voltage (V_{OC}) and fill factor (FF) remain relatively stable. However, a progressive decline in power conversion efficiency (PCE) is observed. This decline is primarily attributed to a reduction in the short-circuit current density (J_{SC}).

The deterioration in JSC can be explained by enhanced recombination within the ETL. A higher concentration of defect states increases the probability of trapping and non-radiative recombination of photogenerated charge carriers, particularly for those generated near the illuminated surface (short-wavelength photons). This process effectively diminishes the collected photocurrent. Furthermore, the presence of these defects can be correlated with an increase in the device's series resistance (Rs), which provides an additional mechanism for the observed performance degradation[11],[12].

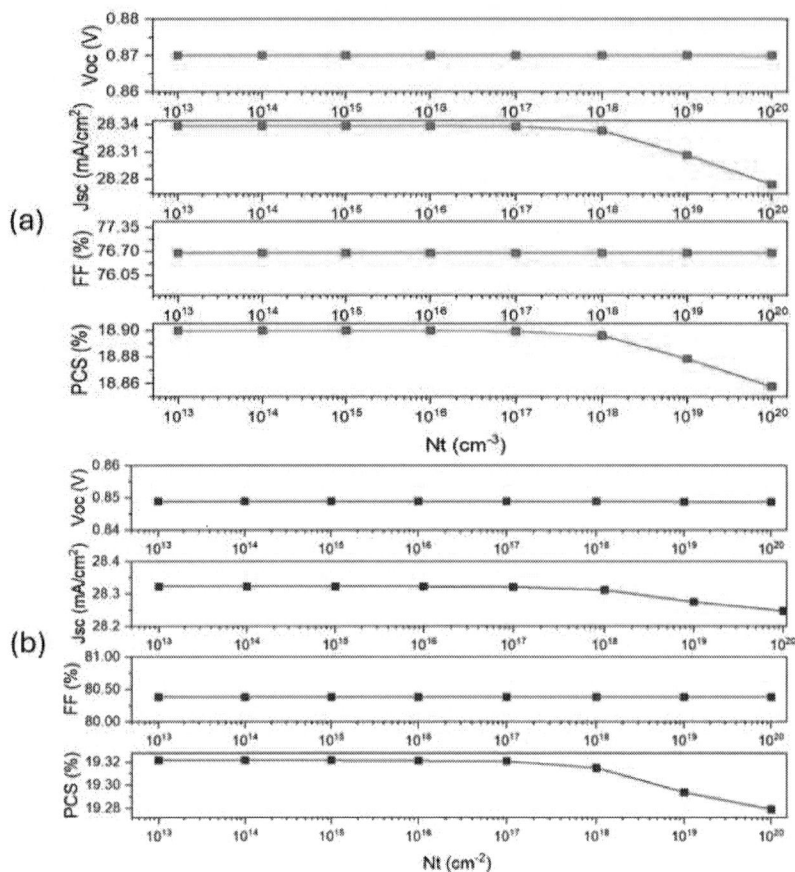

Fig. 4. The effect of (a) TiO$_2$ and (b) ZnO defect density on solar cell performance

D. Effect of interface trap density on solar cell performance

Figure 5 presents the simulated influence of interfacial trap density on the current density-voltage (J-V) characteristics, with the electron transport layer (ETL) thickness fixed at 50 nm. The density of these interfacial trap states was varied from 1×10^7 cm^{-2} to 1×10^{18} cm^{-2}.

The simulation results, as detailed in Figure 5, demonstrate the significant effect of trap state density on photovoltaic output parameters. At low densities (1×10^7 to 1×10^{13} cm^{-2}), the impact on performance is negligible. Conversely, at high densities exceeding 1×10^{15} cm^{-2}, degradation is again minimal. The most pronounced degradation across all four parameters—open-circuit voltage (V$_{OC}$), short-circuit current density (J$_{SC}$), fill factor (FF), and power conversion efficiency (PCE)—is observed within the intermediate range of 1×10^{13} to 1×10^{15} cm^{-2}.

In the low-density regime, the ETL/perovskite interface can be considered relatively defect-free, presenting minimal recombination centers. This allows for the efficient separation and collection of photogenerated charge carriers, resulting in high and stable device performance. In contrast, within the critical intermediate density range, the interface hosts a substantial number of defect states. These traps function as effective non-radiative recombination centers, promoting the recombination of electrons and holes before they can be extracted to the respective contacts, thereby degrading all key performance metrics [13], [14].

When the density of interfacial trap states at the ETL/perovskite junction exceeds a threshold of approximately 1×10^{15} cm^{-2}, the photovoltaic parameters stabilize and become largely invariant. This plateau occurs because interfacial recombination losses have saturated; the density of non-radiative recombination centers is sufficiently high to ensure that the vast majority of photogenerated charge carriers arriving at the interface recombine prior to collection. Consequently, further increases in trap density introduce no additional performance degradation, as the device's current-voltage characteristics are already fundamentally limited by this interfacial recombination. This saturation effect indicates that the interface has transitioned to a "trap-limited" or "recombination-saturated" regime. Within this high-concentration regime, the recombination rate and, by extension, the overall device performance become insensitive to further increments in defect density, rendering subsequent changes in output metrics negligible

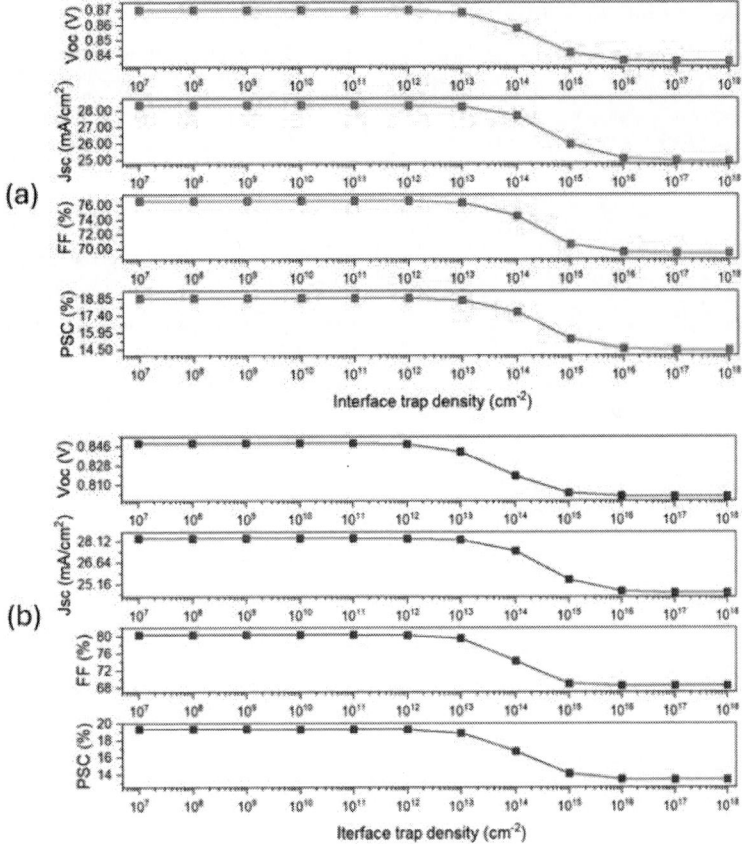

Fig. 5. The effect of (a) TiO2 and (b) ZnO interfacial trap density on solar cell performance

IV. CONCLUSIONS

A comparative analysis of photovoltaic performance was conducted for solar cells utilizing zinc oxide (ZnO) and titanium dioxide (TiO₂) as electron transport layers (ETLs). The effect of changing the parameters on the performance of the solar cell and its parameters for both materials titanium dioxide (TiO₂) and zinc oxide (ZnO) is almost the same, but with a difference in values. The ZnO-based device demonstrated an initial power conversion efficiency (PCE) of 18.9%. For this configuration, systematic variation of the ETL thickness and bulk defect density exhibited a negligible impact on overall performance. However, a positive correlation was observed between n-type doping concentration and efficiency, with PCE increasing from 18.5% to 19.0% across the simulated range. In contrast, the density of defect traps at the ZnO/perovskite interface proved to be a critical parameter. An increase in interfacial trap density from 1×10^{7} cm^{-2} to 1×10^{18} cm^{-2} resulted in a pronounced decrease in PCE from 18.9% to 14.48%, representing a relative degradation of 23.38% from the initial value.

The TiO₂-based solar cell exhibited superior initial performance, achieving a PCE of 19.35%. Its response to parameter variation followed a similar trend to the ZnO-based cell, showing limited sensitivity to changes in ETL thickness, bulk defect density, and doping concentration. However, the device demonstrated greater sensitivity to interfacial quality. Increasing the density of traps at the TiO₂/perovskite interface induced a significant performance decline, with PCE decreasing by 30.5% from its initial value across the same defect density range.

REFERENCES

[1] R. Caspart *et al.*, 'Precise Energy Consumption Measurements of Heterogeneous Artificial Intelligence Workloads', *ISC Work.*, 2022, doi: 10.48550/ARXIV.2212.01698.

[2] A. Goetzberger and V. U. Hoffmann, *Photovoltaic solar energy generation*, vol. 112. Springer Science & Business Media, 2005.

[3] A. Kojima, K. Teshima, Y. Shirai, and T. Miyasaka, 'Organometal halide perovskites as visible-light sensitizers for photovoltaic cells.', *J. Am. Chem. Soc.*, vol. 131, no. 17, pp. 6050–6051, May 2009, doi: 10.1021/ja809598r.

[4] M. A. Green *et al.*, 'Solar Cell Efficiency Tables (Version 66)', *Prog. Photovoltaics Res. Appl.*, May 2025, doi: 10.1002/pip.3919.

[5] P. D. Badillo and A. E. Degterev, 'Transport Layers as a Factor in Perovskite Solar Cells Improvement', in *2024 Conference of Young Researchers in Electrical and Electronic Engineering (ElCon)*, IEEE, Jan. 2024, pp. 698–701. doi: 10.1109/ElCon61730.2024.10468489.

[6] N. R. Pochont and Y. R. Sekhar, 'Numerical Simulation of Nitrogen-Doped Titanium Dioxide as an Inorganic Hole Transport Layer in Mixed Halide Perovskite Structures Using SCAPS 1-D', *Inorganics*, vol. 11, no. 1, p. 3, Dec. 2022, doi: 10.3390/inorganics11010003.

[7] A. Usman and T. Bovornratanaraks, 'Modeling and Optimization of Modified TiO 2 with Aluminum and Magnesium as ETL in MAPbI 3 Perovskite Solar Cells: SCAPS 1D Frameworks', *ACS Omega*, Sep. 2024, doi: 10.1021/acsomega.4c04505.

[8] M. K. Hossain *et al.*, 'An extensive study on multiple ETL and HTL layers to design and simulation of high-performance lead-free CsSnCl3-based perovskite solar cells', *Sci. Rep.*, vol. 13, no. 1, p. 2521, Feb. 2023, doi: 10.1038/s41598-023-28506-2.

[9] M. U. Salman *et al.*, 'Direct correlation between open-circuit voltage and quasi-fermi level splitting in perovskite solar cells: a computational step involving thickness, doping, lifetime, and temperature variations for green solutions', *RSC Adv.*, vol. 15, no. 20, pp. 15618–15629, 2025, doi: 10.1039/D5RA01868D.

[10] K. H. Lee *et al.*, 'Optimized Cu-doping in ZnO electro-spun nanofibers for enhanced photovoltaic performance in perovskite solar cells and photocatalytic dye degradation', *RSC Adv.*, vol. 14, no. 22, pp. 15391–15407, 2024, doi: 10.1039/D4RA01544D.

[11] K. Deepika, A. Singh, and S. Ameen, 'Identifying Recombination Mechanisms in Sn-Based Perovskite/Electron-Transport Layer Interface in Perovskite Device using SCAPS-1D', *Phys. status solidi*, vol. 222, no. 10, May 2025, doi: 10.1002/pssa.202400903.

[12] Y. Ai *et al.*, 'SnO2 surface defects tuned by (NH4)2S for high-efficiency perovskite solar cells', *Sol. Energy*, vol. 194, pp. 541–547, Dec. 2019, doi: 10.1016/j.solener.2019.11.004.

[13] M. M. Haque *et al.*, 'Study on the interface defects of eco-friendly perovskite solar cells', *Sol. Energy*, vol. 247, pp. 96–108, Nov. 2022, doi: 10.1016/j.solener.2022.10.024.

[14] M. Mottakin *et al.*, 'Determination of Suitable Transport Layers in Light of Interface Defect States in MASnX 3 -Based Perovskite Solar Cell', *Phys. status solidi – Rapid Res. Lett.*, vol. 16, no. 11, Nov. 2022, doi: 10.1002/pssr.202200216.

[1] R. Caspart *et al.*, 'Precise Energy Consumption Measurements of Heterogeneous Artificial Intelligence Workloads', *ISC Work.*, 2022, doi: 10.48550/ARXIV.2212.01698.

[2] A. Goetzberger and V. U. Hoffmann, *Photovoltaic solar energy generation*, vol. 112. Springer Science & Business Media, 2005.

[3] A. Kojima, K. Teshima, Y. Shirai, and T. Miyasaka, 'Organometal halide perovskites as visible-light sensitizers for photovoltaic cells.', *J. Am. Chem. Soc.*, vol. 131, no. 17, pp. 6050–6051, May 2009, doi: 10.1021/ja809598r.

[4] M. A. Green *et al.*, 'Solar Cell Efficiency Tables (Version 66)', *Prog. Photovoltaics Res. Appl.*, May 2025, doi: 10.1002/pip.3919.

[5] P. D. Badillo and A. E. Degterev, 'Transport Layers as a Factor in Perovskite Solar Cells Improvement', in *2024 Conference of Young Researchers in Electrical and Electronic Engineering (ElCon)*, IEEE, Jan. 2024, pp. 698–701. doi: 10.1109/ElCon61730.2024.10468489.

[6] N. R. Pochont and Y. R. Sekhar, 'Numerical Simulation of Nitrogen-Doped Titanium Dioxide as an Inorganic Hole Transport Layer in Mixed Halide Perovskite Structures Using SCAPS 1-D', *Inorganics*, vol. 11, no. 1, p. 3, Dec. 2022, doi: 10.3390/inorganics11010003.

[7] A. Usman and T. Bovornratanaraks, 'Modeling and Optimization of Modified TiO 2 with Aluminum and Magnesium as ETL in MAPbI 3 Perovskite Solar Cells: SCAPS 1D Frameworks', *ACS Omega*, Sep. 2024, doi: 10.1021/acsomega.4c04505.

[8] M. K. Hossain *et al.*, 'An extensive study on multiple ETL and HTL layers to design and simulation of high-performance lead-free CsSnCl3-based perovskite solar cells', *Sci. Rep.*, vol. 13, no. 1, p. 2521, Feb. 2023, doi: 10.1038/s41598-023-28506-2.

[9] M. U. Salman *et al.*, 'Direct correlation between open-circuit voltage and quasi-fermi level splitting in perovskite solar cells: a computational step involving thickness, doping, lifetime, and temperature variations for green solutions', *RSC Adv.*, vol. 15, no. 20, pp. 15618–15629, 2025, doi: 10.1039/D5RA01868D.

[10] K. H. Lee *et al.*, 'Optimized Cu-doping in ZnO electro-spun nanofibers for enhanced photovoltaic performance in perovskite solar cells and photocatalytic dye degradation', *RSC Adv.*, vol. 14, no. 22, pp. 15391–15407, 2024, doi: 10.1039/D4RA01544D.

[11] K. Deepika, A. Singh, and S. Ameen, 'Identifying Recombination Mechanisms in Sn-Based Perovskite/Electron-Transport Layer Interface in Perovskite Device using SCAPS-1D', *Phys. status solidi*, vol. 222, no. 10, May 2025, doi: 10.1002/pssa.202400903.

[12] Y. Ai *et al.*, 'SnO2 surface defects tuned by (NH4)2S for high-efficiency perovskite solar cells', *Sol. Energy*, vol. 194, pp. 541–547, Dec. 2019, doi: 10.1016/j.solener.2019.11.004.

[13] M. M. Haque *et al.*, 'Study on the interface defects of eco-friendly perovskite solar cells', *Sol. Energy*, vol. 247, pp. 96–108, Nov. 2022, doi: 10.1016/j.solener.2022.10.024.

[14] M. Mottakin *et al.*, 'Determination of Suitable Transport Layers in Light of Interface Defect States in MASnX 3 -Based Perovskite Solar Cell', *Phys. status solidi – Rapid Res. Lett.*, vol. 16, no. 11, Nov. 2022, doi: 10.1002/pssr.202200216.

MACl-Assisted Improvement of Morphology in Mixed-Halide FAPbBr₂Cl Perovskite Films

Maksim A. Balutin
Department of Photonics
Saint Petersburg Electrotechnical
University "LETI"
St.-Petersburg, Russia
mabalutin@stud.etu.ru

Yuriy E. Isaev
Department of Photonics
Saint Petersburg Electrotechnical
University "LETI"
St.-Petersburg, Russia
yura.isaev.97@bk.ru

Aleksandr S. Tarasov
Department of Photonics
Saint Petersburg Electrotechnical
University "LETI"
St.-Petersburg, Russia
astarasov@etu.ru

Aleksander E. Degterev
Department of Photonics
Saint Petersburg Electrotechnical
University "LETI"
St.-Petersburg, Russia
aedegterev@etu.ru

Mariya M. Degtereva
Department of Photonics
Saint Petersburg Electrotechnical
University "LETI"
St.-Petersburg, Russia
mmromanovich@etu.ru

Ivan A. Lamkin
Department of Photonics
Saint Petersburg Electrotechnical
University "LETI"
St.-Petersburg, Russia
ialamkin@etu.ru

Abstract— Uniform coverage of the absorber layer is crucial for the performance of perovskite-based optoelectronic devices. In this study methylammonium chloride (MACl) was introduced as an additive into the formamidinium lead mixed-halide (FAPbBr₂Cl) perovskite. Perovskite films of pristine FAPbBr₂Cl were synthesized using a traditional one-step spin-coating method under ambient conditions. In contrast, samples with the MACl additive were fabricated using a two-step spin-coating procedure with various annealing times. To evaluate the optoelectronic properties of the fabricated films, which serve as an indicator of film uniformity, photoluminescence, absorption and transmittance spectra were studied. The results indicate that the films containing the MACl additive possess higher uniformity, which can be attributed to enhanced crystallization and higher surface quality. It was confirmed that optimizing the annealing time is a critical factor for achieving high uniformity in perovskite films.

Keywords— *mixed halide perovskite, chloride perovskite, photobrightening, light soaking, photoluminescence spectra*

I. INTRODUCTION

In recent years, mixed-halide wide-bandgap perovskites have attracted increasing attention due to their unique physical properties, compositional diversity, and potential for application as photoactive layers in optoelectronic devices [1]. By varying the material composition, their properties can be finely tuned. Particular focus in this work is placed on chlorine-containing perovskites as promising materials for fabricating visible-blind ultraviolet photodetectors.

Since early reports that Cl⁻ incorporation extends charge-carrier diffusion lengths, chlorine doping has been widely adopted to enhance device efficiency. Studies indicate that Cl improves film crystallinity and compactness [2]. However, the mechanism behind these improvements and the chemical state of chlorine remain debated. Some groups report that Cl is absent in the final film, having served as a volatile processing aid [3], while others suggest it forms a mixed-halide phase or resides preferentially at interfaces [4].

In this study, a series of thin-film perovskite samples of FAPbBr₂Cl were synthesized. Subsequently, the optical absorption and photoluminescence of the obtained samples were measured. The experimental data were then analyzed, and based on this analysis, conclusions were drawn regarding the prospects of using these structures as photoactive layers.

II. EXPERIMENTAL SECTION

A. Preparation of precursor solutions

Lead halide solution (Solution A) was prepared by dissolving 257 mg of PbBr₂ and 83 mg of PbCl₂ in 1 mL of a mixed solvent of N, N-dimethylformamide (DMF) and dimethyl sulfoxide (DMSO) (4:1) at room temperature, followed by stirring on a magnetic stirrer for 2 hours.

Organic salt solution (Solution B) was prepared separately by dissolving 30 mg of FABr (formamidinium bromide) and 12.5 mg of MACl (methylammonium chloride) in 1 mL of isopropyl alcohol at room temperature with stirring for 1 hour. Both solutions were filtered through a hydrophobic syringe filter (0.22 μm pore size) prior to use.

B. Sample preparation

Perovskite thin films were deposited via a two-step spin-coating method (Fig. 1) onto pre-cleaned glass/ITO substrates. All steps were performed under ambient air conditions. 80 μL of Solution A was dispensed onto the substrate, which was then spun at 3000 rpm for 30 seconds.

Immediately after spin-coating, the substrate was thermally treated on a hotplate at 100 °C for 70 seconds.

After cooling to room temperature, 150 μL of Solution B was dispensed onto the formed layer at a spin speed of 1500 rpm for 30 seconds (Fig. 2).

The substrate underwent a final annealing step on a hotplate at 100 °C. The annealing time was systematically varied from 10 to 55 minutes to study crystallization kinetics and optimize film morphology.

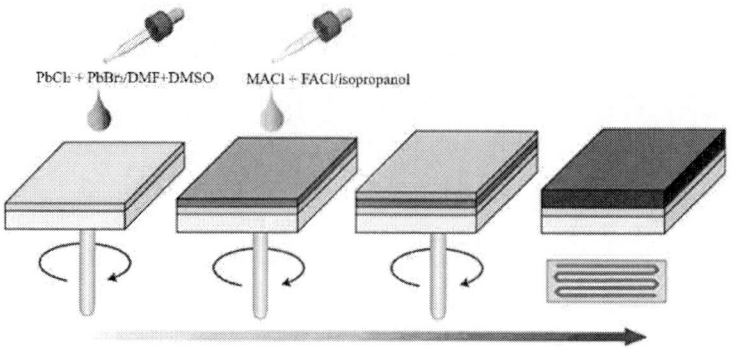

Fig. 1. Two-step spin-coating method

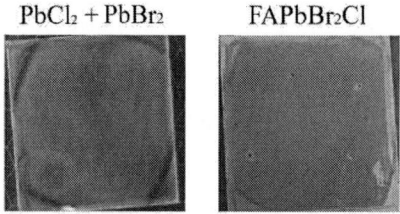

Fig. 2. Photographs of PbCl₂+PbBr₂ (first deposition step) and FAPbBr₂Cl films

III. RESULTS

A. Effect of adding MACl in precursor solution of FAPbBr₂Cl

We first investigated the influence of the MACl additive on the formation and properties of FAPbBr₂Cl perovskite films. Its impact on morphology is profound, as clearly visible in the film photographs presented in Fig. 3.

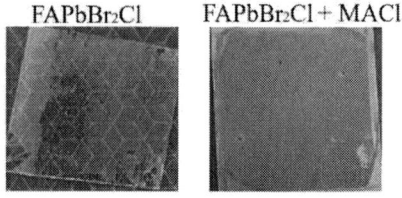

Fig. 3. Photographs of FAPbBr₂Cl films

The addition of MACl to the FABr-based bromide precursor leads to a substantial improvement in the morphology of the resulting perovskite films. Similar to the results obtained for iodide-based systems [5], introducing MACl promotes a more complete conversion of PbBr₂ into the perovskite phase, increases crystallite size, and facilitates the formation of a dense, uniform coating on the substrate. A possible explanation is because chlorine acts as a morphological modifier, improving crystallization kinetics and reducing the density of point defects.

B. Studying the effect of annealing time

Annealing duration is a key processing parameter known to govern the final chemical state and spatial distribution of chlorine in perovskite films, thereby directly dictating their structural and optoelectronic quality. Our investigation revealed that thermal annealing triggers a set of dynamic processes encompassing chlorine redistribution, chemical transformation, and its eventual expulsion from the material. Absorption and transmittance spectra of films with different time of annealing are shown in figures 4 and 5.

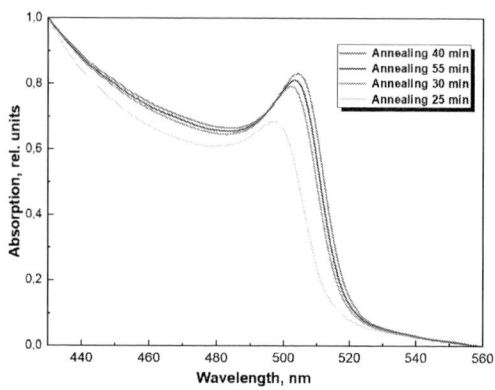

Fig. 4. Absorption spectra of FAPbBr₂Cl + MACl films

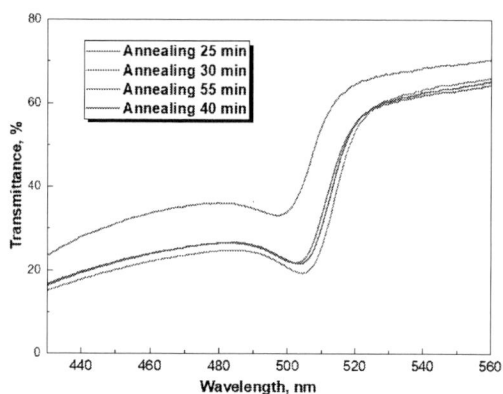

Fig. 5. Transmittance spectra of FAPbBr₂Cl + MACl films

The observed non-monotonic dependence of the film's optical density on annealing time, characterized by a minimum absorption at 10 minutes, a subsequent increase up to 40 minutes, and a decline upon longer processing, reflects the dynamics of structural and phase transformations in the bromide-chloride perovskite system.

The minimum absorption after brief annealing (10 min) may be due to the simultaneous presence of several phases in the film: a wide-bandgap MAPbCl₃ phase, a mixed FA$_x$MA$_{1-x}$PbBr$_y$Cl$_{3-y}$ phase and possible residual amorphous regions. Such a composite structure leads to incomplete absorption in the visible spectral range and increased light scattering at phase boundaries.

The absorption increase in the interval from 10 to 40 minutes is explained by a thermally activated homogenization process and the removal of excess chlorine from the lattice, accompanied by halide redistribution and the formation of a more uniform mixed $FAPbBr_xCl_{3-x}$ phase or, upon complete additive chlorine removal, the $FAPbBr_2Cl$ phase. This process optimizes the bandgap, improves crystallinity, and increases the absorption coefficient in the target spectral region.

The subsequent decrease in absorption after 40 minutes of annealing is likely caused by the onset of thermal degradation of the perovskite. This degradation may include partial decomposition of the material with the loss of volatile components, the formation of defects and pinholes, and possible phase segregation. These processes deteriorate the film's structural quality, increase recombination losses, and reduce the effective optical density.

Thus, the non-monotonic behavior of the optical properties results from the competition between structural ordering and homogenization processes on one hand, and thermally induced degradation on the other.

C. Effect of light soaking time on photoluminescence spectra

$FAPbBr_2Cl$ perovskite demonstrates a photobrightening effect when exposed to laser radiation. In this work, we used a 405 nm diode laser as an excitation source.

Fig. 6 shows the relative PL intensity as a function of light soaking time. PL measurements were taken every 10 seconds. A significant increase in photoluminescence intensity is observed under prolonged laser illumination. This is attributed to the photo-induced passivation of non-radiative recombination centers. Photo-excited carriers can facilitate the migration of ions, leading to the annihilation of vacancy-interstitial pairs or the neutralization of charged defect states that act as traps. Moreover, localized light absorption can drive a reversible redistribution of halide ions, potentially smoothing compositional inhomogeneities and reducing the density of defect clusters that form at halide-rich interfaces.

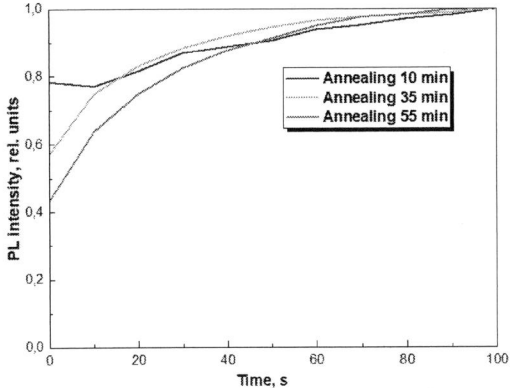

Fig. 6. Photoluminescence intensity as a function of light soaking time

Fig. 7 shows the PL peak wavelength as a function of light soaking time. A red shift is observed under continuous illumination. This is due to the photoinduced halide segregation as well as the photothermal heating of the perovskite film, leading to bandgap reduction.

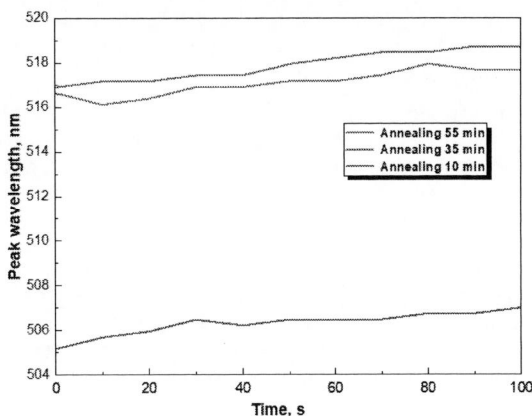

Fig. 7. Photoluminescence peak wavelength as a function of light soaking time

IV. CONCLUSIONS

This work investigates the influence of the MACl addition on the morphological and optical properties of mixed-halide perovskite films $FAPbBr_2Cl$. It was shown that the introduction of MACl into the precursor significantly improves film homogeneity, promotes an increase in crystallite size, and reduces the density of point defects, as evidenced by an increase in photoluminescence intensity and a change in the optical absorption pattern.

The MACl additive acts not only as a source of chlorine but also as a dynamic agent, temporarily stabilizing intermediate phases during crystallization. This slows down the kinetics of crystal growth, resulting in denser and more uniform films with improved optoelectronic characteristics. It was established that the annealing duration greatly affects the properties of $FAPbBr_2Cl$ films: short annealing times (less than 10 minutes) lead to coexistence of multiple phases, minimizing the optical absorption. After 10 minutes, a thermal homogenization and halide redistribution occur, optimizing the film's bandgap and crystallinity leading to higher optical absorption. Prolonged thermal exposure induces decomposition and defect formation. Therefore, choosing the appropriate annealing time is crucial for the creation of high-performance perovskite films. The nonlinear dependence of optical properties on annealing time indicates complex processes of halide redistribution, thermal homogenization, and subsequent degradation. The optimal annealing time (approximately 40 minutes) corresponds to a balance between the completion of phase transformations and the onset of thermal decomposition of the perovskite phase.

Photoluminescence studies reveal that light soaking triggers two competing processes in $FAPbBr_2Cl$ perovskite films. First, a pronounced photobrightening effect occurs, where photoluminescence intensity increases due to the photoinduced passivation of non-radiative defects. The observed photobrightening effect is associated with light-induced defect passivation, indicating the potential of light treatment as a method for post-synthesis film optimization. Concurrently, a progressive red shift of the emission peak is observed, resulting from laser-induced halide segregation and photothermal heating, which narrows the effective bandgap. The simultaneous red shift of the PL peak indicates competition between defect passivation and light-induced halide segregation, a key factor in the instability of mixed perovskites. This duality reflects a kinetic competition: initial illumination enhances radiative efficiency through defect healing, while prolonged exposure promotes halide segregation and phase separation. Together, these processes define the material's evolving optoelectronic stability under operating conditions.

These results pave the way for the creation of stable wide-bandgap perovskite films for UV photodetectors and tandem solar cells. Controlling the morphology through MACl-type additives and optimizing thermal annealing allows for the management of not only the optical, but also the transport properties of the layers. Thus, this work demonstrates that combining chemical modification (MACl) and precise control of thermal processing allows for the manipulation of the microstructure and functional properties of mixed-halide perovskites, which is key to their integration into high-performance optoelectronic devices.

REFERENCES

[1] P. D. Badillov *et al.*, "Perspectives on Perovskite Solar Cells Under the Glass of Characterization and Model-based Research," in *2023 XXVI International Conference on Soft Computing and Measurements (SCM)*, May 2023, pp. 277–280. doi: 10.1109/SCM58628.2023.10159103.

[2] B. Park *et al.*, 'Enhanced Crystallinity in Organic–Inorganic Lead Halide Perovskites on Mesoporous TiO2 via Disorder–Order Phase Transition', *Chem. Mater.*, vol. 26, no. 15, pp. 4466–4471, Aug. 2014, doi: 10.1021/cm501541p.

[3] A. E. Williams, P. J. Holliman, M. J. Carnie, M. L. Davies, D. A. Worsley, and T. M. Watson, 'Perovskite processing for photovoltaics: a spectro-thermal evaluation', *J. Mater. Chem. A*, vol. 2, no. 45, pp. 19338–19346, Oct. 2014, doi: 10.1039/C4TA04725G.

[4] S. Colella *et al.*, 'Elusive Presence of Chloride in Mixed Halide Perovskite Solar Cells', *J. Phys. Chem. Lett.*, vol. 5, no. 20, pp. 3532–3538, Oct. 2014, doi: 10.1021/jz501869f.

[5] Y. Sun, H. Chen, T. Zhang, and D. Wang, 'Chemical state of chlorine in perovskite solar cell and its effect on the photovoltaic performance', *J Mater Sci*, vol. 53, no. 19, pp. 13976–13986, Oct. 2018, doi: 10.1007/s10853-018-2571-2.

Acetone Detection via ZnO/Zn$_2$SnO$_4$ Impedance Spectroscopy Analysis

Cong Doan Bui
Department of micro- and nanoelectronics
Saint Petersburg Electrotechnical University "LETI"
Saint Petersburg, Russia
congdoan6997@gmail.com

Svetlana S. Nalimova
Department of micro- and nanoelectronics
Saint Petersburg Electrotechnical University "LETI"
Saint Petersburg, Russia
sskarpova@list.ru

Abstract— ZnO/Zn$_2$SnO$_4$ core–shell nanorod arrays were prepared via spin-coated ZnO seeds, hydrothermal ZnO growth, and a hydrothermally deposited Zn$_2$SnO$_4$ shell. Impedance spectroscopy at 200 °C in the presence of acetone with different concentrations (0–3000 ppm; 1 kHz–500 kHz) indicates single Nyquist semicircles whose diameters change monotonically with concentration. A parallel (R – C) model with a resistance, that drops from 28.1 kΩ (air) to 16.7, 10.4, and 7.1 kΩ at 1000, 2000, and 3000 ppm, and a constant capacitance of 19 pF was proposed. As a result, $\tau = RC$ drops from 0.53 to 0.14 µs and f_c shifts from 298 kHz to 1.18 MHz as the acetone concentration increases. The response is controlled by an *n*-type mechanism. The results provide compact and dependable sensor design parameters.

Keywords— *impedance spectroscopy, ZnO/Zn$_2$SnO$_4$ core/shell nanorods, acetone gas sensing, equivalent-circuit modeling*

I. INTRODUCTION

Metal oxide semiconductors (MOSs) are widely used gas sensor materials for their stability, relatively low cost, and sensitivity [1]. Zinc ortho-stannate (Zn$_2$SnO$_4$) is known as an *n*-type ternary oxide semiconductor with a relatively large bandgap energy range (typically between 3.46 eV and 3.6 eV), as well as relatively large electron mobility, and so on [2]. This material, together with zinc oxide (ZnO), has widespread applications in transparent conducting electrode layers for solar cells and gas sensors or other kinds of sensors [3, 4]. The intentional combination between ZnO and Zn$_2$SnO$_4$ within heterostructure-based materials provides a promising approach towards improving gas sensitivity for using such materials, instead of using them in their pure form as MOSs. This is attributed to their ability to create *n-n* heterostructural interfaces, resulting in additional electron depletion layers that are very sensitive to gas adsorption on respective layers [5].

One of the most important aspects of composite design is controlling the nanoscale architecture in order to maximize charge transfer and surface reaction rates [6]. The current study focuses on a specific high-performance structure: a ZnO/Zn$_2$SnO$_4$ core-shell composite, with ZnO nanorods serving as the core and a Zn$_2$SnO$_4$ layer forming the exterior shell. This architecture allows the ZnO core to create an improved pathway for electron transport, while the Zn$_2$SnO$_4$ shell acts as an energy barrier, significantly limiting the recombination of charge carriers at the interface between adsorbed molecules and the sensor [7].

Sensitive and stable detection of volatile organic compounds (VOCs) like acetone is important for medical diagnosis and environmental protection. The possibility of using Zn$_2$SnO$_4$-based materials as gas sensors of ethanol, NO$_2$, H$_2$S, and acetone vapor has already been indicated by previous investigations. However, it requires advanced characterization to unravel the intricate sensing mechanism of improved core-shell nanostructures [8].

In this study, Electrochemical Impedance Spectroscopy (EIS) is utilized to provide a complete understanding of electrochemical processes and the behavior of charge carriers in the fabricated composite [9 – 11]. Separate resistive components, like grain bulk resistance (R_g) and grain boundary resistance (R_{gb}), of the complex impedance spectrum can be measured and assess how these components vary with higher acetone gas concentrations [12]. Preparation and systematic study of the acetone gas sensing mechanism of the new ZnO nanorod core/Zn$_2$SnO$_4$ shell composite is revealed in this work by thorough EIS analysis to identify synergistic effects of such a heterostructure.

II. EXPERIMENT

A. Synthesis seed layer ZnO

The seed layer of ZnO nanoparticles was deposited by the spin-coating method. The coating solution was prepared by dissolving zinc acetate dihydrate (Zn(CH$_3$COO)$_2$.2H$_2$O) in 50 mL of distilled water. After stirring for 10 minutes, the resulting solution was applied onto pre-cleaned *BI2* substrates with pre-deposited gold electrical contacts (sensor platform) at a rotation speed of 3000 rpm for 15 seconds, with three successive coatings. The samples were then annealed in air at 500 °C for 15 minutes to form ZnO nanoparticles on the surface of the sensor platform and to remove organic residues.

B. Synthesis nanorod core ZnO

The ZnO nanorod array was synthesized on the substrates using the ZnO nanoparticle seed layer by the hydrothermal synthesis process. The growth solution was synthesized by dissolving hexamethylenetetramine (HMTA) and zinc nitrate hexahydrate (Zn(NO$_3$)$_2$.6H$_2$O) in 50 mL of distilled water. The solution was stirred for 10 minutes and then transferred to a 50 mL autoclave. The substrates were then soaked in the prepared solution. Lastly, the growth solution was kept at 85 °C for 1 hour in a laboratory thermostatic bath. The resulting samples were cleaned with deionized water and ethanol and then annealed in air at 500 °C for 15 minutes.

C. Synthesis shell Zn$_2$SnO$_4$

Zn$_2$SnO$_4$ shell was prepared using the hydrothermal method and then annealed using a muffle furnace. The coating solution was initially derived by dissolving potassium stannate trihydrate (K$_2$SnO$_3$·3H$_2$O) and urea (CH$_4$N$_2$O) in an alcohol-water mixture. The above-prepared solution was again loaded in a 50 mL autoclave with immersed substrates. The solution was then kept in an oven, heated to 170 °C and maintained for 30 minutes. Lastly, the samples were annealed at 500 °C for 15 minutes in air to create a crystalline Zn$_2$SnO$_4$ shell layer on the surface of the ZnO nanorod array.

D. Impedance characterization

Impedance spectroscopy is a widely used technique because it has technical and programmatic tools available, as well as the ability to investigate phase boundaries. Hence, it was decided to provide the laboratory setup with the facility to measure the complex impedance of samples at different frequencies employing an Elins Z-500P impedance meter (Elins, Russia). The device provides possibility to change the amplitude of the alternating signal in the range 0–255 mV and the range of frequency from 15 mHz to 500 kHz. Z-500P has advanced functions, such as approximation impedance spectrum and high impedance measurement accuracy for a wide range of values (from 0.05 Ω to 50 MΩ).

III. RESULTS AND DISCUSSION

The impedance spectra of the sample were measured at a temperature of 200 °C. The acetone vapor generation system allows adjustment of the concentration in the range from 0 to 3000 ppm. The Nyquist plots in Fig. 1 are semicircular arcs with diminishing diameters as the acetone concentration is increased from 0 to 3000 ppm.

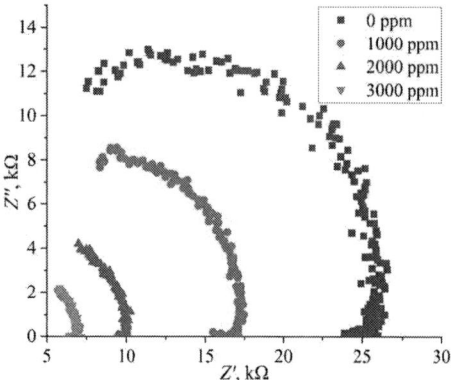

Fig. 1. Nyquist plots of the ZnO–Zn$_2$SnO$_4$ sensor under various acetone vapor concentrations

The trend denotes reduction of the sum of impedance of the sensor with an increase in gas concentration. Reduction in real part of the impedance (Z') and reduction in the radius of the semicircular points towards increased mobility of the charge carrier and enhanced surface conductivity of the sensing film when exposed to acetone vapor.

This kind of response is characteristic for *n*-type semiconductor oxides, including ZnO and Zn$_2$SnO$_4$. When acetone molecules react with adsorbed oxygen species on the surface of the sensor, they form oxidation products, giving back electrons to the conduction band. This causes the depletion layer to decrease in thickness and, therefore, decrease sensor resistance. The regular and periodic change of the impedance arcs with the concentration of gases confirms the high sensitivity and good reproducibility of the sensor response at 200 °C, validating the sensor to be appropriate for detecting acetone in the low-ppm level.

Earlier research has proven that the sensing action of ZnO and its heterostructures directly relates to the adsorption of oxygen species on the material surface. This causes changes in the impedance of the sensing layer due to chemical and electronic reactions between the analyte gas molecules adsorbed on the surface. The sensor response strongly depends on the extent of oxygen adsorption, and the latter is itself highly dependent upon the operating temperature. Adsorption of acetone and oxygen with varying conditions of temperature can be represented as follows [13]:

$$O_{2(g)} \leftrightarrow O_{2(ad)} \tag{1}$$

$$O_{2(ad)} + e^- \leftrightarrow O_{2(ad)}^- \quad (T < 150°C) \tag{2}$$

$$O_{2(ad)}^- + e^- \leftrightarrow 2O_{(ad)}^- \quad (150\ °C < T < 400°C) \tag{3}$$

$$O_{2(ad)} + 2e^- \leftrightarrow O_{(ad)}^{2-} \quad (T > 400°C) \tag{4}$$

$$CH_3COCH_{3(g)} + 8O_{(ad)}^- \leftrightarrow 3CO_2 + 3H_2O + 8e^- \quad 150\ °C < T < 400°C \tag{5}$$

Particularly for ZnO/Zn$_2$SnO$_4$ core–shell nanorod structure, oxygen molecules adsorb on Zn$_2$SnO$_4$ shell surface upon exposure to air and get ionized into O_2^-, O^- and O^{2-} species by trapping the free electrons of the heterostructure. It is the capture of an electron that results in the creation of a surface depletion layer, causing the decrease in concentration and mobility of electrons and increase in material resistance when exposed to air. The steps for these reactions are described by Equations (1) to (4) above.

As a reducing agent, acetone oxidizes assisted by adsorbed oxygen at a working temperature of 200 °C, giving out electrons to flow in the conduction band as described by Equation (5). This actually lowers the height of the Schottky barrier along with the thickness of the depletion layer and consequently lowers the overall resistance. In this case, the Zn$_2$SnO$_4$ shell interacts with acetone molecules while the ZnO core is used as a supporting material to improve electronic properties and sensing performance of the material.

An equivalent circuit model was proposed based on the measured impedance spectra of the sample, consisting of a resistor (R) connected in parallel with a capacitor (C). The model is responsible for processes at grain boundaries and within the grain bulk [14, 15]. The result of the equivalent circuit simulation, fitted to the experimental impedance data measured in the absence of acetone vapor at an operating temperature of 200 °C, is shown in Fig. 2.

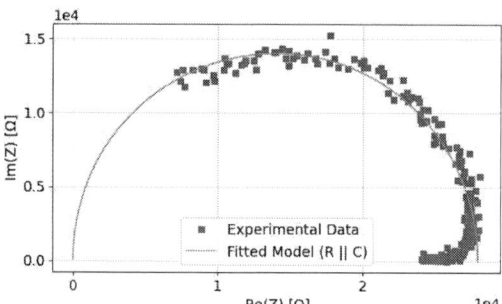

Fig. 2. Nyquist plot of the experimental impedance data and the fitted equivalent circuit model (R – C)

The semicircular curve of the plot shows that the electrical response of the sensor in air is determined by a single relaxation process related to charge transport and storage at the grain boundaries and the bulk sensing layer. The good agreement between the experimental data and the model fit indicates that the suggested (R – C) circuit fairly describes the impedance characteristics of the ZnO/Zn2SnO4 heterostructure without acetone.

The value of the obtained resistance accounts for adding bulk and interfacial resistance of the sensing material in ambient air conditions, which is high because of oxygen adsorption at the surface. This adsorbed oxygen on the surface captures free electrons in the conduction band, creating a surface depletion region and thus adding to the total impedance. Table 1 shows the calculated resistor and capacitor values according to the equivalent circuit above.

At the $1\ kHz - 500\ kHz$ frequency range, the Nyquist semicircular arc is cut off at the high-frequency end because the characteristic relaxation frequency is above the 500 kHz ceiling of the instrument. The semicircle then fails to extend all the way to its high-frequency intercept, and the resulting "missing half" is a property of the instrument and not the product of a second electrochemical reaction.

TABLE I.   EXTRACTED EQUIVALENT-CIRCUIT PARAMETERS

Concentration, ppm	R, kΩ	C, pF
0	28.1	19
1000	16.7	19
2000	10.4	19
3000	7.1	19

The fitted resistance R monotonically varies from 28.1 kΩ (air) to 7.1 kΩ (3000 ppm), with response factors R_{air}/R_{gas} 1.68, 2.70, and 3.96 at 1000, 2000, and 3000 ppm, respectively. With capacitance $C \sim 19$ pF, the relaxation time $\tau = RC$ goes down from 0.534 μs (0 ppm) to 0.317 μs, 0.198 μs, and 0.135 μs, moving the characteristic frequency $f_c = 1/(2\pi RC)$ from 298 kHz to 502 kHz, 805 kHz, and 1.18 MHz. These values rationalize the shorting of the Nyquist semicircles at high frequency: even at 0 ppm the top is close to the analyzer limit of 500 kHz, and at $\geq$1000 ppm f_c is out of range for the instrument. The fact that C is not varying significantly means acetone is modulating the resistive (carrier-depletion) component mainly, and the dielectric/space-charge capacitance and microstructure are little perturbed over $0 - 3000$ ppm.

IV. CONCLUSIONS

We synthesized ZnO/Zn$_2$SnO$_4$ core–shell nanorod arrays through spin-coating a ZnO seed layer, hydrothermal growth of the ZnO nanorods, and hydrothermal deposition/annealing Zn$_2$SnO$_4$ shell. Impedance spectra at 200 °C in acetone (0 – 3000 ppm) show Nyquist arcs that narrow and shift as predicted with concentration. A parallel (R – C) circuit is a reasonable fit for the response: resistance is monotonically decreasing from 28.1 kΩ (air) to 7.1 kΩ (3000 ppm), and capacitance is roughly constant at 19 pF. This suggests that acetone largely sensitizes the resistive component through the traditional n-type mechanism – adsorbed oxygen consumption, electron release into the conduction band, depletion region narrowing, and Schottky barrier reduction.

The subsequent relaxation time $\tau = RC$ decreases from 0.53 μs to 0.14 μs, moving the characteristic frequency from 298 kHz to 1.18 MHz. In the 1 kHz–500 kHz measurement range then, the upper cut-off of the semicircle is truncated; the "missing half" is thus a bandwidth effect of the instrument.

These findings confirm that the ZnO/Zn$_2$SnO$_4$ core–shell nanostructure provides a reproducible, stable, and concentration-dependent impedance response to acetone at 200 °C with R-dominated sensitivity.

REFERENCES

[1] A. A. Ryabko, A. A. Bobkov, S. S. Nalimova, A. I. Maksimov, V. A. Moshnikov, and E. I. Terukov, "Gas sensitivity of nanostructured coatings based on zinc oxide nanorods under combined activation," *Technical Physics*, vol. 68, suppl. 1, pp. S13–S18, Dec. 2023, doi: 10.1134/S106378422390053X.

[2] S. S. Nalimova, Z. V. Shomakhov, D. A. Kozodaev, A. A. Rybina, S. S. Buzovkin, C. D. Bui, I. A. Novikov, and V. A. Moshnikov, "VOC gas sensors based on zinc stannate nanoparticles decorated with silver," *Nanomaterials*, vol. 14, no. 24:1993, Dec. 2024, doi: 10.3390/nano14241993.

[3] A. A. Ryabko, S. S. Nalimova, N. V. Permyakov, A. A. Bobkov, A. I. Maksimov, V. M. Kondratev, K. P. Kotlyar, M. K. Ovezov, A. S. Komolov, E. F. Lazneva, V. A. Moshnikov, and A. N. Aleshin, "Architectonics of zinc oxide nanorod coatings for adsorption gas sensors," *Technical Physics*, vol. 69, pp. 2103–2110, Jul. 2024, doi: 10.1134/S1063784224070387.

[4] R. Zhapakov, D. Murzalinov, M. Begunov, T. Seredavina, A. Gagarina, Y. Spivak, V. Moshnikov, E. A. Dmitriyeva, P. Osipov, and A. Kemelbekova, "Investigation of Complex ZnO–Porous Silicon Structures with Different Dimensions Obtained by Low-Temperature Synthesis," Processes, vol. 13, no. 7, p. 2099, 2025, doi: 10.3390/pr13072099.

[5] C. D. Bui, S. S. Nalimova, Z. V. Shomakhov, A. M. Guketlov, S. S. Buzovkin, and A. A. Rybina, "ZnO/Zn$_2$SnO$_4$ nanorod heterostructure coatings for effective detection of acetone," *Physical and chemical aspects of the study of clusters, nanostructures and nanomaterials*, no. 16, pp. 794–804, 2024, doi: 10.26456/pcascnn/2024.16.794.

[6] S. C. Lims, S. Divya, M. Manivannan, T. Arumanayagam, R. Robert, and M. Jose, "Investigation of structural, temperature- and frequency-dependent dielectric behavior of Zn$_2$SnO$_4$@amorphous SiO$_2$ core–shell nanocomposites," *Chemical Physics Impact*, vol. 8, no. 100485, Jun. 2024, doi: 10.1016/j.chphi.2024.100485.

[7] R. Dridi, I. Saafi, A. Mhamdi, A. Yumak, M. H. Lakhdar, A. Amlouk, K. B. Boubaker, and M. Amlouk, "Structural, optical and AC conductivity studies on alloy ZnO–Zn$_2$SnO$_4$ (ZnO–ZTO) thin films," *Journal of Alloys and Compounds*, vol. 634, pp. 179–186, Jun. 2015, doi: 10.1016/j.jallcom.2015.02.009.

[8] N. H. Hanh, M. N. Trinh, V. D. Lai, M. H. Chu, V. D. Nguyen, and D. H. Nguyen, "A comparative study on the VOCs gas sensing properties of Zn$_2$SnO$_4$ nanoparticles, hollow cubes, and hollow octahedra towards exhaled breath analysis," *Sensors and Actuators B: Chemical*, vol. 343, p. 130147, Sep. 2021, doi: 10.1016/j.snb.2021.130147.

[9] V. M. Kondratev, E. A. Vyacheslavova, T. Shugabaev, D. A. Kirilenko, A. Kuznetsov, S. A. Kadinskaya, Z. V. Shomakhov, A. I. Baranov, S. S. Nalimova, V. A. Moshnikov, A. S. Gudovskikh, and A. D. Bolshakov, "Si nanowire-based Schottky sensors for selective sensing of NH$_3$ and HCl via impedance spectroscopy," *ACS Applied Nano Materials*, vol. 6, no. 13, pp. 11513–11523, Jul. 2023, doi: 10.1021/acsanm.3c01545.

[10] A. Bobkov, V. Luchinin, V. Moshnikov, S. Nalimova, and Y. Spivak, "Impedance spectroscopy of hierarchical porous nanomaterials based on por-Si, por-Si incorporated by Ni and metal oxides for gas sensors," *Sensors*, vol. 22, no. 4, no. 1530, Feb. 2022, doi: 10.3390/s22041530.

[11] V. Balasubramani, S. Chandraleka, T. Subba Rao, R. Sasikumar, M. R. Kuppusamy, and T. M. Sridhar, "Review – Recent advances in electrochemical impedance spectroscopy based toxic gas sensors using semiconducting metal oxides," *Journal of The Electrochemical Society*, vol. 167, Art. no. 037572, 2020, doi: 10.1149/1945-7111/ab77a0.

[12] A. Ch. Lazanas and M. I. Prodromidis, "Electrochemical impedance spectroscopy – A tutorial," *ACS Measurement Science Au*, vol. 3, pp. 162–193, Mar. 2023, doi: 10.1021/acsmeasuresciau.2c00070.

[13] S. Ma, L. Shen, S. Ma, J. Wen, and J. Xu, "Emerging zinc stannate and its application in volatile organic compounds sensing," *Coordination Chemistry Reviews*, vol. 490, p. 215217, Sep. 2023, doi: 10.1016/j.ccr.2023.215217.

[14] Nalimova, S. S.; Kononova, I. E.; Moshnikov, V. A.; Dimitrov, D. T.; Kaneva, N. V.; et al. "Investigation of the vapor-sensitive properties of zinc oxide layers by impedance spectroscopy," *Bulgarian Chemical Communications*, 49(1), 121–126 (2017).

[15] G. M. G. da Silva, P. M. Faia, S. R. Mendes, and E. S. Araújo, "A review of impedance spectroscopy technique: Applications, modelling, and case study of relative humidity sensors development," *Applied Sciences*, vol. 14, no. 13, p. 5754, 2024, doi: 10.3390/app1413575.

Characterization of Nanostructured Ternary Oxide Multisystems for Gas Sensors

Sergey Buzovkin
Department of micro- and nanoelectronics
Saint-Petersburg Electrotechnical University "LETI"
Saint-Peresburg, Russia
sergey.bu2015@gmail.com

Arina Rybina
Department of micro- and nanoelectronics
Saint-Petersburg Electrotechnical University "LETI"
Saint-Peresburg, Russia
arinasvg02@gmail.com

Roman Kryukov
Department of micro- and nanoelectronics
Saint-Petersburg Electrotechnical University "LETI"
Saint-Peresburg, Russia
rom1999@yandex.ru

Svetlana Nalimova
Department of micro- and nanoelectronics
Saint-Petersburg Electrotechnical University "LETI"
Saint-Peresburg, Russia
sskarpova@list.ru

Vyatcheslav Moshnikov
Department of micro- and nanoelectronics
Saint-Petersburg Electrotechnical University "LETI"
Saint-Peresburg, Russia
vamoshnikov@mail.ru

Abstract— **In this work, ternary semiconducting oxides were synthesized and subsequently investigated for their sensitivity to volatile organic compounds. The produced structures were also examined using various characterization techniques, such as Kelvin probe measurements using atomic force microscopy, X-ray photoelectron spectrocopy, Scanning electron microscopy and Energy-dispersive spectroscopy.**

Keywords—gas sensor, ternary oxide, zinc stannate, zinc ferrite

I. INTRODUCTION

Over the past century, the amount of various toxic emissions released into the atmosphere has increased many times. These substances affect not only the environment but also have a strong impact on the human body. Air-polluting compounds begin to influence a person from early childhood, and it is our responsibility to protect nature and future generations so that they inherit a clean world. Gas sensors are necessary for monitoring the concentration of harmful substances in the ambient air. They can also be used for detecting fires and identifying leaks in industrial facilities. Toxic gases that are frequently encountered in industrial environments include ammonia, hydrogen sulfide, carbon dioxide, carbon monoxide, and volatile organic compounds such as methanol, acetone, isopropanol, and others. In our study, a semiconducting gas sensors based on ternary metal oxides and their responses to several volatile organic compounds will be examined.

II. PREPARATION OF STRUCTURE

A. Characteristics of gas sensing materials

Zinc stannate is a promising multicomponent oxide widely considered for use in semiconducting gas sensors. It is an n-type semiconductor with a band gap in the range of 3.3–3.7 eV, which determines its optoelectronic properties as well as its high chemical and thermal stability. This material exhibits enhanced charge-carrier mobility and good electrical conductivity, as well as a developed system of oxygen vacancies and surface active sites that strengthen gas adsorption [1,2].

Zinc ferrite is also one of the promising materials used in gas sensors. It likewise possesses high chemical and thermal stability and a well-developed defect structure, which ensures efficient adsorption of target gas molecules. The band gap of $ZnFe_2O_4$ is approximately 1.9–2.1 eV, which enables it to operate effectively at lower working temperatures compared to wide-band-gap oxides. In addition, the material exhibits pronounced catalytic properties, high sensitivity to reducing gases, and good resistance to humidity, making it attractive for gas-sensing applications [3].

B. Preparation of gas sensor

Samples of zinc oxide nanorods, zinc stannate, and zinc ferrite were prepared in this work. The initial step involved obtaining zinc oxide nanoparticles on the surface of the substrate. These nanoparticles were produced using spin-coating. Next, to synthesize zinc oxide nanorods, the previously prepared substrates with the ZnO nanoparticle structure were placed at the bottom of an autoclave. The autoclave was filled with a solution obtained by mixing 148 mg of zinc nitrate in 20 mL of distilled water and 70 mg of HMTA in 20 mL of distilled water. The hydrothermal synthesis was conducted at 85 °C. After the synthesis, the substrates were rinsed with distilled water to remove reaction products and dried at 40 °C, after which they were placed in a furnace for annealing at 500 °C. The zinc stannate and zinc ferrite structures were obtained by modifying the hydrothermally synthesized ZnO nanorods [4,5].

To obtain zinc stannate, a solution was prepared by dissolving 14.33 mg of potassium stannate in 10 mL of 40% isopropyl alcohol, and another solution was prepared by dissolving 93.6 mg of urea in 10 mL of 40% isopropyl alcohol. The solutions were mixed using a magnetic stirrer at room temperature for 10 minutes. The substrates with the deposited ZnO nanorods were placed at the bottom of the autoclave and filled with the prepared solution to two-thirds of its volume, then placed in an oven at 170 °C for 30 minutes. Afterward, the substrates were rinsed with distilled water, dried, and annealed at 500 °C for 15 minutes.

To obtain $ZnFe_2O_4$, a solution was prepared by dissolving 347.5 mg of iron sulfate in 50 mL of distilled water. The resulting solution was placed in an ultrasonic bath for 5 minutes, after which the substrates were immersed in this solution for 30 minutes. After half an hour, the substrates were rinsed with water to remove excess precipitate and then placed in a furnace at 500 °C for 15 minutes.

The structure containing zinc stannate was obtained by hydrothermal synthesis, whereas the structure containing zinc ferrite was produced by a precipitation method.

III. INVESTIGATION OF STRUCTURE

The investigation of the surface morphology and the surface potential distribution of the sample containing zinc ferrite was carried out using an NTEGRA ACADEMIA scanning probe microscope (NT-MDT, Russia). Topographical data were obtained in the semi-contact mode of atomic force microscopy, while the surface potential

distribution was examined using scanning Kelvin probe microscopy. The obtained results show a nonuniform distribution of the surface potential with regions of sharp local variations (Figure 2). Such nonuniformity may be associated with the partial decomposition of zinc ferrite into zinc oxide (ZnO) and iron oxide (Fe_2O_3). The formation of heterojunctions at the interface between these oxides leads to local charge redistribution, which is explained by the difference in the work function values of the materials when the interface is formed between two nanostructures. As a result, local electric fields arise, which can significantly influence the catalytic and sensing properties of the system [6,7].

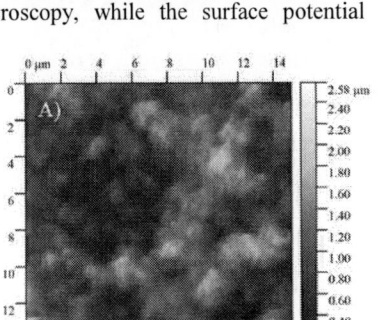
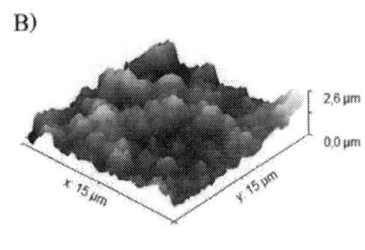

Fig. 1. AFM image of the sample morphology

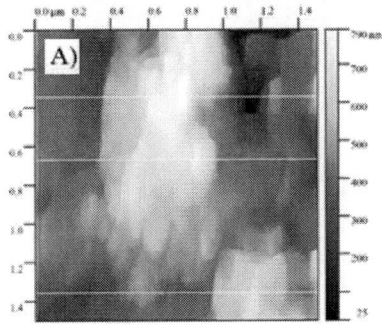
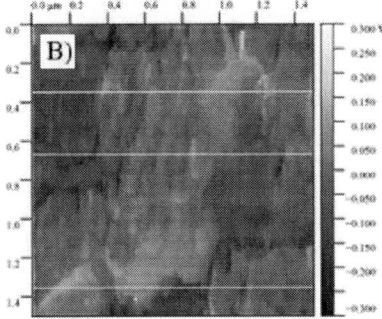

Fig. 2. Kelvin probe microscopy results for the ZnO–Fe_2O_3 sample

Using scanning electron microscopy, images of all the obtained structures were acquired. The Figure 3 shows that the diameter of the synthesized nanorods is approximately 50 nm, and it can also be seen that the structure represents an array of nanorods formed as a result of growth from numerous nanoparticle seeds [8]. In Figure 4 (A), an SEM image of the structure containing zinc ferrite can be observed; the image shows that the nanorods are coated with a thick layer of zinc ferrite. In contrast, in Figure 4 (B), the zinc stannate synthesized on the rods has almost no effect on their appearance and size. The influence of these factors on the gas-sensing behavior will be further examined.

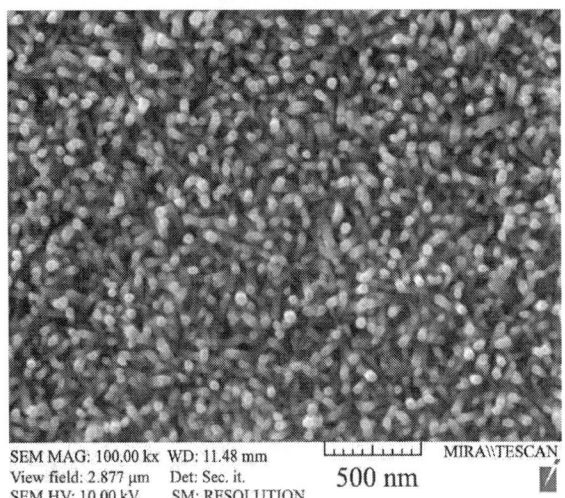

Fig. 3. SEM image of ZnO nanorods

Fig. 4. SEM image of Zn-Fe-O (A), Zn-Sn-O (B)

For our EDS measurements, an energy-dispersive detection system was used. The Figure 5 shows that the chemical elements are distributed fairly uniformly across the entire surface of the sample, which indicates the successful formation of the desired structure.

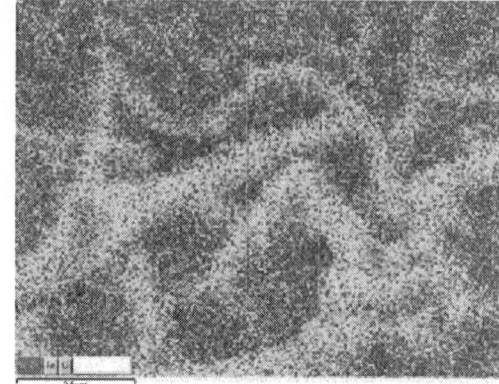

Fig. 5. EDS analysis of the chemical composition of $ZnFe_2O_4$

The Figure 6 also shows the obtained spectrum, which indicates that, in addition to the synthesized compound, silicon is present. This is due to the fact that the analyzed structure was deposited on a silicon substrate; however, the silicon signal is relatively low compared to the other elements, which suggests that the surface is densely covered by the formed structures. The spectrum also shows that iron is present in higher amounts than zinc, which is caused by the partial dissolution of the zinc oxide nanorods.

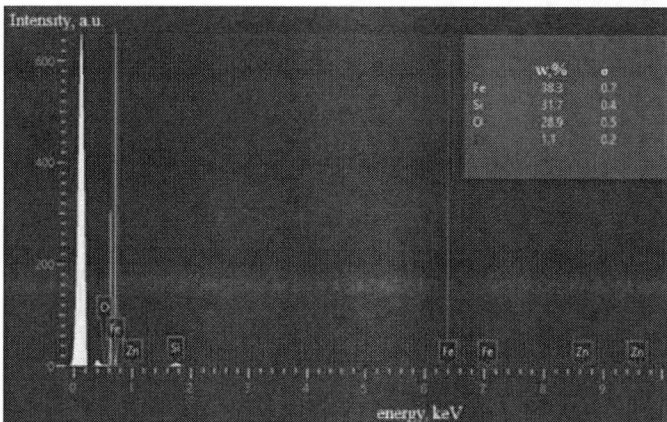

Fig. 6. Spectrum of the analyzed sample containing $ZnFe_2O_4$

During the examination of the samples containing Zn_2SnO_4, an elemental distribution map was also obtained, showing that the zinc oxide nanorods are covered with a layer containing tin and oxygen (Figure 7). The obtained spectrum indicates that the surface of the sample contains a layer predominantly composed of tin, while the zinc oxide nanorods are located beneath it.

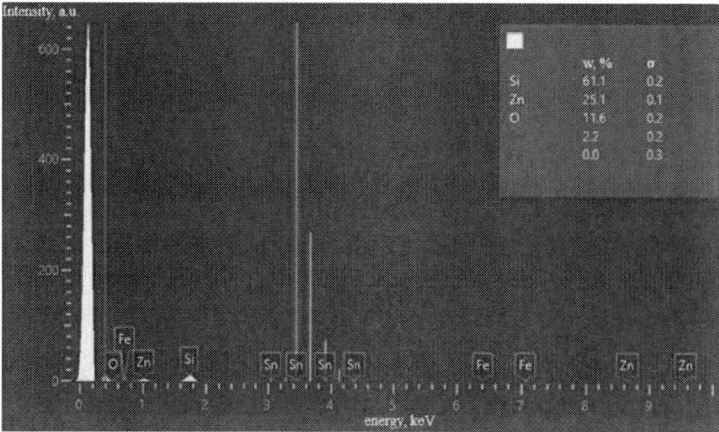

Fig. 7. Spectrum of the analyzed sample containing Zn_2SnO_4

The next method used in this work to investigate the obtained structures was XPS analysis. Initially, the XPS spectrum of the sample with the deposited zinc oxide nanorods was acquired (Figure 8), and the spectrum showed only the elements corresponding to the metal oxide.

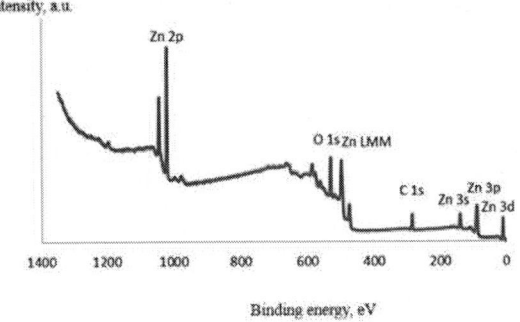

Fig. 8. XPS spectrum of the sample with deposited ZnO

The surfaces of the samples containing zinc stannate and zinc ferrite (Figures 9 and Figure 10) were also examined, and all spectra demonstrated precise correspondence to the expected compounds. Aside from the elements of the synthesized materials, only a carbon-related peak was observed.

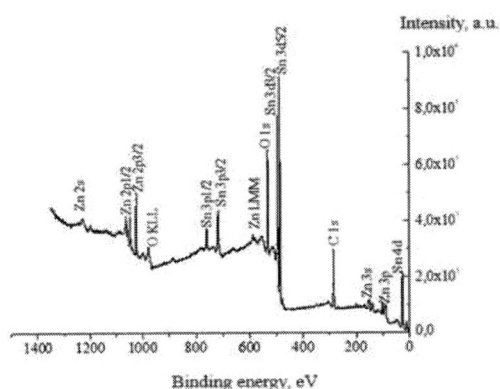

Fig. 9. XPS of the surface of the Zn_2SnO_4-containing sample

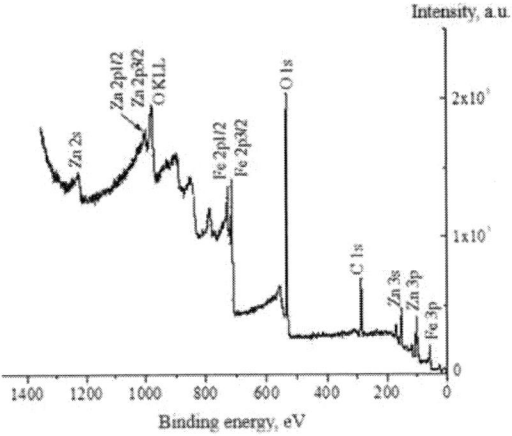

Fig. 10. XPS of the surface of the $ZnFe_2O_4$-containing sample

The gas-sensing measurements of the samples were performed using a specially designed experimental setup.

Rotameters were employed to precisely control both the gas flow rate and its concentration. A desiccant was used to eliminate excess moisture, which varies seasonally in the ambient air.

Experiments were conducted on pure zinc oxide as well as on other structures deposited onto zinc oxide nanorods. The gas-sensing performance of the samples is summarized in Table 1

TABLE I. RESULTS OF GAS-SENSING MEASUREMENTS

Sensing layer	VOC's	Response	t_{Res}, s	t_{Rec}, s
ZnO	*Isopropanol*	*1,7*	*90*	*470*
$ZnO/ZnFe_2O_4$	Ethanol	2,5	19	64
	Isopropanol	2,8	19	67
	Acetone	2,5	153	233
ZnO/Zn_2SnO_4	Methanol	3,3	74	368
	Isopropanol	7,3	58	463
	Acetone	3,6	117	421

IV. RESULTS AND DISSCUSION

Analysis of the results indicates that the sample containing zinc stannate exhibits pronounced selectivity toward isopropyl alcohol. While this sample shows slightly reduced response and recovery times compared to pure zinc oxide, the difference is relatively modest. Conversely, the sample containing zinc ferrite does not demonstrate selectivity toward any of the tested gases; however, it exhibits markedly faster response and recovery times than both pure zinc oxide and the zinc stannate sample. Notably, both modified samples show a higher response to isopropyl alcohol vapors than unmodified zinc oxide.

Surface characterization confirmed that the chemical composition of the samples aligns with the intended structures. Topographical analysis further revealed that zinc stannate forms a thin layer on the surface of zinc oxide nanorods, whereas zinc ferrite forms a comparatively thicker layer.

REFERENCES

[1] Tharsika T. et al. Gas sensing properties of zinc stannate (Zn2SnO4) nanowires prepared by carbon assisted thermal evaporation process //journal of Alloys and Compounds. – 2015. – T. 618. – C. 455-462.

[2] DHAHRI R. et al. A Review of Zinc Stannate (Zn2SnO4) Resistive Gas Sensors //ECS Journal of Solid State Science and Technology. – 2025.

[3] Wu K., Li J., Zhang C. Zinc ferrite based gas sensors: A review //Ceramics International. – 2019. – T. 45. – № 9. – C. 11143-11157.

[4] Nalimova S. S. et al. Investigation of hierarchical gas-sensing ZnFe2O4 nanostructures //Journal of Surface Investigation: X-ray, Synchrotron and Neutron Techniques. – 2023. – T. 17. – № Suppl 1. – C. S416-S422.

[5] Shomakhov Z. V. et al. Changes in the energy of surface adsorption sites of ZnO doped with Sn //Journal of Surface Investigation: X-ray, Synchrotron and Neutron Techniques. – 2023. – T. 17. – № 4. – C. 898-902.

[6] Govind A. et al. Highly sensitive near room temperature operable NO2 gas-sensor for enhanced selectivity via nanoporous CuO@ ZnO heterostructures //Journal of Environmental Chemical Engineering. – 2023. – T. 11. – № 4. – C. 110056.

[7] Cheng L. et al. In-situ growth of CdS QDs on ZnO porous microrods for highly sensitive detection of TEA at lower temperature //Vacuum. – 2023. – T. 212. – C. 112003.

[8] Bobkov A. A. et al. Fabrication of oxide heterostructures for promising solar cells of a new generation //Semiconductors. – 2017. – T. 51. – № 1. – C. 61-65.

Technology of Vacuum Deposition of Combined Thin-film Precision Strain Gauges with Temperature Self-compensation

Sergey A. Gurin
Department of Information and Measurement Technology and Metrology
Penza State University
Penza, Russia
teslananoel@rambler.ru

Anastasia E. Shepeleva
Department of Information and Measurement Technology and Metrology
Penza State University
Penza, Russia
anastasiya.shepeleva.01@mail.ru

Maksim D. Novichkov
Department of Information and Measurement Technology and Metrology
Penza State University
Penza, Russia
novichkov1998maks@gmail.com

Ekaterina A. Pecherskaya
Department of Information and Measurement Technology and Metrology
Penza State University
Penza, Russia
pea1@list.ru

Dmitry V. Agafonov
Department of instrument engineering
Penza State University
Penza, Russia
dmitryagafonov@list.ru

Vadim S. Volkov
Department of instrument engineering
Penza State University
Penza, Russia
distorsion@rambler.ru

Abstract — **The paper presents technological solutions for combined thin-film precision strain gauges with temperature self-compensation TSC to a value of no more than ±3 ppm based on new highly stable materials. Technological modes of deposition of a strain gauge structure based on the Si-C-Cr composition with a negative TSC have been obtained by magnetron sputtering of SiC and Cr targets and a temperature compensation structure based on thin-film nickel with a positive TSC. For the structures obtained a laser correction technique, calculating compensation for resistance changes due to temperature and achieving the required resistance, which allows minimizing the use of suspended compensation elements, has been developed. A prototype has been manufactured and studies, according to the results of which the temperature coefficient of resistance of the strain gauges is within ±3 ppm, and the output signal drift was about -3 ppm have been carried out.**

Keywords — *thin-film microelectronics technology, high-temperature pressure sensor, thin-film strain gauge sensor, temperature coefficient of resistance, resistance instability*

I. INTRODUCTION

Integral thin-film strain gauge pressure sensors (ITSS) are widely used in various industries, including space, mining, metallurgy, and energy, to record information about controlled and measured pressures [1, 2]. The quality, accuracy, and stability of ITSS largely determine the technical level of information-measuring and control systems (IMS) of expensive objects, and also ensure their safety [3, 4].

The operation of these sensors is based on changing the resistance of a strain gauge material obtained by deposition in a vacuum, under the influence of mechanical stresses [1 - 5]. They have high conversion linearity, low inertia, are distinguished by simple design solutions, and are reliable in operation [3 - 5]. However, at operating objects, the sensor, along with the impact of the main measured parameter (pressure), is simultaneously exposed to the influence of a large number of destabilizing factors (temperature, thermal shocks, pressure drops, vibration, etc.), which can distort the true information about the state of the object and take the value of the output signal beyond the limits specified in the technical conditions [6, 7]. The main reason for unreliable information is temperature, which leads to degradation processes in thin-film structures [8-10]. In addition, temperature errors make up 60% or more of the total additional error from all influencing conditions during operation [11-13]. For these reasons, as well as in connection with constant miniaturization, promising technologies are increasingly being considered and new materials and nanostructures are being studied [14, 15]. The aim of the article is to develop heat-resistant thin-film piezoresistive structures with self-compensation of TSC up to $\pm 3 \times 10^{-6}$ 1/°C and to create a piezoresistive sensor capable of operating without additional temperature compensation and balancing resistors on their basis.

II. THEORETICAL PART

The accuracy, time stability and reliability of the ITSS operation are determined primarily by the quality of the thin-film piezoresistive structure of the sensing element (SE) [16]. Studies of the sensors in operation show that in most cases the instability of the bridge circuit is caused by different intensities of change in the resistance of the compression and tension strain gauges. Because of this, the imbalance of the strain gauge circuit bridge $R_1R_4=R_2R_3$ occurs, which under normal climatic conditions is described by the formula [17]:

$$U_0 = U_s \frac{R_1 R_4 - R_2 R_3}{(R_1 + R_3)(R_2 + R_4)} \qquad (1)$$

where the U_s is supply voltage, $U_0 \approx 0$ V at $R_1 \approx R_2 \approx R_3 \approx R_4$.

The cause of the uneven change in resistance is diffusion processes in the thin-film piezoresistive structure of the SE, occurring under the influence of non-stationary heat flows, and thermomechanical stresses developing during film deposition and sensor operation [17]. As a result, gradual (parametric) failures are observed, reflecting only the properties inherent in the materials and the ITSS structures (solid-phase reactions, mutual diffusion, aging), without changing the appearance of the thin-film circuit, which cannot be excluded at the early stages of production. The achieved level of error in the time stability of the initial output signal (U_0) of the ITSS is from ±2% to ±25% during operation, while no more than ±1% is required [18, 19]. Such parameters are due to the fact that digital technologies and software are being widely introduced in all industries. When functionally integrating microprocessor technology with elements of modern IMS based on processor processing of signals taken from pressure sensors, the task of creating technologies for the formation of precision thin-film measuring circuits capable of ensuring long-term stability of output parameters, primarily U_0, arises ,since progressive errors associated with instability ("drift") of U_0 or its abrupt change, which are of a non-stationary random nature over time, cannot be eliminated by microprocessor processing [20, 21].

Additional complications of the ever-growing requirements for the stability of ITSS thin-film structures create requirements for miniaturization and weight and size indicators. The number of sensors in many complex multifunctional objects already amounts to tens of thousands, and this, for example, plays a key role in the case of the space industry.

The weight and size indicators can be estimated based on the design of the mechanical link of a membrane-type pressure transducer (Fig.1).

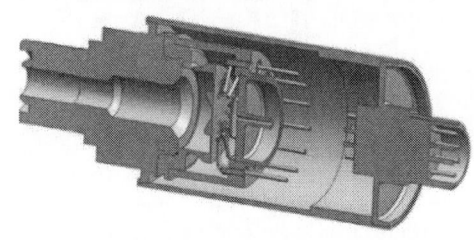

Fig. 1. Graphic representation of the structure in section, consisting of a nozzle 1, a membrane 2, a contact block 3, which also acts as a hermetic housing with a cover 4, an outer housing 5 and a connector 6

The main ITSS unit is the SE, which is a rigidly clamped membrane. A thin-film piezoresistive measuring circuit is formed on the membrane surface using vacuum deposition and photolithography methods. The membrane thickness ensures the SE design flexibility, the sensitivity required for the specified pressures and possible deformation that does not cause the layers destruction. The sensor housing is developed taking into account the possibility of placing all the necessary hanging electronic components.

Fig. 2 shows the electrical schematic diagram of a such ITSS.

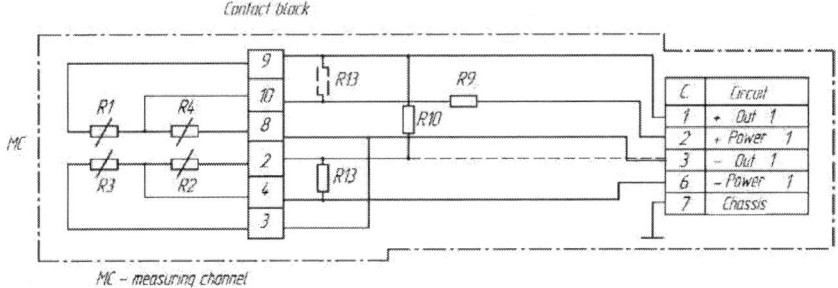

Fig. 2. Electrical schematic ITSS diagram

From Fig. 2 it is evident that the ITSS have a measurement channel (MC) that converts pressure into an electrical signal. Strain gauges R_1, R_2, R_3, R_4 form a bridge measuring circuit sensitive to the controlled pressure of the working medium. The hanging resistor R_{10} is used to adjust U_0, R_9 - to adjust the nominal output signal U_{out}, R_{13} - to adjust temperature compensation U_0. The dashed line shows a possible option for connecting resistor R_{13} to other arms of the bridge circuit. It is clear that the presence of hanging temperature compensation and balancing resistors significantly increases the weight and size characteristics of the ITSS. But since the developed SE have unacceptable TSC values (±50 ppm), it is impossible to do high-precision measurements without them.

The last thing that is important to note is the special sensitivity of thin films to electrical overvoltages. As a result, a significant role is played by failures caused by the strain gauge destruction when a high-density direct current passes through due to electromigration processes (Fig. 3) [16, 18, 19].

Fig. 3. Photograph of the burnout of a part of the resistive film, taken on a «Hirox 7700» optical microscope with a magnification of 150×

When the sensor operates under normal conditions, the current density through the strain gauge can be from $0,3 \cdot 10^5$ to $0,4 \cdot 10^6$ A/cm² [20, 21]. In polycrystalline films, at such a current density, electromigration over a long period of time can already lead to film damage, as can be seen from Fig. 3. This process accelerates under electrical overloads and elevated temperatures.

979-8-3315-4784-4/26 $31.00 © 2026 IEEE   101

As a result of the analysis of the main causes of the discrepancy between the ITSS parameters, it was shown that solutions related to changing the thin-film piezoresistive SE structure, the characteristics of which will determine the characteristics of the sensor as a whole, are relevant for improving thin-film pressure sensors and obtaining a new IMS class. This concerns the inclusion of temperature self-compensation of the TSC in the SE to a value of no more than ±2 ppm based on thin-film strain gauges with a negative TSC and thin-film temperature-compensating resistors with a positive TSC, excluding suspended compensation resistors. Development of methods for laser fitting taking into account the calculation of self-compensation of resistance changes from temperature and achieving the required resistance. Implementation and development of technologies for the synthesis of thin-film strain gauges based on new thermally stable materials that provide the necessary time stability and resistance to high-density currents.

III. MATERIALS, DESIGN AND TECHNOLOGIES

To achieve the set goals for the development of new thermally stable strain gauges with a negative TSC value, the Si-C-Cr composition, obtained in [17 - 19] by joint sputtering of SiC and Cr targets using the magnetron method was chosen. The advantages of SiC-based materials have been well studied [9, 10, 22, 23] and are widely used in products for extreme electronics [24, 25]. Therefore, the task is to develop a technology for synthesizing strain gauges with a consistently obtained negative TSC value (in the range from minus $80 \cdot 10^{-6}$ 1/°C to minus $5 \cdot 10^{-6}$ 1/°C before preliminary temperature and laser correction of resistance), based on a mechanical mixture of SiC and Cr target materials, sputtered in a single vacuum cycle, which can be attributed to the group of cermets, For self-compensation of the temperature dependence of the strain gauge film, it was decided to use a series-connected resistive layer of nickel Ni sputtered by magnetron sputtering with a titanium sublayer Ti, which has a positive TSC significantly greater in modulus (from $4000 \cdot 10^{-6}$ to $5000 \cdot 10^{-6}$ 1/°C). Film nickel with a titanium sublayer is known as a contact layer for sensors used at elevated temperatures.

The technology is implemented on an automated vacuum installation SAVTECH U-PVD-211, equipped with one high-frequency magnetron (RF) and three direct current (DC) magnetrons. The process modes with the specified parameters are listed in Table 1.

TABLE I. Technological modes for obtaining thin-film ITSS structures

Layer	Gas flow rate	Parameters of target sputtering	Substrate temperature, C
Composition «Si–C–Cr»	2.5	Cr: 470 V; 3 A; 1,4 kW SiC: 720 W	380
NiTi	2.2	Ti: 2,2 kW; 4 A; Ni: 2,2 kW; 4 A	320

Fig. 4 shows the topology of a thin-film measuring strain gauge circuit consisting of self-compensating strain gauges and designed for 10 kOhm, where the resistance of the strain gauge based on the Si–C–Cr composition R_{scr} is many times greater than the resistance of the nickel resistor R_{Ni}.

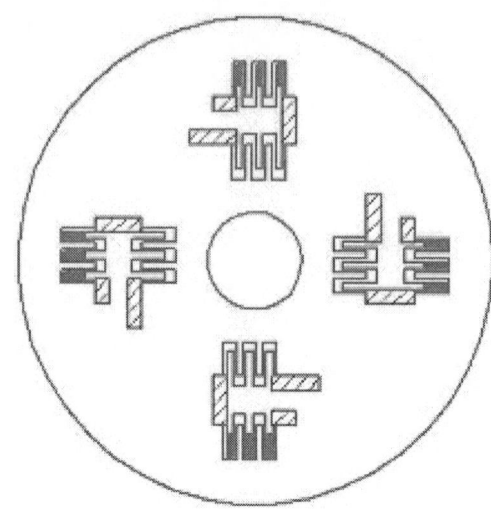

Fig. 4. Topology of the SE ITSS based on combined strain gauges

The strain gauges R_{scr} are located in the zone of greatest deformation and are sensitive to the applied pressure. Therefore, its relative change in resistance from the applied pressure should be as large as possible. The strain gauge R_{scr} and the temperature-compensating resistor R_{Ni} have a shape containing adjustable sections for "rough" and fine adjustment of the parameters, both for changing the nominal resistance and the temperature coefficient of resistance. The temperature-compensating resistors R_{Ni} are located on the periphery of the SE.

The technology presented in Table 1 should provide the following parameters of the deposited thin-film structures after photolithography and preliminary heat treatment:

1. The resistance R_{scr} should be within 70 to 90% of the required total resistance of the proposed combined resistive structure;

2. The resistance R_{Ni} should be within 0.4 to 0.6% of the required total resistance of the proposed combined resistive structure;

3. The TSC R_{scr} should be within $\alpha = -(80 \div 5)$ ppm, determined by the technology;

4. The TSC R_{Ni} should be within $\alpha = (4000 \div 5000)$ ppm, determined by the technology.

The functional laser correction technique is carried out using the calculated values of compensation for resistance changes both from temperature and achieving the required resistance. First, the resistances of the strain gauges R_{scr} and temperature compensation resistors R_{Ni} are measured at temperatures of 0 °C and 100 °C, and the TSC values are calculated using the obtained values using the formula

$$\alpha_R = \frac{R_t - R_0}{R_0 \Delta T},$$ (2)

where α_R is the TCR; R_t is the Resistance of R_{scr} and R_{Ni} at a temperature of 100 °C; R_0 is the Resistance of R_{scr} and R_{Ni} at a temperature of 0 °C.

Next, the resistance value of each temperature compensation resistor R_{Ni} is determined to perform functional laser correction in order to eliminate the effect of temperature (TSC correction) based on the condition:

$$-\Delta R_{scr} = \Delta R_{Ni},$$ (3)

where $\Delta R_{scr} = R_{scrt} - R_{scr0}$ is the value of the change in resistance of the strain gauge R_{scr} when heating it from 0 °C to 100 °C;

$\Delta R_{Ni} = R_{Nit} - R_{Ni0}$ is the value of the change in resistance of the strain gauge R_{Ni} when heating it from 0 °C to 100 °C.

The formula shows that to compensate for the TSC, the values of ΔR_{scr} and ΔR_{Ni} must be equal in magnitude and have opposite signs of the TSC. The required resistance value R'_{Nit} itself is determined from the expression:

$$R_{Ni} = R_{scr0}\frac{\alpha R_{scr}}{\alpha R_{Ni}},\qquad (4)$$

According to the resistance values of the temperature-compensating resistors R_{Ni} obtained by formula (4), their functional laser correction is carried out, thereby eliminating the effect of temperature on the strain gauges R_{scr}.

At the next stage, the resistance values of the series-connected strain gauge R_{scr} and the temperature-compensating resistor R_{Ni} are calculated with subsequent functional laser correction to achieve the required value of their total resistance, so that the TSC compensation is maintained. To do this, the following condition must be met:

$$\frac{\Delta R_{scr}}{R_{scr}} = \frac{\Delta R_{Ni}}{R_{Ni}}\qquad (5)$$

where ΔR_{scr} and ΔR_{Ni} are the change in resistance during laser correction of R_{scr} and R_{Ni}, respectively.

As a result, the following system of equations is solved to find the required values:

$$\begin{cases}\dfrac{\Delta R_{scr}}{R_{scr}} = \dfrac{\Delta R_{Ni}}{R_{Ni}},\\ R = R_{scr}' + R_{Ni}'\end{cases}\qquad (6)$$

where R'_{scr}, R'_{Ni} are the resistances of R_{scr} and R_{Ni} after laser correction, respectively.

After performing algebraic transformations, we obtained:

$$R'_{scr} = \frac{R_{scr}}{R_{Ni}}\frac{R}{\left(1 + \dfrac{R_{scr}}{R_{Ni}}\right)},\qquad (7)$$

$$R'_{Ni} = R - R'_{scr},\qquad (8)$$

where R'_{scr}, R'_{Ni} are the resistances of R_{scr} and R_{Ni} after laser correction, respectively.

After calculating the resistances R_{scr} and R_{Ni}, the second functional laser correction is performed.

IV. EXPERIMENTAL PROCEDURE

The testing of the obtained results on the production of heat-resistant strain gauges with temperature self-compensation based on the developed spraying technology and the calculation method for the parameters for functional laser correction was carried out on the SE ITSS, designed for a pressure of P = 120 bar.

The SE appearance is shown in Figure 5.

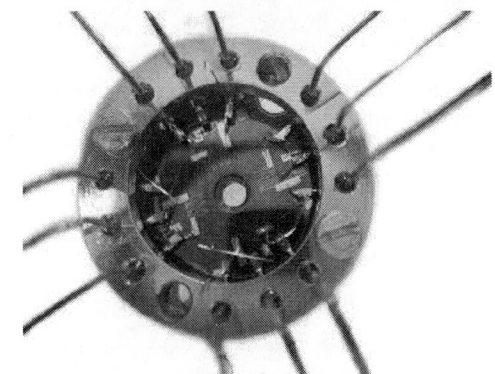

Fig. 5. SE ITSS based on combined strain gauges

The layers of the strain gauge measuring circuit of the SE ITSS were obtained on a SAVTECH U-PVD-211 vacuum deposition installation using the film deposition process modes corresponding to those showed in Table 1.

The bridge measuring circuit was formed by direct photolithography with selective liquid and plasma etching. The contact pads were obtained by selective etching of Au in a solution of distilled water, potassium iodide, iodine and V in hydrogen peroxide. The Rscr strain gauges were obtained by ion-beam etching in an inert gas environment on a Caroline E12 installation, and the RNi temperature-compensating resistors were obtained in an etching solution for nickel based on acetic, orthophosphoric and nitric acids. In the photolithography process, an EM-5026AM installation was used to align and expose the circuit pattern. After photolithography, the SE was heat-treated in a vacuum at a temperature of 300±10 °C for 4 hours.

After preliminary heat treatment, the obtained resistance values Rscr and RNi were measured using the ammeter-voltmeter "V7-16A" ATD2.710.000 TU by placing the measuring probes of the device on the contact pads of the SE at a temperature of 25 °C and 125 °C. Based on the obtained data, the correction resistances R'scr, R'Ni were calculated using the developed technique for functional laser adjustment of resistances.

Figure 6 shows the graphs of the dependence of the resistance values of strain gauges $R_1..R_4$ on temperature.

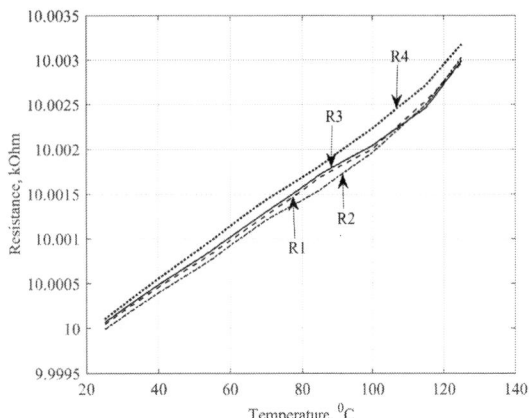

Fig. 6. Temperature dependence of strain gauge resistance

The analysis of tables 2 and 3 shows that the TCR of the strain gauges after adjustment decreased by an order of magnitude from values of about 30 ppm to values of about 3 ppm.

It is evident from the figure that the resistance changes of strain gauges R1. R4 from temperature have a fairly linear characteristic with the declared value of TSC.

The calculation of the output signal and the bridge circuit according to formula (1) is shown below, while

$$R_{1,4} = R_0\big(1 + \alpha(T_{max} - T_{min})\big)(1 + k\varepsilon), \qquad (9)$$

$$R_{2,3} = R_0\big(1 + \alpha(T_{max} - T_{min})\big)(1 - k\varepsilon), \qquad (10)$$

In formulas (9) and (10), R_0 is the resistance of the corresponding strain gauges after correction, α is the TSC of the strain gauges. The value of k is the strain sensitivity coefficient, the minimum value of which, according to reference data, is 5. The value of ε is the strain of the strain gauges under the action of the measured pressure, the value of which, for a linear elastic characteristic of the membrane, we will take to be equal to 0.005. The calculation will be performed for a supply voltage of 6 V.

At a supply voltage of 6 V, the nominal output signal will be equal to 15 mV. The output signal of the strain gauge circuit, calculated on the basis of the parameters obtained after correction (Table 3), is shown in Figure 7.

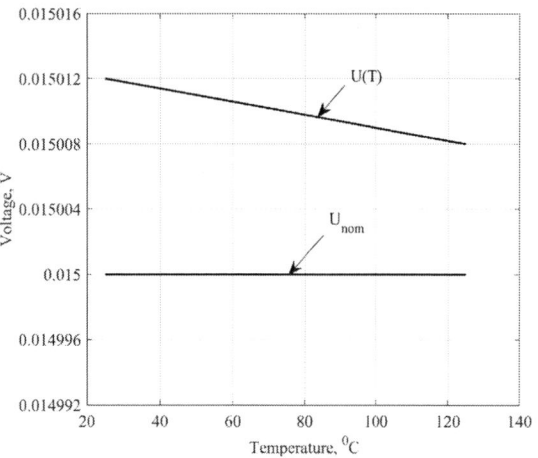

Fig. 7. Output signal of the strain gauge circuit at a supply voltage of 6 V

V. CONCLUSIONS

A pressure sensor based on a sensitive element made of a rigidly clamped membrane, on the surface of which a thin-film structure is formed, including strain gauge thin films of the "Si-C-Cr" composition with a negative TSC value and an additional functional layer for compensating temperature effects made of nickel with a positive TSC sign, is manufactured.

Thin films were formed by magnetron sputtering in a single vacuum cycle, while the "Si-C-Cr" composition was synthesized by joint sputtering of SiC and Cr targets at a power of 720 W and 1400 W, respectively. Ni sputtering was performed at a power of 2200 W with preliminary sputtering of the Ti sublayer at the same power.

The presented modes make it possible to achieve the TSC of the strain gauge structure in the range from -80 to -5 ppm and from +4000 to +5000 ppm for the compensation film. For the obtained parameters of thin-film structures, a laser correction technique was developed with the calculation of

compensation for changes in resistance values from temperature and subsequent achievement of the required resistance value.

The presented technologies, materials and techniques for creating SE allow creating ITSS with the declared characteristics of stability, accuracy and TSC value. The achieved TSC results of strain gauges were ±3 ppm, the drift coefficient of the nominal output signal β was -3 ppm.

ACKNOWLEDGMENT

The work was supported by the grant of the Ministry of Science and Higher Education of the Russian Federation «Synthesis and research of promising nanomaterials, coatings and electronics devices» (# 124041700069-0).

REFERENCES

[1] García-Alonso, E Castaño, I Obieta, J Garcia, FJ Gracia, Thin film technology applied to the development of a multilayer pressure sensor device, Vacuum, Volume 45, Issues 10–11, 1994, 1103-1105, Doi: 10.1016/0042-207X(94)90035-3.

[2] Yang JieHua, Di Ping, Li Qiao, Wang Xi, Ding Hao, Bai YunFeng, Zhu ShiGen. (2021). Characterization of a Commercial Thin Film-based Pressure Sensor. Journal of Physics: Conference Series. 1790. 012041. Doi:10.1088/1742-6596/1790/1/012041.

[3] Yoshiharu Kakehi, Yoshiharu Yamada, Yusuke Kondo, Taizo Oguri, Kazuo Satoh, Electrical and piezoresistive properties of titanium oxycarbide thin films for high-temperature pressure sensors, Vacuum, Volume 193, 2021, 110550, Doi:10.1016/j.vacuum.2021.110550.

[4] Pecherskaya E., Gurin S., Novichkov M. (2022). Combined Thin-Film Resistive and Strain-Resistant Structures with Temperature Self-Compensation. Journal of Surface Investigation: X-ray, Synchrotron and Neutron Techniques. 16. 1074-1080. Doi: 10.1134/S1027451022060209.

[5] Zhao Yang, Liu Jin, Ying Yuhuang, Chen Hongyu, Wang Wenxuan, Zhang Sijie, Hai Zhenyin, Sun Daoheng. (2024). Temperature Self-Compensation Thin Film Strain Gauges Based on nano-SiO2/AgNPs Composite. Journal of Materials Chemistry C. 12. Doi: 10.1039/D4TC01645A.

[6] Shchegolkov Aleksei, Shchegolkov Alexandr, Kaminskii Vladimir, Chumak Maxim. (2025). Polymer Composites with Nanomaterials for Strain Gauging: A Review. Journal of Composites Science. 9. 8. 10.3390/jcs9010008.

[7] Afanasev A.V., Ilyin V.A., Luchinin,V.V. Ion Doping of Silicon Carbide in the Technology of High-Power Electronic Devices (Review). Semiconductors 56, 472–486 (2022). Doi: 10.1134/S1063782622130024.

[8] Marsi Noraini, Majlis Burhanuddin, Hamzah Azlan, Mohd-Yasin Faisal. (2014). The Mechanical and Electrical Effects of MEMS Capacitive Pressure Sensor Based 3C-SiC for Extreme Temperature. Journal of Engineering. 2014. 1-8. Doi:10.1155/2014/715167.

[9] Keshyagol Kiran, Hiremath Shivashankar, H M Vishwanatha, Kini U., Naik Nithesh, Hiremath Pavan. (2024). Optimizing Capacitive Pressure Sensor Geometry: A Design of Experiments Approach with a Computer-Generated Model. Sensors (Basel, Switzerland). 24. Doi:10.3390/s24113504.

[10] Alekseyev Nikolay, Khmelnitskiy Ivan, Aivazyan Vagarshak, Broyko Anton, Korlyakov Andrey, Luchinin Victor. (2021). Ionic EAP Actuators with Electrodes Based on Carbon Nanomaterials. Polymers. Doi: 13. 4137. 10.3390/polym13234137.

[11] Luchinin, V.V., Panov, M.F., Pavlova, M.V. et al. Optical Control of the Parameters of Substrates in Silicon-Carbide Epitaxial Structures. Semiconductors 56, 455–461 (2022). Doi:10.1134/S1063782622130073.

[12] Bogdanov V., Volobuyev V., Gorbushin Anton. (2009). Research on thermal dynamics of strain-gauge balance and development of methods for its temperature error reduction. TsAGI Science Journal. 40. 619-629. Doi: 10.1615/TsAGISciJ.v40.i5.80.

[13] Zhang Guangan, Lu Zhibin, Jibin Pu, Wu Guizhi, Wang Kaiyuan. (2013). Structure and Thermal Stability of Copper Nitride Thin Films. Indian Journal of Materials Science. 2013. 1-6. Doi: 10.1155/2013/725975.

[14] Zhao Hong, Zheng Zhong, Sun Lixian, Liu Hongwei, Tsoutas Kostadinos, Akhavan Behnam, Liu Yanping(Martin), Bilek Marcela,

Liu Zongwen. (2023). Introducing a new heterogeneous nanocomposite thin film with superior mechanical properties and thermal stability. Materials & Design. 234. 112333. Doi: 10.1016/j.matdes.2023.112333.

[15] Gurin Sergey, Pecherskaya Ekaterina, Novichkov Maksim, Safronova Olga. (2022). Multilayer thin-film resistive structures with temperature self-compensation for super-precision resistors and strain gauges. Journal of Physics: Conference Series. 2373. 032028. Doi:10.1088/1742-6596/2373/3/032028.

[16] Wrbanek, John & Fralick, Gustave & Gonzalez III, Jose. (2006). Developing Multilayer Thin Film Strain Sensors With High Thermal Stability. Collection of Technical Papers - AIAA/ASME/SAE/ASEE 42nd Joint Propulsion Conference. 4. Doi: 10.2514/6.2006-4580.

[17] Volokhov I.V., Gurin S.A., Vergazov I.R. Study of the Properties of High-Sensitivity Thermally-Stable Thin-Film Resistance Strain Gauges for Integral Pressure Sensors. Meas Tech 59, 80–86 (2016). Doi: 10.1007/s11018-016-0921-5.

[18] Qian Long, Linqing Wang, Weijie Yu, Weijiu Huang, Li Wang, Structural and mechanical properties of amorphous Si–C-based thin films deposited by pulsed magnetron sputtering under different sputtering powers, Vacuum,Volume 191, 2021,110319, Doi: 10.1016/j.vacuum.2021.110319.

[19] Wang Xiaoyan, Lim Eng, Hoettges Kai, Song Pengfei. (2023). A Review of Carbon Nanotubes, Graphene and Nanodiamond Based Strain Sensor in Harsh Environments. C. 9. 108. Doi: 10.3390/c9040108.

[20] Zhang Congchun, Kang Zhipeng, Zhao Nan, Lei Peng, Yan Bo. (2023). A Bilayer Thin-Film Strain Gauge With Temperature Self-Compensation. IEEE Sensors Journal. PP. 1-1. Doi: 10.1109/JSEN.2023.3238328.

[21] Pang Yawen, Zhang Congchun, Lei Peng, Wang Yusen, Lv Zhenjie. (2021). Effects of Thermal Annealing on the Electrical Properties and Stability of Pt Thin Film Resistors with Ti and PtxOy Interlayers. Journal of Physics: Conference Series. 2002. 012002. Doi: 10.1088/1742-6596/2002/1/012002.

[22] Novichkov M.D., Gurin S.A., Pecherskaya E.A., Shepeleva J.V., Grishchenko V.I. Self-temperature compensated thin-film resistor. AIP Conference Proceedings, 2024, 3102(1), 020013, Doi: 10.1063/5.0199900.

[23] Jiang Zhuangde, Zhao Yulong, Zhao Libo, Tingzhong Xu. (2018). High Temperature Silicon Pressure Sensors. Doi: 10.1007/978-981-10-5945-2_16.

[24] Chabi S, Kadel K. Two-Dimensional Silicon Carbide: Emerging Direct Band Gap Semiconductor. *Nanomaterials*. 2020; 10(11):2226. Doi:10.3390/nano10112226.

[25] Cui Hongzhi, Bai Lu, Cao Xiangpeng, Shiheng YU. (2024). Experimental study of the in-situ fabrication of strain sensor with conductive adhesive to monitor the concrete structure. Construction and Building Materials. 452. 139000. Doi: 10.1016/j.conbuildmat.2024.139000.

Remote Temperature Monitoring Method Based on Tapered Fiber Sensors

Vyacheslav G. Nesterov
The Bonch-Bruevich Saint-Peterburg
State University of Telecommunications
Saint Petersburg, Russia
nesterov.vyacheslav7@gmail.com

Sergei A. Shagako
The Bonch-Bruevich Saint-Peterburg
State University of Telecommunications
Saint Petersburg, Russia
s.shagako15@yandex.ru

Diana D. Shagako
The Bonch-Bruevich Saint-Peterburg
State University of Telecommunications
Saint Petersburg, Russia
diana.diana.tsyganova@mail.ru

Abstract — The article examines various temperature-monitoring systems based on optical fibers used under different conditions, enabling continuous temperature control in diverse media or objects. The operating principle of such systems relies on the change in the optical properties of the fiber material as temperature varies. Limitations that arise during long-term operation of existing sensors are highlighted. It is shown that when measuring temperature varying over wide ranges and across many zones, it becomes necessary to simplify the sensor design while maintaining accuracy. The use of optical fibers with a tapered core profile in fiber-optic sensor designs is proposed. A dependence is established between the variation of optical power losses of the detected signal and the temperature T to which the fiber is exposed. The obtained results make it possible to implement remote temperature monitoring using tapered fibers at distances exceeding 120 km from the laser source, photodetector module, and processing device. This significantly expands the applicability of remote temperature monitoring via optical fibers in open-air environments, for example, along power transmission lines.

Keywords — fiber-optic sensor, temperature, monitoring, tapered fiber, laser radiation, Rayleigh scattering coefficient, optical power.

I. INTRODUCTION

Temperature control is one of the most important requirements in industrial production, environmental monitoring, and many other areas [1-5]. The development of new high-precision industries characterized by automated technological processes has led to a demand for new types of sensors featuring high reliability, stability, compactness, and low weight [5-9]. These sensors must consume minimal energy, be compatible with electronic processing systems, have low manufacturing cost, and be easy to fabricate. Fiber-optic sensors represent one of the most promising directions in monitoring systems [8, 10-14]. They may be classified according to the optical wave parameter used for measurement: optical power, phase, polarization state, spectral characteristics, or mode composition [8, 10-13, 15-18].

In this work, the measurement of temperature T is based on detecting changes in the optical power of the transmitted signal. Physical quantities such as temperature, pressure, or tensile force can affect a local region of the fiber, altering its properties. This leads to attenuation of optical power, which is detected by a photodetector. Prior calibration of the relationship between optical-power change and the physical quantity (e.g., temperature T) enables remote measurements—an essential capability for many applications [7-9, 14, 15, 19-21].

Temperature sensors are used nearly everywhere. Any domain where the temperature of an object or environment affects production quality or safety requires continuous monitoring. A key development trend in fiber-optic temperature sensors is the increase of sensitivity, directly influencing measurement accuracy [1-4, 9, 14, 19-23].

This is particularly relevant for sensors in which temperature is measured through optical-power attenuation. These sensors are simple to operate, highly reliable, and allow measurements at large distances from the main equipment, unlike other types of fiber sensors [8, 10-13, 15-18, 21].

To ensure high sensitivity and compactness, this study considers a temperature sensor based on tapered fiber with short waist sections. Implementing such structures requires accounting for several factors to ensure stable operation. Addressing these aspects constitutes the primary objective of this work.

II. FEATURES OF USING TAPERED FIBER IN TEMPERATURE SENSORS

The primary effect of temperature T on the optical signal occurs in the tapered region of the fiber. To measure temperature at a specific point or surface, the tapered waist must be small; otherwise, the sensor will record temperature gradients rather than the local temperature. On the other hand, the tapered region must not be too short, as this may cause excessive optical loss or even mechanical failure. Therefore, reproducible fabrication of tapered fibers with optimized geometric parameters is necessary for each sensing application.

A separate challenge when measuring temperature T over long distances is ensuring compatibility between standard transmission fiber (e.g., 30–50 km spans) and the tapered-fiber sensor. Fibers must have identical refractive-index profiles and originate from the same manufacturer; otherwise, significant coupling losses or complete disappearance of the optical signal may occur [3, 24-26].

III. FABRICATION OF TAPERED OPTICAL FIBER

Tapered-fiber fabrication requires heating a local section of fiber to reduce material viscosity, enabling stretching. Both direct and indirect heating methods exist, including flame heating, focused-laser heating, electric-arc heating, and microwave-oven heating. Each method has advantages and limitations. The heating temperature must remain below 1500 °C to avoid quartz crystallization into the β-phase (occurring around 1550 °C), which leads to high optical losses.

During fabrication, the fiber is mounted in a translation system providing controlled linear motion. The stretching speed is programmed and typically below several millimeters per second. To fabricate tapered fibers with the smallest possible waist diameter, it is necessary to heat the minimum possible cross-section of the fiber within a short period of time. Achieving minimal waist diameters requires highly localized heating using either a focused laser beam or an electric arc. These methods allow heating fiber sections smaller than 1 mm in length. When using other techniques, heating typically begins over regions of 2–3 mm or more. Heating with a focused laser beam is a rather complex and specialized process. Moreover, tapering quartz optical fiber using a laser requires an expensive high-power infrared laser and dedicated infrared optics. In contrast, electric-arc heating is simple and cost-effective.

The method of fabricating tapered fibers using electric-arc heating has been known for a long time; however, this technology has not yet been studied in sufficient detail. Standard commercially available devices that employ electric arcs for fiber heating are designed solely for splicing quartz optical fibers. In such devices, the ends of two fibers are positioned close to each other, heated, and rapidly fused. This process is automatic, although its parameters can be optimized. In contrast, the fiber for tapering must be stretched rather than compressed. These factors necessitated the development of dedicated equipment required for producing short tapered sections of quartz optical fibers. Several process parameters must be varied to optimize the fabrication conditions of short tapered fibers, including arc power, its supply time, arc position relative to the fiber, and the heating duration of the quartz fiber.

Each section of the fiber tapers during every oscillation cycle. As the flame begins to heat the fiber, it is gently stretched in opposite directions, as illustrated in Fig. 1.

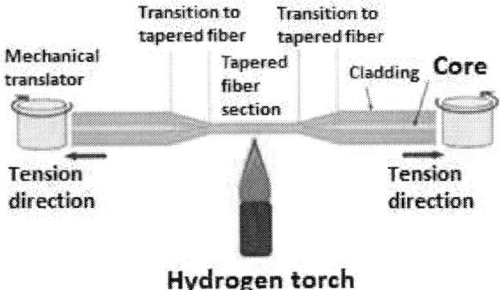

Fig. 1. Schematic illustration of the flame-based approach used for the fabrication of tapered optical fibers.

In our experiments, the flame oscillation speed (v_0) was set to approximately 6 mm/s, and the oscillation amplitude L_0 was 10 mm. The translation speed of each sliding stage (v) was approximately 2.4 mm/min. The total stretch distance z is the sum of the distances traveled by both stages; in this experiment, z = 10 mm. The core diameter was 4.5 μm. All the above parameters were precisely controlled by a computer program. The experimental fiber used in this work was custom-fabricated by NPO "Volokno" to address several specific tasks. It was manufactured on the basis of a standard telecom single-mode fiber compliant with ITU-T G.652, featuring a numerical aperture of 0.14 and a cutoff wavelength of 1260 nm (core diameter 9 μm, cladding diameter 125 μm). Commercially available single-mode fibers with a pure-quartz core — such as Corning SMF-28e (manufactured in India) — may also be used. Similar single-mode fibers can additionally be obtained on special order from the Central Research Institute "Electropribor".

IV. EXPERIMENTAL SETUP AND RESULTS OF RESEARCH OF THERMAL EFFECT ON OPTICAL FIBER

In single-component glasses, optical losses due to Rayleigh scattering from density fluctuations are described by:

$$\alpha_{ray,\rho} = \frac{8\pi^3}{3\lambda^4}(n^2 - 1)\beta k T_f, \quad (1)$$

where λ is wavelength, n is refractive index, k is Boltzmann constant, β is isothermal compressibility at the fictive temperature T_f, which depends on glass composition and thermal history.

The fictive temperature may differ significantly from the glass-transition temperature and depends on both the glass composition and its thermal history. The spectral dependence of optical losses due to Rayleigh scattering remains unchanged

$$\alpha_{ray} = \frac{A_{ray}}{\lambda^4}, \quad (2)$$

where A_{ray} is the Rayleigh scattering coefficient (RSC).

Temperature variation T is monitored by measuring changes in the detected optical-signal power. Figure 2 shows the experimental setup used to verify the performance of the tapered-fiber-based temperature sensor. This setup has been successfully used to study the influence of temperature T on the variation of optical losses α(λ) in single-mode fibers [26].

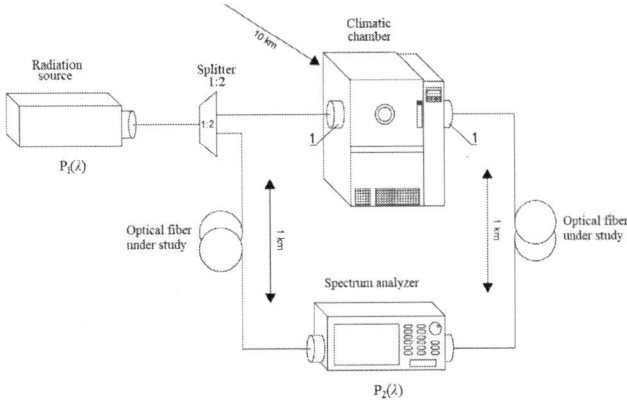

Fig. 2. Schematic diagram for measuring the dependence of optical losses at the wavelength of 1.55 μm on the temperature T: 1 - thermal outputs.

The measurement of optical losses $\alpha(\lambda)$ is carried out as follows. A temperature sensor based on a 10-km tapered fiber is placed inside a climate chamber in which the temperature T is controlled. The optical power level $P_2(\lambda)$ of the signal transmitted from the laser source to the receiver is then measured through a 10-km fiber coil located in the chamber, as well as through an additional 1-km fiber segment located outside the chamber. An optical signal from the splitter, serving as the reference channel, propagates over an equivalent distance, providing the reference power $P_1(\lambda)$. The number of connector interfaces in both channels is identical. The total loss of the two splice joints is 0.25 dB, which is also accounted for when determining $\alpha(\lambda)$, calculated in this experiment according to formula (3):

$$\alpha(\lambda) = -10\frac{\log(P_2(\lambda)\,/\,P_1(\lambda))}{L}, \qquad (3)$$

where L is the length of the optical fiber over which the losses are evaluated (in this setup, L=10 km).

Figure 3 presents the results of the study on how variations in temperature T affect the attenuation of the optical signal, expressed through changes in the transmitted power P2(λ), at a wavelength of λ=1550 nm in the tapered fiber with a quartz core. The output power of the laser source is 5 mW.

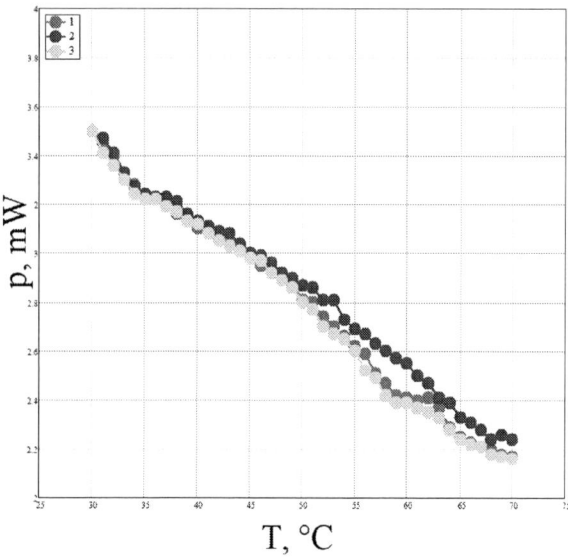

Fig. 3. Dependence of the laser-radiation power variation at a wavelength of 1.55 μm transmitted through the tapered optical fiber on temperature T; 1 – core diameter 4.5 μm; 2 – core diameter 4.9 μm; 3 – core diameter 4.2 μm.

Analysis of the obtained data demonstrated good resolution in detecting temperature variations T within less than one degree, which is sufficient for a number of technological processes or monitoring tasks performed under challenging conditions. Given the simplicity and reliability of the sensor design, a system comprising multiple optical sensors can be implemented using a single laser source, enabling more accurate temperature assessment at a specified location—for example, within an airflow after cooling, and in other similar applications.

The established dependence of increasing optical losses $\alpha(\lambda)$ with rising temperature T offers a new approach to the fabrication of tapered optical fibers for temperature fiber-optic sensors. This enables achieving a steeper $P_2(T)$

response, allowing such sensors to be used effectively under various operating conditions.

Tapered optical fibers with a reduced core diameter demonstrate significantly increased sensitivity when used as elements of fiber-optic sensors. This is due to the fact that the main influence of temperature exposure is localized in the taper region, and the reduced core enhances the contribution of changes in optical losses resulting from mode transformations and enhanced Rayleigh scattering. This facilitates the detection of even minor temperature fluctuations that cause a noticeable change in the power of the transmitted signal, thereby achieving high measurement accuracy.

However, such sensitivity is achieved at the expense of significant manufacturing complexity: to obtain a stable and reproducible taper, it is necessary to maintain strictly controlled local heating and fiber stretching while observing temperature constraints to avoid quartz crystallization and the growth of internal losses. In addition, at small core diameters, there is an increased risk of mechanical destruction and unwanted losses during subsequent power input into the standard fiber, which requires high precision of technological processes and matching of fiber profiles.

V. CONCLUSION

It has been established that the reason for the high level of optical losses in high-alloyed optical fibers at decreasing temperature is additional losses localized in the region of the core-shell boundary and in the region of the central failure of the RIP. The results confirm the effectiveness of the tapered-fiber fabrication technology developed at NPO "Volokno." The sensors allow detecting temperature fluctuations below 1 K in remote-monitoring applications while remaining simple, low-cost, and easy to replace.

Future work will explore how taper-geometry variations affect sensor sensitivity and determine optimal taper lengths L. Placing the tapered section in coils or on dedicated spools will increase sensor distance from the main equipment and allow more sensors to operate from a single source.

REFERENCES

[1] N. Popovskiy, "Proceedings of the 2024 Conference of Young Researchers in Electrical and Electronic Engineering, ElCon 2024," pp. 741–744, 2024.

[2] D. Isaenko, B. Reznikov, S. Rodin, N. Popovskiy, D. Vakorina, "Proceedings of the 2022 International Conference on Electrical Engineering and Photonics, EExPolytech 2022," pp. 315–319, 2022.

[3] V. Davydov, B. Reznikov, V. Dudkin, "Energies," vol. 16, no. 3, pp. 1040, 2023.

[4] E. M. Dianov, V. M. Mashinsky, V. B. Neustruev, O. D. Sazhin, A. N. Guryanov, V. F. Khopin, N. N. Vechkanov, "Optical Fiber Technol," vol. 3, pp. 77-86, 1997.

[5] A. V. Moroz, V. V. Davydov, "Journal of Physics: Conference Series," vol. 1410, no. 1, pp. 012212, 2019.

[6] E. V. Isupova, A. P. Valov, "Scientific and Technical Bulletin of St. Petersburg State Polytechnic University. Physical and Mathematical Sciences," vol. 17, no. 3.2, pp. 368–372, 2024.

[7] A. S. Podstrigaev, A. S. Lukiyanov, A. V. Smolyakov, A. N. Shishkov, V. V. Davydov, A. P. Glinuchkin, E. A. Sinicyna, "Journal of Physics: Conference Series," vol. 1410, no. 1, pp. 012155, 2019.

[8] D. Wu, T. Zhu, K. S. Chiang, M. Deng, "J. Lightw. Technol," vol. 30, no. 5, pp. 805–810, 2012.

[9] D. S. Dmitrieva, V. M. Pilipova, R. V. Davydov, E. I. Andreeva, V. Y. Rud, "Lecture Notes in Computer Science (including subseries Lecture Notes in Artificial Intelligence and Lecture Notes in Bioinformatics)," vol. 12526 LNCS, pp. 348-356, 2020.

[10] P. Lu and Q. Chen, "Asymmetrical fiber Mach-Zehnder interferometer for simultaneous measurement of axial strain and

temperature," IEEE Photon. J., vol. 2, no. 6, pp. 942–953, December 2010.

[11] P. Lu, L. Men, K. Sooley, and Q. Chen, "Tapered fiber Mach-Zehnder interferometer for simultaneous measurement of refractive index and temperature," Appl. Phys. Lett., vol. 94, no. 13, pp. 131110- 1–131110-3, April 2009.

[12] Y. Geng, X. Li, X. Tan, Y. Deng, and Y. Yu, "High-sensitivity MachZehnder interferometric temperature fiber sensor based on a waist-enlarged fusion bitaper," IEEE Sensors J., vol. 11, no. 11, pp. 2891–2894, Nov. 2011.

[13] C. Chen, Y. S. Yu, R. Yang, L. Wang, J. C. Guo, Q. D. Chen, and H. B. Sun, "Monitoring thermal effect in femtosecond laser interaction with glass by fiber bragg grating," J. Lightw. Technol., vol. 29, no. 14, pp. 2126–2130, July 2011.

[14] V. V. Davydov, N. M. Grebennikova, K. Y. Smirnov, "Measurement Techniques," vol. 6, pp. 37-43, 2019.

[15] D. M. Hernandez, A. M. Rios, I. T. Gomez, and G. S. Delgado, "Compact optical fiber curvature sensor based on concatenating two tapers," Opt. Lett., vol. 36, no. 22, pp. 4380–4382, Nov. 2011.

[16] J. Villatoro, V. P. Minkovich, and D. Monzón-Hernández, "Temperature-independent strain sensor made from tapered holey optical fiber," Opt. Lett., vol. 31, no. 3, pp. 305–307, Feb. 2006.

[17] Z. Tian and S. S. H. Yam, "In-line abrupt taper optical fiber MachZehnder interferometric strain sensor," IEEE Photon. Technol. Lett., vol. 21, no. 3, pp. 161–163, Feb. 2009.

[18] M. Hatta, Y. Semenova, Q. Wu, and G. Farrell, "Strain sensor based on a pair of single-mode-multimode-single-mode fiber structures in a

ratiometric power measurement scheme," Appl. Opt., vol. 49, no. 3, pp. 536–541, Jan. 2010.

[19] D. S. Dmitrieva, V. M. Pilipova, V. I. Dudkin, R. V. Davydov, V. V. Davydov, "Lecture Notes in Computer Science (including subseries Lecture Notes in Artificial Intelligence and Lecture Notes in Bioinformatics)," 13158 LNCS, pp. 230–239, 2022.

[20] F. Esposito, A. Stancalie, A. Srivastava, M. Śmietana, R. Mihalcea, D. Neguṭ, S. Campopiano, A. Iadicicco, J. Light. Technol., vol. 41, pp. 4389–4396, 2023.

[21] S. A. Filatova, A. E. Fale, V. A. Kamynin, A. I. Fedoseev, V. B. Tsvetkov, "Journal of Lightwave Technology," vol. 41, no. 19, pp. 6400–6407, 2023.

[22] A. A. Petrov, D. V. Zaletov, V. V. Davydov, D. V. Shapovalov, "Radio Engineering and Electronics," vol. 66, no. 3, pp. 285-290, 2021.

[23] E. V. Isupova, A. S. Budnikov, V. V. Davydov, A. P. Valov, A. A. Petrov, "Scientific Notes of the Physics Department of Moscow University," no. 4, pp. 2241201, 2022.

[24] N. I. Popovskiy, V. V. Davydov, V. Yu. Rud, "St. Petersburg State Polytechnical University Journal. Physics and Mathematics," vol. 16, no. S3.2, pp. 81-86, 2023.

[25] E. I. Andreeva, D. P. Andreev, M. A. Orlov, A. I. Isupov, "St. Petersburg State Polytechnical University Journal. Physics and Mathematics," vol. 17, no. S3.1, pp. 252-256, 2024.

[26] V. G. Nesterov, V. V. Davydov, V. V. Naumova, "Scientific Notes of the Physics Department of Moscow University," no. 6, pp. 2460401, 2024.

Effect of Micro- and Nanostructure of Carbon Anode on the Performance of Biophotovoltaic Cell

Rakshaev Bair Ts.
Department of Electronics
St. Petersburg Electrotechnical
University LETI
St. Petersburg, Russia
rakshaev00@mail.ru

Konditerov Pavel B.
Department of Electronics
St. Petersburg Electrotechnical
University LETI
St. Petersburg, Russia
pavel.konditerov@mail.ru

Snarskaya Dina D.
St. Petersburg University,
Resource Center
St. Petersburg, Russia
dina.snarskaya@spbu.ru

Kononova Svetlana V.
Institute of Macromolecular
Compounds, RAS
St. Peterburg, Russia
scisecretary@hq.macro.ru

Fallahzade Peyman Z.
Department of Electronics
St. Petersburg Electrotechnical
University LETI
St. Petersburg, Russia
peyman.fallahzadeh@gmail.com

Zimina Tatiana M.
Department of Electronics
St. Petersburg Electrotechnical
University LETI
St. Petersburg, Russia
tmzimina@etu.ru

Abstract— **Massive energy consumption is driving a growing energy crisis, increasing the need for renewable solutions. One promising approach is the development of biophotovoltaic systems (BPhVS), which use cyanobacteria to convert sunlight into electricity. The dependence of the electrical performance of BPhVSs on the texture and morphology of carbon-based anode materials was investigated. The study was conducted on planar BPhV cells assembled using printed electronics technologies. Carbon fabric and carbon felt with different fiber densities were used as anodes and bonded to the substrate with carbon nanotube ink, while the cathode was coated with a 0.2 mm Nafion film. The photosynthetic component consisted of a Synechocystis sp. 1823 biofilm applied from an aqueous suspension. The electrolytic zone was formed from polyethylene oxide and agar-agar-based hydrogels, ensuring the system's integrity and moisture retention. Experiments demonstrated that carbon fabric outperformed felt, and optimized anode porosity significantly improved photocurrent generation. Overall, this work provides a systematic comparison of carbon cloth and carbon felt anodes fabricated by printed electronics methods under identical biofilm and electrolyte conditions, demonstrating that microporosity optimization, rather than maximization of total surface area, is critical for achieving stable photovoltage generation in cyanobacteria-based BPhV cells.**

Keywords—biophotovoltaic cell, cyanobacteria, carbon anode, microstructure, porosity, nanostructured carbon, printed electronics, electron transfer

I. Introduction

The tendency of non-renewable energy sources (gas and oil) depletion is causing concern among the population all over the world. While at present time, more than two-thirds of the global energy demand is met by non-renewable energy derived from natural resources. [1]. The development of solar and wind energy cannot replace these sources yet [2 – 6]. The growing need for renewable clean energy can be met with achievements of so called biophotovoltaics (BPhV), or a power generation technology including a stage with photosynthetic microorganisms, such as cyanobacteria (or blue-green algae), transforming light radiation energy into electrical energy [7 – 10].

In biophotovoltaic systems, the interface between the biological component and the electrode plays a key role, since the efficiency of photogenerated electron transfer directly depends on the structure, chemical composition, and electrical conductivity of the anode material. Carbon materials are widely used as anodes in BPhV cells due to their high chemical stability, biocompatibility and the possibility of structural modification.

The BPhV system contains two main components: an anode and a cathode which can be separated by a proton exchange membrane (PEM). In BPhV system, the output power or current is mainly affected by the design of the working electrodes. Thus, in [11] authors identified tin oxide as an effective semiconductor for generating electrons initiated by photosynthesis because of water oxidation due to a conduction band suitable for receiving electrons originating from water oxidation by photosynthesis. For BPhV devices, indium tin oxide (ITO) is considered as the best anode coating material. When comparing the output power of Pseudanabaena limnetica biofilm on different substrates such as carbon paper, stainless steel, fluorene-coated glass fiber coated with conductive polyaniline, and indium tin oxide-coated polyethylene terephthalate (ITO-PET), it was shown that the best light/dark ratio and the highest photo response were achieved using stainless steel and ITO-PET electrodes [12, 13].

However, carbon-based materials became more commonly employed in the 2000s because of initial difficulties in introducing nano and micron structures in inorganic materials: carbon cloth [14]; carbon nanotubes [15]; graphite [16]; reduced graphene oxide (rGO) [17]. Thus, in the previous studies it has been demonstrated that high electrical conductivity, reliability, versatility, and chemical inertness can be achieved at a low cost by using carbon-based anode materials. Compared with simple carbon anode materials, high output currents in BPhV cells can be achieved by using different coatings or surface modifications of the electrodes, including using cyanobacteria as whole-cell catalysts. Since the anode in BPhV cells serves as a sink for the reducing elements of photosynthetic cyanobacteria, the electrode material should have a large surface area to support their living conditions and ensure the attachment of the biofilm to the electrode.

In this paper, the effect of the microstructure and porosity of carbon anodes based on carbon fabric modified with nanostructured carbon ink on the electrical characteristics of planar biophotovoltaic cells is investigated. The cyanobacterium Synechocystis sp. 1823 was used as a photosynthetically active component. The main attention is paid to the relationship between the microstructural parameters of the anode, biofilm formation and the efficiency of electric energy generation.

The aim of this study is to improve the anode design for BPhV cell using cyanobacteria Synechocystis and hybrid technologies including printing, and to establish the role of process kinetics on the power output of the BPhV cell.

The contribution of this work includes:

(1) a comparative study of carbon cloth and carbon felt anodes in a planar printed BPhV architecture under identical biofilm and electrolyte conditions;

(2) evaluation of porosity and effective specific surface area through weighing of dry and wetted samples;

(3) demonstration that optimal microporosity improves voltage stability and specific power output, whereas excessive anode densification reduces light accessibility and limits the active biofilm volume.

II. MATERIALS AND DESIGN

A. Materials and Design of Carbon Anodes

Carbon-based anodes were manufactured using a hybrid approach combining nanostructured carbon inks with three-dimensional carbon substrates in the form of carbon fabric. The nanocarbon ink served as a conductive interfacial layer, while the carbon fabric provided a mechanically stable three-dimensional architecture and microporosity. This hierarchical design made it possible to independently control the electrical properties of the anode at the nanoscale and the structural parameters of the anode at the microscale.

Carbon fabric is a three-dimensional network of intertwined carbon fibers with characteristic diameters in the micrometer range, forming a developed porous structure. This architecture provides a combination of mechanical strength, high electrical conductivity, and an accessible surface necessary for the formation of a cyanobacterial biofilm. To systematically study the effect of the microstructure of the anode on the characteristics of BPhV elements, carbon fabrics with different surface densities were used. Changing the density of the fabric made it possible to vary the pore geometry and fiber packing density with the same chemical composition and the same composition of the nanocarbon inks used to form the conductive interfacial layer.

The conductive ink was produced as a graphene-based nanocomposite using polymethylmethacrylate (PMMA) as a binder and N,N-dimethylformamide (DMFA) as a solvent. Graphene flakes (0.05 g) and PMMA (0.5 g) were dispersed in DMFA to form a homogeneous suspension suitable for precipitation. PMMA was used primarily to provide adhesion to the surface and mechanical strength, while the electrical conductivity of the layer was determined by a percolating mesh of graphene flakes. The ink composition made it possible to form a continuous conductive carbon surface after evaporation of the solvent.

Ink was applied to polymer substrates using printing electronics technologies that made it possible to precisely control the thickness of the layer and the uniform distribution of the nanocarbon phase. After application, the samples were dried at room temperature for a long time (up to 20 hours) to ensure gradual evaporation of DMF and stabilization of the carbon layer. In some cases, moderate heat treatment was used to accelerate solvent removal and improve the mechanical stability of the film. After drying, the nanocarbon layer formed a strong conductive surface with a stable electrical resistance of the order of several kilohms per meter, indicating the formation of a continuous conductive network.

To minimize the possible effect of the electrolyte on biological activity, the anodes were treated before biofilm formation by exposure to ambient air and subsequent contact with an aqueous electrolyte. Thanks to this procedure, biological experiments were conducted on stabilized solid carbon surfaces, rather than on liquid or semi-liquid ink films. The resulting anode surfaces were mechanically strong and retained their electrical properties during subsequent processing and application of the biofilm.

The print-based manufacturing approach allowed geometry to be quickly introduced into production and allowed for a fully customized graphical interface for Electrode users. This reproducibility was crucial for determining the effect of microscale structural parameters (fiber packing density, porosity, available surface area) on the electrical characteristics of biophotovoltaic cells while maintaining a constant composition of nanocarbons in all samples.

B. Formation of Cyanobacterial Biofilm

The cyanobacterium Synechocystis sp. 1823 was used as the photosynthetically active biocatalyst. Cultivation was carried out in nutrient medium No. 6 under standard laboratory conditions. The cyanobacterial suspension was deposited onto the anode surface to form a biofilm.

Biofilm formation was achieved by sequential deposition of multiple layers of the cyanobacterial suspension followed by controlled drying. This approach enabled gradual increase in cell density and promoted uniform surface coverage. Due to the porous structure of the carbon anodes, cyanobacterial cells partially penetrated the pore network, forming a three-dimensional biofilm interacting with both micro- and nanoscale carbon surface features. Such spatial organization is essential for maximizing the effective contact area between microorganisms and the anode while maintaining adequate access to light and nutrients.

C. Assembly of Biophotovoltaic Cells

A schematic diagram of the developed biophotovoltaic (BPhV) cell is shown in Fig. 1. The device had a planar layered architecture and was assembled on a polymer substrate. A polymethyl methacrylate (PMMA) sheet with dimensions of 1.5 mm × 40 mm × 40 mm was used as the supporting base. Alternatively, a 500 μm thick polypropylene film was employed to provide mechanical flexibility of the structure.

The rectangular anode and cathode electrodes, each measuring 9 × 40 mm, were arranged in the same plane and separated by a distance of several millimeters. Spatial separation of the electrodes was achieved using an 80 μm thick polyvinyl chloride (PVC) spacer film with an adhesive layer patterned by laser ablation. This configuration allowed precise control of the interelectrode gap geometry and reduced the likelihood of internal short circuits.

The background electrode layer was prepared using carbon ink deposited from the electrode intended to function as the anode, which was necessary to enable the formation of a cyanobacterial biofilm on the sample. Under illumination, the photosynthetic activity of the biofilm resulted in electron generation, which was collected by the anode and transported through the external electrical circuit.

To improve the operational stability of the device, the cathode was coated with a thin Nafion layer, providing selective proton transport and spatial separation of anodic and cathodic processes. A 5.2 wt.% aqueous Nafion solution was used and applied directly to the cathode surface.

A hydrogel electrolyte was placed between the electrodes, simultaneously serving as an ionic conductor, a moisture-retaining medium, and a matrix supporting the viability of the photosynthetically active biofilm. The use of a hydrogel increased the compactness and sealing of the device and ensured stable operating conditions during prolonged illumination.

During BPhV operation under illumination, photosynthetic water oxidation reactions occurred within the cyanobacterial biofilm, accompanied by electron generation. The generated electrons were transferred to the anode and subsequently flowed through the external electrical circuit to the cathode, where the electrochemical loop was completed. The use of a proton-exchange membrane and spatial separation of the electrodes contributed to the reduction of parasitic currents and enhanced the stability of the electrical characteristics of the BPhV cell.

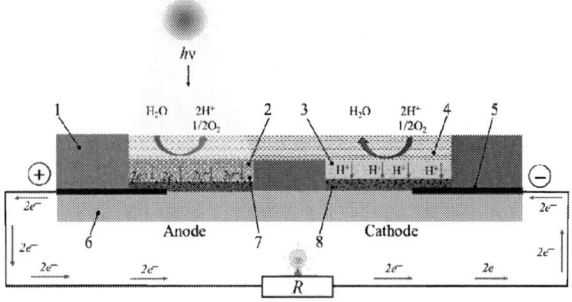

Fig. 1. Schematic presentation of BPhC design: 1 – PVC film planar casing, 2 – biofilm of cyanobacteria, 3– proton-exchange layer (nafion), 4 – hydrogel with nutrient electrolyte, 5 – conductor, 6 – base, 7 – carbon fiber layer, 8 – back-ground carbon ink layer of anode and cathode, R – load.

To assess the effect of the microstructure of the anode on the characteristics of BPhV, samples of carbon tissue and various weights were used. The mass of dry samples, carbon ink-coated samples, and moistened samples was determined using high-precision electronic scales. The amount of added water was selected depending on the mass of the anode and was 40, 50 and 80 µl for samples of different densities. The obtained mass values of the studied samples are shown in Table 1.

TABLE I.   MASS OF THE STUDIED MATERIALS FOR ANODE MODIFICATION

	Sample number		
	1	2	3
Sample mass, m g	0,0004	0,0040	0,0210
Mass of dry sample with ink, m_{dry}, g	0,0045	0,0089	0,0294
Wet sample mass with ink, m_{wet}, g	0,0448	0,0643	0,1074

Experimental verification of the proposed architecture and materials was carried out by measuring the electrical characteristics of laboratory samples of biophotovoltaic cells.

III. RESULTS AND DISCUSSION

A. Influence of Microstructure on Biofilm–Anode Interaction

SEM analysis showed that carbon fabric anodes consist of fibers with a diameter of approximately 5-6 microns forming interconnected networks of pores. It is not an easy task to calculate the exact volume occupied by the biofilm,

however we can make estimations. To determine the porosity and specific surface area of the three-dimensional electrode, it is necessary to know not only the dimensions of the electrode formed by the system of fibers, but also to measure their mean radius and length. SEM images of cross-section of carbon fiber anode layers are presented in Figure 2 (a, b, c). The carbon fibers demonstrated sufficiently uniform diameter of the mean value $<d_{CF}> = 5.74$ µm.

Modified anodes with carbon felt will not be considered because it was not possible to take pictures of the fiber structure using SEM. Accordingly, the calculation of the specific surface area using this method is impossible.

Another important parameter of the carbon fiber anode is the porosity, ϕ, which is the dimensionless quantity defined as a ratio of pore volume to the total volume of the sample (5):

$$\phi = V_P/V_A, \tag{5}$$

where: V_P – pore volume, V_A – total volume of carbon fiber anode.

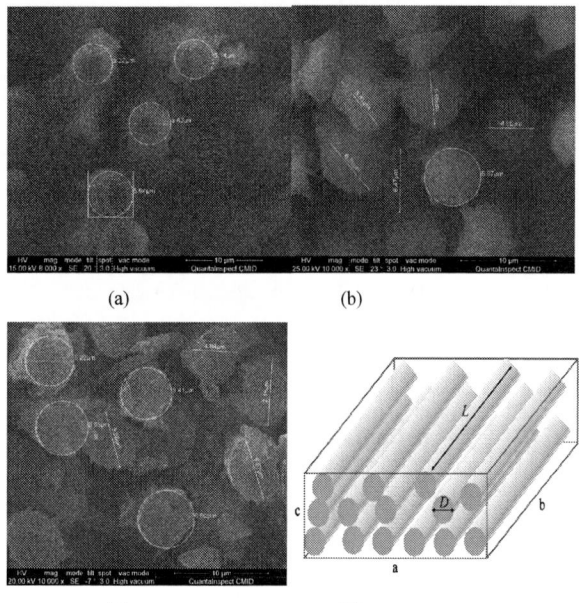

Fig. 2. SEM image of cross sections of carbon fibers in the fabric of the samples of BPhV cells Nos 1, 2 and 3 (a, b, c correspondingly). Diameters of the fibers are presented on the images and are all within the range of 5 – 6 µ. The schematic of porosity and specific surface area calculation (d)

To evaluate the porosity of the anode carbon fiber layer it is necessary to determine the pore volume and total volume of the sample. The overall thickness of the carbon fiber anode layer is 10-4 m. Total volume (Figure 9 d) will be V_A = a × b × c = 0.8 mm × 1.5 mm × 0.1 mm = 0.12 mm3.

The total weight of fiber, WF, of certain volume could be determined as follows:

$$W_F = (\pi D^2/4) \times L \times P_S \tag{6}$$

where: P_S is the specific weight of carbon fiber – 1.7 g/cm³, L - total length of fibers

Then, L could be determined as (7):

$$L = W_F/[(\pi D^2/4) \times P_S] \tag{7}$$

and the volume of fibers in the sample will be

$$V_F = L\pi D^2/4$$

Then the porosity of the sample 1 will be

The specific surface area (SSA) for the sample 1 could be estimated according to expression (8)

$$SSA = \pi DL/W_F \tag{8}$$

Another approach for determination of porosity and SSA is the experimental measurement of the dry and wetted samples (Table 2).

TABLE II. EVALUATION OF CARBON FIBER ANODE POROSITY BY WEIGHTING DRY AND WET SAMPLES.

		Sample Number		
		1	2	3
Sample mass, g	Dry	0.0045	0.0089	0.0294
	Total	0.0448	0.0643	0.1074
	Liquid	0.0403	0.0634	0.0780
Volume, cm³	Dry	0,0026	0.0052	0.017
	Liquid	0.0403	0.0634	0.0780
	Total	0.0429	0.0686	0.095
φ		0.939	0.924	0.81

B. Role of Nanostructures in Electron Transfer

Although the anode microstructure governs biofilm formation and light accessibility, the electrical power output of BPhV cells is ultimately limited by the efficiency of charge extraction at the biofilm–electrode interface. In this context, nanostructured carbon layers play a secondary but important role by reducing interfacial charge transfer resistance rather than increasing the total number of photogenerated electrons.

Nanocarbon inks form a percolating conductive network on the anode surface, enabling localized charge collection from cells located within the porous structure. This effect is particularly important for electrons generated in the inner regions of the biofilm, where transport through the extracellular polymeric matrix may be rate-limiting.

However, the presented results indicate that the contribution of nanostructuring is strongly constrained by the underlying microstructure. In densely packed anodes, reduced pore size and limited light penetration suppress photosynthetic activity in deeper biofilm layers, preventing effective utilization of the enhanced conductivity provided by the nanocarbon network. As a result, nanostructuring does not compensate for unfavorable microstructural characteristics.

These observations demonstrate that nanostructured carbon layers primarily function as interfacial charge-collection layers, the effectiveness of which depends on the spatial accessibility of active biological components rather than on the total geometric surface area of the anode.

C. Electrical Characteristics of BPhV Cells

BPhV cells with low-mass carbon cloth anodes exhibited the highest output voltages (up to ~0.35 V) and the most stable current generation (Fig. 3). Anodes of intermediate mass showed moderate performance, whereas high-mass anodes produced only negligible power.

Cells employing carbon felt anodes displayed unstable electrical behavior, which can be attributed to the hydrophobic nature of the felt, limiting electrolyte access and biofilm hydration.

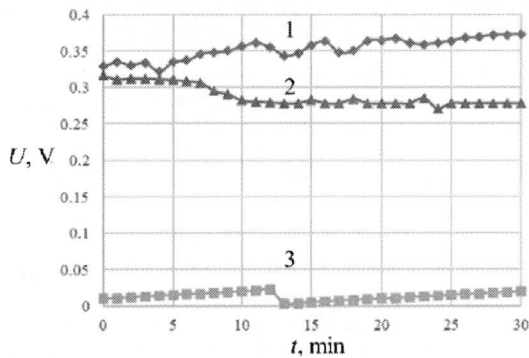

Fig. 3. Relationships of output voltage versus time for BFC numbers 1, 2 and 3 with Synechocystis sp. 1823 with carbon fiber

Analysis showed that increasing porosity and accessible specific surface area generally promotes higher specific power output of BPhV cells. However, this effect is non-monotonic, and further increases in anode mass and fiber packing density lead to a deterioration of electrical performance despite an increase in geometric surface area.

The obtained results indicate the existence of an optimal range of anode microstructural parameters at which maximum electrical efficiency of BPhV cells is achieved. Within this range, the average pore size is comparable to the characteristic dimensions of Synechocystis cells, simultaneously satisfying three key conditions: (i) effective penetration and uniform distribution of the biofilm, (ii) sufficient light permeability of the active layer, and (iii) minimal losses during electron transport.

Thus, increasing the anode specific surface area alone is not a determining factor for enhancing power output, and BPhV anode design should be based on microstructural optimization rather than maximization.

From a phenomenological standpoint, the specific power output can be considered a function of both porosity and pore size, exhibiting a maximum at intermediate values rather than a monotonic increase.

IV. CONCLUSION

In this paper, the effect of the microstructure and porosity of carbon anodes on the characteristics of biophotovoltaic cells with the cyanobacterial biofilm Synechocystis sp. 1823 is experimentally investigated. It was shown that anodes with a lower tissue mass had larger pores and higher porosity, which allowed cyanobacterial cells to partially penetrate the structure while remaining open to light.

On the contrary, a higher mass of tissue led to a dense packing of fibers, a reduction in pore size and a restriction of light penetration into deeper layers of biofilm. Although such structures provided a large total surface area, a significant part of this area was not effectively used by active cyanobacteria.

TABLE III. RESULTS OF SPECIFIC SURFACE AREA AND SPECIFIC POWER CALCULATION IN BPhV CELLS

Sample number	Sample porosity, $p_\text{п}$	Specific surface area, m²/m³	Specific power, W·m²/m³
1.1	0,8388	$0,606 \cdot 10^5$	$8,3 \cdot 10^{-9}$
1.2	0,829	$0,605 \cdot 10^5$	$5,9 \cdot 10^{-9}$
1.3	0,896	$0,365 \cdot 10^5$	$0,5 \cdot 10^{-9}$

It has been established that an excessive increase in the mass and density of carbon tissue, despite an increase in specific surface area, leads to a deterioration in light transmission and a limitation of photosynthetic activity of microorganisms, which reduces the electrical output power of the BPhV cell. Nanostructuring anodes using carbon ink helps to reduce contact resistance and increase the efficiency of electron collection, but its positive effect is fully manifested only with an optimal microstructure of the anode.

The results obtained confirm that the maximum efficiency of biophotovoltaic cells is achieved by matching the micro- and nanostructural characteristics of the anode with the biological and optical requirements of the photosynthetic system. The presented approach, based on printed electronics technologies, opens up opportunities for scalable production and further optimization of biophotovoltaic devices.

Acknowledgment

The authors acknowledge SPBU Research Park, Centre for Culture Collection of Microorganisms (a research project 124032000041-1) for providing strains CALU 1823.

The authors would also like to thank Sonya Zenkova for her help and the work she has done.

References

[1] 1. Pang, L., Liu, L., Zhou, X., Hafeez, M., Ullah, S. and Sohail, M.T., 2024. How does natural re-source depletion affect energy security risk? New insights from major energy-consuming coun-tries. Energy Strategy Reviews, 54, p.101460.

[2] Khan A., Hussain I., Khan N. Impact of Energy Crisis on Developed and Developing Countries: A Comparative Review //Energy Crisis and Its Impact on Global Business. – 2024. – C. 288-315.

[3] Einolander J., Kiviaho A., Lahdelma R. Detecting changes in price-sensitivity of household electricity consumption: The impact of the global energy crisis on implicit demand response behavior of Finnish detached households //Energy and Buildings. – 2024. – T. 306. – C. 113941.

[4] Fearn G. The end of the experiment? The energy crisis, neoliberal energy, and the limits to a socio-ecological fix //Environment and Planning E: Nature and Space. – 2024. – T. 7. – № 1. – C. 212-233.

[5] Lawrence B. et al. Stock market connectedness during an energy crisis: Evidence from South Africa //Emerging Markets Review. – 2024. – C. 101194.

[6] Zhou Y. et al. Evaluating the social benefits and network costs of heat pumps as an energy crisis intervention //Iscience. – 2024. – T. 27. – № 2.

[7] Hassan Q. et al. The renewable energy role in the global energy Transformations //Renewable Energy Focus. – 2024. – T. 48. – C. 100545.

[8] Sharma V. Integrating renewable energy with building management systems: Pathways to sustainable infrastructure //Journal of Waste Management & Recycling Technology. – 2024. – T. 2. – № 1.

[9] Ani E. C. et al. Renewable energy integration for water supply: a comparative review of African and US initiatives //Engineering Science & Technology Journal. – 2024. – T. 5. – № 3. – C. 1086-1096.

[10] Mirza Z. T. et al. A thematic analysis of the factors that influence the development of a renewable energy policy //Renewable Energy Focus. – 2024. – C. 100562.

[11] H. Ochiai, H. Shibata, Y. Sawa, M. Shoga, S. Ohta, Appl. Biochem. Biotechnol. 1983, 8, 289–303.

[12] Mccormick A. J., Bombelli P., Scott A. M. [et al.]. Photosynthetic biofilms in pure culture harness solar energy in a mediatorless bio-photovoltaic cell (BPV) system // Energy Environ. – 2011. – Vol. 4. – P. 4699–4709.

[13] Chin, J.C., Khor, W.H., Chong, W.W.F., Wu, Y.T. and Kang, H.S., 2022. Effects of anode materials in electricity generation of microalgal-biophotovoltaic system-part I: natural biofilm from floating microalgal aggregation. Materials Today: Proceedings, 65, pp.2970-2978.

[14] A. Cereda, A. Hitchcock, M. D. Symes, L. Cronin, T. S. Bibby, A. K. Jones, PLoS One 2014, 9, 91484.

[15] N. Sekar, Y. Umasankar, R. P. Ramasamy, Phys. Chem. Chem. Phys. 2014, 16, 7862–7871

[16] K. Hasan, V. Grippo, E. Sperling, M. A. Packer, D. Leech, L. Gorton, ChemElectroChem 2017, 4, 412–417

[17] F. L. Ng, M. M. Jaafar, S. M. Phang, Z. Chan, N. A. Salleh, S. Z. Azmi, K. Yunus, A. C. Fisher, V. Periasamy, Sci. Rep. 2014, 4, 7562.

[18] Zimina, T. M., Mandrik, I. V., Pudova, A. V., Gataullin, A. O., & Snarskaya, D. D. (2023). Biophotovoltaic Energy Sources Based on Cyanobacteria. Nanobiotechnology Reports. – 2023. – T. 18. - № 1, P. S156-S164.

[19] T. M. Zimina, N. O. Sitkov, Y. A. Volkov et. al, Comparative study of biosensing principles: fet, impedimetric, light scattering, with regard to operational efficiency in antibiotic resistance testing. AIP Conference Proceedings. 2019. C. 020083.

Resent Advances in the Production and Research of Composite Materials Based on Graphene and Polymers

Pavel F. Samsygin
Department of Micro- and Nanoelectronics
Saint-Petersburg Electrotechnical University "LETI"
Saint Petersburg, Russia
pfsamsygin@stud.etu.ru

Svetlana S. Nalimova
Department of Micro- and Nanoelectronics
Saint-Petersburg Electrotechnical University "LETI"
Saint Petersburg, Russia
sskarpova@list.ru

Abstract—Graphene-polymer mixes have become popular as useful materials. They combine graphene's good electrical, heat, and strength features with the flexibility of polymers. This paper organizes the recent progress in how to make, change, and the applications of graphene-polymer mixes between 2021 and 2025. It looks at both basic ways to make them, like mechanical and liquid-phase separation, chemical vapor placement, melt mixing, and in situ polymerization. Special attention is given to how graphene is spread out, how well it sticks to the polymer, and how its modification controls the mechanical, electrical, heat, and barrier qualities of the mixes. Modern ways to examine the microstructure and filler spacing are discussed, along with new ways to analyze things in place and in three dimensions. Main uses like flexible and wearable electronics, energy storage, membrane tech, gas sensing, and 3D printing are checked. Current problems with graphene rules, clumping, how easy it is to make, and cost are noted. Also, future research to make things more reliable and easier to put into use in industry is suggested.

Keywords—graphene–polymer composites, graphene dispersion, functionalization, electrical conductivity, barrier properties, flexible electronics, additive manufacturing, nanocomposites

I. Introduction

The carbon-nanostructure-based composites have been gaining a great deal of research interest owing to their extraordinary trade-off between mechanical strength and stability, combined with the ability to tune the corresponding materials' physicochemical properties in a controlled manner. One special case in this regard amongst them is the graphene-polymer combinations, which combines functional attributes of graphene and technologically versatile polymer matrix.The combination of the excellent electrical and thermal conductivities, as well as the chemical inertness of graphenes has prompted opportunities for the application of such composites to a broad spectrum of engineering applications [1, 2].

However, exploiting the potential of graphene requires technology development. The problem is the agglomeration of particles and nonuniform distribution of particles in the volume of matrix that cause a degradation of mechanical and transport properties of nanocomposite. Research focusing on the analysis of filler dispersion reports the dramatic influence of structural homogeneity on the final features of nanocomposite materials [3]. An analogous relation can be seen from the literature investigating in thermally conductive composites, where the strength of interfacial heat transfer is thereby directly influenced by the crossover between "graphene–polymer" adhesion [4].

A new wave of interest in the field is largely provided by the progress in scalable graphene synthesis methods, the development of functional polymers, and the growing demand from the flexible electronics industry and membrane technologies. For example, conductive fillers with polymers have been the most efficient and reliable solution in gas sensor development [5]. It is important to mention that modern approaches to computational modeling also play a major role in the field since they enable scientists to precisely predict the conductivity of composites, even at the design stage [6].

The aim of the present work is to systematize contemporary approaches to the synthesis and modification of graphene-polymer composites. Particular attention is given to analyzing the latest achievements in controlling their structure and functional properties. The review is based on a critical analysis of publications from the 2021–2025 period, allowing for the identification of key trends and vectors for further technological development.

II. Methods for Producing Graphene-Polymer Composites

A. Top-down: Mechanical and Liquid-Phase Exfoliation of Graphite

In the Top-down approach, graphene is obtained by destroying the large-scale structure of graphite; this is achieved through its exfoliation—mechanical, chemical, or liquid-phase. Mechanical exfoliation is most often carried out using ultrasonic treatment (bath or probe), which promotes the effective breakdown of weak van der Waals bonds between graphite layers, or via micromechanical cleavage [6].

The advantage of this method is its comparative low cost, technological simplicity, and capability for scalability, making it sought after for industrial production [1]. However, the main limitation of the method is the non-uniformity of the resulting graphene particles in thickness, the presence of structural defects, and the inevitable formation of agglomerates. These factors critically reduce the mechanical and transport properties of the final composite [7].

III. The good news is that a way forward has been mapped out in recent years in the face of these difficulties. Namely, efficient ways of stabilizing graphene suspensions with surfactants have been formulated, the ultrasonic treatment has been perfected to provide gentler yet highly effective exfoliation, and numerous "green" solvents (e.g., ethanol or water) are widely accepted to lower the production's environmental impact. The directions mentioned above fully align with the general trends of graphene composite fabrication methods that have been discussed in several review articles [1].

A. Bottom-up: CVD Graphene and its Integration with Polymers

The Chemical Vapor Deposition (CVD) method implements the Bottom-up approach, allowing for the synthesis of mono- or few-layer graphene with high uniformity and quality. Atomic-level control over the process ensures the production of material with a minimal number of structural defects.

The key technological difficulty is associated with the transfer of the synthesized CVD films from the catalytic substrate to the target polymer matrix. Mechanical damage—folds, tears, and defects—frequently arises during this process [7].

Nevertheless, significant improvements have emerged in this area over the past few years: roll-to-roll approaches have been developed to ensure scalable transfer, mechanical and thermal damage during transport has been reduced, and methods for direct graphene growth on polymer substrates at lower temperatures are being actively promoted, minimizing the risk of damage to the thermosensitive matrix. These strategies are discussed in detail in reviews dedicated to composite preparation and their electrical and thermal properties [8].

B. Compounding and Melt Blending

The melt blending (compounding) approach is one of the most popular in industry, as it avoids the use of solvents and is easily scalable for mass production.

Polymer melt is of utmost importance since it determines the formation of conductive networks, the percolation threshold, and, in the end, the functional properties of the material [8].

Recent achievements in this direction include: the use of twin-screw extruders to improve graphene distribution, and the application of functionalized graphene (e.g., GO or rGO) to enhance its compatibility with the polymer. Such strategies help to lower the percolation threshold, as functional groups on the filler surface strengthen the interfacial interaction and facilitate the formation of denser conductive networks [4].

C. In Situ Polymerization

The in situ polymerization method involves mixing the monomer (or low-molecular-weight precursor) with graphene before the polymerization reaction begins. This allows the polymer to "grow" directly near or on the surface of the graphene layers.

This results in excellent adhesion between the matrix and the filler and a very high degree of graphene dispersion.

Modern studies demonstrate the active use of controlled radical polymerization (e.g., ATRP, RAFT) on the graphene surface, allowing for the production of hybrids with improved morphology and high stability.

Thanks to the strong interface and uniform graphene distribution, composites obtained in situ form stable electroconductive structures and exhibit improved operational characteristics. This makes them especially promising for high-performance membranes and sensor systems [7].

IV. PROPERTIES AND STRUCTURE OF GRAPHENE-POLYMER COMPOSITES

A. Mechanical Properties

The introduction of graphene into a polymer matrix traditionally leads to a noticeable increase in the stiffness and strength of the composites, even at low filler content. The observed increase in Young's modulus and tensile strength limits is directly explained by the exceptional intrinsic mechanical characteristics of the graphene sheets, as well as their critical contribution to load transfer through the "filler–matrix" interfacial interaction [1].

The key factor determining the reinforcement efficiency is not the absolute graphene content but the quality of its dispersion and the number of layers. It is established that homogeneously distributed mono- or few-layer particles provide the maximum reinforcing effect, while the presence of large agglomerates and multi-layer stacks sharply reduces the efficiency of stress transfer [4].

Recent works actively investigate the practical applications of these effects: compositions specifically adapted for 3D printing are being developed, where optimization of dispersion and graphene modification significantly increase layer adhesion and the mechanical integrity of the final products. Flexible reinforced structures, required in wearable devices and structural elements, are also being created [1].

Finally, fractographic studies are becoming increasingly widespread for a detailed understanding of fracture processes. Analysis of the fracture surface morphology (microfractography) allows for establishing a direct correlation between the type of crack formation and the scale of filler dispersion, as well as the nature of the interface. This, in turn, provides valuable practical recommendations for improving fatigue resistance and impact toughness [4].

B. Electrical and Thermal Properties

A critical property for functional applications is the formation of conductive networks in the polymer matrix, which occurs upon reaching the percolation threshold. With a sufficient number of effective contacts between graphene sheets, a sharp, non-linear increase in the composite's specific conductivity is observed [2].

Functionalization of graphene (e.g., oxidized derivatives or reduced graphene oxide, rGO) is a powerful tool for property regulation, as it changes both inter-particle contacts and interfacial resistances. This allows for fine-tuning the conductivity at a fixed filler mass fraction [1].

Current research directions focus on developing composites for flexible electrodes in wearable electronics, creating thermal interface materials with improved thermal conductivity for efficient heat dissipation, and using graphene-polymer layers for electromagnetic interference (EMI) shielding [5].

C. Barrier and Chemical Properties

The exceptional impermeability of individual graphene sheets provides a significant improvement in the barrier characteristics of composites. Even at relatively low filler concentrations, 2D graphene sheets substantially increase the diffusion path of molecules through the polymer matrix, leading to a reduction in gas and vapor permeability [1].

This quality is actively applied in the development of membrane technologies for water and gas purification:

hybrids of graphene oxide (GO) with polymers demonstrate enhanced selectivity and efficiency in separating components during filtration [5].

Furthermore, in a number of systems, due to graphene's heat-dissipating and protective properties at the interface, an increase in chemical stability and enhanced flame retardancy is noted [1].

D. Microstructure and Analysis Methods

Characterization of the morphology and interface remains a key component of research. The standard set of methods includes Atomic Force Microscopy (AFM), Scanning and Transmission Electron Microscopy (SEM, TEM), as well as Raman Spectroscopy and X-ray Photoelectron Spectroscopy (XPS) for a comprehensive assessment of the filler's dispersion, defectiveness, and surface chemical state [4].

Among recent innovations are in situ methods for observing layer distribution directly during the shaping process, and the correlation of local morphology with electrical properties (conductivity mapping). Separate attention is paid to the use of 3D tomography for three-dimensional analysis of nanofase distribution in the matrix. These advanced approaches allow for a more accurate establishment of the relationship between microstructure and macroscopic properties, which is necessary for optimizing composite production technologies [5, 6].

V. APPLICATION AREAS OF GRAPHENE-POLYMER COMPOSITES

A. Flexible Electronics and Wearable Devices

In recent years, graphene-polymer composites have been actively positioned as a key foundation for flexible electronic systems. As noted in several reviews [8], graphene films and conductive inks provide a unique combination of high conductivity and mechanical flexibility. This makes them ideal candidates for creating strain sensors, pressure sensors, and highly sensitive breath sensors. Importantly, polymer matrices significantly enhance the stability of the response under multiple cyclic bending.

The latest experimental works demonstrate successful prototypes of wearable biosensors, where graphene conductive networks maintain stable operation and high sensitivity to humidity and gas impurities even under significant stretching or compression. Researchers actively use the combination of graphene with elastomers, enabling the creation of soft sensor platforms for monitoring human movements and physiological signals.

B. Energy Storage

In the field of energy applications, the main direction remains the use of these composites as flexible conductive layers for supercapacitors and batteries. Analytical reviews confirm that graphene contributes to the formation of stable transport networks for both electrons and heat dissipation.

Studies show that polymer electrolytes with the addition of oxidized forms of graphene exhibit improved ionic mobility, which is explained by the modification of the interfacial interaction. Authors also emphasize the significant prospect of enhancing electrode cycle stability due to the reinforcing role of graphene sheets, which effectively reduce the mechanical destruction of the structure during multiple charge-discharge cycles. A recent work on the numerical modeling of carbon nanocomposites clearly described how the formation of conductive pathways critically affects the overall electrical conductivity of the hybrids [6].

C. Membranes and Filtration Systems

Graphene and GO-polymer hybrids are the subject of intensive research in the field of membrane technologies. As highlighted in reviews [8], oxidized graphene (GO) not only creates additional transport pathways but also paradoxically increases separation selectivity due to controlled interlayer spacing.

Such composites are successfully applied in water purification, where the modified surface chemistry of the graphene sheets provides high resistance to biofouling and simultaneously high permeability. Furthermore, active development is underway for new multilayer structures where graphene sheets are incorporated into polyamide and other polymer matrices, allowing for improved barrier properties while reducing resistance to water or gas flow.

D. Composites for 3D Printing

One of the most dynamically developing areas is graphene-polymer materials adapted for additive technologies (3D printing). Reviews focusing on graphene dispersion in nanocomposites indicate that the addition of reduced graphene oxide (rGO) significantly improves layer adhesion during layer-by-layer printing and minimizes the number of internal defects in the structure.

Materials based on popular polymers such as PLA and ABS, with rGO filler, demonstrate a noticeable increase in mechanical strength and elastic modulus, which is attributed to improved interaction between the graphene planes and the polymer chain. Moreover, the possibility of forming stable conductive pathways even at relatively low filler content is reported, opening the door for creating printed conductive parts and integrating simple sensor elements directly into printed products. These achievements confirm the high promise of graphene hybrids as functional materials for modern additive technologies.

VI. CURRENT CHALLENGES AND FUTURE DIRECTIONS

Despite significant achievements in the field of graphene-polymer composites, their widespread commercial implementation is still constrained by a number of systemic problems. First and foremost, the issue of graphene standardization remains unresolved: variations in filler thickness, number of layers, defect level, and degree of oxidation lead to a substantial scatter in the properties of the final composites, even when an identical compounding technology is used [1].

Closely linked to this is the problem of functionalization methods, which remain disparate; the lack of unified, reproducible protocols makes it extremely difficult to compare results obtained in different research laboratories. Furthermore, issues with the uniform distribution of graphene within the polymer matrix remain relevant. The presence of agglomerates is a key factor limiting the achievement of maximum mechanical, electrical, and barrier characteristics [4].

An equally important constraint is the issue of scalability and cost of technologies for producing both high-quality graphene itself and the finished composites. This is particularly true for methods like CVD synthesis and some top-down approaches, where effective defect control and the film transfer procedure require expensive equipment and complex multi-stage processes [9].

In the coming years, the scientific community is expected to focus efforts on developing quality standards and control

979-8-3315-4784-4/26 $31.00 © 2026 IEEE

methods for graphene, which should ultimately ensure the predictability and reproducibility of composite properties [8]. The research emphasis is likely to shift towards integration with biopolymers and soft matrices to create a new generation of wearable devices and biosensors. Another key direction will be the search for new, more effective ways of chemical adhesion of graphene to the polymer, including controlled functionalization and direct in situ polymerization [6].

The implementation of these approaches will enable the creation of graphene-polymer composites with predictable and stable characteristics, paving the way for their mass production for critical applications in flexible electronics, energy, and membrane technologies.

VII. CONCLUSION

Graphene-polymer composites firmly retain their position as one of the most promising areas of modern materials science, demonstrating a synergistic combination of graphene's unique characteristics with the technological flexibility of polymer matrices. Currently, functionalization methods aimed at ensuring improved interfacial adhesion and homogeneous distribution of graphene layers are being actively refined. In parallel, substantial progress is observed in approaches to scalable production, including the refinement of CVD synthesis, the optimization of top-down exfoliation, and the increased efficiency of compounding.

The main contemporary research trends are focused on flexible electronics, wearable sensors, membrane and filtration systems, and energy applications, where the high conductivity, mechanical strength, and stability of hybrid materials are critically needed. The progress achieved in recent years suggests a gradual and confident transition from purely laboratory developments to the creation of industrial platforms. This provides grounds to believe that graphene-polymer composites may soon become standard functional materials for a wide range of advanced technological solutions.

ACKNOWLEDGMENT

The authors would like to express their gratitude to the Scanning Microscopy Center at Saint Petersburg Electrotechnical University "LETI" for providing the equipment.

REFERENCES

[1] D. Niyobuhungiro and L. Hong, "Graphene polymer composites: Review on fabrication method, properties and future perspectives," Adv. Sci. Technol. Res. J., vol. 15, no. 1, pp. 248–259, 2021. DOI: 10.12913/22998624/132605

[2] V. K. Samoei and A. H. Jayatissa, "A mini review of conductive graphene/polymer composites," J. Compos. Sci., vol. 7, no. 1, p. 19, 2023. DOI: 10.3390/jcs7010019

[3] S. J. Lee, S. J. Yoon, and I.-Y. Jeon, "Graphene/polymer nanocomposites: Preparation, mechanical properties, and application," Nanomaterials, vol. 12, no. 1, p. 77, 2022. DOI: 10.3390/nano12010077

[4] P. Govindaraj, A. Sokolova, N. Salim, S. Juodkazis, F. K. Fuss, B. Fox, and N. Hameed, "Distribution states of graphene in polymer nanocomposites: A review," Compos. Part A Appl. Sci. Manuf., vol. 150, p. 106606, 2021. DOI: 10.1016/j.compositesa.2021.106606

[5] P. Ak, M. Muthiah, Y. Venkatraman, and S. Sagadevan, "Recent advances and synergistic interactions in graphene-based polymer nanocomposites for enhanced gas sensing applications," Chemosphere, vol. 351, p. 141121, 2024. DOI: 10.1016/j.chemosphere.2023.141121

[6] B. Mohammadian and M. Ghasemi, "A highly efficient numerical method to investigate the conductivity of CNT/polymer composite," Comput. Mater. Sci., vol. 193, p. 110398, 2021. DOI: 10.1016/j.commatsci.2021.110398

[7] S. Perumal, R. Atchudan, and I. W. Cheong, "Recent studies on dispersion of graphene–polymer composites," Polym. Rev., vol. 63, no. 3, pp. 696–740, 2023. DOI: 10.1080/15583724.2022.2101966

[8] A. Tarhini and A. R. Tehrani-Bagha, "Advances in preparation methods and conductivity properties of graphene-based polymer composites," Curr. Opin. Chem. Eng., vol. 35, p. 100778, 2022. DOI: 10.1016/j.coche.2021.100778

Formation of Low-Dimensional Halide Perovskite Crystals in Radiation-Resistant Phosphate Glass

Tatyana Sedegova
Peter the Great St. Petersburg Polytechnic University
Saint Petersburg, Russia
TatianaSedegova@yandex.ru

Abstract — **Addressing the global energy demand requires improving the efficiency of photovoltaic devices, which can be achieved using functional composite halide perovskite coatings. A promising approach is to convert the high-energy part of the solar spectrum into the spectral range of the solar cell. Halide perovskites exhibit high absorption coefficients and quantum yields of luminescence (64%), enabling the utilization of previously unexploited solar energy. However, their instability toward moisture and oxygen limits practical application. To enhance stability, perovskites were embedded in a phosphate glass matrix doped with cerium, which acts as a radiation-protective component. This composite demonstrated high luminescence quantum yields, while cerium doping effectively suppressed the formation of radiation-induced defects, improving the material's radiation resistance. The spectral and luminescent properties of the composites were characterized, confirming the feasibility of using cerium-doped phosphate glass as a protective matrix for halide perovskites in photovoltaic applications.**

Keywords—halide perovskites, luminescence, phosphate glass, ionizing radiation.

I. INTRODUCTION

The growth in global energy consumption necessitates an increase in electricity generation capacity. Today in this context renewable energy sources are becoming increasingly important. Photovoltaics is a particularly promising area. An urgent scientific task is to increase the efficiency of photovoltaic converters [1, 2]. One possible approach to solving this problem is the use of functional composite coatings based on halide perovskites.

In recent years halide perovskites have established themselves as significant luminescent materials for modern optoelectronics. They have found application as phosphors, active media for lasers and for radiation detectors. These crystals have a high quantum yield luminescence, a multi-stage and inexpensive production technology, which justifies their potential for practical application. The disadvantage of perovskites is their low resistance to water, temperature changes and low chemical stability [3-5].

To protect perovskites from external influences, that are formed in porous structures and solid-state matrices. One promising type of protective matrix is glass. Glass not only provides effective insulation of crystals from aggressive environmental factors, but also allows additional properties, that useful for their application in special devices to be imparted to the material by introduction alloying additives [6, 7]. The composites obtained in this manner can be milled into a powder and dispersed in a flexible polymer matrix, enabling the fabrication of functional composite films.

Standard silicon photovoltaic converters most efficiently convert visible-range radiation into electrical energy [8]. However, the solar radiation spectrum is broader than the visible radiation spectrum. Consequently, one strategy to increase solar panel efficiency is to expand the usable light energy spectrum. Halide perovskites re-emit ultraviolet radiation into the visible range, enabling them to supply additional energy to the solar cell and thereby enhance its operating efficiency [3].

An important operational requirement for materials in space and terrestrial energy applications is their radiation hardness. Lead halide perovskites themselves demonstrate high stability against ionizing radiation [9]. The radiation stability of halide-based composites has been previously demonstrated for borosilicate glass matrices containing $CsPbBr_3$ nanocrystals, which preserved their luminescence even after exposure to high doses of gamma and electron irradiation [10]. When using glass as a matrix, it is possible to increase its inertness to the effects of ionizing radiation by adding protective additives that suppress the formation of color centers in the glass. One of the most effective additives of this type are cerium ions (Ce^{3+}/Ce^{4+}), which, due to their variable valence, act as traps for electron-hole pairs, thereby preventing the degradation of the glass's optical properties [11, 12]. Glassy matrices with high solubility for additives are employed for incorporating protective and functional additives. For the composite to be usable in solar cells, the matrix must be transparent in the visible and ultraviolet ranges. Based on the outlined criteria, a glassy phosphate matrix was selected for these investigations [6, 13].

Therefore, this study presents an approach based on the nucleation of halide perovskites within a cerium-doped glassy matrix, with subsequent pulverization of the obtained glass-ceramic material and its integration into a polymer composite. This strategy is expected to yield a functional film capable of enhancing solar cell efficiency while maintaining stability under elevated radiation exposure. The primary objective of this study is to elaborate a synthesis methodology for $CsPbBr_3$ and $CsPbCl_3$ perovskite crystals embedded within cerium-doped glassy phosphate matrix, and to investigate their spectro-luminescent properties.

II. MATERIALS AND METHODS

A. Synthesis technique of glass

The object of this study was a glass system with the composition $8.16Al_2O_3$-$30.4Na_2O$-$59P_2O_5$-$(2.45-x)La_2O_3$-$xCeO_2$. To form perovskite crystals within the glass matrix, precursors were added for the formation of $CsPbBr_3$ (NaBr, $PbBr_2$, $CsCO_3$) and $CsPbCl_3$ (NaCl, $PbCl_2$, $CsCO_3$). Reagents with chemical purity categories of "especially pure" and "pure for analysis" were used. The crucibles used for glass synthesis were cleaned with distilled water prior to use and then thermally treated at 100 °C for 15 min. All batch components were weighed on analytical balances with an accuracy of four decimal places. To ensure the homogeneity of the batch the mixture was ground in an agate mortar for 15 min.

979-8-3315-4784-4/26 $31.00 © 2026 IEEE

The glass synthesis was carried out in an electric resistance furnace under ambient air in two stages. The melt was first held at 1100 °C for 1 h and then at 1200 °C for 30 min. The melt was poured into a preheated steel mold, and the primary annealing was performed at 480 °C. In the case of perovskite nucleation within the matrix, the melt was cooled to 1100 °C, and the precursors required for crystal nucleation were introduced. The melt was stirred thoroughly and then kept in the furnace at 1100 °C for an additional 15 min. To minimize the loss of volatile halides, the crucibles were covered with lids during synthesis. The resulting melt was poured into a preheated brass mold and annealed in a muffle furnace at 360 °C. The crystallization degree of the obtained glasses was assessed visually. The electron irradiation was made on Accelerator RTE - 1V.

B. Characterization

For X-ray diffraction (XRD) analysis, the samples were ground in an agate mortar to a particle size of approximately 10 µm. The measurements were carried out using a «Bruker D8-Advance» (Germany). Optical transmission and absorption spectra were recorded with an «Analytik Jena Specord 40» (Germany) with a step 1 nm. The absolute quantum yield was measured using a calibrated multichannel spectroscopic system «Hamamatsu Photonics C9920-03» (Japan). The luminescence spectra were measured by Avantes-2048 fiber spectrometer (Netherlands) with a semiconductor laser with a peak wavelength of 405 nm. The mathematical processing of the data was carried out in the Origin Pro software.

III. RESULTS AND DISCUSSION

As part of this research work, a series of phosphate glasses doped with cerium was synthesized; their compositions are listed in Table 1. According to the available literature, there are no systematic studies describing the dependence of the spectral characteristics of these glasses on their composition and the irradiation dose. Therefore, in order to select an optimal matrix for radiation-resistant composites, these dependencies were investigated in the present study.

TABLE I.   COMPOSITIONS (MOL. %) AND DESIGNATIONS OF THE SYNTHESIZED GLASS SAMPLES

№	Name	Compositions				
		Al_2O_3	Na_2O	La_2O_3	P_2O_5	CeO_2
1	Ce2,45	8,16	30,4	0	59,00	2,45
2	Ce1,78	8,16	30,4	0,67	59,00	1,78
3	Ce1,20	8,16	30,4	1,25	59,00	1,20
4	Ce0,78	8,16	30,4	1,67	59,00	0,78
5	Ce0	8,16	30,4	2,45	59,00	0

A.   Protective effect of cerium on phosphate glasses

To demonstrate the protective effect of cerium in the selected glass system, absorption spectra of the glasses with different Ce contents were recorded before and after electron-beam irradiation (Fig. 1). It can be seen that in Ce-containing samples the fundamental absorption edge is shifted toward longer wavelengths. However, it is sharper compared to that of the glass without the protective additive. After irradiation with an electron fluence of $5*10^{13}$ e⁻ cm⁻² color centers appeared in the Ce-free samples at 430 nm and 510 nm corresponding to POHC-type defects formed in phosphate glasses due to hole trapping by PO groups [14].

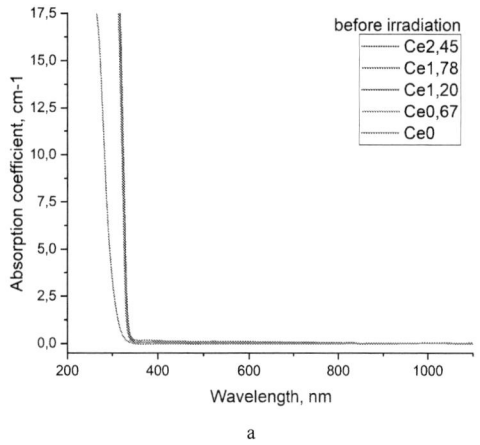
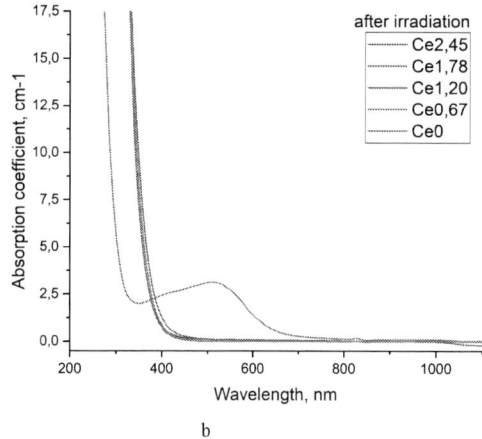

a   b

Fig. 1.   Absorption spectra of glasses with different Ce contents (a) before and (b) after electron irradiation with a fluence of $5*10^{13}$ e⁻ cm⁻²

It should be noted that, when irradiated with the same dose of ionizing radiation, the glasses with different cerium contents exhibit induced absorption of different magnitudes. The variation of optical transmittance as a function of irradiation is shown in Fig. 2 (a). It can be seen that, with increasing Ce content in the glass, the change in integral transmittance at the same irradiation dose becomes smaller.

An important parameter of glasses intended for operation under conditions of increased radiation exposure is their relaxation time. The dependence of the integral transmittance of the glasses on the time elapsed after irradiation with an electron fluence of $1.15*10^{15}$ e⁻ cm⁻² was measured (Fig. 2(b)). In constructing the plots the integral transmittance of the glass immediately after irradiation was normalized to 100 %. During relaxation, the transmittance an increase in the cerium concentration results in a larger recovery of transmittance within the same relaxation period.

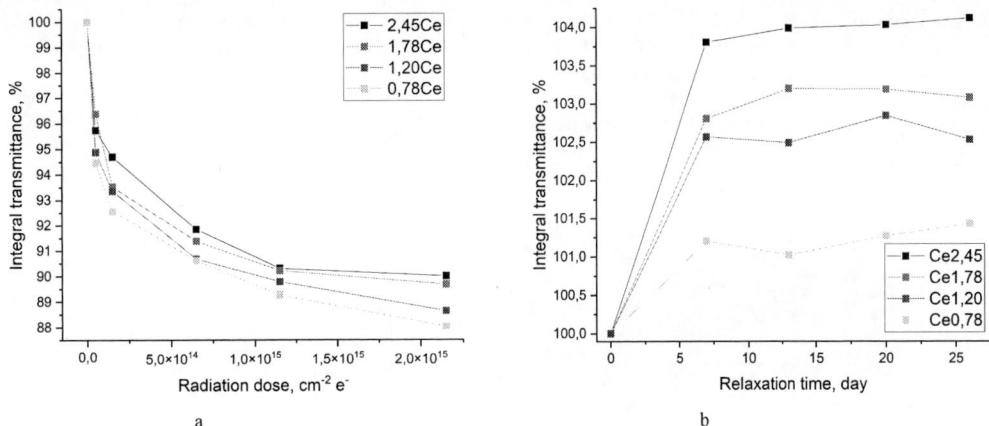

a

b

Fig. 2. Dependence of the integral transmittance of the glass samples on (a) the electron irradiation dose and (b) the time elapsed under normal conditions after irradiation with an electron fluence of $1.14*10^{15}$ e⁻ cm⁻²

Thus, an increase in the cerium concentration in the glass enhances the material's resistance to ionizing radiation. With increasing irradiation dose, the integral transmittance of the glass decreases to a lesser extent, while the relaxation occurs to a greater degree. Moreover, within the Ce concentration range of 0.78-2.45 mol. %, the position of the fundamental absorption edge varies only slightly, as can be seen in Fig. 1. These dependencies can be explained by the saturation of cerium-related traps at higher irradiation doses, which results in a smaller decrease in transmittance with increasing Ce content. The greater relaxation observed at higher cerium concentrations is associated with a larger number of trapped charge carriers capable of recombination.

B. Glass with perovskites

To substantiate the potential of phosphate glass as a matrix material, components required for the formation of perovskite crystals were introduced into the glass matrix without cerium oxide. The spectral and luminescent properties of the obtained samples were investigated. To identify the crystalline phase, XRD analysis was performed. The diffraction patterns of the glass-ceramic samples, together with the reference cards of the corresponding crystalline phases from the PDF database, are presented in Fig.3. Comparison of the diffraction patterns with the reference data clearly confirms the formation of the desired crystalline phase in the glass.

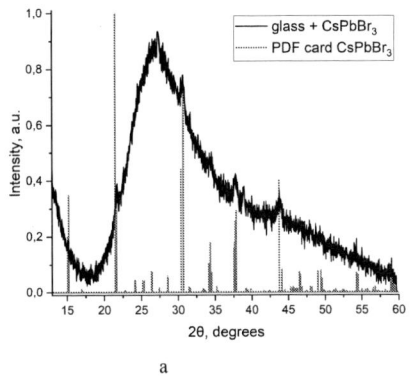

a

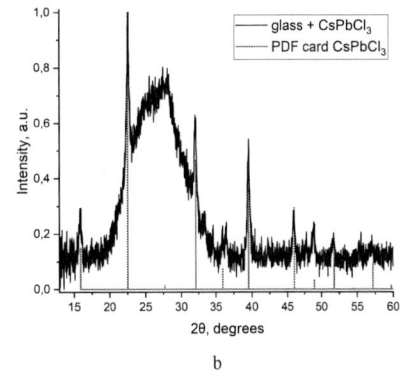

b

Fig. 3. XRD patterns of glass-ceramic samples containing perovskite phases: (a) CsPbBr3 and (b) CsPbCl3. Peaks are indexed according to the reference PDF cards

The luminescence spectra of the glass samples are shown in Fig. 4. The luminescence of $CsPbBr_3$ perovskite was excited using a 405 nm laser, while that of $CsPbCl_3$ perovskite was excited using a UV lamp with an emission peak at 365 nm. As can be seen from the figure, the obtained chloride and bromide perovskite crystals exhibit luminescence maxima at 410 nm and 508 nm respectively.

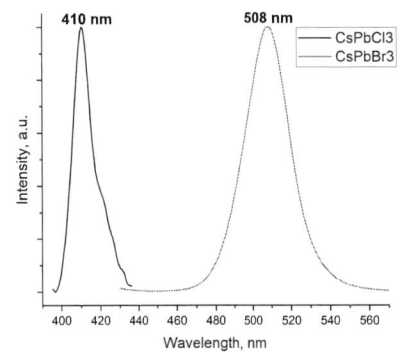

Fig. 4. Luminescence spectra of glass samples containing $CsPbBr_3$ and $CsPbCl_3$

The photoluminescence quantum yield of the samples was also measured. For the chloride perovskite it was found to be 0.23 %, while for the bromide perovskite it reached 64 %. This significant difference is most likely associated with the lower defects tolerance and lower decomposition energy of CsCl and PbCl2, which makes the decomposition of the chloride perovskite phase energetically favorable and renders it less stable under external influences [15, 16].

To demonstrate the high photoluminescence quantum yield of the glass containing bromide perovskite, its luminescence spectrum was recorded under solar illumination as shown in Fig.5.

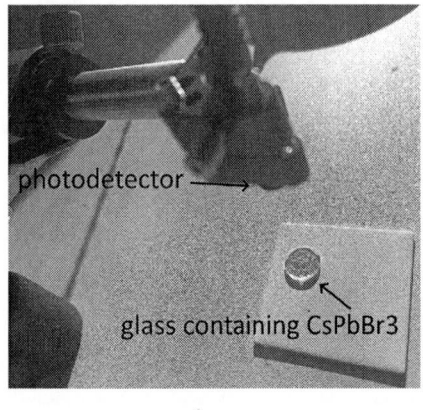

a

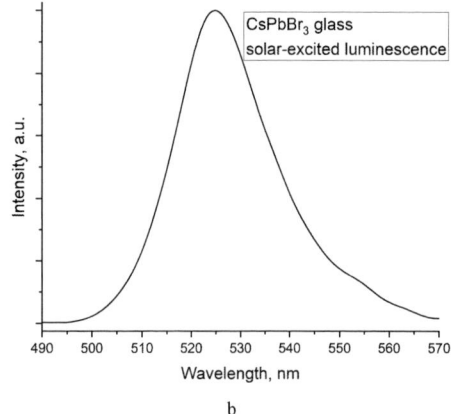

b

Fig. 5. Excitation of the CsPbBr3-containing glass under solar illumination (a) luminescence measurement process; (b) luminescence spectrum

Thus, the phosphate glass with the composition 8,16Al$_2$O$_3$-30,4Na$_2$O-59P$_2$O$_5$-2,45La$_2$O$_3$ can serve as a matrix for the nucleation of CsPbBr$_3$ and CsPbCl$_3$ perovskites.

C. Radiation-resistant glasses with perovskites

In the sample containing 1.78 mol% cerium oxide, a balance between cerium content and its effectiveness was maintained; therefore, this sample was subsequently used for the incorporation of components for CsPbBr3 and CsPbCl3 formation. The changes in the spectral characteristics of the glass after exposure to ionizing radiation were analyzed for sample containing bromide perovskite (Fig. 6). The absorption spectrum of the glass in the visible region exhibited no color centers, which indicates the absence of radiation-induced defects significant for material operation. After irradiation, the fundamental absorption edge shifted toward longer wavelengths. Moreover, the defects responsible for this effect showed almost no relaxation over time, as evidenced by the negligible shift of the transmission edge during storage of the glass under ambient conditions.

IV. CONCLUSION

In this work, a composition of radiation-resistant phosphate glass containing halide perovskites was developed, and a high photoluminescence quantum yield was achieved for the composite with CsPbBr$_3$ perovskite. The obtained glass withstands electron irradiation up to approximately $5*10^{15}$ e$^-$ cm^{-2}. The resulting films, in the case of bromide perovskite, exhibit low transparency in the visible range, while the films with chloride perovskite demonstrate low photoluminescence quantum yield. Therefore, neither material is suitable for enhancing the efficiency of photovoltaic elements. To achive the goal set at the beginning of this study, it is necessary to synthesize more complex structures based mixed perovskites. Furthermore, the conducted analysis demonstrated that the incorporation of cerium ions effectively suppresses the formation of radiation-induced defects in phosphate glass and provides a significant enhancement of its radiation resistance, confirming the feasibility of using Ce as a protective component of the matrix.

ACKNOWLEDGMENT

The study was carried out with the financial support of the Ministry of Science and Higher Education of the Russian Federation under the Strategic Academic Leadership Program " Priority 2030" (Agreement No. 075-15-2025-210 dated April 4, 2025).

REFERENCES

[1] Lewis N. S. Research opportunities to advance solar energy utilization //Science. – 2016. – T. 351. – №. 6271. – C. aad1920.

[2] Al-Ezzi A. S., Ansari M. N. M. Photovoltaic solar cells: a review //Applied System Innovation. – 2022. – T. 5. – №. 4. – pp. 67.

[3] Yuan X. et al. Thermal degradation of luminescence in inorganic perovskite CsPbBr 3 nanocrystals //Physical Chemistry Chemical Physics. – 2017. – T. 19. – №. 13. – C. 8934-8940.

[4] Zhou Y. et al. Nonlinear optical properties of halide perovskites and their applications //Applied Physics Reviews. – 2020. – T. 7. – №. 4.

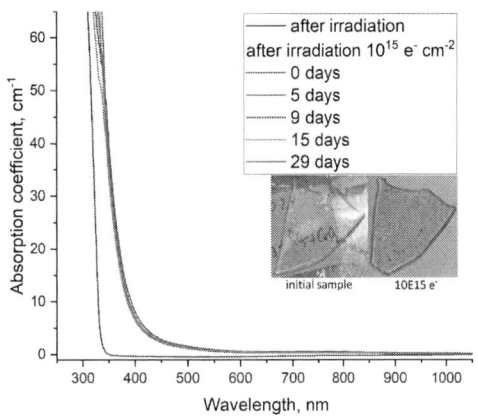

Fig. 6. Absorption spectra of the glass containing Ce and CsPbBr$_3$ before, immediately after, and some time after irradiation with ionizing radiation

[5] Schulz P., Cahen D., Kahn A. Halide perovskites: is it all about the interfaces? //Chemical reviews. – 2019. – T. 119. – № 5. – C. 3349-3417.

[6] Samiei S. et al. Exploring CsPbX3 (X= Cl, Br, I) perovskite nanocrystals in amorphous oxide glasses: innovations in fabrication and applications //Small. – 2024. – T. 20. – № 17. – C. 2307972.

[7] Li S. et al. CsPbX3 (X= Cl, Br, I) perovskite quantum dots embedded in glasses: Recent advances and perspectives // Chemical Engineering Journal. – 2022. – vol. 434. – P. 134593.

[8] Akinoglu B. G., Tuncel B., Badescu V. Beyond 3rd generation solar cells and the full spectrum project. Recent advances and new emerging solar cells //Sustainable Energy Technologies and Assessments. – 2021. – T. 46. – C. 101287.

[9] Svanström S. et al. X-ray stability and degradation mechanism of lead halide perovskites and lead halides //Physical Chemistry Chemical Physics. – 2021. – T. 23. – № 21. – C. 12479-12489.

[10] Klinkov V. et al. Effect of Gamma and Electron Irradiation on the Spectral and Luminescent Properties of a Composite Film with Halide Perovskite Nanocrystals //2024 International Conference on Electrical Engineering and Photonics (EExPolytech). – IEEE, 2024. – C. 429-431.

[11] Xinjie F. U., Lixin S., Jiacheng L. I. Radiation induced color centers in cerium-doped and cerium-free multicomponent silicate glasses //Journal of Rare Earths. – 2014. – T. 32. – № 11. – C. 1037-1042.

[12] McGrath B., Schönbacher H., Van de Voorde M. Effects of nuclear radiation on the optical properties of cerium-doped glass //Nuclear Instruments and Methods. – 1976. – T. 135. – № 1. – C. 93-97.

[13] Brow R. K. The structure of simple phosphate glasses //Journal of Non-Crystalline Solids. – 2000. – T. 263. – C. 1-28.

[14] Petit L. Radiation effects on phosphate glasses //International Journal of Applied Glass Science. – 2020. – T. 11. – № 3. – C. 511-521.

[15] Evarestov R. A. et al. First-principles comparative study of perfect and defective CsPbX 3 (X= Br, I) crystals //Physical Chemistry Chemical Physics. – 2020. – T. 22. – № 7. – C. 3914-3920.

[16] Peters J. A. et al. Defect levels in CsPbCl3 single crystals determined by thermally stimulated current spectroscopy //Journal of Applied Physics. – 2022. – T. 132. – № 3.

979-8-3315-4784-4/26 $31.00 © 2026 IEEE

Formation of a Contact Pad in the Ceramic Structure of an Electrostatic Chuck

Viktor S. Traktirshchikov
JSC NPP ESTO
Zelenograd, Moscow, Russia
National Research University of Electronic Technology
Zelenograd, Moscow, Russia
traktirshikoff@yandex.ru

Maksim E. Shiryaev
Head of the Department of Innovative Developments
LLC "Estika"
Zelenograd, Moscow, Russia

Viktor V. Kalugin
National Research University of Electronic Technology
Zelenograd, Moscow, Russia

Ivan A. Korotkevich
LLC "Estika"
Zelenograd, Moscow, Russia

Abstract - **The article presents the results of the search for the optimal way to form an electrical contact on the ceramic structure of an electrostatic chuck. We have considered four ways of forming an electrical contact. A ceramic mock-up was made to assess the quality and suitability of the formed areas of electrical contact. The analysis of the obtained contact pads allowed us to determine the optimal way to form electrical contacts. The method of filling thin channels with subsequent metallization of the contact area provides the most reliable contact pad.**

Keywords— electrostatic chuck, electrostatic chuck, electric contact, conductive paste

I. INTRODUCTION

The electrostatic fastening device (ESC) is a layered flat structure comprising insulating, electrode and working layers. The first and last are made of dielectric materials, in turn, the electrode layer is a thin layer of metallization (10-50 microns) deposited on one of the dielectric layers. The dielectric layers can be made as separate discs (if the material is solid). The layers are joined by various processes, such as gluing or splicing [1].

The principle of operation of the device, in the general case, is to hold the wafer placed on the ESC's working surface by an electrostatic field. The electrostatic field is formed by applying a high voltage to the electrode layer [2].

It is necessary to provide for the possibility of an electrical connection of the ESC with a high-voltage power supply unit for transmitting high voltage to the electrode layer of the ESC. Due to the fact that ESC is a complex in-chamber technological equipment on which nanostructured wafers are placed, the highest requirements are placed on the electrical connection, as well as on the device itself. For example, an electrical contact must have small overall dimensions, be well insulated from the conductive elements of the installation, have a reliable connection to the power supply, and so on [3].

The connection of the ESC electrode layer to the high-voltage power supply system can be carried out through a detachable or non-removable connection. In this case, the ESC design must provide an output, a contact pad, that is electrically connected to the electrode layer.

In order to determine the optimal implementation option for the contact pad of an electrostatic fastening device, it was decided to produce a mockup.

II. MATERIALS AND METHODS

Structurally, the mockup repeats the layered structure of the ESC and consists of a pair of flat aluminum oxide Al2O3 plates, one of which simulates an insulating layer with a thickness of 1.5 mm, the other simulates a working layer with a thickness of 0.3 mm. At the same time, there are many holes in the insulating ceramics for creating contact areas with diameters of 2, 3 and 4 mm, which allows us to consider several options for implementing electrical contacts of various sizes. An electrode layer (20-30 microns) was applied to the insulating ceramic. Areas of the electrode layer (20-30 microns) were applied to the working ceramics in places designated for the "hole without filling" method. After that, the Al2O3 plates were glued together with a special adhesive compound (the plates are oriented with an electrode layer inside). After gluing the plates, the provided holes were filled according to the methods under consideration (Figure 1).

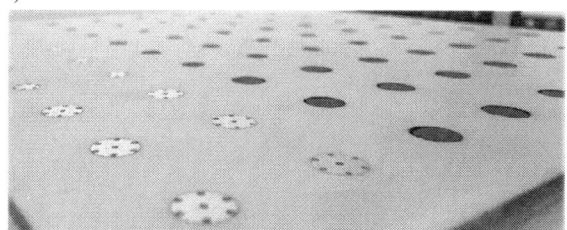

Fig. 1. Mockup with various types of contact pads

The mockup provides 4 versions of the contact pads (Figure 2):

1. *The hole without filling.* The method consists of creating through holes in the insulating (lower) layer of the ESC, which provides direct access to the electrode layer. The presence of the hole can also facilitate the centering of the mating contact part.

2. *Partial filling of the hole.* The method consists of creating through holes in the insulating layer and then partially filling them with a conductive paste, which thickens the electrode layer in the contact area to increase reliability.

979-8-3315-4784-4/26 $31.00 © 2026 IEEE

The presence of a partially filled hole can also facilitate the centering of the mating contact part.

3. Complete filling of the hole. The method consists of creating through holes in the insulating layer and then filling them with a conductive paste to the full depth, which thickens the electrode layer in the contact area to increase reliability and aligns the contact plane with the base plane of the ESC.

4. Complete filling of thin channels, followed by the application of a metal contact area. The method consists in creating a group of thin through channels in an insulating layer and then filling them with a conductive paste to the full depth. Next, a metallization layer is applied on top of the obtained channels to combine the channels into a common group.

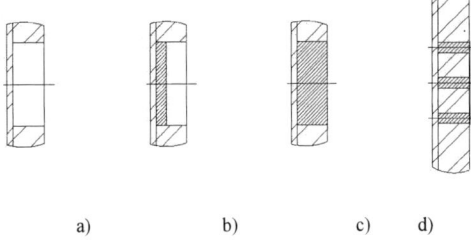

a)   b)   c)   d)

Fig. 2. Implementation options for contact pads: a) a hole without filling; b) partial filling of the hole; c) full filling of the hole; d) full filling of thin channels with subsequent application of a metal contact area.

III. RESULTS

The resulting mockup allows us to evaluate the ways of forming contact pads. Using a multimeter in the mode of checking the integrity of the electrical circuit, it was found that all the obtained contact pads have electrical contact with the inner electrode layer. Figure 3 shows the appearance of the contact pads.

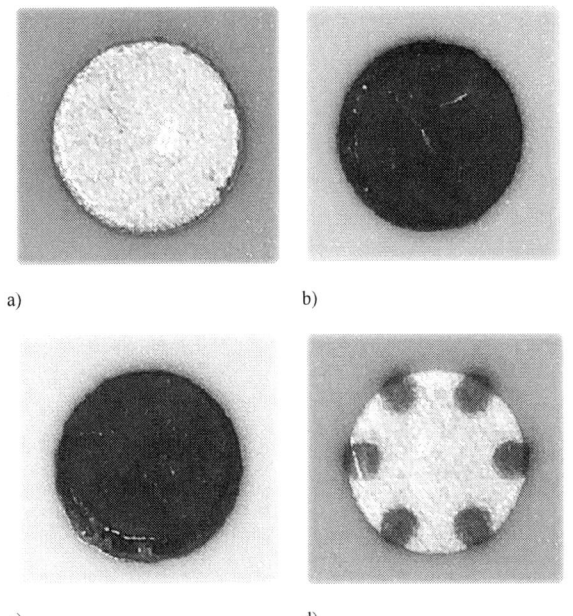

a)

c)   d)

Fig. 3. Appearance of the formed contact pads: a) holes without filling; b) partial filling of holes; c) full filling of holes; d) filling of thin channels with metallization Note: the option of partially filling the hole to a given depth turned out to be unfeasible in our conditions.

Despite the presence of an electrical contact, a number of contact pads have significant defects that exclude the normal operation of the device or reduce the reliability of the application.

The list of defects includes the following defects (Figure 4):

- peeling of the paste as a result of shrinkage;

- influx of adhesive composition on the contact surface;

- exceeding the filling level of the contact hole.

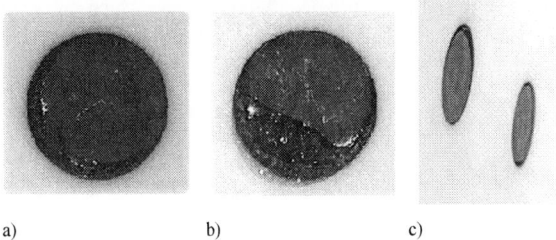

a)   b)   c)

Fig. 4. Defects of the contact pads: a) peeling of the paste due to shrinkage; b) influx of adhesive composition; c) excess of the filling level of the hole

With each method of forming contact pads, contact areas with diameters of 2, 3 and 4 mm were made. For greater clarity of the results, all contact pads and the presence of defects in them are compared in Table 1. The defect columns indicate the defect coefficient, which varies from 0 to 1, where 0 is the absence of defects in the group under consideration, and 1 is the presence of defects in each site of the group under consideration.

TABLE I.   DEFECTS OF THE OBTAINED CONTACT PADS

№	The method of forming contact pads	Diameter of the area, mm	Peeling off the paste	Influx of adhesive composition	Exceeding the filling level
1	*Holes without filling*	2	0	1	0
		3	0	0.7	0
		4	0	0.2	0
2	*Partial filling of holes*	2	0.9	0	1
		3	1	0.2	1
		4	1	0.1	1
3	*Full filling of the holes*	2	0.9	0	0
		3	1	0.2	0
		4	1	0.1	0
4	*Filling of thin channels with metallization*	2	0	0	0
		3	0	0	0
		4	0	0	0

IV. DISCUSSIONS

Analyzing the results obtained, the following conclusions can be drawn:

1. Holes without filling:

- During the bonding of ceramic plates, the adhesive composition seeps into the area of the contact pad, which can prevent the formation of electrical contact. At the same time, as the diameter of the holes increases, this type of defect decreases.

- Due to the fact that the contact area is applied to the ESC working layer, the thickness of which can vary from several tens of microns to hundreds of microns, there is a danger of destruction of the upper layer due to the mechanical impact of the counter contact part. In the case of using this method, it is necessary to take into account the impact of the mating part on the contact area.

2. Holes with full filling:

- - Filling holes with a diameter of 2-4 mm with a conductive paste is associated with significant shrinkage of the material, resulting in the detachment of the newly formed conductive area from the walls of the holes. As the size of the hole to be filled decreases, the degree of peeling decreases.

- - In the process of gluing ceramic plates, the adhesive composition seeps into the gap formed between the wall and the settled conductive paste, which can lead to an influx of glue onto the contact pad, but also additional fixation of the newly formed conductive body.

3. Partially filled holes:

- The same as with full-fill holes.

- It is practically difficult to fill holes with a diameter of 2-4 mm to a certain depth with a conductive paste. To achieve this goal, it is necessary to improve the method of applying the paste.

4. Complete filling of thin channels, followed by application of a metal contact area:

- - The manufacture of contact pads by the specified method was carried out without the above-mentioned defects.

V. CONCLUSIONS

In this paper, four methods of forming contact pads in the ceramic ESC structure were considered: direct use of the electrode layer, partial filling of the contact holes, complete filling of the contact holes, filling of a group of thin channels with subsequent application of a metal contact area. To carry out the assessment, a mock–up was made, repeating the layered structure of the ESC, for the formation of contact pads of various sizes - from 2 to 4 mm in diameter.

The entire array of contact pads obtained has an electrical contact with the electrode layer of the ESC. At the same time, some of the contact pads have defects that worsen or hinder their normal operation.

According to the results obtained, the use of partial and complete filling of holes with diameters from 2 to 4 mm is associated with significant defects, the largest of which is the peeling of the paste due to shrinkage.

The method of direct application of the ESC working layer or the method of holes without filling, which consists in creating through holes in the ESC insulating layer providing direct access to the electrode layer, has a high defect (influx of adhesive composition on the contact area) with hole sizes from 2 to 3 mm. As the size of the holes increases, the defect is significantly reduced. At the same time, in the case of using this method, it is necessary to take into account the force of the mechanical impact of the mating part on the contact area, since the ESC working layer has a small thickness and can be destroyed by minor mechanical impact.

All contact pads obtained by filling thin channels and then combining them by metallizing the contact area were made without defects for all the sizes of the contact areas under consideration.

In the case of the formation of contact pads from 2 to 4 mm in size in solid structures by filling holes with conductive pastes, it is recommended to use the method of filling of thin channels with metallization.

REFERENCES

[1] Asano, K., Hatakeyama, F., & Yatsuzuka, K. (2002). "Fundamental study of an electrostatic chuck for silicon wafer handling." IEEE Transactions on Industry Applications, 38(3), 840-845.

[2] Sun, Y., Cheng, J., Lu, Y., Hou, Y., & Ji, L. (2015). "Design space of electrostatic chuck in etching chamber." Journal of Semiconductors, 36(8), 084004.

[3] G. Kalkowski, S. Risse, S. Müller, G. Harnisch "Ultraplanare Elektrostatische Chucks für Next-Generation-Lithographie." Fraunhofer IOF Jahresbericht. 2005

Investigation of the Effect of Laser Annealing on the Structural Properties of BaSnTiO₃ Thin Films

Oleg E. Zaytsev
Department of Physical Electronics and Technology
Saint Petersburg Electrotechnical University "LETI"
Saint Petersburg, Russia
Ninbyarki1@gmail.com

Igor N. Zakasovsky
Department of Physical Electronics and Technology
Saint Petersburg Electrotechnical University "LETI"
Saint Petersburg, Russia
Igorzakasovskij1@gmail.com

Abstract—**Thin ferroelectric BaSnTiO₃ films were deposited on polycrystalline Al₂O₃ substrates by RF magnetron sputtering and subsequently annealed using a pulsed ytterbium fiber laser. A model for laser radiation absorption in semitransparent media and heat dissipation within the film was developed. The annealing parameters were optimized based on numerical modeling performed in COMSOL Multiphysics. The resulting X-ray diffractograms reveal a significant dependence of the film's structural properties on the laser annealing parameters, establishing a correlation between the modeled temperature and thermal annealing temperatures.**

Keywords— *thin films, ferroelectric, laser annealing, COMSOL*

I. INTRODUCTION

Ferroelectric materials in the microwave frequency range possess high nonlinearity of dielectric permittivity, relatively low losses, radiation resistance, and fast switching times (~10^{-11} s) [1–6].

Among ferroelectric materials, barium titanate-stannate (BTS) is one of the most promising. It exhibits a high dielectric constant at room temperature (2×10^4) and, due to the substitution of Ti atoms with more stable Sn atoms in the crystal lattice, demonstrates low dielectric losses [6]. This combination of properties makes BTS a highly attractive material for microwave electronics.

To improve the structural and, consequently, electrical properties of BTS films, thermal annealing with uniform and prolonged heating of the entire sample is required [6]. An alternative method, laser annealing, allows for localized and rapid heating, minimizing the thermal impact on the substrate by focusing energy onto the ferroelectric film [7]. However, the effect of laser annealing on thin barium titanate-stannate films has not been reported in the literature.

This work is devoted to studying the process of annealing BTS films with an ytterbium pulsed laser, based on data obtained from modeling the heating process of the ferroelectric material.

II. EXPERIMENT

Thin BaTi₀.₈Sn₀.₂O₃ films, deposited by high-frequency magnetron sputtering in an Ar/O₂ atmosphere (75/25 ratio) at a gas pressure of 2 Pa and a substrate temperature of 1200 K, were annealed in air using a Microset laser marker with an ytterbium radiation source. The laser parameters are listed in Table 1.

The annealing process was simulated using the COMSOL Multiphysics software package, employing the "Heat Transfer with Radiative Beam in Absorbing Media" module. The modeled sample consisted of a 1 μm thick BTS layer on a alumina substrate with a thickness of 50 μm. The scanning speed (the velocity of laser movement across the film surface) and the pulse repetition rate (pulses per second) were fixed at the maximum values for the setup (see Table 1). The absorption of laser radiation by the BTS film was accounted for using the Beer-Lambert law, with an absorption coefficient of 4×10^5 m^{-1} [8]. The simulation continued until the film reached its maximum temperature. A Gaussian distribution of the radiation power on the film surface was considered, with the standard deviation equal to the laser beam radius. The total power of the distribution corresponds to the source's operating mode power. Figure 1 shows the temperature distribution profile on the film surface at the end of the laser pulse.

The structure of the ferroelectric film was analyzed by X-ray diffraction (XRD) using a DRONE-6 diffractometer with monochromatic CuKα₁ radiation ($\lambda = 1.5406$ Å).

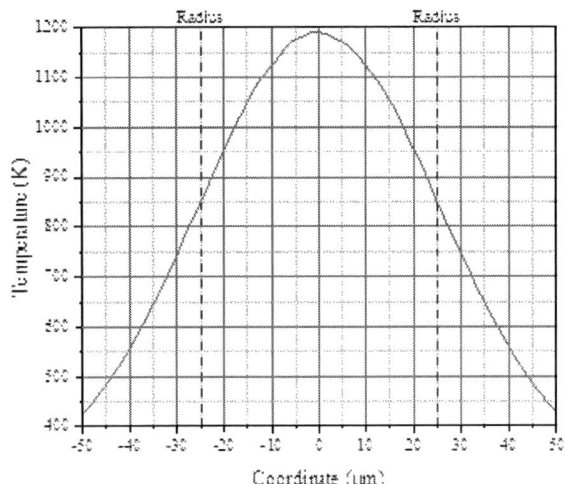

Fig. 1. Heat distribution at film surface

TABLE I. LASER PARAMETERS

Power, W	Pulse Time, ns	Laser beam diameter, μm	Scan speed, mm/s	Wavelength, nm	Line density, 1/mm	Pulse repetition rate, kHz
2-20	2-200	50	0-5000	1064	100-500	0-1000

III. RESULTS

The dependence of film temperature on pulse time and laser beam power was investigated through modeling; the results are presented in Fig. 2.

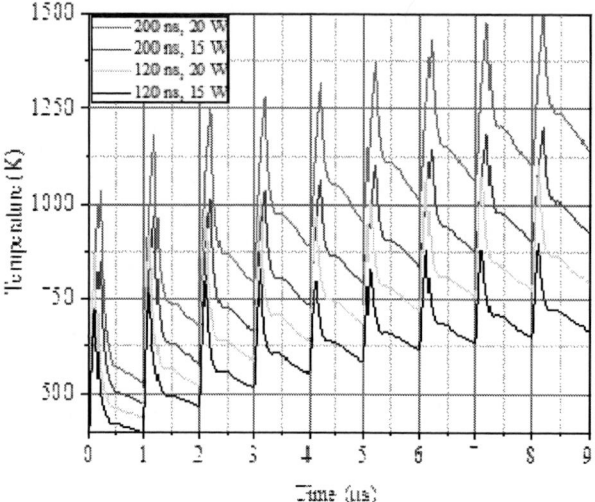

Fig. 2. Simulation results of temperature versus time under laser irradiation

The simulated annealing regimes cover a wide temperature range (800–1500 K), providing a basis for determining optimal thermal processing conditions for the films.

Table 2 presents the parameters of the experimental annealing regimes for thin BTS films, selected based on the simulation results.

TABLE II.   ANNEALING MODES

№	Power, W	Pulse Time, ns	Line density, 1/mm	Repetition count	Fluence, mJ/cm²
1	20	200	100	1	203
2	20	120	100	1	122
3	15	200	100	1	152
4	15	120	100	1	91

Figure 3 presents the XRD analysis results for samples 1–4, illustrating the influence of laser annealing temperature on the position and intensity of the (110) peak.

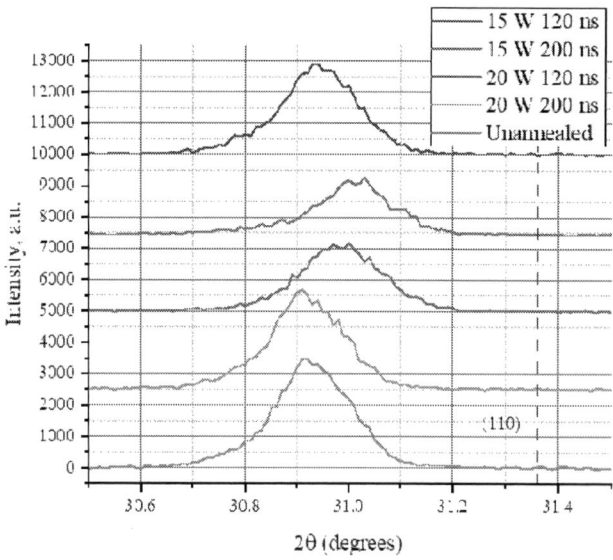

Fig. 3. Diffractograms of annealed films №1-4

The graph (Fig. 3) shows that the greatest influence on peak position is exerted by the 15 W, 200 ns regime, corresponding to a maximum simulated temperature of 1200 K. Analysis of the fluence reveals that the most significant results were achieved with pulses having close energy density values (120–150 mJ/cm²). The dashed line indicates the reference value for the (110) orientation in the $BaTi_{0.8}Sn_{0.2}O_3$ film.

To further improve film properties, it was decided to vary the number of beam interactions with the film, based on existing studies for other materials [7].

Parameters determining the number of times the laser beam passes over the same area (line density and number of repeated anneals) do not change the peak heating temperature. This is confirmed by Fig. 4, which shows that the area cools down before the next pass of the laser beam. Nevertheless, these parameters influence the structural properties due to repeated impacts on the annealed sample.

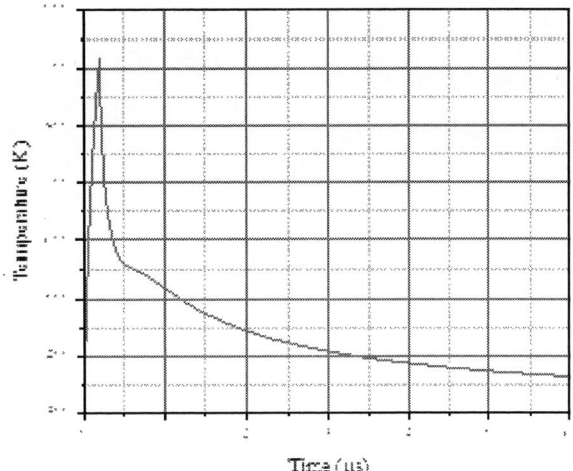

Fig. 4. Cooling curve for a single 20W 200ns pulse

To study the influence of line density and number of repetitions on the position of the (110) peak, the regime with the greatest shift, namely 15 W, 200 ns, was selected. The annealing regimes are listed in Table 3. The diffractograms of the samples are shown in Fig. 5.

The annealing results show a shift from the reference values with increasing line density. In the sample set for the number of repeated anneals, an approach to the reference value is observed at 10 anneals, while at 100 anneals, the values return to those of the unannealed sample.

TABLE III. ANNEALING MODES

№	Power, W	Pulse Time, ns	Line density, 1/mm	Repetition count
5	15	200	100	10
6	15	200	100	100
7	15	200	300	1
8	15	200	500	1

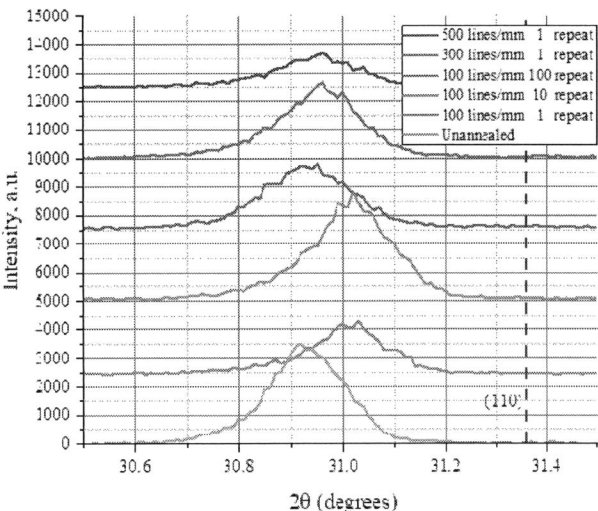

Fig. 5. Diffractograms of annealed films №3 and №5-8

IV. CONCLUSIONS

Based on the obtained results, laser annealing at fluence values of 120–150 mJ/cm² has a positive effect on the structural properties of BTS films. No influence on the diffractogram was detected when exceeding the threshold of 200 mJ/cm² or failing to reach 100 mJ/cm². Increasing the number of repeated anneals to 10 shifts the position of the (110) texture closer to the reference. Further impacts returned the peak to its initial position. Achieving a line density of 300 lines/mm increased the peak intensity, but 500 lines/mm led to a deterioration of characteristics compared to unannealed samples. Further research is required to improve the structure of BTS.

REFERENCES

[1] Vendik, O.G.; Zubko, S.P. Ferroelectrics as constituents of tunable metamaterials. in theory and thenomena of metamaterials; CRC Press: Boca Raton, FL, USA, 2017; p. 33-1.

[2] Crunteanu, A.; Muzzupapa, V.; Ghalem, A.; Huitema, L.; Passerieux, D.; Borderon, C.; Gundel, H.W. Characterization and performance analysis of BST-based ferroelectric varactors in the millimeter-wave domain. Crystals 2021, Vol. 11, p. 277.

[3] Aymen, S.; Mascot, M.; Jomni, F.; Carru, J.C. High tunability in lead-free Ba0.85Sr0.15TiO3 thick films for microwave tunable applications. Ceram. Int. 2019, Vol. 45, pp. 23084–23088.

[4] Pronin, I.P.; Kaptelov, E.Y.; Tarakanov, E.A.; Afanas'ev, V.P. Effect of annealing on the self-poled state in thin ferroelectric films. Phys. Solid State 2002, Vol. 44, pp. 1736–1740.

[5] Firsova, N.Y.; Mishina, E.D.; Sigov, A.S.; Senkevich, S.V.; Pronin, I.P.; Kholkin, A.; Yuzyuk, Y.I. Femtosecond infrared laser annealing of PZT films on a metal substrate. Ferroelectrics 2012, Vol. 433, pp. 164–169.

[6] Tumarkin A., Sapego E., Gagarin A., Senkevich S., Enhanced tunability of BaTixSn1− xO3 films on dielectric substrate //Applied Sciences 2021, Vol. 11, №. 16., p. 7367.

[7] Queraltó A. et al. Ultraviolet pulsed laser crystallization of Ba0.8Sr0.2TiO3 films on LaNiO3-coated silicon substrates //Ceramics International, 2016. Vol. 42, №. 3. pp. 4039-4047. DOI: 10.1016/j.ceramint.2015.11.075

[8] Kraft V. et al. Influence of tin concentration on the electronic structure and ferroelectric behavior of barium titanate: Experimental and first-principles insights //Journal of Applied Physics, 2025 Vol. 138, №. 9. DOI: 10.1063/5.0277365

Study of a Resistive-Type Superconducting Fault Current Limiter (SFCL) for Fault Protection in a Microgrid

Qusai K. Al Naimi
Institute of Energy
Peter The Great Saint-Petersburg
Polytechnic University
Saint-Petersburg , Russian Federation
ORCID: 0009-0008-5223-872X

Abdulhadi Haj Ahmad
Institute of Energy
Peter The Great Saint-Petersburg
Polytechnic University
Saint-Petersburg , Russian Federation
abdulhadi.hajahmad@mail.ru

Hayder J. Mohammed
Institute of Energy
Peter the Great saint Petersburg
Polytechnic University
Saint Petersburg , Russian Federation
hdr_jsm@uomisan.edu.iq

Abstract— Short circuit faults, which can result in excessive currents, damage to equipment, and destabilize the system, are becoming more widespread in small electrical grids with a high penetration of renewable energy sources. Typical protection mechanisms frequently find it difficult to react quickly enough to these unexpected incidents. The application of a resistive-type superconducting fault current limiter (SFCL) as a microgrid fault mitigation solution is examined in this work. The SFCL was modeled using MATLAB/Simulink, considering the critical current, thermal dynamics, and transition properties of the superconductor. Both with and without the SFCL, the microgrid was simulated under a range of fault scenarios. As demonstrated by the results, the SFCL improves grid stability and maintains sensitive equipment by efficiently limiting fault currents, lowering voltage sag, and speeding up system recovery. These results show that resistive SFCLs present a quick, reliable, and effective way to protect small electrical grids from fault occurrences.

Keywords— *Superconducting Fault Current Limiter, Resistive-Type SFCL, Micro Grid, High-Temperature Superconductors (HTS), Grid Stability Enhancement, Renewable Energy Integration, Transient Fault Analysis.*

I. INTRODUCTION

The integration of distributed generation (DG) resources, such as solar photovoltaic (PV) farms, wind turbines, and fuel cells, into modern power systems has given rise to the concept of the microgrid (MG) [24-28]. A microgrid is a localized group of DGs, energy storage systems, and loads that can operate in both grid-connected and islanded (stand-alone) modes [1, 29, 31]. This concept enhances power system resilience, improves power quality, facilitates the use of renewable energy, and can reduce transmission losses [2]. However, the transition from a traditional, centralized radial distribution network to a bidirectional, active network with multiple generation sources introduces significant engineering challenges. Among these, ensuring reliable and selective protection is paramount. The dynamic nature of fault currents—which vary considerably between grid-connected and islanded modes of operation—renders conventional overcurrent protection schemes ineffective, often leading to miscoordination, nuisance tripping, or failure to operate [3]. Consequently, there is a pressing need for advanced protective devices and adaptive schemes to safeguard microgrid infrastructure and ensure system stability. The Superconducting Fault Current Limiter (SFCL) has emerged as a promising technology to address these challenges, offering a fast, self-triggering, and effective means of controlling fault current levels without compromising normal operation.

The protection of microgrids is fundamentally complicated by their structural and operational characteristics. Unlike passive radial distribution networks, microgrids feature bidirectional power flow due to the presence of multiple DG units at various nodes [4]. This bidirectional flow disrupts the unidirectional fault current assumption upon which traditional directional overcurrent relays are coordinated. Furthermore, the magnitude of fault current is highly variable. When connected to the main grid, fault currents are substantial due to the high fault contribution from the utility source. In contrast, in islanded mode, fault currents are significantly lower—often only 2–3 times the rated current for inverter-interfaced DGs—which may fall below the pickup thresholds of relays set for grid-connected operation [5, 6]. This disparity creates a protection blind spot and necessitates either adaptive relay settings or supplementary devices.

Existing protection strategies can be broadly categorized into communication-based, local measurement-based, and external device-based methods. Communication-based schemes, such as differential protection or adaptive relaying using centralized controllers, offer precision but add cost, complexity, and a vulnerability to communication failure [7, 8]. Local measurement-based techniques, including voltage-based protection [9] or distance relaying, reduce dependency on communications but may lack sensitivity or be highly dependent on specific network configurations [10]. Therefore, the deployment of external devices like Fault Current Limiters (FCLs) presents a compelling alternative to modify the network's fault response directly, thereby simplifying the protection coordination problem.

The Superconducting Fault Current Limiter (SFCL) is a revolutionary protective device that leverages the unique properties of high-temperature superconducting (HTS) materials. Its core operational principle is based on the nonlinear transition of a superconductor from a zero-resistance state to a high-resistance state when subjected to fault conditions.[23]

Basic Operation: Under normal operating conditions, the current, temperature, and magnetic field affecting the HTS element remain below their critical thresholds (Ic, Tc, Bc). In this superconducting state, the SFCL presents negligible impedance (virtually a short circuit) to the system, resulting in zero power loss [11]. When a fault occurs in the system, the fault current surge rises above the critical current of the superconductor and also the current density. This initiates an sudden transition state known as "quenching", that known as the state of losing superconductivity state and the transition to normal conducting state of the superconductor. This instant increase in impedance, which can occur within the

979-8-3315-4784-4/26 $31.00 © 2026 IEEE

first quarter-cycle of the fault effectively limits the prospective fault current to a controllable level. Afterwards, the fault current removed from the system using protection devices and the superconductor temperature will decrease below the critical temperature value leading to regaining the superconducting state and ready for next operation [12].

Numerous key SFCL technologies have been developed, each with different operational principles. The most common type is the Resistive SFCL (R-SFCL), where the superconductor is positioned directly in series with the power line. This design offers advantages in simplicity, fast response, and efficiency, often incorporating a parallel shunt resistor or inductor to mitigate excessive voltage rise and localized heating, or "hot spots," during its quenching phase [13]. In contrast, the Inductive SFCL (I-SFCL) operates on a transformer principle. Its primary winding is series-connected to the line, while the secondary winding consists of a short-circuited high-temperature superconducting (HTS) coil. During normal operation, this coil provides a shielding effect, resulting in low impedance. When a fault occurs and the HTS coil quenches, high impedance is introduced into the primary circuit as magnetic flux couples with the core. While this design reduces thermal stress on the superconductor itself, it typically results in a larger and heavier apparatus [14]. More advanced designs, known as Hybrid SFCLs, seek to optimize performance by combining superconducting elements with power electronic switches or saturable cores. This approach aims to enhance performance characteristics, reduce the volume of expensive superconducting material required, and improve the management of energy dissipation during fault events [15].

II. RESISTIVE-TYPE SFCLs

The resistive-type Superconducting Fault Current Limiter (R-SFCL) is described by its simple design and obvious suitability for power systems. It operates based on the real behavior of High-Temperature Superconductor (HTS) materials, which demonstrate negligible electrical resistance when maintained below critical thresholds of temperature (Tc) and current (Ic). During normal operation, the HTS component conducts load current without losses, maintaining grid efficiency.

Upon a fault that surpasses Ic, the HTS element rapidly undergoes a "quench" — transitioning to a normal resistive state. This sudden increase in resistance inserts substantial impedance into the circuit within the first fault cycle, effectively limiting the peak fault current and safeguarding equipment like circuit breakers from damage [20].

Advantages of R-SFCLs include a more compact and lightweight construction compared to inductive types, easing integration into existing networks. The use of HTS tapes further improves performance due to their high current density and swift self-recovery once the fault clears and conditions return to normal [21]. To enhance reliability during quenching, a parallel bypass branch (resistive or inductive) is typically incorporated to manage localized heating and overvoltages [22].

Within micro grids, where distributed generation can raise fault levels, the R-SFCL provides a fast, autonomous current-limiting solution, making it particularly valuable for protecting distribution networks and micro-grid interconnection points. Subsequent case studies illustrate its practical performance in grids with renewable integration.

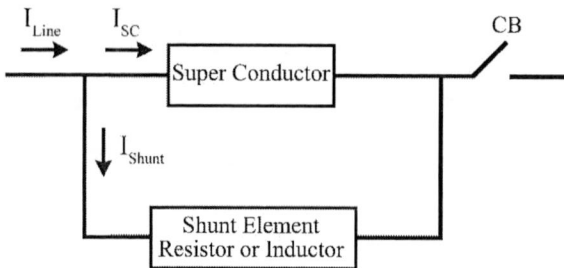

Fig. 1. Basic structure of resistive superconducting fault current limiter

In this paper, a three-phase R-SFCL is modeled with the following key parameters, as outlined in Table I:

TABLE. I. R-SCFL PARAMETERS

Parmeter	Value	Unit
l_{sc}	50	m
d_{sc}	0.004	M
T_a	77	K
Tc	95	K
α	6	-
β	3	-
J_c (77 K)	1.5e7	A/m²
E_0	0.1	V/m
ρ	1e-6	Ωm
K	1.5e3	W/Km²
C_v	1.0e6	J/Km³

III. POWER SYSTEM MODEL

The power system model expresses a micro grid isolated power system that includes 3 buses where a synchronous generator was attached to the first bus and a wind turbine to the second bus and a load were attached as well to the third bus. As presented in the figure (2).

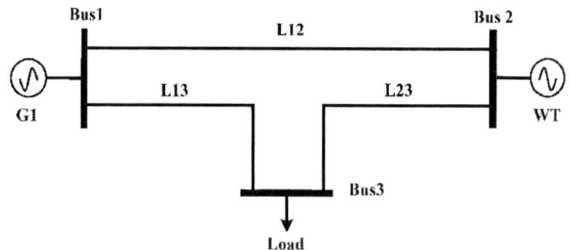

Fig. 2. Basic structure of the studied micro grid power system

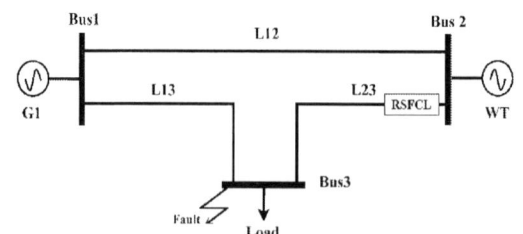

Fig. 3. Fault on bus (3) with SCFL connected to L (2-3)

IV. RESULTS AND DISCUSSION

The simulation results, summarized in the table, demonstrate the significant impact of implementing a Resistive-Type Superconducting Fault Current Limiter (SFCL) on the microgrid's response to a three-phase fault at Bus 3. For the simulation, four scenarios have been tested in order to study the impact of the SCFL on the micro grid. The first scenario was implemented to study the three-phase fault at the bus (3) without the integration of SCFL in the micro grid as presented in figure (2). While, the second scenario is showing three-phase fault on the third bus while the SCFL was implemented in line between the second and the third bus as presented in figure (3). This scenario will explain the impact of the implementation of the SCFL on this line and the other lines in the grid.

The third scenario was studied about the implementation of the SCFL on the output of the wind turbine in bus (2) as presented in figure (4). The last scenario was studied about the occurrence of the three phase fault also on bus (3) as the SCFL connected to the output of the generator on the bus (1) as presented in figure (5). All the values of the fault current in each line and the voltage drops also were presented in Table II. The simulation results clearly demonstrate the effectiveness of the resistive-type Superconducting Fault Current Limiter (SFCL) in enhancing the protection scheme of the studied microgrid. The analysis focuses on a three-phase fault at Bus 3 under different SFCL placement scenarios.

Fault Current Limitation: Without the use of SFCL, the fault currents in lines L12, L13, and L23 are extensive (ranging from 2797 A to 3370 A). The installation of an SFCL significantly reduces these currents, with the most proven effect observed on the line where the SFCL is directly installed. For instance, placing SFCL on line L23 suppresses its fault current from 2797 A to a simple 203.6 A, a remarkable reduction of over 93%. Correspondingly, installing the SFCL on the Wind Turbine (WT) or Generator G1 output leads to a significant and balanced decline in fault currents across all lines (to approximately 945-1942 A), indicating a system-wide benefit. This applicable decrease directly mitigates the risk of thermal and mechanical damage to cables, transformers, and other connected equipment. The figures (6)(7) show the significant degradation of the fault current on the line due to the installation of SCFL.

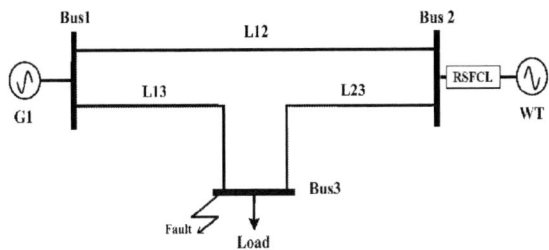

Fig. 4. Fault on bus (3) with SCFL connected to WT output

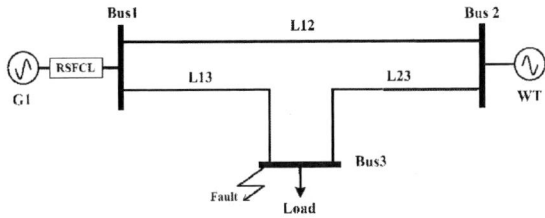

Fig. 5. Fault on bus (3) with SCFL connected to G1 output

Voltage Profile Improvement: During the fault, the bus voltage at the faulted bus (Bus 3) experiences a severe sag to 0.1 pu without SFCL. The SFCL's impedance during the fault state significantly improves this voltage profile. With the

SFCL on L23, the voltage at Bus 3 during the fault recovers to **0.2253 pu**. More particularly, when the SFCL is placed at the WT or G1 output, the voltage during the fault is maintained at significantly higher levels (**0.2040 pu to 0.4856 pu**). This reduction in voltage sag is critical for maintaining the stability of other healthy sections of the microgrid and preventing the tripping of sensitive loads.

Impact of SFCL Placement: The results underscore that the strategic placement of the SFCL is crucial. Locating it on the faulted feeder (L23) provides the strongest localized protection for that line but offers less improvement for fault currents coming from other sources. Conversely, installing the SFCL at the output of major generation sources (WT or G1) provides a more complete protection scheme. It not only limits the fault current contribution from that specific source but also facilitates a more balanced current distribution from other paths, leading to a better overall voltage profile during the fault, as seen in the higher and more uniform voltage values (0.1060-0.2040 pu).

Figure (8) shows the change in the resistance of the SCFL in normal conditions for a value equals zero, and in fault condition as the resistance of the SCFL rises to limit the fault current and try to quench it later.

TABLE. II.   RESULTS OF THE CURRENT AND VOLTAGE VALUES DURING IN ALL CASES

		Fault current (A)	Voltage during Fault (Vp)
Fault on bus (3) without SCFL	L (1-2)	0.1	2937
	L (1-3)	3370	2937
	L (2-3)	2797	2937
Fault on bus (3) with SCFL connected to L (2-3)	L (1-2)	2115	4856
	L (1-3)	4628	4856
	L (2-3)	203.6	225.3
Fault on bus (3) with SCFL connected to WT output	L (1-2)	945.9	2040
	L (1-3)	1942	2040
	L (2-3)	1006	1060
Fault on bus (3) with SCFL connected to G1 output	L (1-2)	945.9	1060
	L (1-3)	1006	1060
	L (2-3)	1941	2040

979-8-3315-4784-4/26 $31.00 © 2026 IEEE

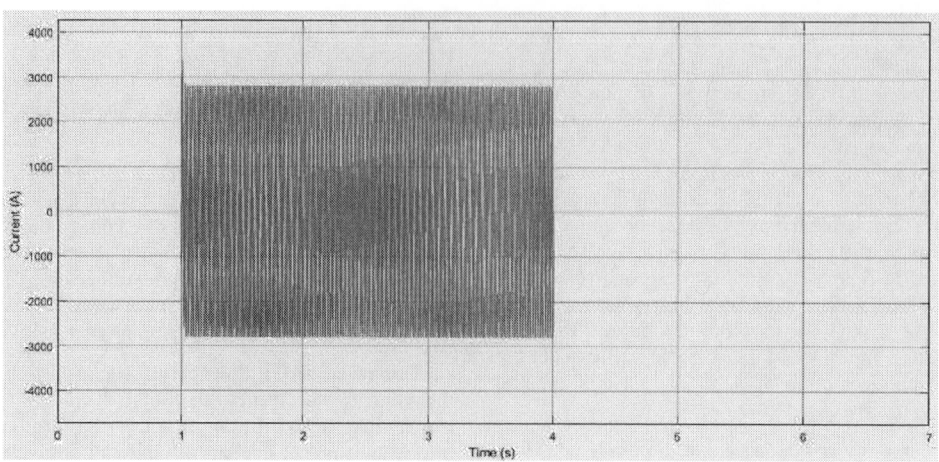

Fig. 6. Fault current on L (2-3) during fault without the integration of SCFL

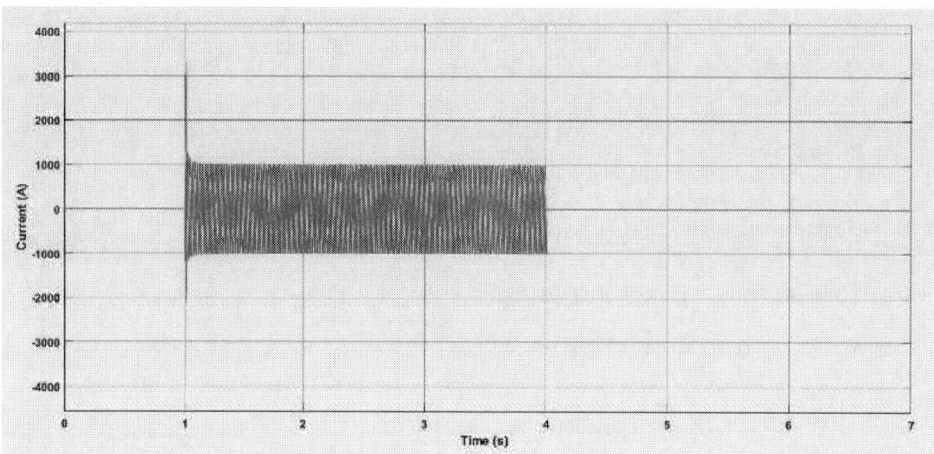

Fig. 7. Fault Current on L (2-3) after the integration of SCFL

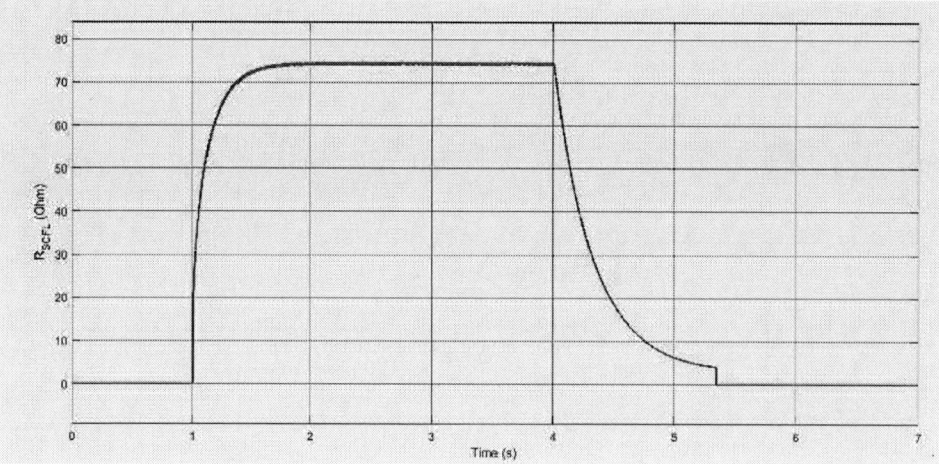

Fig. 8. The behavior of resistance of SCFL during fault condition

V. CONCLUSION

This study demonstrates the significant potential of resistive-type Superconducting Fault Current Limiters (SFCLs) as a robust and efficient solution for fault protection in renewable-rich microgrids. After preforming simulation for many cases, SCFL improves its ability to mitigate the excessive fault currents and its ability to improve the voltage stability during fault occurrence. Based on the results, SCFL significantly reduced the fault currents in the micro grid by over 90%, thus protecting sensitive equipment in the grid from thermal and electrodynamic stress. Concurrently, it substantially improves voltage sags at the faulted bus, enhancing the overall stability and ride-through capability of the microgrid. Furthermore, the analysis emphasizes that the SFCL's performance is well dependent on its placement within the network. While installation on a critical feeder offers targeted protection, positioning it at the output of distributed generators provides a more balanced and system-wide improvement in both current limitation and voltage support. The resistive SFCL presents a fast, reliable, and adaptable technological option to strengthen microgrid protection schemes. Its integration can facilitate the safe and stable operation of modern electrical grids with high penetration of renewable resources, ensuring compliance with protection coordination and paving the way for more resilient distributed energy systems.

REFERENCES

[1] Mirsaeidi, S., Said, D. M., Mustafa, M. W., Habibuddin, M. H., & Ghaffari, K. (2014). Progress and problems in micro-grid protection schemes. *Renewable and Sustainable Energy Reviews*, *37*, 834-839.

[2] Sung, B. C., Park, D. K., Park, J. W., & Ko, T. K. (2009). Study on a series resistive SFCL to improve power system transient stability: modeling, simulation, and experimental verification. *IEEE transactions on industrial electronics*, *56*(7), 2412-2419.

[3] Laaksonen, H. J. (2010). Protection principles for future microgrids. *IEEE Transactions on power electronics*, *25*(12), 2910-2918.

[4] Dewadasa, M., Ghosh, A., & Ledwich, G. (2011, September). Protection of microgrids using differential relays. In *AUPEC 2011* (pp. 1-6). IEEE.

[5] Nikkhajoei, H., & Lasseter, R. H. (2007, June). Microgrid protection. In *2007 IEEE Power Engineering Society General Meeting* (pp. 1-6). IEEE.

[6] Memon, A. A., & Kauhaniemi, K. (2015). A critical review of AC Microgrid protection issues and available solutions. *Electric Power Systems Research*, *129*, 23-31.

[7] Zeineldin, H. H., El-Saadany, E. F., & Salama, M. M. A. (2006, March). Distributed generation micro-grid operation: Control and protection. In 2006 Power Systems Conference: Advanced Metering, Protection, Control, Communication, and Distributed Resources (pp. 105-111). IEEE.

[8] Swathika, O. G., & Hemamalini, S. (2016). Prims-aided Dijkstra algorithm for adaptive protection in microgrids. *IEEE Journal of Emerging and Selected Topics in Power Electronics*, *4*(4), 1279-1286.

[9] Redfern, M. A., & Al-Nasseri, H. (2008, March). Protection of micro-grids dominated by distributed generation using solid state converters. In IET 9th International Conference on Developments in Power Systems Protection (DPSP 2008) (pp. 669-673). Stevenage UK: IET.

[10] Dewadasa, M., Ghosh, A., & Ledwich, G. (2009, January). An inverse time admittance relay for fault detection in distribution networks containing DGs. In *TENCON 2009-2009 IEEE Region 10 Conference* (pp. 1-6). IEEE.

[11] Young, M., & Hassenzahl, W. (2009). Superconducting fault current limiters: Technology watch 2009. *Electric Power Research Institute (EPRI), Palo Alto*.

[12] Tixador, P. (2002). Superconducting current limiters-some comparisons and influential parameters. *IEEE transactions on applied superconductivity*, *4*(4), 190-198.

[13] Lim, S. H., Choi, H. S., Chung, D. C., Jeong, Y. H., Han, Y. H., Sung, T. H., & Han, B. S. (2005). Fault current limiting characteristics of resistive type SFCL using a transformer. *IEEE transactions on applied superconductivity*, *15*(2), 2055-2058.

[14] Didier, G., Bonnard, C. H., Lubin, T., & Lévêque, J. (2015). Comparison between inductive and resistive SFCL in terms of current limitation and power system transient stability. *Electric Power Systems Research*, *125*, 150-158.

[15] Ebrahimpour, M., Vahidi, B., & Hosseinian, S. H. (2013). A hybrid superconducting fault current controller for DG networks and microgrids. *IEEE transactions on applied superconductivity*, *23*(5), 5604306-5604306.

[16] Khan, U. A., Seong, J. K., Lee, S. H., Lim, S. H., & Lee, B. W. (2010). Feasibility analysis of the positioning of superconducting fault current limiters for the smart grid application using simulink and simpowersystem. *IEEE Transactions on Applied Superconductivity*, *21*(3), 2165-2169.

[17] Khan, U. A., Seong, J. K., Lee, S. H., Lim, S. H., & Lee, B. W. (2010). Feasibility analysis of the positioning of superconducting fault current limiters for the smart grid application using simulink and simpowersystem. *IEEE Transactions on Applied Superconductivity*, *21*(3), 2165-2169.

[18] Ghanbari, T., & Farjah, E. (2012). Unidirectional fault current limiter: An efficient interface between the microgrid and main network. *IEEE Transactions on Power Systems*, *28*(2), 1591-1598.

[19] Noe, M., & Steurer, M. (2007). High-temperature superconductor fault current limiters: concepts, applications, anddevelopment status. *Superconductor science and technology*, *20*(3), R15.

[20] Kozak, S., Janowski, T., Kondratowicz-Kucewicz, B., Kozak, J., & Wojtasiewicz, G. (2005). Experimental and numerical analysis of energy losses in resistive SFCL. *IEEE transactions on applied superconductivity*, *15*(2), 2098-2101.

[21] Kim, C. H., Lee, K. M., & Ryu, K. W. (2006). A numerical study on temperature increase in the resistive SFCL element due to the quench condition. *IEEE transactions on applied superconductivity*, *16*(2), 636-641.

[22] Pei, X., & Smith, A. C. (2016). Experimental testing and development of improved modelling for multistrand resistive SFCL. *IEEE Transactions on Applied Superconductivity*, *26*(4), 1-5.

[23] V. M. Govor, A. G. Kalimov, S. Bagan, E. N. Kobzar and E. R. Mannanov, "Numerical Simulation of Frequency-dependent AC Transport Losses in HTS 2G Tape with Copper Stabilizer," *2022 Conference of Russian Young Researchers in Electrical and Electronic Engineering (ElConRus)*, Saint Petersburg, Russian Federation, 2022, pp. 1174-1178, doi: 10.1109/ElConRus54750.2022.9755604.

[24] Hayder J. Mohammed, Mushtaq A. Al-Furaiji, Qays A. ALI, Ahmed M. Al-Antaki, H.A. Issa, Study of the MPPT for PV Systems Using Simulation in MATLAB/Simulink. 2023 Seminar on Industrial Electronic Devices and Systems (IEDS), Saint Petersburg, Russian Federation, 2023, pp. 181-185, doi: 10.1109/IEDS60447.2023.10425824.

[25] Qays Adnan Ali, Mushtaq A. Al-Furaij, Hayder J. Mohammed, Naji Abdullah Mezaal, Using Solar Systems for the Power Supply of Baghdad City in Iraq, 2023 Seminar on Electrical Engineering, Automation & Control Systems, Theory and Practical Applications (EEACS), Saint Petersburg, Russian Federation, 2023, pp. 24-30, doi: 10.1109/EEACS60421.2023.10397422.

[26] Hayder J. Mohammed, Qays A. Ali, Bishro H. Rasool, Asra H. Al-Denenawe, Analysis of the Operating Mode for the Photovoltaic Module by Experimental Methods. 2024 Conference of Young Researchers in Electrical and Electronic Engineering (ElCon), Saint Petersburg, Russian Federation, 2024, pp. 831-834, doi: 10.1109/ElCon61730.2024.10468463.

[27] Hayder J. Mohammed, Mohammed S. Hasan, Faeza Mahdi Hadi, Korovkin Nikolay V., Study the Possibility of Application of Wind Turbine under the Climatic Conditions of Iraq, 2022 Conference of Russian Young Researchers in Electrical and Electronic Engineering (ElConRus), Saint Petersburg, Russian Federation, 2022, pp. 1234-1238, doi: 10.1109/ElConRus54750.2022.9755597.

[28] Bishro H. Rasool, Hayder J. Mohammed, Qays A. Ali., Response of 100Kw Grid Connected PV System at Slava City – Iraq in the Event of Short Circuit and Loss of Grid, 2024 Conference of Young Researchers in Electrical and Electronic Engineering (ElCon), Saint Petersburg, Russian Federation, 2024, pp. 795-799, doi: 10.1109/ElCon61730.2024.10468473.

[29] H. J. Mohammed and N. V. Korovkin, "Improving Energy Demand Management Through Applying a Smart Electricity Grid for Energy Sector of Iraq," 2025 XXVIII International Conference on Soft Computing and Measurements (SCM), Saint Petersburg, Russian Federation, 2025, pp. 237-242, doi: 10.1109/SCM66446.2025.11060200.

[30] H. J. Mohammed and N. V. Korovkin, "Study of Artificial Intelligence Methods and Systems in Renewable Energy for Iraq: A Review," 2025 XXVIII International Conference on Soft Computing and Measurements (SCM), Saint Petersburg, Russian Federation, 2025, pp. 357-362, doi: 10.1109/SCM66446.2025.11060409.

[31] Mohammed, H. J., & Vladimirovich, K. N. (2024). Optimization Performance PV system by using Theoretical and Practical Methods to Improve Energy Sector of Iraq, The 7th International Conference on Renewable Energy and Environment Engineering (REEE 2024), Nantes, France, 28-30 August, 2024, In E3S Web of Conferences (Vol. 572, p. 01005). EDP Sciences, https://doi.org/10.1051/e3sconf/202457201005.

979-8-3315-4784-4/26 $31.00 © 2026 IEEE

Plasma-Enhanced Atomic Layer Deposition of WO₃ Using Tungsten Hexacarbonyl and Activated Oxygen

Vladislav E. Zamoshets
Peter the Great Saint-Petersburg Polytechnic University
Saint-Petersburg, Russia
Zamoshets.ve@edu.spbstu.ru

Abstract — **Tungsten trioxide thin films have gathered attention due to their versatile applications in electrochromic devices, gas sensors, photocatalysts, and photoelectrochemical cells. However, conventional deposition methods face significant challenges. That is why some researches shifted to more precise method of deposition – atomic layer deposition, that is featured by uniform, conformal deposition with a precise thickness. Additionally, coatings on complex three-dimensional structures can be deposited. In this paper, plasma-enhanced atomic layer deposition of tungsten trioxide in the reagent system tungsten hexacarbonyl-plasma activated oxygen was used. The mentioned reagent system, most likely, was not presented in other papers. The lowest substrate temperature allowing reproducible conformal growth is in between 80 °C and 120 °C, the highest – in between 200 °C and 225 °C. A tungsten trioxide coating on indium tin oxide whiskers as a test sample was obtained. Test sample demonstrated electrochromic properties, turning blue when voltage was applied.**

Keywords — tungsten trioxide; WO3; atomic layer deposition; ALD; plasma-enhanced ALD; tungsten hexacarbonyl; indium tin oxide; thin coatings; electrochromic materials.

I. INTRODUCTION

Tungsten trioxide WO3 is a semiconductor with a wide bandgap and with a density of 7.16 g/cm³, which in its ordinary form appears as a light-yellow crystalline powder. The width of the band gap in tungsten trioxide film primarily depends on the crystal structure and film thickness. For example, an amorphous film of tungsten trioxide is characterized by a band gap of approximately 3.25 eV, while for a monoclinic structure at room temperature the value of bandgap is 2.62 eV [1]. The wide bandgap of tungsten trioxide is actively utilized in optoelectronics, as the material interacts excellently with visible and ultraviolet light. These optical properties of tungsten trioxide are applied in the creation of photocatalysts, photoanodes, and photodetectors. Tungsten trioxide is also known for its gas-sensing properties, and gas sensors based on it are manufactured for dangerous gases such as nitrogen dioxide, ammonia, hydrogen sulfide, and hydrogen [2]. It is also worth mentioning the applications based on the electrochromic properties of tungsten trioxide. Among all electrochromic materials, tungsten trioxide is still one of the most studied and widely applied materials.

Electrochromic materials are materials whose optical properties can be modulated by applying an electric field. To the human eye, the electrochromic effect manifests as a change in the material's color. Electrochromic devices based on these materials can be used as smart windows, displays, anti-glare mirrors for automobiles, self-tinting sunglasses, near-infrared filters, and other applications. Among all these uses, smart windows attract the most significant interest, primarily due to their potential for energy savings in air conditioning and heating, as well as their thermal regulation capabilities, which can be particularly valuable in spacecraft.

There are several methods for synthesizing tungsten trioxide, ranging from physical and chemical vapor deposition methods to hydrothermal method. As devices and their prototypes continue to advance, there is an increasing demand for innovative strategies to enhance their performance characteristics. For instance, in photocatalysts, photoanodes, gas sensors, and electrochromic devices based on tungsten trioxide, researchers are pursuing nanostructured films to increase the surface area. A larger surface area promotes the greater active layer. In gas sensors, this increases the number of active sites available for gas adsorption, which ultimately improves sensitivity and signal output during gas detection. For electrochromic films, an advanced surface morphology plays a particularly important role in improving device performance. A well-developed surface morphology can be achieved through several approaches, with the creation of a porous surface being one of the most effective. This approach has already been explored for films prepared by magnetron sputtering, where the surface morphology exhibited a columnar structure [3].

Various works presented a relatively novel approach to developing the surface morphology of films, which can be primarily aimed at applications in electrochromic devices. The method is based on creating a core-shell structure, wherein a film is deposited onto a substrate with a complex morphology. The core-shell structure formation involves synthesizing a film that, to some extent, replicates the substrate's structure. This approach has previously been demonstrated in research [4] where a nickel oxide film was deposited onto an indium tin oxide (ITO) substrate using spray pyrolysis. However, the authors did not achieve significant improvements in electrochromic properties. One reason for this is the insufficiently developed surface morphology of the substrate, as well as the difficulty in obtaining a conformal film through spray pyrolysis. Given these limitations, atomic layer deposition (ALD) — a more precise method — is being considered for creating the core-shell structure.

The primary advantage of atomic layer deposition for creating core-shell structures lies in the self-limiting nature of the chemical reaction between the precursor and the functional groups on the deposition surface. This enables conformal film growth even on substrates with a developed surface [5]. Because of that feature, ALD can be used for coating synthesis that faithfully replicates the substrate's morphology with high fidelity. Achieving such results with other conventional methods is challenging, as air-filled pores form in the film when depositing onto highly porous substrates. Another advantage of atomic layer deposition is its low deposition temperatures (usually does not exceed 300 °C) [6, 7] and its excellent control over film thickness, which prevents the porous structure from being filled due to excessive film growth. However, despite these benefits, atomic layer deposition has a significant drawback related to its slow deposition rate, high cost of equipment and high energy costs [8]. The deposition rate is primarily determined by both the deposition conditions and the choice of precursor system. Various combinations of precursors continue to be investigated.

Numerous precursor systems are known for atomic layer deposition of tungsten trioxide, employing both organic and inorganic compounds of tungsten. Precursor systems based on organic compounds, such as $WH_2(Cp)_2$ (Cp = cyclopentadienyl group), demonstrate high deposition rates (0,7 Å/cycle) [9], however, the high cost of these compounds limits cost reduction of the films production. Deposition using the inorganic compound, such as tungsten fluoride exhibits satisfactory growth rates (0,45 Å/cycle) [10], but at elevated deposition temperatures, etching of the growing film occurs. Furthermore, the precursor itself and its vapors are extremely hazardous and possess high corrosive properties. Another inorganic compound does not suffer from the drawbacks associated with tungsten fluoride. This compound is tungsten hexacarbonyl $W(CO)_6$, which is the most thermally stable among metal carbonyls and has very low cost. Among scientific research, only a small fraction of studies were focused on atomic layer deposition using this compound. In published works, hydrogen peroxide and ozone have been employed as oxidants [11, 12], but no studies directed at plasma-enhanced deposition of tungsten trioxide using tungsten hexacarbonyl have been found.

This work examines atomic layer deposition using a precursor system of tungsten hexacarbonyl and plasma-activated oxygen. The research findings will provide potentially valuable insights into the characteristics of atomic layer deposition employing this understudied precursor system. Furthermore, the objective of this work is to obtain a core-shell structure with a tungsten trioxide film, utilizing an ITO whisker layer as the substrate with well-developed surface morphology. As part of the investigation, the electrochromic properties of the resulting film deposited on the ITO whisker layer substrate will also be examined.

II. MATERIAL AND METHODS

The deposition of WO_3 using the $W(CO)_6$-O_2^* precursor system was conducted on a laboratory apparatus equipped for atomic layer deposition (Fig.1). Inside the reactor chamber, a graphite disk is positioned on a pedestal to hold the substrate. The substrate is heated by a heater located within the pedestal, with temperature monitored by a thermocouple. To minimize heat loss during heating, a copper shield is employed. The graphite disk also serves as an electrode, enabling plasma ignition between it and another electrode. The heating of the reactor chamber walls and the temperature control of the evaporator are managed by an external furnace. Since the temperature of the external furnace for heating the evaporator, which contains tungsten hexacarbonyl powder, is considerably lower than that of the substrate heater, the reactor can be classified as a cold-wall reactor. Precursors are supplied to the reactor from a gas distribution system through a gas shower in the manifold assembly via a feedthrough.

Wilson seals, located in both the reactor lid and the lower section of the reactor chamber, are used to ensure proper sealing.

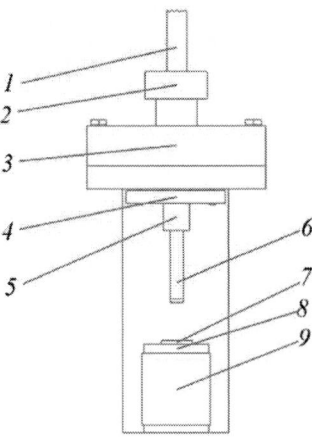

Fig. 1. Simplified reactor block scheme: 1 – rod; 2 – nut; 3 – lid with seal; 4 – system with bushings; 5 – evaporator; 6 – electrode; 7 – sample; 8 – pedestal; 9 – copper screen.

Tungsten carbonyl in powder form was used for deposition of the tungsten trioxide film. Monocrystalline silicon substrates with a native oxide layer and borosilicate glass with an ITO whisker layer served as substrates. The conditions for obtaining the ITO layer on borosilicate glass substrates are the same as in the presented reference [13]. The silicon substrates had an area of approximately 375 mm². Prior to deposition, the substrates were treated in an acetone vapor. Deposition was conducted at a pressure of 100 Pa and in a temperature range of 80–225 °C. The pulse times for $W(CO)_6$ and O_2^* were 20 and 6 seconds, respectively, with purge times of 20 seconds for both. The flow rates of argon and O_2 were 55 and 30 mL/min, respectively. The discharge current was set at 5 mA.

Film thickness of WO_3 coatings was measured using a null ellipsometer LEM-3M, with thickness calculations performed using the "LEF-72" software. Transmission spectra of the samples were acquired using an AnalytikJena SPECORD 40 spectrophotometer. Surface morphology analysis of the tungsten trioxide film on ITO whiskers was conducted using scanning electron microscopy (SEM). A Carl Zeiss SUPRA 55VP scanning electron microscope was employed to obtain images of the scanned surfaces.

To characterize the electrochromic properties of the tungsten trioxide sample on the substrate with ITO whiskers, chronoamperometry was performed. The three-electrode cell used for chronoamperometry consisted of two reservoirs: an optical cuvette and a chemical beaker. The optical cuvette and chemical beaker were filled with lithium perchlorate solution diluted with polycarbonate (PC) to a concentration of 0.5 M ($LiClO_4$/PC) and 0.1 M potassium chloride (KCl), respectively. The working electrode was the WO_3 coating on a glass substrate with deposited ITO whiskers, while a platinum wire served as the auxiliary electrode. The sample and platinum wire were immersed in the optical cuvette, and a Ag/AgCl electrode was immersed in the chemical beaker. The reservoirs containing the solutions were connected by a salt bridge. Sample with electrochromic coating acted as the working electrode, and the platinum wire as the counter electrode. Potential control applied to the working electrode was managed using the Ag/AgCl reference electrode. All aforementioned electrodes were connected to a potentiostat.

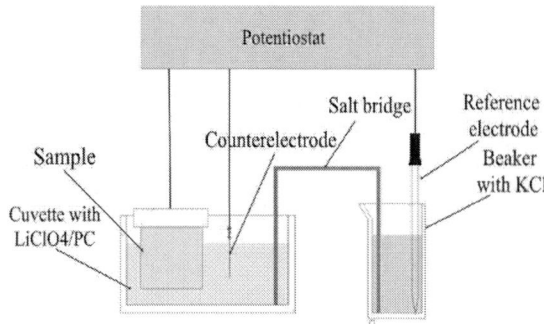

Fig. 2. Scheme of three electrode system

III. RESULTS AND DISCUSSION

A. Investigation of the characteristics of WO₃ deposition by atomic layer deposition using the $W(CO)_6$–O_2* precursor system.

A series of experiments was carried out to study ALD of WO₃ using the $W(CO)_6$–O_2* precursor system, aimed at determining the dependence of the growth rate at various temperatures. To eliminate random error, the experiments were performed with a random choice of deposition temperature. In each experiment, 30 deposition cycles were conducted over a temperature range of 80–225 °C (Table 1). Based on the obtained data, the temperature dependence of the growth rate was plotted (Fig. 3). In the 125–200 °C range, only a slight change in growth rate is observed, which suggests that this region may correspond to the ALD temperature window. The upper and lower boundaries of this window lie in the 200–225 °C and 80–125 °C regions, respectively, and their refinement requires additional experiments. The increase in growth rate observed when the temperature decreases from 125 to 80 °C may be associated with condensation of the precursor vapor on the substrate surface. This can be explained by the fact that during deposition the substrate temperature was lower than the temperature of the crucible containing the precursor. An increase in growth rate is also observed when the temperature is raised from 200 to 225 °C, which is attributed to precursor pyrolysis.

To evaluate the contribution of pyrolysis to the coating growth rate, an experiment was carried out at 225 °C, in which molecular oxygen and $W(CO)_6$ were supplied simultaneously for 10 minutes, corresponding to 23 deposition cycles. Silicon substrates coated with a uniform WO₃ layer were used in the experiment. Substrates with a native SiO₂ layer were not used, since during chemical vapor deposition the film grows with grains, which distorts the thickness analysis by ellipsometry. An increase in the WO₃ film thickness on the WO₃-coated substrate after the experiment was confirmed by ellipsometric measurements. As a result, the contribution of pyrolysis during deposition at

225 °C was found to be 15%. However, when analyzing the experimental data, structural changes in the WO₃ film must be taken into account, and therefore a more detailed testing procedure is required to accurately evaluate the pyrolysis contribution.

TABLE I.   PARAMETERS AND RESULTS OF THE DEPOSITION SERIES IN VARIOUS TEMPERATURES.

Deposition temperatures, °C	Mean thickness, Å	Mean growth per cycle, Å/cycle
80	16,04±2,86	0,53±0,09
125	3,36±2,14	0,11±0,07
140	3,34±1,14	0,11±0,04
175	2,87±1,03	0,09±0,03
200	5,02±0,90	0,17±0,03
225	77,52±5,02	0,78±0,05

Another experiment was also carried out to determine the dependence of film thickness on the number of deposition cycles (Fig. 3). The experiment was performed at a deposition temperature of 200 °C. As can be seen from the figure, the resulting dependence is linear. In addition, extrapolation of the dependence intersects the origin, which indicates the absence of retarded film growth during the initial deposition cycles. The absence of such retarded growth may be related to a 40-second surface functionalization step in oxygen plasma performed prior to deposition, however, additional experiments are required to clarify the underlying cause.

B. Investigation of the electrochomic properties of WO₃ coating deposited by atomic layer deposition on ITO whiskers

The electrochromic film under analysis was synthesized at a deposition temperature of 200 °C, which falls within the ALD temperature window and offers a high film growth rate (0,17 Å/cycle) (Fig. 3). The highest film growth in the ALD temperature window is slightly (for around 15-27%) lower than in the classic thermal ALD with $W(CO)_6$ [11, 12]. The same ALD cycle was employed. To obtain a WO₃ film with a thickness of 9.2 ± 0.7 nm, 540 deposition cycles were performed. SEM images of the ITO whisker layer, acquired before and after WO₃ deposition, are shown in Fig. 4. The whisker diameter visibly increased post-deposition (Fig. 4a, b), confirming successful WO₃ layer formation. On the sample cross-section image, slight filling of the space between ITO whiskers by the WO₃ layer is discernible (Fig. 4c), indicating potential and rationale for depositing thicker coatings in the future. The transmission spectrum of the as-deposited sample was measured (Figure 5). The WO₃ film exhibits optical transmission of about 82% in the visible range and slightly lower in the near-IR.

979-8-3315-4784-4/26 $31.00 © 2026 IEEE

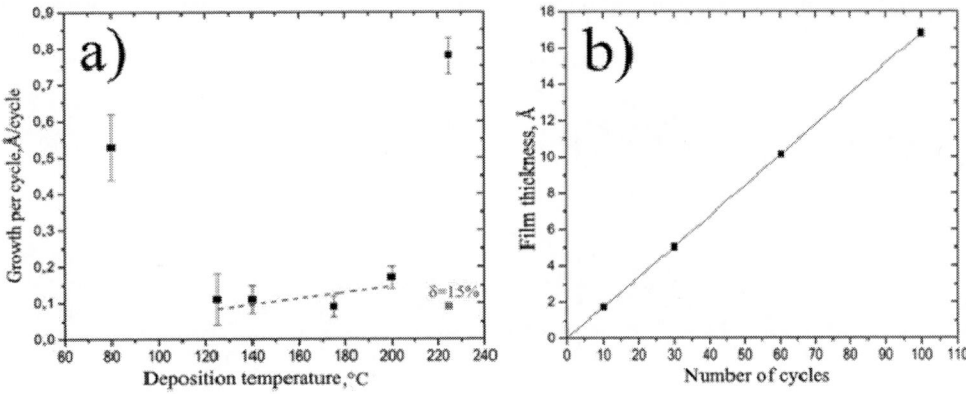

Fig. 3. Dependencies of deposition parameters: film growth per cycle versus deposition temperature (a) and film thickness versus number of deposition cycles (b).

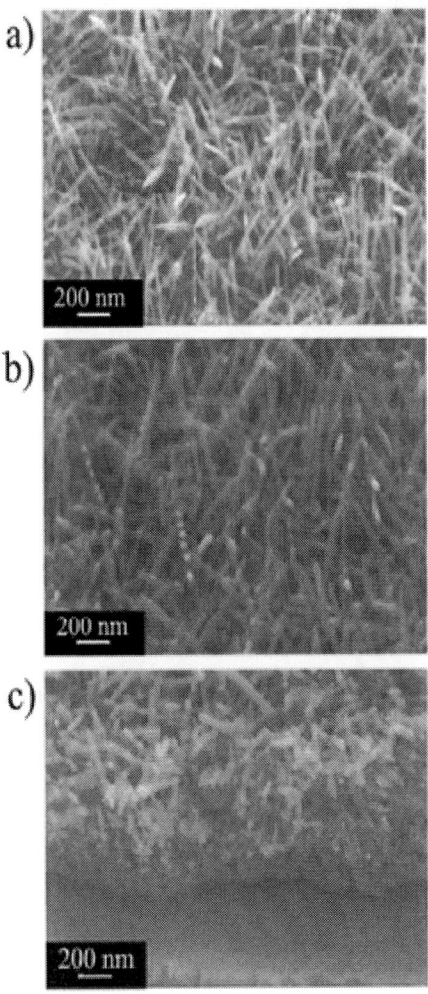

Fig. 4. SEM images of the ITO whisker layer: a) surface before WO₃ deposition; b) surface after WO₃ deposition; c) surface near the fracture after WO₃ deposition.

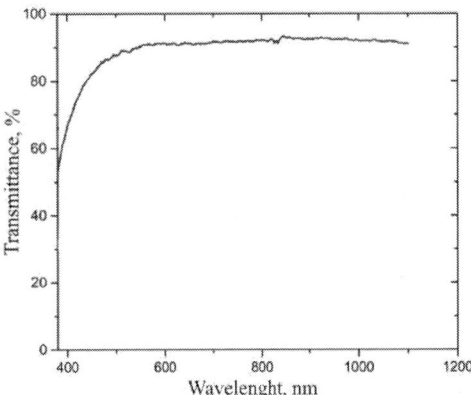

Fig. 5. Sample transmittance spectra

After spectral analysis, the sample underwent chronoamperometric testing. A voltage of -1 V was applied for coloring and +1 V for bleaching, each voltage was applied for 20 seconds. During testing, the sample colored to a light blue with low density of color (Fig. 6). Judging by the transmittance spectra from another work [14], in which the color of the sample was similar to the color of the sample in this work, the transmission modulation of the deposited sample might be around 40,8 %. The reason of low coloring density may be because of the insufficient film thickness, which affects the optical path length based on Lambert-Beer law. It could also suggest limited sites for electrolyte ion intercalation, though this seems unlikely given the high surface area. Overall, ten chronoamperometry cycles were conducted, during which sample coloring was observed. Response time was determined from chronoamperometry data (Fig. 7), defined as the time required for the measured signal to change by 90% during coloring and bleaching, as indicated on the figure. The resulting coloring and bleaching times were 16.0 and 0.8 seconds, respectively. Slow coloring process on chronoamperogram indicates poor charge transfer, which should be investigated in the future with optical measurements. After sample testing, a SEM image of its surface was obtained (Figure 20) to visually assess changes in the film that occurred during analysis of electrochromic properties. The image shows that the appearance of the ITO whiskers with the deposited WO₃ layer changed after testing. Some whiskers appear damaged, and the presence of the tungsten trioxide film on the whiskers is less visible compared to the SEM image acquired immediately after

979-8-3315-4784-4/26 $31.00 © 2026 IEEE   139

deposition. It is likely that the thickness of the WO₃ film on the ITO whiskers decreased during the process of testing. This change is probably attributable to film etching during water rinsing.

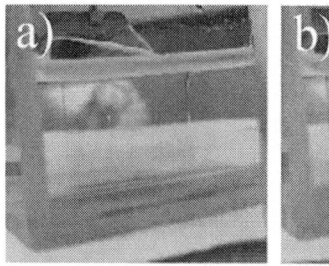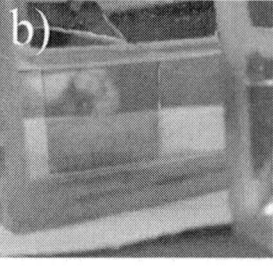

Fig. 6. Sample appearance: a) in the bleached state; b) in the colored state.

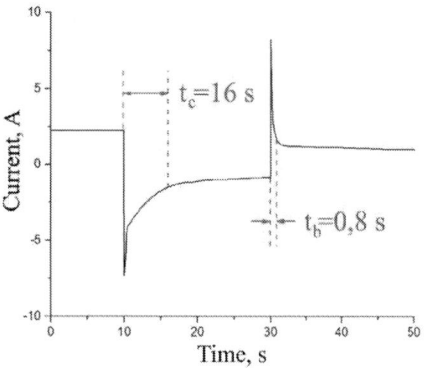

Fig. 7. Graph of current dependence vs measurement time for the coating, with time intervals corresponding to coloring (t_c) and bleaching (t_b) durations.

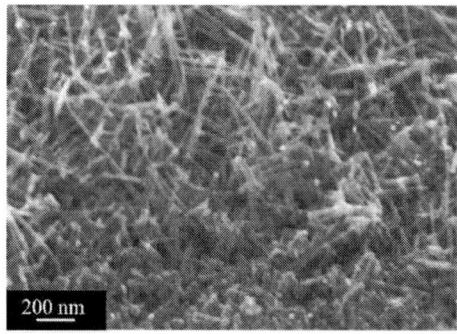

Fig. 8. SEM image of the sample surface after testing.

IV. CONCLUSIONS

During the investigation of ALD characteristics for synthesis of WO₃ coatings using the novel $W(CO)_6$–plasma-activated oxygen precursor system, several experimental series were conducted to study the dependence of growth rate versus temperature and film thickness versus number of deposition cycles. It was determined that the growth rate remains constant in the 125–200 °C range. When temperature is increased above 200 °C, an increase in growth rate is observed, which is attributed to precursor pyrolysis. At 225 °C, the contribution of pyrolysis to the increased growth rate is 15%. When temperature is decreased below 125 °C, an increase in growth rate is also observes, which is, most likely, attributed to condensation of precursor vapor. The dependence of growth rate on the number of cycles was also established. The growth rate exhibits a linear dependence on

cycle number, and no retarded film growth was observed during initial deposition cycles, which warrants further detailed investigation in the future.

The electrochromic properties of WO₃ coating deposited on the ITO whisker layer were investigated using chronoamperometry. The sample with WO₃ obtained from the ALD using $W(CO)_6$–O_2* precursor system exhibited light blue coloring during ten chronoamperometry cycles. Based on chronoamperometric measurements, the coloring and bleaching times were determined to be 16.0 and 0.8 seconds, respectively. The prolonged coloring time is likely attributed to poor charge transfer. The film underwent changes following testing, as confirmed by SEM images of the coating acquired after the experiment, which should be investigated in the future. Optical modulation was determined by comparing the sample's coloring with a similar sample from another study. The optical modulation value was found to be approximately 40.8%. Future work will include conducting independent optical measurements to determine optical modulation, enabling clear comparison with samples from other researches. Additionally, it would be beneficial to deposit thicker tungsten trioxide films on the ITO whiskers.

REFERENCES

[1] Gillet M. [et al.] The structure and electrical conductivity of vacuum-annealed WO3 thin films // Thin Solid Films. – 2004. – Vol. 467. – P. 239-246.

[2] Guanglu L. [et al.] Thin films of tungsten oxide materials for advanced gas sensors // Sensors and Actuators: B. Chemical. – 2021. – Vol. 341. – P. 129996.

[3] Z. Luo, L. Liu, X. Yang, X. Luo, P. Bi, Z. Fu, A. Pang, W. Li, Y. Yi, Revealing the charge storage mechanism of nickel oxide electrochromic supercapacitors, ACS Appl. Mater. Interfaces 12 (2020) 39098–39107

[4] Lin S. [et al.] Electrochromic properties of nano-structured nickel oxide thin film prepared by spray pyrolysis method // Applied Surface Science. – 2008. – Vol. 254. – P. 2017-2022

[5] Tong Z. [et al.] Self-supported one-dimensional materials for enhanced electrochromism // Nanoscale Horizons. Royal Society of Chemistry. – 2018. – Vol. 3. – № 3. – P. 261–292.

[6] Nematzadeh M. [et al.] Investigation of the Atomic Layer Deposition of the Titanium Dioxide (TiO2) Film as pH Sensor Using a Switched Capacitor Amplifier // Chemosensors. – 2022. – Vol. 10. – № 7. – P. 274.

[7] Koshtyal Y. [et al.] Atomic layer deposition of NiO to produce active material for thin-film lithium-ion batteries //Coatings. – 2019. – Vol. 9. – №. 5. – P. 301.

[8] Oviroh P. O. [et al.] New development of atomic layer deposition: processes, methods and applications //Science and technology of advanced materials. – 2019. – Vol. 20. – №. 1. – P. 465-496.

[9] Romanov R. [et al.] Radical-Enhanced Atomic Layer Deposition of a Tungsten Oxide Film with the Tunable Oxygen Vacancy Concentration // The Journal of Physical Chemistry C – 2020. – Vol. 124. – P. 18156-18164.

[10] Strobel A. [et al.] Room temperature plasma enhanced atomic layer deposition for TiO2 and WO3 films // Journal of Vacuum Science & Technology A – 2016. – Vol. 34. – P. 01A118.

[11] Dunn B. [et al.] Tungsten Hexacarbonyl and Hydrogen Peroxide as Precursors for the Growth of Tungsten Oxide Thin Films on Titania Nanoparticles // AIChE Journal. – 2014. – Vol. 60. – P. 1278-1286.

[12] Malm J., Sajavaara T., Karpinnen M. Atomic Layer Deposition of WO3 Thin Films using W(CO)6 and O3 Precursors // Chemical Vapor Deposition. – 2012. – Vol. 18. – P. 1-4.

[13] Filatov L. [et al.] Concept of atomic layer deposition application in electrochromic device fabrication approved on ITO@ NiO whisker layers // Materials Today Communications. – 2025. – Vol. 44. – P. 112116.

[14] Tuna O. [et al.] Electrochromic properties of tungsten trioxide (WO3) layers grown on ITO/glass substrates by magnetron sputtering // Vacuum. – 2015. – Vol. 120. – P. 28-31.

979-8-3315-4784-4/26 $31.00 © 2026 IEEE

Biconical Antenna for Use in Systems with High Radiated Power

Alexander A. Borisov
MIREA – Russian Technological University
Moscow, Russia

Valeriy V. Demshevsky
MIREA – Russian Technological University
Moscow, Russia
demshevskyv.v@yandex.ru

Igor A. Bogachev
MIREA – Russian Technological University
Moscow, Russia

Andrey D. Bazhenov
MIREA – Russian Technological University
Moscow, Russia

Vyacheslav V. Lobodin
MIREA – Russian Technological University
Moscow, Russia

Stanislav S. Sidorenko
MIREA – Russian Technological University
Moscow, Russia

Abstract—**The paper considers one of the main problems of antennas when emitting useful high-power signals: high heating temperatures at the points of antenna structure, which lead to a decrease of the real emitted power due to a decrease in the efficiency of antennas as well as the destruction of their parts such as connectors, coaxial cables and soldered joints.**

The paper provides a study of a biconical antenna used in systems with high radiated power. Analytical calculations of the antenna were performed, and its parameters and characteristics were improved, including the expansion of its operating frequency band and radiation efficiency.

Keywords—efficiency, radiated power, radiation efficiency, wide-band antenna

I. INTRODUCTION

In modern systems there is an increasing need to enlarge the range of transmission and reception of a useful signal, while maintaining the directional characteristic. Such requirements entail an increase in the transmitted signal power, which will provide a sufficient signal level over long distances. At the same time there is an increasing need to limit the mass and size of such systems, reduce their design complexity and cost of materials and components.

Therefore, in order to ensure the relevance of using systems with high radiated power, it is important to study the antennas used in such type of systems. Ensuring their mass and size characteristics and reliability of materials and components used in antenna design for high-power radiation is a priority task for ensuring the functionality for the entire system with the specified characteristics.

One of the main problems when using antennas that do not have the ability to emit high-power useful signals is the high temperatures that can be reached at certain points in the structure. This can lead to a decrease in the efficiency of radiation due to a reduction in their efficiency as well as to destruction of certain parts such as connectors, coaxial cables and soldered joints. Taken together, these factors lead to a critical decrease in the transmission or reception range, determined by the level of the useful signal transmitted power.

The study was based on an ultra-wideband biconical antenna. Due to its symmetry and geometry a biconical antenna has an omnidirectional radiation pattern in the horizontal plane, which makes it suitable for transmitting and receiving signals in almost all directions.

Biconical antennas can be implemented in various variations and geometric sizes and some of their modifications during the development process allow for customization of their characteristics to meet specific requirements and frequency bands. Several configurations of biconical antennas are shown in Fig. 1.

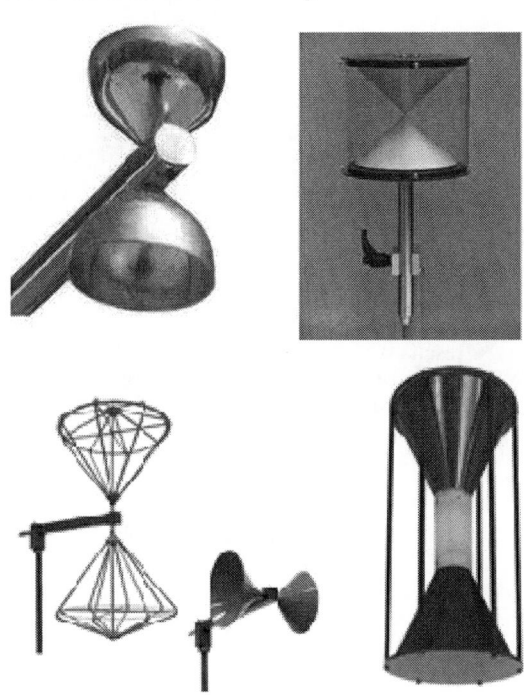

Fig. 1. Types of biconical antennas

Such antennas have a number of advantages, such as a frequency-independent radiation pattern (RP), relatively small size and weight, simplicity of design and broad-band.

II. STATEMENT OF PROBLEM

To strictly evaluate the functionality of antenna under study, an electrodynamic model was developed and calculated, and a prototype was manufactured based on the results of calculations. The paper presents the results of electrodynamic and thermal analysis, manufacturing, assembly and measurement of radiation characteristics of a broadband biconical antenna, capable of emitting more than 50 W of power without significant heating.

979-8-3315-4784-4/26 $31.00 © 2026 IEEE

III. THE RESULTS OF ELECTRODYNAMIC MODELING

The solution of the set tasks is based on a developed strict electrodynamic model of a biconical antenna, which includes metal cones, a section of coaxial cable, a dielectric housing a metal basis and a cover. During the calculations the shape of the cones as well as the width and height of the dielectric spacer between the cones were selected to achieve the maximum possible operating frequency range with minimal antenna dimensions. Fig. 2 shows the appearance of the antenna.

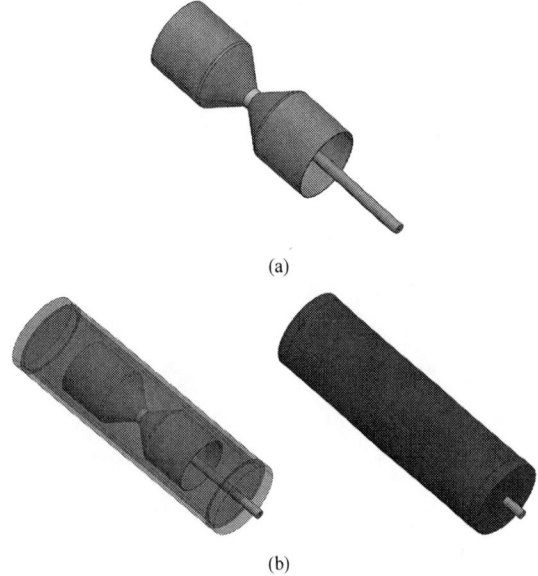

(a)

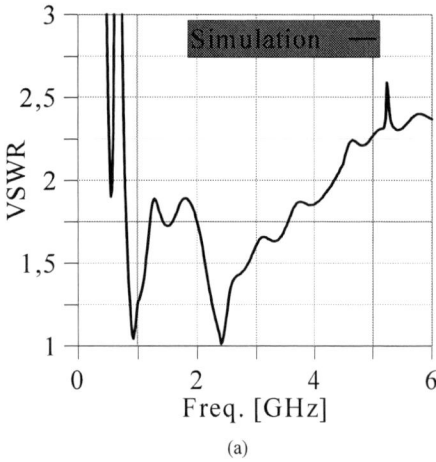

(b)

Fig. 2. The appearance of the biconical antenna (a) electrodynamic model; (b) model in a dielectric case

As a result of the electrodynamic model analysis, the VSWR characteristics in the operating frequency band and a radiation pattern at various points of the operating range were obtained (Fig. 3). The VSWR shows that the antenna operates in the range 600 to 4200 MHz.

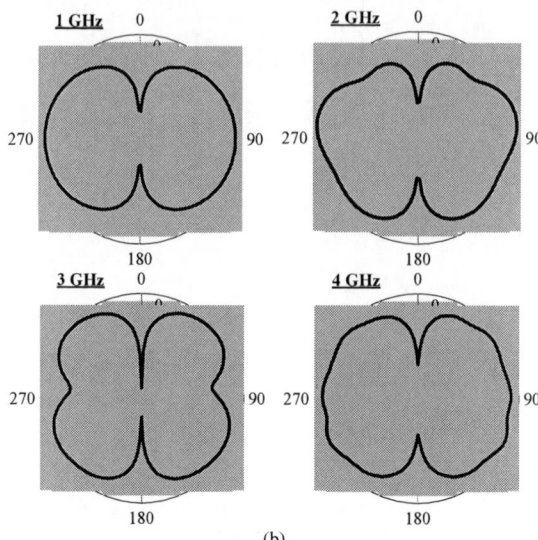

(b)

Fig. 3. Characteristics of electrodynamic model (a – VSWR; б – RP at different frequencies)

The second key task in the antenna development is to assess the heating temperature during high-power applications, in order to select technical solutions at the modeling stage to ensure heat dissipation from the most heated points and increase the antenna radiation efficiency. In the process of studies calculations were made of antenna heating when 50 W (Fig. 4) and 100 W (Fig.5) were supplied to its input. According to the calculation results, even when a 100 W signal is applied to the antenna input, it does not exceed 55 degrees Celsius.

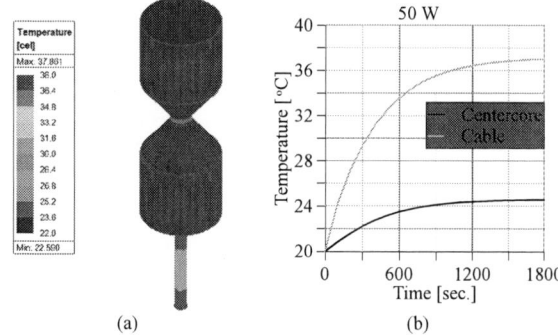

(a)   (b)

Fig. 4. Estimation of the effect of antenna heating when 50 W is applied (a) the model appearance; (b) – heating temperature over time, where 1 – heating temperature of the coaxial cable, 2 – heating temperature of the radiation point)

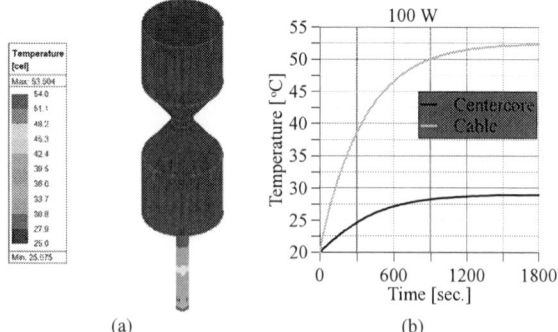

(a)   (b)

Fig. 5. Estimation of the effect of antenna heating when 100 W is applied (a) the model appearance; (b) heating temperature over time, where 1 – heating temperature of the coaxial cable, 2 – heating temperature of the radiation point)

IV. THE RESULTS OF MEASUREMENTS

Based on the developed model a prototype was manufactured in order to conduct further measurements and verify the calculated characteristics in practice Fig. 6(a). During the assembling process in order to reach high efficiency special attention was paid to the type and quality of the cable assembly used, as well as to the losses it introduces and its impact on the output characteristics of the biconical antenna, including the use of a microwave connector.

(a)

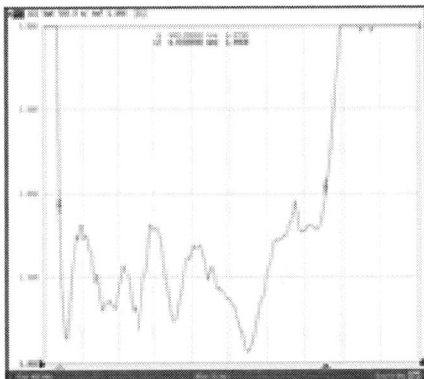

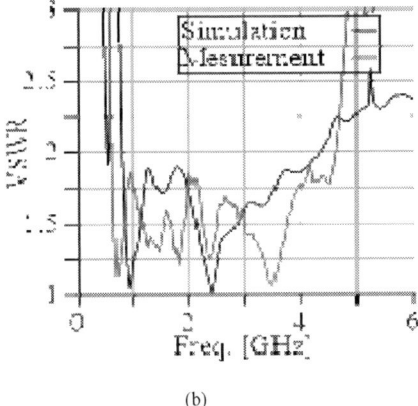

(b)

Fig. 6. Prototype model of a biconical antenna (a) photo of the antenna; (b) measured antenna VSWR and comparison with calculated data

The characteristics of the manufactured prototype of a biconical antenna were also tested in laboratory conditions by applying a constant signal with a power of 90-100 W to its input. Fig. 7 shows photos of the antenna connected to a powerful signal source and the signal spectrum at various frequencies.

(a)

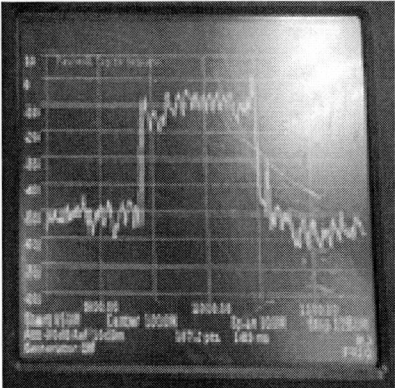

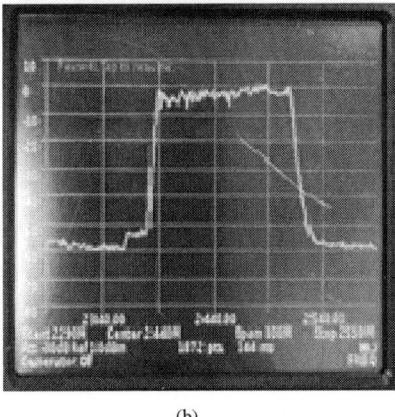

(b)

Fig. 7. Measurement of antenna characteristics when a 100 W interference signal is applied to the antenna (a) antenna connection diagram; (b) – antenna frequency characteristics

The antenna demonstrated its ability to maintain its performance when receiving a constant 100 W signal for an extended period of time without distortion or overheating.

Besides, using Harald Friis's equation (1), the radiation efficiency of the developed antenna was evaluated based on the power, received by the receiving antenna in an ideal environment, at a specific distance from the transmitting antenna, when a 100 W signal was transmitted.

$$P_r = P_t \times G_t \times G_r \times \left(\frac{\lambda}{4 \times \pi} \right)^2 \qquad (1)$$

According to the calculations performed, the use of the antenna developed during the work as part of systems with high radiated power, increases the power by 60%, compared with the analogs considered due to the achievement of high radiation efficiency of 80-85%.

V. Conclusions

During the work theoretical calculations of the antenna were performed, and its parameters were upgraded and improved. The model's performance was evaluated during prolonged periods of high-power operation. Based on the data obtained during the modeling process, a prototype antenna was manufactured and assembled, and its characteristics were measured in laboratory conditions. The radiation efficiency of the developed antenna was evaluated, and according to analytical calculations it showed a 60% increase in power.

References

[1] Pat. RU2022428C1

[2] R. Kudpik, K. Meksamoot, N. Siripon, "Design of a compact biconical antenna for UWB application, " 2011 Int. Symp. On Intelligent Signal Processing and Communications Systems (ISPACS), Chiang Mai, Thailand, 2011, pp. 1-6;

[3] U. Keda, "Biconical antenna with dielectric lens, " Information radio systems and radio technologies, Minsk, 2020, pp. 337-341;

[4] A.M. Bobreshov, A.S. Zhabin, E.A. Seregina, G.K. Uskov "Biconical antenna whith inhomegeneous dielectric lens for UWB applications," in Electronics Letters, vol. 56, no 17, pp. 857-859, 2020, doi: 10.1049/el.2020.1098;

[5] Y. Zhieheng, R. Pengshan, M. Debgpan "Design of a compact biconical antenna with asymmetric configuration," 2018 UEEE 4th Int. Conf. on Computer and Communications (ICCC), Chengdu, People`s Republic of China, 2018, pp. 1022-1026;

Clustered Antenna Array for AESA with Wide Angles of Electronic Scanning

Valeriy V. Demshevsky
MIREA – Russian Technological University
Fryazino, Russia
demshevskyv.v@yandex.ru

Grigoriy S. Anikin
MAI – Moscow Aviation Institute
Moscow, Russia

Stanislav S. Sidorenko
MIREA – Russian Technological University
Moscow, Russia

Ilya A. Tsygitsa
MAI – Moscow Aviation Institute
Moscow, Russia

Dmitri V. Bagno
MAI – Moscow Aviation Institute
Moscow, Russia

Eugene V. Iliin
MAI – Moscow Aviation Institute
Moscow, Russia

Abstract—**The increasing requirements for the weight and size characteristics of electronic equipment and products inevitably lead to structural difficulties in the layout of elements and modules inside the products themselves. When developing an AESA the main design challenge is to accommodate simultaneously a great number of transceiver modules and control buses in a limited space which can lead to a shortage of space for cooling systems and other components. The paper considers the possibility of developing the clustered antenna array for AESA with a ±60° electronic angle scanning sector in the main planes.**

Keywords—active electronically scanned array, antenna array, clustering, electronic angle scanning, radiation pattern.

I. INTRODUCTION

The number of discrete elements (the number of individual channels) determines not only the capabilities of the active electronically scanned array (AESA), but also its cost. In theory, each radiator on the antenna array should have its own T/R-module with an independent phase shifter and attenuator. From an economic point of view the cost of such system will definitely be high. There is another practical reason for reducing the number of T/R-modules in the AESA – to reduce the complexity and number of discrete connections in the high-frequency distribution system, to reduce the number of power supply and control system bundles, and to ensure heat dissipation from the transceivers. In other words, the developers are required to ensure simultaneously both a reduction in the price of the final product and preservation of the tactical and technical requirements, while reducing the number of individual parts as much as possible and simplifying the assembly of the product.

For example, to build a monopulse AESA based on antenna array with a size of 16x14 elements, it is estimated that 224 T/R-modules will need to be placed, more than 900 microwave connectors will need to be connected, with 600 bullets or cable assemblies, 450 power supply pins or cable assemblies, and 230 control buses will need to be soldered.

Therefore, at the moment the following tasks are coming to the fore:

1) Reducing the number of emitters and/or the number of used T/R-modules without significant reducing the directivity and increasing the side lobes level;
2) Simplification of the power supply and control scheme;

These tasks are successfully solved to some extent by using sparse or clustered antenna array, which have an unconventional architecture [1]. As a rule, when designing such antenna arrays, they try to obtain either a non-equidistant grid of excitation points (when clustering), or a non-equidistant grid of emitter placement (when sparse). In this case it is possible to eliminate the periodicity of the array factor with respect to the generalized angular coordinate, which reduces the magnitude of the diffraction maxima in the structure of the directional characteristic.

Methods of analysis and synthesis of non-equidistant grids are divided [2] into 3 main types:

- sampling methods used when there are few antenna array elements. They partially include methods using genetic algorithms to find the optimal placement of emitters and applied amplitude distributions;

- methods using the approximation of the array factor by series, and reduce the non-equidistant structure to an equivalent (equidistant) structure;

- methods that compare the density of elements position on certain axes lying in the scanning planes with amplitude distributions along a continuous linear emitter.

All the above methods are usually based on some approximations, which make them statistically dependent. In addition, the efficiency and speed of some methods, especially the sampling methods, strongly depend on the size and number of antenna array elements.

II. CLUSTERING OF THE AESA

AESA cluster consists of several emitters on an antenna array that are combined into a single structure, with one output which is connected to one transceiver that has one transmitter, one receiver, one phase shifter, one attenuator. In fact, this is a synphase subgroup of elements excited with equal amplitudes, with one phase which corresponds to the phase of some characteristic cluster point (for instance, phase center point, if it is available), calculated from the phasing equation. A cluster can be considered as a single complex-shaped emitter, which is characterized by its vector complex radiation pattern, phase center (or partial phase center), etc. The simplest methods of clustering allow for electronic scanning either in one plane, or a very small scanning sector in any plane.

The antenna arrays presented in the paper [3] are regular with equidistant elements. Regular clustering of the antenna

979-8-3315-4784-4/26 $31.00 © 2026 IEEE

array was carried out. A rather narrow electronic scanning sector was considered which was ±20° from the normal to the antenna array.

Fig.1 shows an example of clustering a circular AESA given in [3].

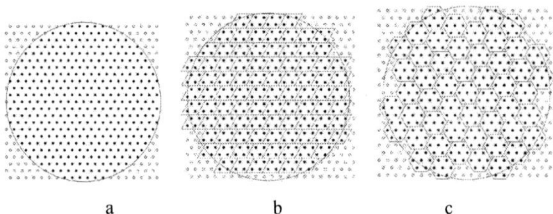

Fig.1 Antenna array. a) regular with discrete elements; b) clustered, with 4 elements cluster size; c) clustered, with 7 elements cluster size

Reference [4] describes a similar problem but it is not about analyzing the directional characteristics of electron scanning, but rather about the effect of the number of clusters on the side lobes level (SLL).

In cases when it is necessary to provide electrical scanning in the entire front hemisphere and in a wider sector of angles, they try to implement an irregular arrangement of clusters in antenna array, i.e. to provide a non-equidistant arrangement of the phase centers of the clusters (or other characteristic points of the clusters if there is no phase center). In this case the grid of emitters on the antenna array remains equidistant and regular, but the grid of excitation points ceases to be as such.

III. ANTENNA ARRAY CLUSTERIZATION USING POLYOMINO

The first clustering method, the polyomino method, is based on dividing AESA into polyomino-shaped synphase subgroups having the appearance of polyomino. In mathematics polyominoes are flat geometric shapes formed by connecting several single-cell squares along their sides [5]. Such geometric shapes can be used as the main elements that form the antenna array. They can be based on single emitters that set the period of antenna array.

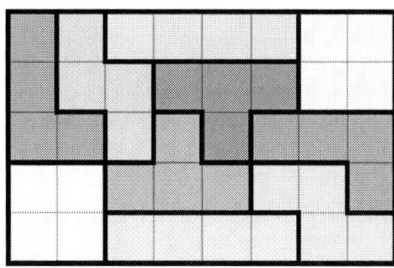

Fig. 2. An example of the structure of a regular rectangular antenna array with an equidistant grid of elements formed based on polyomino joining

References [6,7] consider a method for dividing the antenna array into clusters, which are domino tiles based on circuit optimization and based on Pareto optimization.

Papers [8,9] describe a method for designing clustered flat antenna array, characterized by an irregular arrangement of two-element clusters in the form of dominoes across the antenna array as shown in figures 4, 5. Based on theorems about optimal placement derived from mathematical theory and using an analytical procedure for checking the possibility of filling the aperture with clusters using the domino

principle, two methods of multi-criteria optimization are derived which ensure efficient operation with small and medium/large antenna arrays.

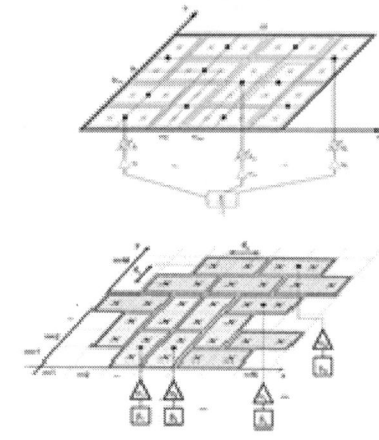

Fig. 3. Clustered antenna array architecture based on dominoes and element excitation scheme

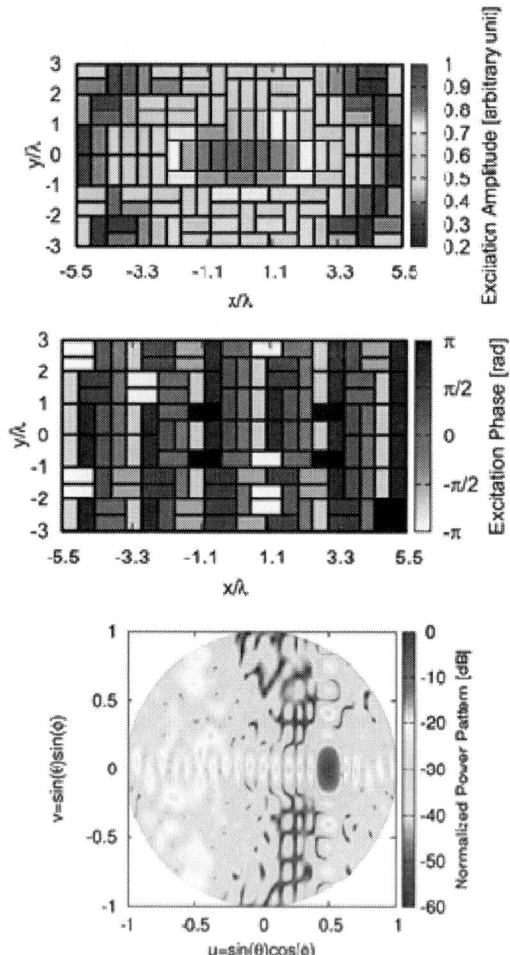

Fig. 4. Numerical simulation of the characteristics of a clustered antenna array based on dominoes, amplitude and phase distribution

The above-mentioned articles also discuss the method of designing antenna array based on polyomino clusters in the form of non-symmetrical polyoimino, as shown in Fig. 6.

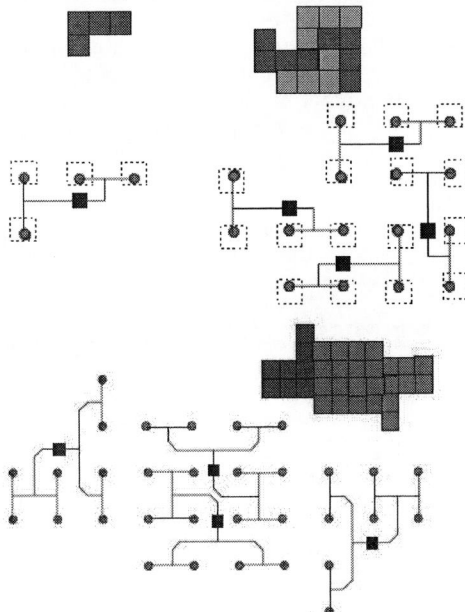

Fig. 5. Layout-design based on clusters that have the shape of polyominoes

The transition from antenna array with a rectangular equidistant grid of element placement (Fig. 7-10) to an irregular one with non-equidistant element placement is shown in reference [10].

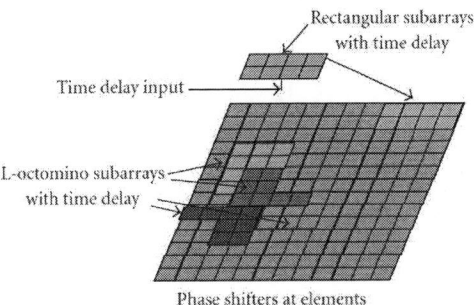

Fig.6. An example of an antenna array with a regular rectangular grid and an irregular cluster layout

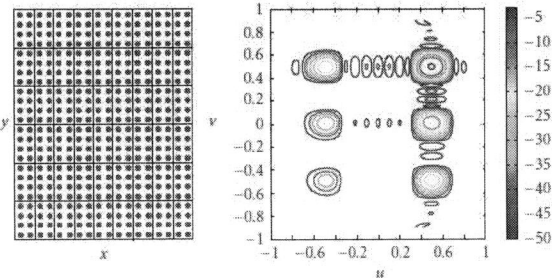

Fig. 7. An example of an antenna array with a rectangular grid, regular arrangement of elements and subarrays and map projection of the amplitude directional characteristic in a map projection when the beam is deflected in a diagonal plane

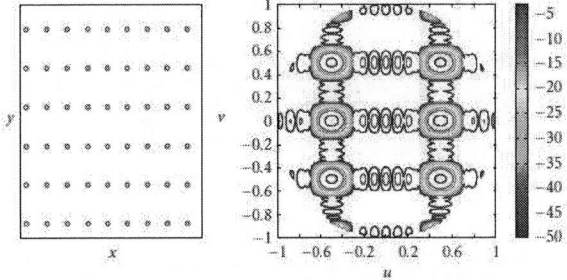

Fig. 8. An example of an antenna array with a rectangular grid and regular arrangement of elements with a large step and map projection of the amplitude directional characteristic in a map projection when the beam is deflected in a diagonal plane

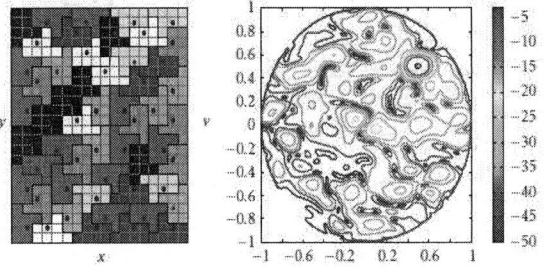

Fig. 9. An example of a clustered antenna array with a rectangular grid and regular layout of elements and map projection of the amplitude directional characteristic when the beam is deflected in a diagonal plane

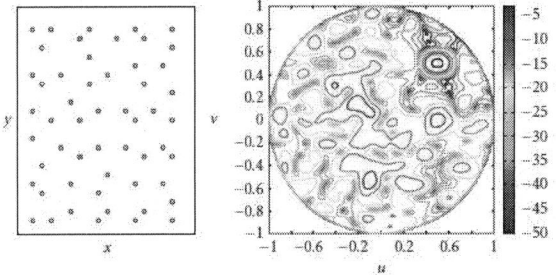

Fig. 10. An example of a sparse antenna array with elements located at the phase centers of asymmetric clusters and map projection of the amplitude directional characteristic when the beam is deflected in a diagonal plane

The advantage of the polyomino method is that the AESA can be divided into clusters of the desired size either by using clusters of the same size and shape or by combining different types of clusters on the same antenna array. This way it is possible to control the partitioning grid of the clustered antenna array.

Let's take a closer look at the clustered antenna array, built on two-element clusters (dominoes) (Fig. 4). In this case there are four possible arrangements of clusters (Fig. 11), which, when paired, effectively form a four-element array, that can have a radiation pattern feature depending on configuration.

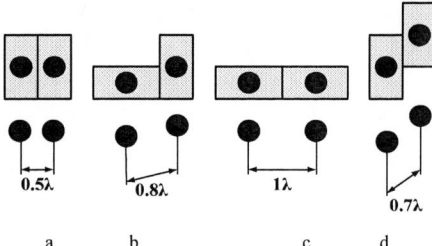

Fig.11. Graphical representation of all possible arrangements of clusters in the X algorithm structure (a) connecting clusters on a wide wall; (b) connecting clusters on a wide and narrow wall; (c) connecting clusters on a narrow wall; (d) connecting clusters on a wide wall with an offset

Let's assume that the minimum distance between the elements in a regular antenna array grid is 0.5λ, and let's also assume, that the phase centers of the clusters coincide with their geometric centers. Make calculations on the distance between the clusters centers. Fig.11 shows that the worst case of cluster arrangement is case «c» – the connection on a narrow wall, in which the distance between the phase centers becomes equal to 1λ. If we consider a cluster as a single element of an antenna curtain with a complex configuration (and a more complex pattern of partial directivity characteristics within the array), then by combining several such "elements" we obtain an antenna array with a large step. It is clear that with this configuration the cluster line does not provide the required wide electronic scanning sector.

To illustrate the above, Fig. 12 shows the step-by-step process of transition from the element grid to the excitation point grid during the clustering of the antenna array.

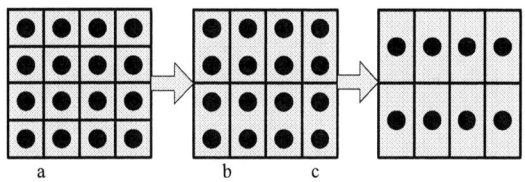

Fig. 12. Step-by-step transition from antenna array with a rectangular grid to clustered antenna array with equidistant arrangement of clusters (a) antenna array with a rectangular grid; (b) distribution of antenna array elements between clusters using uniform clustering; (c) clustered antenna array with marked phase (geometric) cluster centers

As an algorithmic basis for the synthesis of variants of a clustered antenna array the X algorithm was used, proposed by American mathematician D. Knuth [11], who mathematically solved the problem of complete covering a given-size area with figures of the specific type, avoiding overlapping and gaps. In antenna engineering applications the use of this algorithm for clustering the antenna array is described in [12].

Since this algorithm is a solution to a mathematical problem, its result is finding all possible ways to fill the antenna array with clusters. When working with antenna arrays of any large size the number of solutions is on the order of 1×10^7, so the time required to analyze the radio technical characteristics of all clustering options is very long. When creating a clustered antenna array with a wide electronic scanning sector up to $\pm 60°$ in the entire front hemisphere some restrictions were formulated to reduce the load on the computer used for calculations. These restrictions work in conjunction with the algorithm, allowing unsuccessful solutions to be discarded at the preliminary review stage (before calculating the antenna arrays radio technical characteristics). Sequential arrangements of clusters

with a large distance between their geometric centers were excluded; combinations of adjacent clusters were selected to ensure the minimal distance between phase centers, and the number of "unsuccessful" cluster combinations was analyzed. As a result of the antenna array synthesis based primarily on combinations of two-element domino-shaped clusters the antenna array version shown in Fig. 13 was obtained. Figure 14 shows the amplitude direction diagrams in the four main planes when scanning by $60°$. They were calculated both analytically (using the lattice multiplier method), and by electrodynamic analysis of the antenna array model by finite element method.

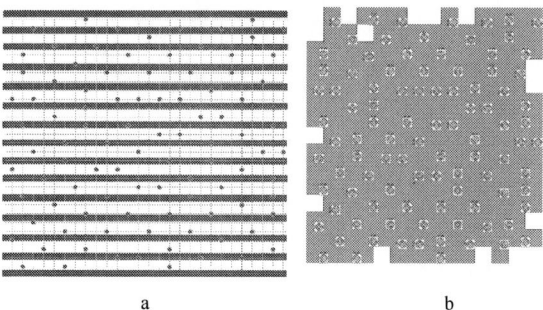

Fig.13 Clustered antenna array (a) antenna array analytical model; (b) rear view of electrodynamic model with microwave connectors located at the phase centers of individual clusters

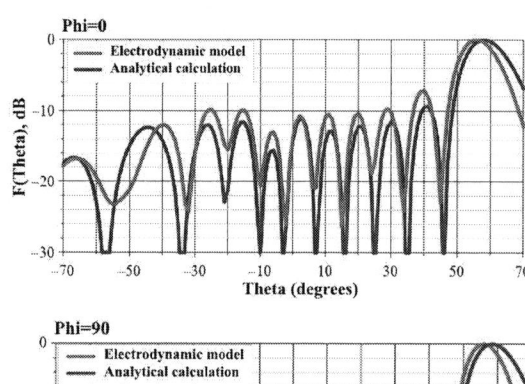

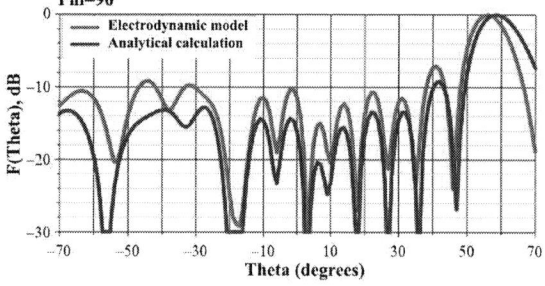

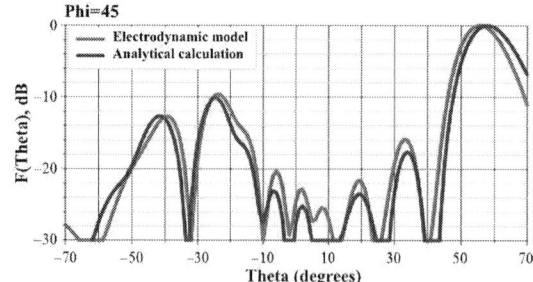

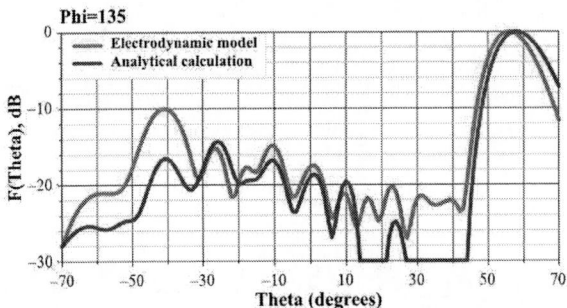

Fig.14 Comparison of the characteristics of electronic scanning of clustered antenna array models calculated by the analytical method and using electrodynamic modeling

IV. CONCLUSIONS

Antenna array clustering is a good method not only to reduce the cost of AESAs and simplify their assembly, but also to ensure that the system meets the requirements for electronic beam scanning. The conducted analytical and electrodynamic calculations demonstrate the similarity of the results both qualitatively (in terms of the envelope shape of the side lobes), and quantitatively (in terms of their level). The possibility of creating a clustered antenna array, capable of providing electronic scanning in a wide range of angles has been demonstrated, simultaneousely reducing the number of T/R-modules by more than 2 times as compared to the original completely filled regular antenna array. Based on the developed electrodynamic model it is possible to implement the antenna array practically.

In further research it is planned to evaluate and compare gain for clustered and regular antenna array, as well as to evaluate the probability of blind spots occurrence in clustered antenna array and to establish the conditions of their occurrence.

REFERENCES

[1] Rocca, P. Unconventional Phased Array Architectures and Design Methodologies – A Review / P. Rocca, G. Oliveri, R. J. Mailloux, A. Massa // Proceedings of the IEEE. – 2016. – Vol. 104, No. 3.;

[2] Antenna arrays. Calculation and design methods (in russian)/ Benenson L.S. and others // Moscow. Publisher: SOV.RADIO - 1966;

[3] D.A. Trifonov Development of recommendations for the design of an antenna system for aerological studies of the atmosphere // Ural Federal University// Ekaterinburg - 2020;

[4] Ghanta, H. A. Realizable Active Array Antenna Configuration For Fighter Aircraft Having Decent Radar Performance / H. Ghanta, S. P., B. Mole K. S., S. Kreenivasulu // 11th International Radar Symposium India - 2017 (IRSI-17). – 2017.;

[5] Solomon W. Golomb Polyominoes / Princeton Univercity Press – 1965;

[6] Nicola Anselmi, Paolo Rocca, Marco Salucci and Andrea Massa//Irregular Phased Array Tiling by Means of Analytic Schemata-Driven Optimization// IEEE TRANSACTIONS ON ANTENNAS AND PROPAGATION, VOL. 65, NO. 9, SEPTEMBER 2017;

[7] Paolo Rocca, Nicola Anselmi, Alessandro Polo and Andrea Massa // Pareto-Optimal Domino-Tiling of Orthogonal Polygon Phased Arrays// IEEE TRANSACTIONS ON ANTENNAS AND PROPAGATION, VOL. 70, NO. 5, MAY 2022;

[8] Roman Chirikov, Paolo Rocca, Luca Manica, Scott Santarelli, Robert J. Mailloux, and Andrea Massa // Innovative GA-based Strategy for Polyomino Tiling in Phased Array Design // 2013 7th European Conference on Antennas and Propagation (EuCAP);

[9] R.J. Mailloux, S.G. Santarelli and T.M. Roberts// Polyomino-shaped subarrays for time delay control of planar arrays// 1-4244-0123-2/06/$20.00 ©2006 IEEE;

[10] R. J. Mailloux, S. G. Santarelli, T. M. Roberts and D. Luu // Irregular Polyomino-Shaped Subarrays for Space-Based Active Arrays // Hindawi Publishing Corporation International Journal of Antennas and Propagation Volume 2009;

[11] D.E Knuth. Dancing links. arXiv:cs/0011047v1 [cs.DS]. 2000. 26 p.;

[12] Xiong, Z.-Y. Subarray Partition in Array Antenna Based on the Algorithm X / Z.-Y. Xiong, Z.-H. Xu, S.-W. Chen, S.-P. Xiao // IEEE Antennas and Wireless Propagation Letters. – 2013. – Vol. 12.

Numerical Modelling of Spin Wave Logic Gates

Roman Haponchyk
Department of Physical Electronics and Technology
St. Petersburg Electrotechnical University
St. Petersburg, Russia
https://orcid.org/0000-0002-8764-7100l

Alexey Ustinov
Department of Physical Electronics and Technology
St. Petersburg Electrotechnical University
St. Petersburg, Russia
https://orcid.org/0000-0002-7382-9210

Abstract—**In this study, an original theory was developed for a frequency response simulation of an exclusive NOR (XNOR) and exclusive OR (XOR) logic gates based on construction of nonlinear spin wave Mach-Zehnder interferometer. It is shown that the logical operations are perform due to the effect of an induced nonlinear phase shift of spin waves in nonlinear phase shifters located in different arms of the interferometer.**

Keywords—magnonics, nonlinear spin waves, logic gate

I. INTRODUCTION

The Mach-Zehnder-type spin wave (SW) interferometers are a fundamental and versatile component in the design of microwave magnonic circuits and devices [1, 2]. Since their invention, the concept of spin wave interference within this architecture has been successfully applied to create a wide range of functional devices. These include magnonic logic gates that perform Boolean operations [3-4], nonlinear directional couplers for signal routing and control [5, 6], and highly sensitive sensors for detecting magnetic fields [7, 8]. A significant advance in the field is the recent development of these interferometers at the nanoscale [9, 10], which has opened up the possibility for their integration with conventional semiconductor electronics.

The physical foundation for the operation of such interferometers lies in the rich nonlinear dynamics exhibited by propagating microwave SW in ferromagnetic films when their amplitude is sufficiently high [11-15]. Among these nonlinear phenomena, the nonlinear phase shift [16-18] is considered particularly promising for the development of magnonic logic circuits. The practical significance of this effect has been substantially enhanced by the recent discovery and demonstration of a giant nonlinear phase shift in engineered ultrathin magnetic film structures [19].

Current research shows that the induced nonlinear phase shift of spin waves in magnetic films and magnonic crystals [20, 21] represents another significant physical effect for future magnonic logic and computing devices. The driving motivation behind the investigation of the induced nonlinear phase shift is a key engineering challenge in magnonics: the development of truly cascadable logic gates. Such gates are essential for constructing complex circuits, as they require the output magnon signal from one logic stage to reliably control or switch the magnon signal in a subsequent stage, a functionality for which this effect is a prime candidate [6]. The purpose of the present work is to show the numerical simulation of nonlinear XNOR and XOR logic gates, whose operating principle is based on the effect of a nonlinear phase shift induced in the surface SW.

II. OPERATING PRINCEPLE

A block diagram of a nonlinear magnonic logic gate is presented in Fig. 1(a). It is based on a Mach-Zehnder interferometer architecture. This architecture represents itself as a two-arm bridge structure where the core elements are SW delay lines (fig. 1 (b)). This two identical, low-loss thin-film SW delay lines are used as a nonlinear phase shifters (NPS1 and NPS2). In the simulation, these elements are placed in a uniform magnetic field oriented tangentially to the film surface. This configuration of the field ensures the excitation and propagation of the surface SW. At the input of the logic gate, a low-amplitude of continuous wave signal at frequency f_s fed through a divider into NPS1 and NPS2. Simultaneously two independent high-amplitude pump pulsed signals with frequencies $f_{p1,2}$ are fed through a combiner into same NPSs. On the output side of the NPSs, the signals pass through auxiliary band-pass filters (BPFs). These BPFs suppress the pump signals at frequencies $f_{p1,2}$.

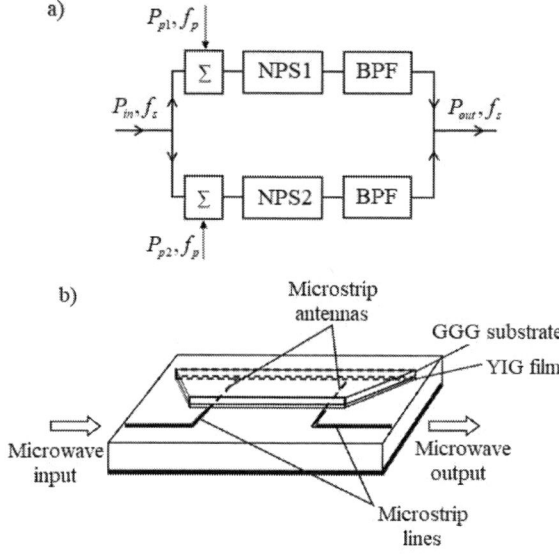

Fig. 1. Block diagram of a nonlinear magnonic logic gate (a) and spin wave delay line (b)

The operation principle of proposed nonlinear XNOR and XOR gates is realized in two steps. First, pump pulsed signals is translated into an equivalent sequence of phase jumps induced through nonlinear effect in the operation of continuous signal wave f_s propagating in NPS1,2 (different arms of the MZI). Finally, the logic signal at the gate output is produced as a result of conventional linear constructive/destructive interference occurring on the output combiner. This mechanism directly implements the XNOR and XOR logic functions.

In the numerical implementation, identical NPSs were modeled in each arm of the logic gate using the following parameters for an yttrium iron garnet (YIG) film. The film had a saturation magnetization of 1980 G, a ferromagnetic resonance linewidth of 0.6 Oe, and a thickness of 5.5 μm. Spin wave excitation and reception in NPS1 and NPS2 were simulated using two microstrip antennas, each 2 mm long and 50 μm wide, separated by a distance of 3 mm, with a

979-8-3315-4784-4/26 $31.00 © 2026 IEEE

characteristic impedance of 50 Ohms. An external magnetic field of 1094 Oe was applied tangentially to the film surface.

III. NUMERICAL SIMULATIONS

In order to describe the nonlinear XNOR and XOR logic gates performance a theoretical formalism of the device outlined below has been developed. This formalism relies upon the theoretical model established in our previous works [20, 22]. As illustrated in Fig. 1(a), the investigated logic gate is fundamentally a Mach-Zehnder interferometer containing two nonlinear spin wave phase shifters (NPS1 and NPS2) and an output power combiner. As outlined earlier, the operating principle of this nonlinear gate involves the deliberate introduction of phase shifts to two continuous-wave microwave signals propagating in NPS1 and NPS2. High-intensity control pulses via nonlinear SW mechanisms induce these shifts. Subsequently, these two phase shifted signals interfere, either constructively or destructively, on a microwave combiner thus forming the output signal corresponding to the targeted logical function.

The formula for the nonlinear SW interferometer (NSWI) power-dependent transmission coefficient has the form:

$$H_{NSWI}(\omega, P_{p1,2}) = \frac{1}{4}[H_{NPS1}(\omega) + H_{NPS2}(\omega) + \qquad (1)$$
$$+2\sqrt{H_{NPS1}(\omega)H_{NPS2}(\omega)}\cos(\Delta\varphi(\omega, P_{p1,2}))],$$

where

$$H_{NPS1}(\omega) = H_{exc1}(\omega)H_{SW1}(\omega)H_{rec1}(\omega) \qquad (2)$$

$$H_{NPS2}(\omega) = H_{exc2}(\omega)H_{SW2}(\omega)H_{rec2}(\omega) \qquad (3)$$

$$\Delta\varphi(\omega, P_{p1,2}) = \varphi_1(\omega) + \varphi_{INL1}(\omega, P_{p1}) - \qquad (4)$$
$$-\varphi_2(\omega) - \varphi_{INL2}(\omega, P_{p2})$$

here H_{NPS1} and $H_{NPS2}(\omega)$ are transmission coefficients of NPS1,2 located in the arms of the interferometer, $H_{exc1,2}(\omega)$ and $H_{rec1,2}(\omega)$ are the coefficients characterizing the efficiency of the SW excitation and reception by microstrip antennas, respectively, $H_{SW1,2}(\omega)$ is the coefficient characterizing the damping of the SW propagating in the ferrite film, $\Delta\varphi(\omega, P_{p1,2})$ is phase difference between SW signals at the output, which is a function of the frequency and power of the pump signals, $\varphi_1(\omega)$ and $\varphi_2(\omega)$ are linear phase shifts in NPSs, $\varphi_{INL1}(\omega, P_{p1})$ and $\varphi_{INL2}(\omega, P_{p2})$ are power dependent induced nonlinear phase shifts.

The induced nonlinear phase shifts $\varphi_{INL1,2}(\omega, P_{p1,2})$ of the SW in the NPSs were calculated using the model developed in our previous work [20]. Below we present an expression only for the phase shift induced in NPS1, a similar expression holds for NPS2:

$$\varphi_{INL1}(z) = -\frac{1}{V_{g1}}\int_0^z \left(N_{11}u_1^2(z) + N_{12}u_2^2(z)\right)dz, \qquad (5)$$

here, $u_{1,2}$ is the dimensionless amplitude of continuous wave and pump SW, V_{g1} is the group velocity, N_{11} and N_{12} are the cubic nonlinear self-interaction and induced interactional coefficients. The model takes into account the nonlinear damping of SWs [5] and shows that in general case each wave accumulates its own nonlinear self-phase shift with an increase in its power as well as the induced phase shift controlled with power of the pump signal.

Using this theoretical formalism, the frequency responses of various magnonic logic gates were modeled. Now let us examine these logic gates. The first logic gate is an element implementing XNOR logic function (equivalence). This function is realized by applying pump powers to NPS1 and NPS2 according to the truth table provided in Table I.

TABLE I.   TRUTH TABLE OF THE XNOR LOGIC GATE

NPS1	NPS2	XNOR
0	0	1
1	0	0
0	1	0
1	1	1

Here and hereafter, we will take the value of the dimensionless squared amplitude of the pumping SW as the pump power $P_{p1,2} \sim |u|_{1,2}^2$. When a dimensionless squared amplitude $|u|_2^2$ of 0.028 arb. units is applied to NPS1, the output of the logic gate yields a logical «1», when $|u|_2^2$ of 0 arb. units is applied, it yields a logical «0». The second phase shifter, NPS2, operates on an identical principle. Figure 2 illustrates the relationship between the induced nonlinear phase shift $\varphi_{INL1,2}$ of the probe signal, accumulated in the nonlinear phase shifters NPS1 and NPS2, and the dimensionless squared amplitude of the control signal. The plot demonstrates that a phase shift of 180 degrees is achieved when $|u|_2^2$ reaches a value of 0.028 arb. units. This specific 180-degree (π) phase shift is the fundamental mechanism enabling the interferometer to operate as XNOR logic gate.

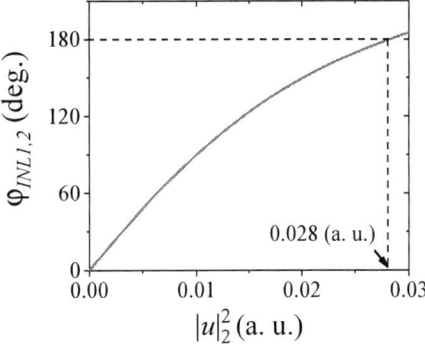

Fig. 2. Induced nonlinear phase shift of the continuous wave signal vs dimensionless squared amplitude in NPS1, 2 for XNOR logic gate

The linear continuous wave signal wave, propagating through the interferometer arms, acquires an additional induced nonlinear phase shift. Subsequently, these two phase shifted signals are combined. Destructive interference produces a weak output signal (logical «0»), while constructive interference produces a relatively strong output signal (logical «1»). The simulated frequency response of this element is shown in Figure 3. It can be seen from the frequency response graph that the logical XNOR operation is implemented in the probe signal frequency f_s range from 5.135 GHz to 5.185 GHz. The pump signal frequency f_{p1} was set at 5.275 GHz.

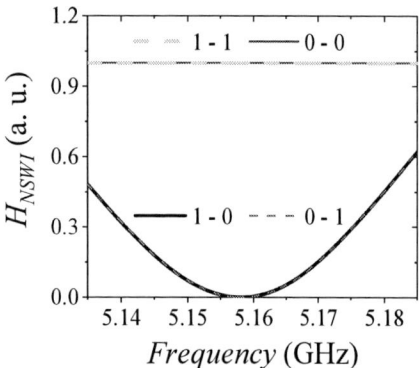

Fig. 3. Modeled frequency response of the XNOR logic gate

The second logic element implements XOR, or logical inequality, function. Its operation is governed by applying specific dimensionless squared amplitude $\left|u\right|_2^2$ levels to the nonlinear phase shifters NPS1 and NPS2, as defined in the truth table provided in Table II. For instance, a dimensionless squared amplitude $\left|u\right|_2^2$ of 0.01 arb. units to NPS1 leads to a logical «1» at the output, whereas a $\left|u\right|_2^2$ of 0.001 arb. units yields a logical «0». NPS2 functions analogously. The dependence of the induced nonlinear phase shift $\varphi_{INL1,2}$ in the probe signal, as it propagates through NPS1 and NPS2, on the dimensionless squared amplitude is shown in Figure 4. According to the data, a phase shift of 90 degrees is attained at an amplitude of 0.01 arb. units, while a shift of 10 degrees corresponds to an amplitude of 0.001 arb. units. This controlled asymmetry in the phase shift between the two interferometer arms is what physically implements the XOR logical gate.

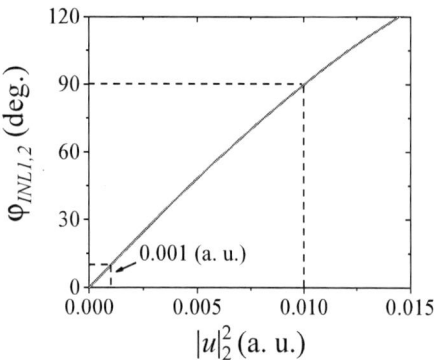

Fig. 4. Induced nonlinear phase shift of the continuous wave signal vs dimensionless squared amplitude in NPS1, 2 for XOR logic gate

TABLE II.   TRUTH TABLE OF THE XOR LOGIC GATE

NPS1	NPS2	XOR
0	0	0
1	0	1
0	1	1
1	1	0

Following this controlled phase shift, the operational mechanism follows the previously described principle. The two phase-shifted signals are interfered in the combiner. Destructive interference results in a low-power output, interpreted as a logical «0». Conversely, constructive interference produces a higher-power output, corresponding to a logical «1». This behavior is illustrated in Figure 5. Analysis of the frequency response on the figure 5 confirms the implementation of the XOR function. The logical operation is defined within the probe signal frequency f_s from 5.119 GHz to 5.175 GHz. In this simulation, the pump frequency for the second control signal f_{p2} was set at 5.275 GHz.

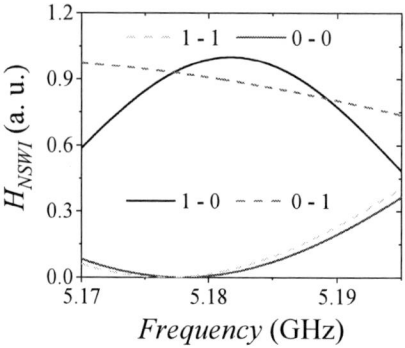

Fig. 5. Modeled frequency response of the XOR logic gate

IV. CONCLUSUIONS

In conclusion, this work has presented a numerical simulation of XOR and XNOR logic gates based on a nonlinear magnonic Mach-Zehnder interferometer. The results show that both fundamental logic functions can be implemented within the Mach-Zegnder interferometer architecture. The operational principle is based on the induced nonlinear phase shift of surface SWs, where the value of nonlinear phase shift is determined by the power level of the pump signal. For the XNOR gate, a π-phase shift, induced at a specific control power corresponding to a dimensionless amplitude of 0.028 arb. units, leads to constructive or destructive interference at the output combiner, representing logic states «1» and «0». Similarly, the XOR function is realized through a deliberate misalignment of phase shifts 90 and 10 degrees for different input states between the arms, enabling XOR functionality.

The numerical models were developed taking into account key physical mechanisms underlying the principle of operation of the device: the nonlinear phase shift and nonlinear damping of spin waves. This numerical model confirms that the required phase shift is achievable with moderate microwave pump power levels. Beyond the specific logic gates studied, the developed theoretical model possesses generality. It is adaptable for describing a wide variety of logic functions and can be extended to alternative interferometer-like configurations, including those utilizing normally magnetized waveguides for forward volume SWs, tangentially magnetized structures, or even magnonic crystals, which offer further control over dispersion and damping.

A feature of the proposed architecture is its cascadability. Since the logic state is encoded in the amplitude of a SW probe signal at a designated frequency, the output from one gate can be directly used as the control input for a subsequent stage. This property, demonstrated through the operational principle of induced phase control, is an essential prerequisite for sequential logic and the development of complex computational circuits. Therefore, this research paves the way for the design of all-magnonic sequential logic systems and highlights the induced nonlinear phase shift as a versatile and promising physical mechanism for wave-based computing. The fundamental concepts explored relying on interference and nonlinear phase shift in a scalable interferometric layout hold strong potential for translation into other wave-based computing paradigms, such as for example integrated photonics.

ACKNOWLEDGMENT

This work was supported by the Ministry of Science and Higher Education of the Russian Federation (grant number No. FSEE-2025-0008).

REFERENCES

[1] Y. K. Fetisov, C. E. Patton, "Microwave bistability in a magnetostatic wave interferometer with external feedback," IEEE Trans. Mag., vol. 35, pp 1024-1036, March 1999. DOI: 10.1109/20.748850.

[2] A.B.Ustinov, B.A.Kalinikos, "Nonlinear microwave spin wave interferometer," Tech. Phys. Lett., vol. 27, i. 10, pp. 403-405, May 2001. DOI:10.1134/1.1376765.

[3] M. P. Kostylev, A. A. Serga, T. Schneider, B. Leven, B. Hillebrands, "Spin-wave logical gates," Appl. Phys. Lett., vol. 87, pp. 153501, October 2005. DOI:10.1063/1.2089147.

[4] A.B. Ustinov, E. Lähderanta, M. Inoue, B.A. Kalinikos, "Nonlinear Spin-Wave Logic Gates," IEEE Magn. Lett., vol. 10, pp. 5508204, October 2019. DOI:10.1109/LMAG.2019.2950638

[5] A.B.Ustinov, B.A. Kalinikos, "Power-dependent switching of microwave signals in a ferrite-film nonlinear directional coupler," Appl. Phys. Lett., vol. 89, pp. 172511, October 2006. DOI: 10.1063/1.2362576.

[6] A. Mahmoud et al., "Fan-out enabled spin wave majority gate," AIP Advances, vol. 10, pp. 035119, March 2020. DOI:10.1063/1.5134690.

[7] Taichi Goto et al., "Spin wave differential circuit for realization of thermally stable magnonic sensors," Appl. Phys. Lett., vol. 106, pp. 132412, April 2015. DOI:10.1063/1.4916989.

[8] M. Balynsky et al., "A Magnetometer Based on a Spin Wave Interferometer," Sci Rep. vol. 7, pp. 11539, September 2017. DOI: 10.1038/s41598-017-11881-y.

[9] J. Chen et.al., "Reconfigurable Spin-Wave Interferometer at the Nanoscale," Nano Lett., vol. 21, i. 14, pp. 6237-6244, July 2021. DOI:10.1021/acs.nanolett.1c02010.

[10] A. A. Grachev , A. V. Sadovnikov , and S. A. Nikitov ," Strain-Tuned Spin-Wave Interference in Micro- and Nanoscale Magnonic Interferometers," Nanomaterials, vol. 12, i. 9, pp. 1520, April 2022. DOI:10.3390/nano12091520.

[11] A. B. Ustinov , A. V. Kondrashov, I. Tatsenko , A. A. Nikitin , and M. P. Kostylev, "Progressive development of spin wave chaos in active-ring oscillators," Phys. Rev. B, vol. 104, pp. L140410, October 2021. DOI:10.1103/PhysRevB.104.L140410

[12] Anastasia S. Bir, et.al, "Direct electric current control of hyperchaotic packets of dissipative dark envelope solitons in a magnonic crystal active ring resonator," Phys Rev. Appl., vol. 21, pp. 044008, April 2024. DOI:10.1103/PhysRevApplied.21.044008.

[13] Alexey B. Ustinov, et.al., "Excitation of fundamental multiple dark solitons from forced biharmonic oscillations in a magnonic active ring," Phys. Rev. B, vol. 110, pp. 174413, November 2024. DOI:10.1103/PhysRevB.110.174413.

[14] Maria A. Morozova, et.al., "Gap solitons in nanoscale YIG magnonic crystals," Phys. Rev. B, vol. 110, pp. 104408, September 2024. DOI: 10.1103/PhysRevB.110.104408.

[15] Nikolaev K.O.,. et al., "Resonant generation of propagating second-harmonic spin waves in nano-waveguides," Nat Commun., vol. 15, pp. 1827, Februaru 2024. DOI:10.1038/s41467-024-46108-y.

[16] A. B. Ustinov, B. A. Kalinikos, "A microwave nonlinear phase shifter," Appl. Phys. Lett., vol. 93, pp. 102504, September 2008. DOI: 10.1063/1.2980022.

[17] U. H. Hansen, V. E. Demidov, S. O. Demokritov, Appl. Phys. Lett., vol. 94, pp. 252502, June 2009. DOI:10.1063/1.3159628.

[18] B. K. Kuanr, et. al., "Monolithic Microwave Nonlinear Phase Shifter," IEEE Magn. Lett., vol. 6, pp. 3500304, October 2015. DOI: 10.1109/LMAG.2015.2406295.

[19] H. Merbouche, et al., "Giant nonlinear self-phase modulation of large-amplitude spin waves in microscopic YIG waveguides," Sci Rep., vol. 12, pp. 7246, May 2022. DOI:10.1038/s41598-022-10822-8.

[20] A. B. Ustinov, et.al., "Induced nonlinear phase shift of spin waves for magnonic logic circuits," Appl. Phys. Lett., vol. 119, pp. 192405, November 2021. DOI:10.1063/5.0074824.

[21] A. B. Ustinov, R. V. Haponchyk, "Nonlinear phase shifts induced by pumping spin waves in magnonic crystals," Appl. Phys. Lett., vol. 122, pp. 212401, May 2023. DOI:10.1063/5.0153392.

[22] A.B. Ustinov, B.A. Kalinikos, "Ferrite-film nonlinear spin wave interferometer and its application for power-selective suppression of pulsed microwave signals," Appl. Phys. Lett., vol. 90, pp. 252510, June 2007. DOI:10.1063/1.2751121.

Tunable Ring Resonator Based on Ferroelectric Capacitors

Kamil Karymsakov
Department of Microwave Electronics
Saint Petersburg Electrotechnical
University "LETI"
Saint Petersburg, Russia
kamakar2003@mail.ru

Kirill Dmitriev
Department of Physical Electronics and
Technology
Saint Petersburg Electrotechnical
University "LETI"
Saint Petersburg, Russia
Ne4ehbka88@gmail.com

Svetlana Zubko
Department of Physical Electronics and
Technology
Saint Petersburg Electrotechnical
University "LETI"
Saint Petersburg, Russia
spzubko@etu.ru

Danyar Alzhanov
Department of Physical Electronics and
Technology
Saint Petersburg Electrotechnical
University "LETI"
Saint Petersburg, Russia
alzhanoff.daniar@yandex.ru

Valeriia Bobrovskaia
JSC "PO «Sevmash"
Saint Petersburg, Russia
bobrovskaaa937@gmail.com

Abstract—**The results of a study of a ring tunable resonator on ferroelectric capacitors are presented. The result was manufactured using microstrip transmission line sections. A comparative analysis of the measured and calculated frequency characteristics demonstrates their correspondence.**

Keywords— resonator, ferroelectric, tunable, resonance.

I. INTRODUCTION

The development of modern telecommunication systems and radars requires the designing tunable microwave resonators with high stability, compact size and low energy consumption. Traditional mechanical tuning methods based on changing the geometry of the resonator have low performance and limited reliability. Ferroelectric materials represent a promising alternative. Its special feature is dependence of the dielectric constant on the external biasing electric field, which allows for rapid electrical restructuring of the capacity of a ferroelectric capacitor without mechanical displacement. The integration of such a capacitor makes it possible to control its equivalent capacitance and, consequently, the resonant frequency, realizing the principle of electrical tuning. Being compared with ferrites and semiconductor the devices based on ferroelectrics might be advantageously distinguished by more simple fabrication technology, smaller controlling circuit energy consumption, and good compatibility with high temperature superconductors. [1].

II. MODELLING

A. Initial data

Fig. 1 shows a model of a ring resonator [2]. The linear dimensions of the resonator are 40x40 mm. The resonator is designed on an FR4 substrate with a thickness of 2 mm, a dielectric constant of 4.5, and tanδ = 0.02. The resonator is a closed metallized ring having 2 slots. Ferroelectric capacitors are placed in gaps. The ring resonator is connected to the external circuit via an inductive-capacitive coupling 1. The controlled elements in the circuit are ferroelectric capacitors 2. To tune the capacitance, dc voltage must be applied to the capacitor plates. To apply voltage, a bias line circuit is used, consisting of a radial stub 3 for decoupling the RF signal and power supply and a thin line 4, which is a low-pass filter [3].

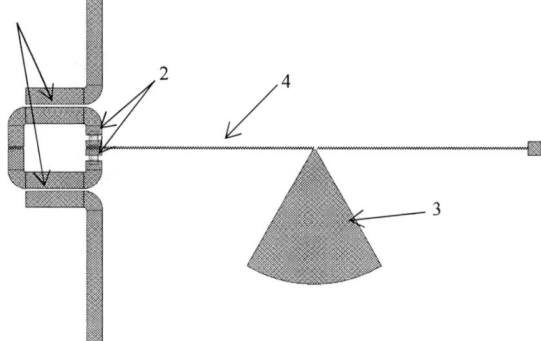

Fig. 1. The layout of the ring resonator

The open radial stub is designed to short-circuit the microwave signal, providing attenuation at the central frequency. The radius of the stub mainly determines the attenuation frequency, while the opening angle affects the bandwidth. It is also worth considering that the radial open stub itself is separated from the line of the microwave signal path at a distance of a quarter of the wavelength.

The capacitors are an interdigital structure shown in fig. 2. The material, used in capacitors, is BSTO. The linear dimensions of the BSTO capacitor are 0.8 x 1.7 mm. Fig. 3 shows the experimental voltage-farad characteristic (VFC) of the capacitor [4].

Fig. 2. Ferroelectric capacitor photograph

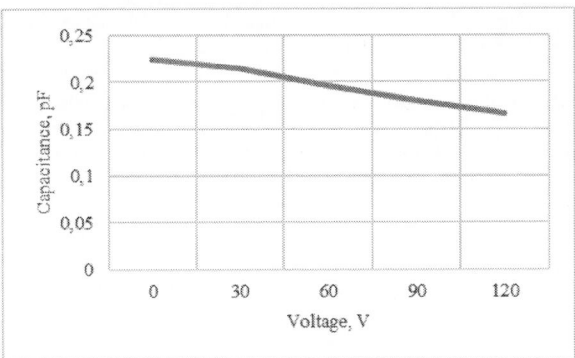

Fig. 3. VFC of ferroelectric capacitor

To excite ring resonators, a power supply method is proposed through connected lines arranged symmetrically (fig. 4 "a", "b" and "d") and asymmetrically (fig. 4 "c") relative to each other. Fig. 5 and 6 show their frequency response.

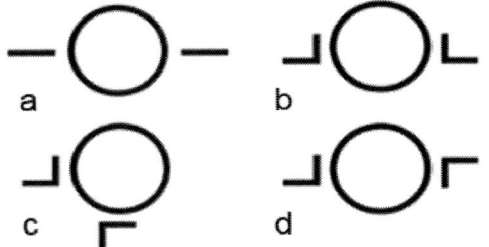

Fig. 4. Configurations of supply lines

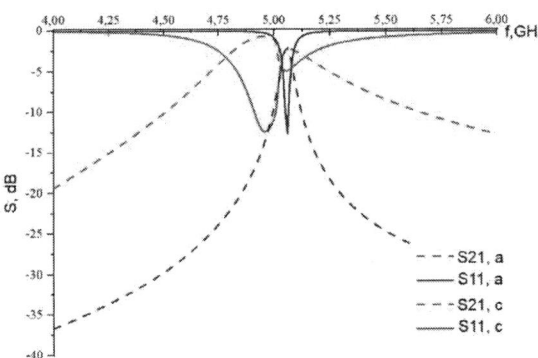

Fig. 5. Frequency response for configurations "a" and "c"

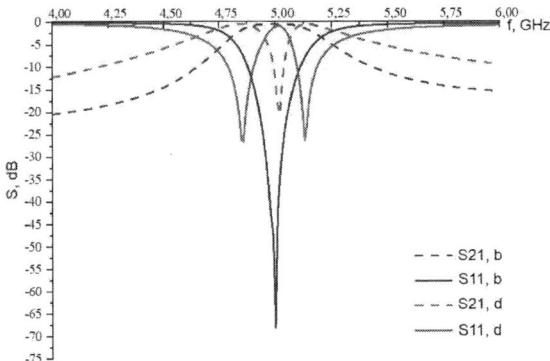

Fig. 6. Frequency response for configurations "b" and "d"

The modeling has shown that in the case of non-tunable structures, the option of an asymmetric structure is preferable, while a symmetrical structure is suitable for tunable structures. In further modeling, a symmetrical method of energy transfer with inductive-capacitive coupling was chosen.

B. Simulation results

Fig. 7 shows the modelled frequency dependence of the reflection coefficient of the ring resonator shown in fig. 1. The arrows indicate the reflection coefficients at bias of 0 V and 120 V. At 0 V the central frequency is 3.35 GHz, at this frequency the reflection coefficient is -17 dB. At a 120V offset, it is 3.51 GHz and the level is -17.4 dB.

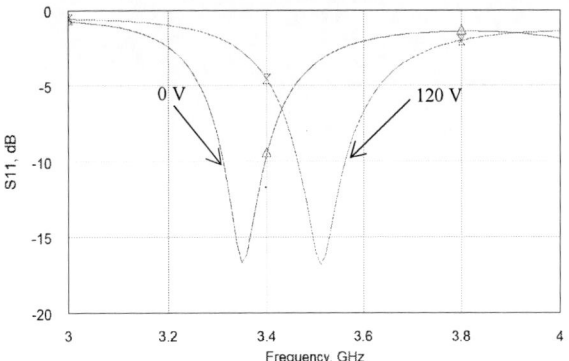

Fig. 7. Modelled dependence of the reflection coefficient on frequency

Fig. 8 shows the modelled dependence of the transmission coefficient on the frequency when the voltage is 0 V and 120 V. The transmission coefficient level at the center frequency at a 0 V offset is -1.7 dB, at a 120 V offset it is -1.8 dB.

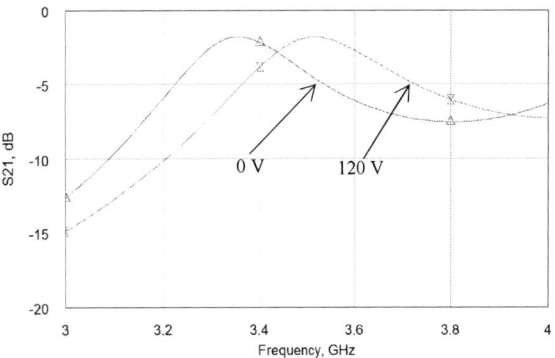

Fig. 8. Modelled dependence of the transmission coefficient on frequency

III. EXPERIMENT

Based on the performed electromagnetic modeling the ring resonator loaded by ferroelectric capacitors was manufactured and experimentally studied. This section presents the results of the experiment.

A. Layout

Photograph of designed tunable resonator is presented in fig. 9. To realized control biasing voltage network via 1 and a bias line 2 were used. The via is metallized and connects the top and bottom electrodes. A positive voltage is supplied through the bias line 2 to the junction of the capacitors. Capacitors separate the "plus" from the "minus".

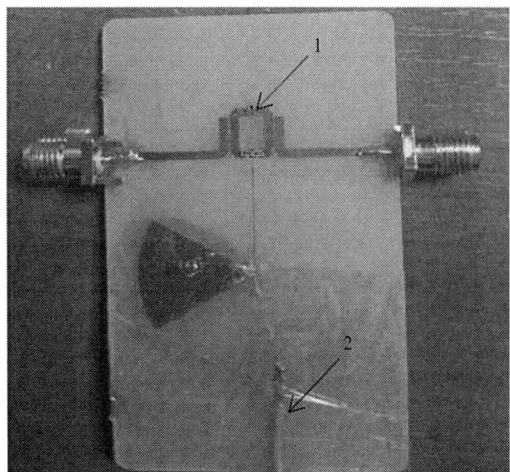

Fig. 9. The layout of the ring resonator

B. Results

Fig. 10 shows the experimentally measured frequency dependence of the reflection coefficient at 0 V and 120 V. This dependence was obtained using a vector circuit analyzer.

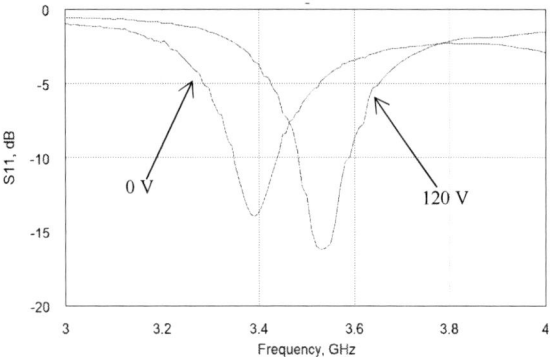

Fig. 10. Experimental dependence of the reflection coefficient on frequency

Fig. 11 shows the experimental frequency dependence of the transmission coefficient at 0 V and 120 V.

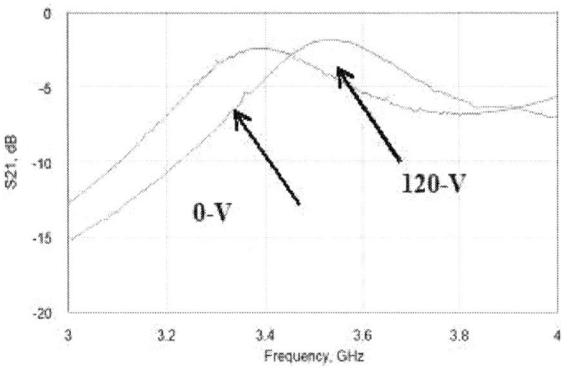

Fig. 11. Experimental dependence of the transmission coefficient on frequency

Compared to the theoretical dependencies, experimentally the center frequencies were shifted relative to the theoretical center frequencies. At U_{DC} = 0 V the experimental center frequency is 3.39 GHz, and the theoretical frequency is 3.35 GHz. At U_{DC} = 120 V the experimental center frequency is 3.53 GHz, and the theoretical frequency is 3.51 GHz. The loss levels have also changed slightly. The experimental value of the transmission coefficient at a 0 V offset at the center frequency was -13.9 dB (the theoretical value of the reflection coefficient at the center frequency at a 0 V offset is -17 dB). At a displacement of 120 V, the experimental value of the reflection coefficient became -16.23 dB (the theoretical value was 17.4 dB). The measured value of the transmission coefficient at a 0 V offset was -2.4 dB, at a 120 V offset it was -1.86 dB. Measured operational frequency range is 140 MHz, modelled operational frequency range is 160 MHz. These changes in frequency characteristics are associated with inaccuracy in the manufacture of the layout, neglect of physical phenomena during modeling, and disregard for external leads.

IV. CONCLUSIONS

The almost complete coincidence of the measured frequency response with the characteristic obtained during the simulation indicates the correctness of the chosen model and the accuracy of calculations. The frequency adjustment was 140 MHz, which confirms the effectiveness of using ferroelectric capacitors as tunable elements in resonant systems. Minor discrepancies can be explained by spurious mounting parameters, element nominal tolerances, and measurement equipment errors. For tunable structures, the option of symmetrical microwave power supply with inductive-capacitive coupling is the most preferable.

REFERENCES

[1] Matthaei G. L., Young L., Jones E. M. Design of microwave filters, impedance-matching networks, and coupling structures. Volume 2. – 1963.

[2] Sarkar D., Moyra T. A low cost electrically tunable bandpass filter with constant absolute bandwidth //AEU-International Journal of Electronics and Communications. – 2017. – T. 77. – C. 156-162.

[3] Makimoto M., Sagawa M. Varactor tuned bandpass filters using microstrip-line ring resonators //1986 IEEE MTT-S International Microwave Symposium Digest. – IEEE, 1986. – C. 411-414..

[4] Miller S. L. et al. Device modeling of ferroelectric capacitors //Journal of applied physics. – 1990. – T. 68. – №. 12. – C. 6463-6471.

Application of X-Ray Fluorescence Analysis for Identification of Paint Pigments in Easel and Wall Paintings

Laleh M. Moazzami Lavasani
Department of Laser and Navigation Systems
Saint Petersburg Electrotechnical University "LETI"
Saint Petersburg, Russia
moazszami@gmail.com

Abstract—**The results of elemental analysis of paint layers of Russian icons, paintings and frescoes of the 18th–19th centuries, performed using a portable X-ray fluorescence spectrometer, are presented. The main attention is paid to the interpretation of the fluorescence spectra recorded at the control points of the analyzed samples, which made it possible to determine the type of pigments of the paint layer and identify the features of their use.**

Keywords—X-ray fluorescence analysis, portable spectrometer, paint layer, pigments, restoration.

I. INTRODUCTION

Restoration and conservation of cultural and historical heritage objects usually require a comprehensive study, including determining the chemical composition of materials or types of surface contamination, the state of its preservation, etc. Currently, optophysical methods and technologies are increasingly used to solve these problems. For example, to study the elemental composition of the paint layer of works of art, modern practices use the method of X-ray fluorescence analysis (XRF).[1]

The relevance of the application of the XRF method in the study of cultural heritage objects is due to the increasing requirements for preserving the integrity of works of art during their scientific study. This necessitates the use of predominantly non-invasive methods when working with historical monuments. XRF meets the key requirements applied to analytical methods in restoration: it is non-invasive, does not require sampling or preliminary preparation of samples, and does not violate the physicochemical properties of the object being studied.[2]

Statement of the Problem

The objective of the experimental research carried out in this work was to identify the paint pigments of works of Russian monumental and stage painting of the 18th–19th centuries on the basis of the St. Petersburg Academy of Arts. Ilya Repin.

Course Materials

The studies were carried out using a portable X-ray fluorescence spectrometer "MetExpert", designed for multi-element analysis of metals and alloys. The device is compact, equipped with an X-ray emitter and is capable of detecting about 80 elements - from sodium (Z=11) to americium (Z=95) - in solid, powder and liquid states. Icons, paintings and frescoes by Russian masters of the 18th–19th centuries were chosen as objects of research.

On each object under study, control points with the most dense and pure pigment were determined. In them, a spectral analysis was performed using the XRF method, which provides information about the chemical composition of the area under study. All experiments were carried out using a portable X-ray fluorescence spectrometer "MetExpert". Information from the analyzer is processed by software and displayed on a PC in the form of a spectrogram (the abscissa axis is the fluorescence energy, the ordinate axis is the radiation intensity).

When conducting XRF studies, the strongest fluorescence peaks are usually taken into account. To avoid errors in identifying a chemical element by the quantitative value of its energy, these values are double-checked against the corresponding tables of photon energy of fluorescent X-ray radiation. Based on the time and place of creation of the object, the resulting set of elements is also checked against tables of basic artistic pigments and fillers. Pigments are identified on these grounds.

II. EXPERIMENTAL RESULTS

This section presents the results of XRF measurements and interpretation of the obtained spectra for the two studied works of painting.

A. Icon "St. Panteleimon", 19th Century

The icon of an unknown author "Saint Panteleimon" was allegedly created in the 19th century. Fig. 1 shows its photograph and the corresponding cartogram, on which the numbers indicate the so-called control points at which XRF measurements were made.

Fig. 1. Icon "Saint Panteleimon" in color and its cartogram

Visually, the state of preservation of this icon could be assessed as unsatisfactory. However, its paint layer was partially lost due to shedding and peeling of paint, as a result of which the number of fragments with sufficient pigment thickness for reliable analysis was small. However, it was possible to conduct an XRF analysis of it, and spectrograms were obtained for each control point.

At each analyzed point, high levels of the element lead (Pb) were found. The presence of this element was characteristic of many oil painting pigments of that era.[3] Also, until the beginning of the 20th century, lead white, that is, white whose main component is lead carbonate, was the most common in tempera painting, so its presence was taken into account when identifying pigments.[3]

As an example, consider point 4, which refers to green pigments. Its spectrum is shown in Fig. 2

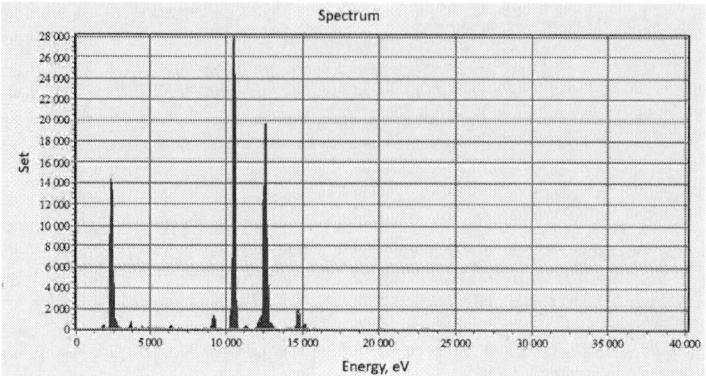

Fig. 2. Spectrum point 4 (green pigment)

In this case, the main fluorescence peaks correspond to the elements: lead (Pb), barium (Ba), calcium (Ca), chromium (Cr), iron (Fe), and zinc (Zn). Taking into account this combination of elements, it was concluded that this pigment is green cinnabar, which is a mixture in various proportions of Zn, Cr, Fe, Ba, Pb, Ca.

B. "Portrait of a young man in a red camisole."

Another object of our research was a painting by an unknown artist "Portrait of a Young Man in a Red Jacket." During its examination, a writing was discovered, presumably made after the painting was painted, and it was thanks to X-ray fluorescence measurements that not only its elemental composition was studied and the pigment was determined, but also the time period in which it was left was revealed. The estimated time of its writing dates back to the 18th century. Fig. 3 shows the picture in color, and Fig. 4 shows its cartogram with marked control points at which measurements were taken.

Fig. 4. Portrait cartogram

The most interesting point is point 5, corresponding to the area supposedly containing whitewash, since it was there that the recording was discovered. Thanks to X-ray fluorescence analysis, the spectrum presented in Fig. 5 was obtained.

Fig. 3. Painting "Portrait of a young man in a red camisole"

979-8-3315-4784-4/26 $31.00 © 2026 IEEE

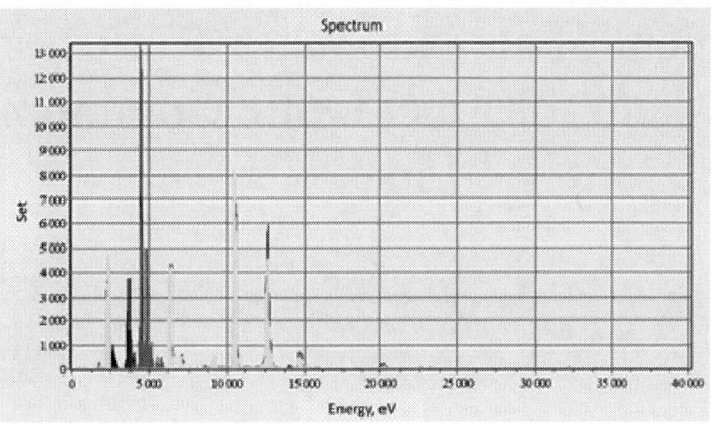

Fig. 5. Spectrum of point 5 (white pigment)

The main elements in the pigment composition at control point 5 are Ba (barium), Pb (lead), Ti (titanium). It is worth noting that titanium began to be used as part of colorful pigments only in the 20s of the 20th century, and therefore, the artist himself could not use them in this work. After comparing this set of elements with the table of pigments and fillers, it was concluded that this pigment is titanium white or Blanfix. The latter is used to partially replace titanium dioxide in the production of pigments, and is a synthetic barium sulfate. This suggests that the two pigments were mixed.

III. CONCLUSION

During the study, using X-ray fluorescence analysis, more than 30 cultural heritage objects undergoing restoration at the St. Petersburg Academy of Arts were studied. Ilya Repin. The data obtained made it possible to identify the pigments of all the studied works with the most accurate probability. The results obtained confirm that XRF is an effective method for the non-destructive study of works of easel and monumental painting of the 18th – 19th centuries.

IV. ACKNOWLEDGMENTS

The authors express their gratitude to the staff of the Department of Painting Restoration of the St. Petersburg Academy of Arts named after Ilya Repin for the opportunity to conduct the research described in this article. Also, the authors would like to express their gratitude to V. A. Parfenov (St. Petersburg Institute of History of the Russian Academy of Sciences, St. Petersburg Electrotechnical University "LETI") for his scientific guidance and constant support throughout this study.

V. REFERENCES

[1] Chernorukov N. G., Nipruk O. V. Theory and practice of X-ray fluorescence analysis // Nizhny Novgorod: Nizhny Novgorod State University, 2012. – 57 p.

[2] G. Revenko, V. A. Revenko Application of the X-ray spectral method of analysis for the study of cultural heritage materials (Review) // Methods and objects of chemical analysis, 2007, Vol. 2, No. 1, p. 4 – 29.

[3] Grenberg, Yu. I. Technology and study of works of easel and wall painting: a textbook for students of art universities and art schools / Yu. I. Grenberg; State Research Institute of Restoration. - Moscow: State Research Institute of Restoration, 2000. - 179 p.

Development of an Automated Reflectometer Control System for Fiber Optic Communication Lines

Evgeny S. Minin
The Bonch-Bruevich Saint Petersburg
State University of Telecommunications
Saint Petersburg, Russia
eminin636@gmail.com

Sergey S. Gryzulev
The Bonch-Bruevich Saint Petersburg
State University of Telecommunications
Saint Petersburg, Russia
sgryz.sut@gmail.com

Bogdan K. Reznikov
The Bonch-Bruevich Saint Petersburg
State University of Telecommunications
Saint Petersburg, Russia
rznkff@net.sut.ru

Timofey A. Kotov
The Bonch-Bruevich Saint Petersburg
State University of Telecommunications
Saint Petersburg, Russia
kotoff201211@gmail.com

Nikolay Yu. Kolybelnikov
The Bonch-Bruevich Saint Petersburg
State University of Telecommunications
Saint Petersburg, Russia
ya.nikolai-kolyb@yandex.ru

Dmitry I. Isaenko.
The Bonch-Bruevich Saint Petersburg
State University of Telecommunications
Saint Petersburg, Russia
isaenko-d@mail.ru

Abstract — **Optical Time Domain Reflectometers (OTDR) are widely used in optical communications at all stages of the life cycle. Examples of use include: incoming inspection (measurement of the cable section prior to installation), measurements during construction (inspection of splice installation using the loopback method), acceptance testing (line measurement before commissioning), ongoing monitoring (measurement during line operation), line modernization and reconstruction (identification of lines for reinstallation in joints — so-called re-soldering). The article is devoted to reviewing methods for monitoring optical transport networks and developing a monitoring system.**

Keywords — Fiber Monitoring, OTDR, Reflectometer, DAS, Laser, Raman Scattering, Network Management System, NMS

I. CLASSIC OTDR: GENERAL PRINCIPLES

Measuring the distribution of optical losses along the length of an optical cable using the backscatter method is based on recording the backscattered radiation in the optical fiber of the cable being measured as an optical pulse passes through it and measuring the dependence of the intensity (power) of this radiation on time. The method is suitable for determining the distribution of optical losses along the length of the cable, cable attenuation, distributed and local inhomogeneities such as breaks, splice locations, and distances to inhomogeneities, measuring the value of losses at inhomogeneities, as well as fiber length and distances to break locations. Figure 1 shows Classic OTDR block diagram and typical reflectogram.

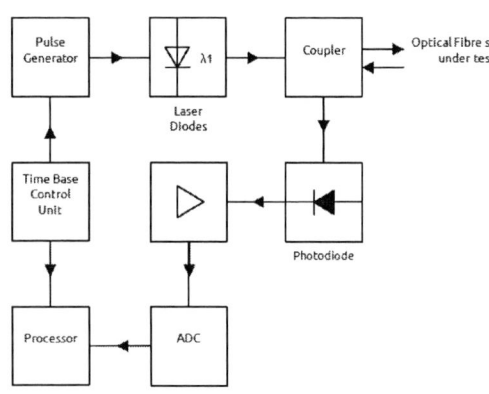

a

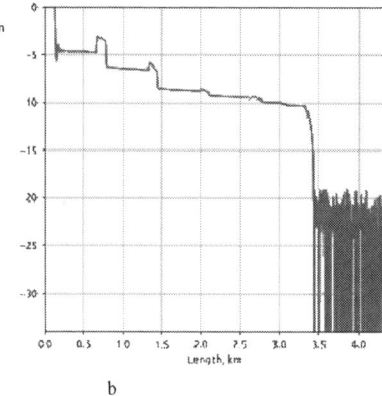

b

Fig. 1. a – Classic OTDR block diargam. b – typical reflectogram.

An OTDR is a device designed to analyze long optical lines based on reflected and scattered light. It works by recording Rayleigh backscatter and Fresnel point reflections that occur at fiber irregularities. The device sends short optical pulses into the fiber and measures the dependence of the reflected signal power on time. After converting time into distance, a reflectogram is formed — a graph reflecting the change in optical losses along the line.

Several key metrological characteristics are important for a classic OTDR [2]:

- The operating wavelength of the optical radiation at the reflectometer output;

- The length measurement range and its basic absolute error;

- The dynamic range of attenuation measurements (Figure 2 a);

- The basic absolute error in attenuation measurements;

- The duration of the probe pulse (Figure 2 b);

The dead zone when measuring attenuation and the position of the heterogeneity (Figure 2 c)

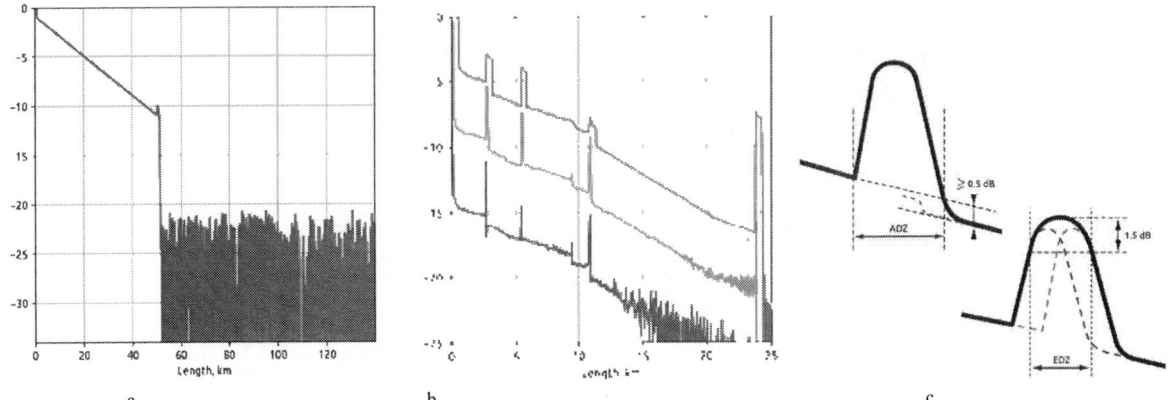

Fig. 2. a – Dynamic range of OTDR. b – Reflectogram of a line with different probe pulse durations. c – Attenuation dead zone and event dead zone.

Typical events can be identified from characteristic sections of the reflectogram: splices, connectors, bends, sections with increased losses, as well as the exact distance to breaks.

The classic OTDR is the main tool for diagnosing fiber optic lines during construction and maintenance, but its use in DWDM transport networks is limited due to the need to temporarily shut down traffic and its sensitivity to active elements (amplifiers and filters). It is these limitations that have stimulated the development of more advanced monitoring methods, described below.

LIMITATIONS OF CLASSIC OTDR IN TRANSPORT NETWORKS

Despite its widespread use in the construction and maintenance of fiber optic communication lines, classic OTDR has a number of limitations that significantly reduce its suitability for modern DWDM transport networks.

Complexity of operation during traffic transmission. Classic OTDR injects powerful pulses into the fiber, which interfere with active DWDM channels and can cause errors and interruptions in equipment operation. Therefore, measurement usually requires shutting down the line or dedicated wavelength, which is not always possible on a line under traffic.

Limited compatibility with optical network elements (ONE). EDFAs, Raman amplifiers, ROADM nodes, and WSS filters are located between transport network nodes. These devices change the pulse shape, introduce amplified spontaneous emission (ASE) noise, or block the reflected signal. As a result, the reflectogram becomes incorrect or unavailable.

Difficulties with long distances and high losses. The distances between line amplifiers can reach 80–120 km, and the total losses can be tens of decibels. A conventional OTDR does not always have sufficient dynamic range to "break through" such sections without breaking the route.

Low sensitivity to small loss increments. Heterogeneities in tenths or hundredths of a decibel (e.g.,

weld degradation) are often below the detection threshold. This is not sufficient for early detection of problems.

The appearance of artifacts and dead zones. Strong reflections (connectors, coils) create an event dead zone—the inability to distinguish events immediately after the reflection; attenuation dead zone—the inability to measure smooth losses over short distances. With a large number of connections, such zones can hide real defects.

Limited monitoring frequency. Each measurement takes time (averaging), so it is difficult to monitor in real time and track rapid changes or vibration effects.

Together, these limitations make the classic OTDR inconvenient for continuous monitoring of trunk lines. This has prompted the development of alternative methods — in-service OTDR, coherent reflectometers, and distributed sensing systems — that can operate without interrupting traffic and provide more information about the condition of the line.

DISTRIBUTED LINE MONITORING (DAS/DTS/DSS)

In addition to classic and integrated OTDR methods, distributed fiber optic sensors are increasingly being used in transport networks to measure physical parameters along the entire fiber in real time. These technologies use light scattering phenomena (Rayleigh, Raman, Brillouin) and provide information not only about losses, but also about external influences on the cable.

Phase-sensitive OTDR — DAS (Distributed Acoustic Sensing). Phase OTDR uses a coherent source and analyzes the phase of the scattered Rayleigh signal [3]. The features of DAS include high sensitivity to acoustic and vibration effects, the ability to detect earthworks, vehicle movement, attempts to open the cable, and the length of the monitored section — tens of kilometers. Such systems are used to protect highways, facility perimeters, and monitor pipelines, but are increasingly being implemented in operators' fiber optic communication lines for early damage detection. Figure 3 shows DAS block diagram and typical reflectogram.

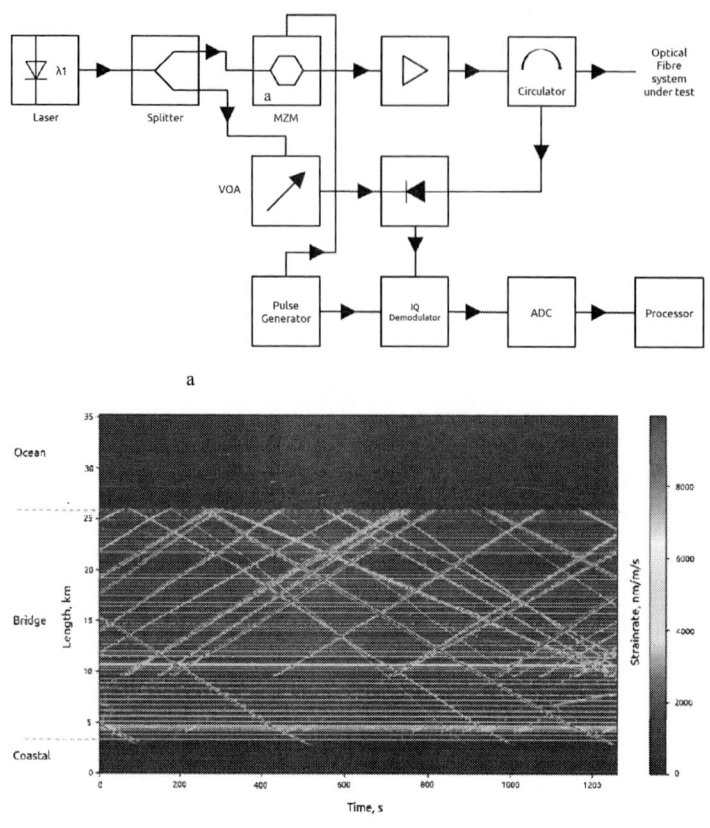

b

Fig. 3. a – DAS block diargam. b – typical DAS reflectogram.

Raman distributed temperature sensor — DTS (Distributed Temperature Sensing). The method is based on analyzing the ratio of the anti-Stokes and Stokes components of Raman scattering, which depends on the fiber temperature

[4]. Key features of DTS include temperature measurement along the entire line, with a typical resolution of 1–2 m. Figure 4 shows DTS block diagram and typical reflectogram.

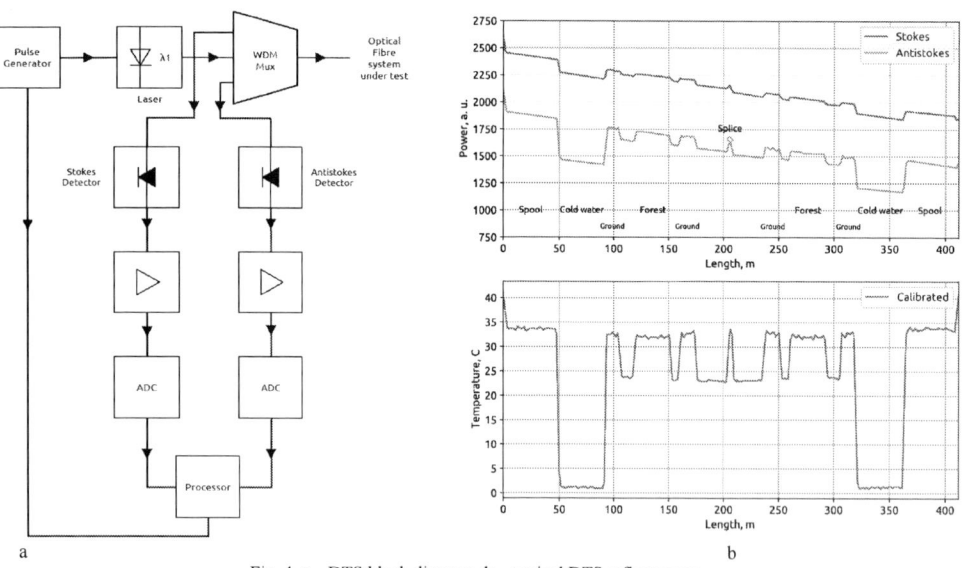

Fig. 4. a – DTS block diargam. b – typical DTS reflectogram.

It can be used to detect overheating, fires, areas with abnormal ground temperatures, and to monitor overhead lines. In DWDM trunk lines, DTS allows hidden emergency processes to be identified, such as damage to the sheath and heating of the fiber.

Brillouin reflectometry systems (BOTDR) — DSS (Distributed Strain Sensing). This method uses the Brillouin scattering effect, the frequency of which depends on the mechanical stress and temperature of the fiber [5]. DSS allows you to measure strain along the entire length of the line, identify areas of cable tension, soil displacement, and bends, and monitor the mechanical stability of underground sections. In practice, Brillouin systems can combine two channels: DSS (strain) and DTS (temperature), which allows you to distinguish between thermal and mechanical effects. Figure 5 shows DSS block diagram and typical reflectogram.

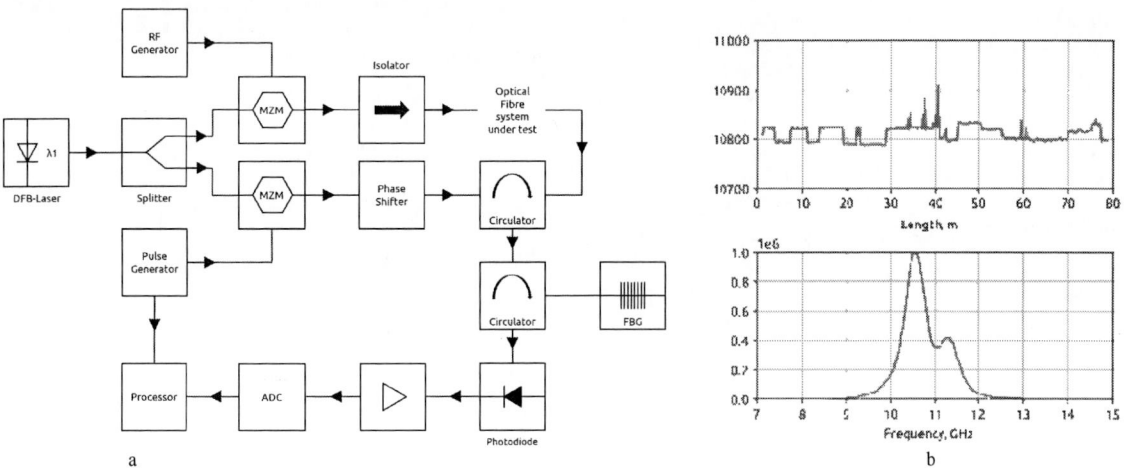

Fig. 5. a – DSS block diargam. b – typical DSS reflectogram and Brillouin scattering spectrum.

II. OPTIONS FOR INTEGRATED OTDR MONITORING IN TRANSPORT SYSTEMS

To ensure continuous monitoring of the condition of trunk fibers in DWDM transport networks, specialized OTDR monitoring methods are used that are adapted to operate without interrupting traffic and are compatible with active elements of the optical line. These approaches combine classic reflectometry principles with optical filtering, coherent reception, and dedicated service channels.

OTDR at a dedicated wavelength (out-of-band monitoring). In this option, a special wavelength is used in the line that does not intersect with the working DWDM channels. Probing pulses pass through a separate OADM/ROADM passband, do not interfere with useful traffic, and allow measurements to be taken while the line is in operation. The disadvantage is the need to allocate a channel, ensure its passage through all filters, and match the power with amplifiers.

In-service OTDR (low-power OTDR in the presence of traffic). The method involves injecting very low-power pulses that do not cause errors in DWDM channels. The technology includes narrowband filters that isolate backscatter, matched powers for passing through EDFA/ROADM, and periodic measurement without disconnecting the line. This is a compromise option: it works over long distances, but its sensitivity is lower than that of traditional measurement.

Coherent OTDR (COTDR, ϕ-OTDR). Instead of short optical pulses, a modulated probe signal (often FMCW or pseudo-noise sequence) is used, and reception is performed by a coherent detector. The advantages include significantly higher sensitivity, better noise immunity, and compatibility with amplifiers and filters. COTDR is currently one of the most promising methods for linear monitoring and fiber optic path status.

Pseudorandom and coded OTDR signals (coded OTDR). Probing is performed not with single pulses, but with sequences: complementary Galois sequences, Barker codes, pseudorandom bit sequences. After correlation, a high signal-to-noise ratio and long measurement range are achieved without increasing power. This approach is more commonly used in coherent or integrated monitoring modules.

OFDR (Optical Frequency-Domain Reflectometry). The method is based on slow laser frequency tuning and interference spectrum analysis. Its advantages include very high resolution (mm–cm) and high sensitivity. The disadvantage is limited range (usually up to several kilometers), which is why OFDR is more often used in very short lines (data centers, optical modules, photonic integrated circuits, short sections of DWDM paths).

Inline monitoring as part of DWDM equipment. Many modern platforms (Cisco, Ciena, Nokia, ADVA) include built-in OTDR units and/or optical switches that distribute probing pulses across cable fibers. Key features: centralized monitoring of multiple spans, automatic measurement initiation in case of failure, integration with NMS/SDN controllers. This is the most practical and common option in transport networks. Figure 6 shows inline in-band monitoring system block diagram.

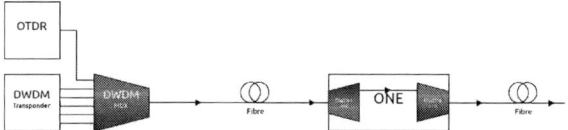

Fig. 6. Inline in-band monitoring system block diargam.

III. THE ROLE OF DISTRIBUTED METHODS IN DWDM TRANSPORT NETWORKS

Although DAS/DTS/DSS do not replace the classic OTDR, they complement it by providing real-time monitoring of external influences, early detection of faults before significant optical losses occur, and the ability to perform predictive analysis of the condition of the cable infrastructure. Together, these methods enable a shift from reactive maintenance to proactive monitoring of optical backbones.

In-line monitoring using both classic OTDR and distributed monitoring systems such as DAS, DTS, DSS, and BOTDR allows expensive equipment located in one place to be used to monitor multiple lines in different directions. Using these methods in combination with NMS, predictive analytics, and artificial intelligence technologies makes it possible to make such systems fully automatic, which will reduce their operating costs.

979-8-3315-4784-4/26 $31.00 © 2026 IEEE

IV. DEVELOPMENT OF AN AUTOMATIC MONITORING SYSTEM FOR OPTICAL COMMUNICATION LINES BASED ON IN-LINE OUT-OF-BAND OTDR

Modern DWDM transport networks require continuous monitoring of fiber status without interrupting user traffic. One of the most practical approaches is to use in-line out-of-band OTDR, in which reflectometric probing is performed at a dedicated wavelength that does not coincide with the working data transmission channels. This approach allows for automatic monitoring of the physical condition of the line in real time [6-10]. Figure 7 shows In-line out-of-band monitoring system block diagram.

How the system works. The system is based on an OTDR module connected to the line via optical combiners and filters, which ensure that the probe signal is injected and extracted outside the DWDM traffic band. A service channel or a specially allocated spectral interval is usually used as the probe wavelength.

OTDR pulses travel along the entire line, including optical amplifiers, and backscatter and reflections are recorded and analyzed without affecting the useful signal.

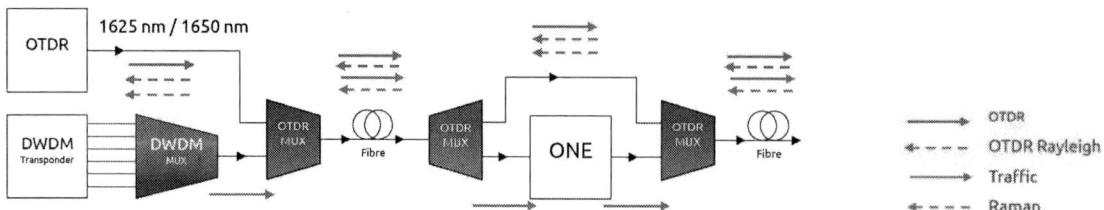

Fig. 7. Inline out-of-band monitoring system block diargam.

Automated monitoring system architecture. The system under development includes the following functional blocks.

- OTDR measurement module, which generates pulses and records the return signal;

- optical input/output interface, consisting of WDM combiners, filters, and optical switches;

- measurement control unit, which implements scheduling, triggers, and OTDR parameter adaptation;

A data processing module that performs trace analysis, comparison with reference profiles, and deviation detection.

An interface for integration with the network management system (NMS/OSS).

This architecture provides scalability and the ability to centrally control a large number of spans. Figure 8 shows In-line out-of-band monitoring system in DWDM-system block diagram and OTDR NMS example.

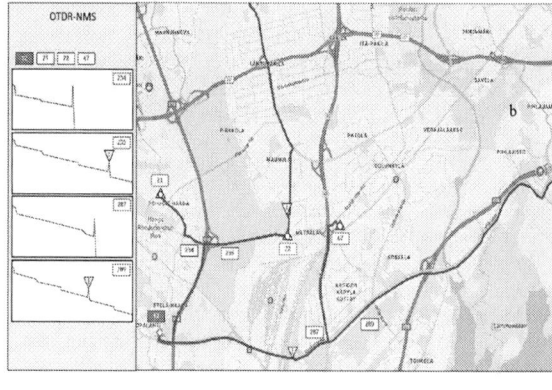

b

Fig.8. a – Inline out-of-band monitoring system block diagram in DWDM-system. b –OTDR NMS example

Automatic monitoring algorithms. The system is automated through the use of algorithms:

- periodic polling of lines with adaptive selection of measurement parameters (pulse width, averaging time);

- comparison of the current OTDR trace with the reference trace formed when the line was put into operation;

- event detection (local losses, reflections, breaks) based on the analysis of differential curves;

- classification of degradations by the nature of the trace changes (smooth losses, sharp jumps, reflection growth).

When the set thresholds are exceeded, the system automatically generates emergency notifications.

Features of operation in a DWDM environment. When developing the system, it is necessary to take into account the influence of active elements of the transport line:

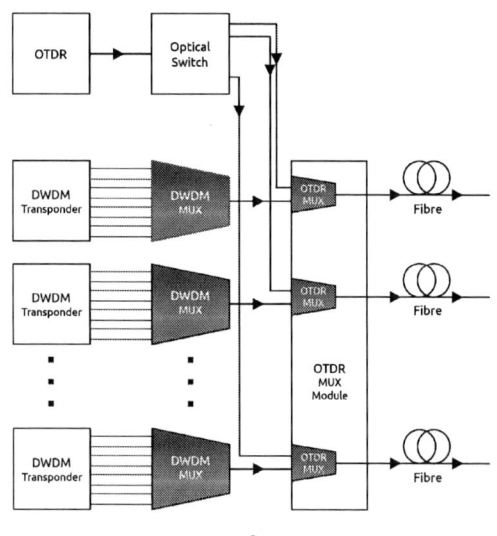

a

979-8-3315-4784-4/26 $31.00 © 2026 IEEE   164

- optical amplifiers introduce amplified spontaneous emission (ASE) noise, which degrades the OTDR signal-to-noise ratio;

- ROADM nodes and WSS filters require bandwidth matching for the service wavelength;

- the power of probing pulses must be limited to avoid nonlinear effects.

The correct choice of spectral position and power of the out-of-band channel is a critical factor for stable system operation.

Integration and operational benefits. The automatic monitoring system based on in-line out-of-band OTDR can be integrated into existing transport network management platforms and provides:

- continuous fiber status monitoring without traffic interruption;

- accurate localization of faults and degradations;

- reduction of mean time to repair (MTTR);

- transition from reactive to predictive maintenance of backbone networks.

Thus, in-line out-of-band OTDR provides an effective basis for building intelligent monitoring systems for the physical parameters of DWDM optical communication lines.

REFERENCES

[1] GOST 26814–86. Optical cables. Methods for measuring parameters. Moscow, USSR: Izdatel'stvo standartov, 1986.

[2] GOST R 50.2.071–2009. State system for ensuring the uniformity of measurements. Optical reflectometers. Verification procedure. Moscow, Russia: Standartinform, 2011.

[3] Shang, Y.; Sun, M.; Wang, C.; Yang, J.; Du, Y.; Yi, J.; Zhao, W.; Wang, Y.; Zhao, Y.; Ni, J. Research Progress in Distributed Acoustic Sensing Techniques. Sensors 2022, 22, 6060. https://doi.org/10.3390/s22166060

[4] Ukil, A., Braendle, H., & Krippner, P. (2011). Distributed temperature sensing: Review of technology and applications. IEEE Sensors Journal, 12(5), 885-892.

[5] Arsenault, Tyler J., et al. "Development of a FBG based distributed strain sensor system for wind turbine structural health monitoring." Smart Materials and Structures 22.7 (2013): 075027.

[6] Urban, Patryk J., et al. "Fiber plant manager: An OTDR-and OTM-based PON monitoring system." IEEE Communications Magazine 51.2 (2013): S9-S15.

[7] Davydov, V.; Reznikov, B.; Dudkin, V. New Optical System for Long Distance Control of Electrical Energy Flows. Energies 2023, 16, 1040. https://doi.org/10.3390/en16031040

[8] Reznikov, Bogdan K., et al. "Narrowband Lengthy Analogue Fiber Optic Communication Line for Telemetry Data Transmission." 2023 Seminar on Networks, Circuits and Systems (NCS). IEEE, 2023.

[9] Antil, Reena, S. Beniwal Pinki, and Sonal Beniwal. "An overview of DWDM technology & network." Int. J. Sci. Technol. Res 1.11 (2012): 43-46.

[10] Kartalopoulos, Stamatios V. DWDM: networks, devices, and technology. John Wiley & Sons, 2002.

Development of a Simulator for a Transceiver Device for Long-Distance Fiber Optic Communication Lines

Evgeny S. Minin
The Bonch-Bruevich Saint Petersburg
State University of Telecommunications
Saint Petersburg, Russia
eminin636@gmail.com

Sergey S. Gryzulev
The Bonch-Bruevich Saint Petersburg
State University of Telecommunications
Saint Petersburg, Russia
sgryz.sut@gmail.com

Bogdan K. Reznikov
The Bonch-Bruevich Saint Petersburg
State University of Telecommunications
Saint Petersburg, Russia
rznkff@net.sut.ru

Timofey A. Kotov
The Bonch-Bruevich Saint Petersburg
State University of Telecommunications
Saint Petersburg, Russia
kotoff201211@gmail.com

Nikolay Yu. Kolybelnikov
The Bonch-Bruevich Saint Petersburg
State University of Telecommunications
Saint Petersburg, Russia
ya.nikolai-kolyb@yandex.ru

Dmitry I. Isaenko
The Bonch-Bruevich Saint Petersburg
State University of Telecommunications
Saint Petersburg, Russia
isaenko-d@mail.ru

Abstract — **The necessity of developing a simulator for testing the operation of blocks that generate sequences of control command codes, monitor various switching systems, and transmit measured environmental parameters and process them after transmission via an optical channel is justified. The design of a simulator using an air-optical communication channel for transmitting information in the form of a sequence of command codes is presented. The features of using a sequence of command codes when transmitting analog signals in an optical communication channel have been established. Confirmation has been obtained of the legitimacy of using the proposed method of forming analog optical signals for transmitting them over long distances.**

Keywords — *analog optical signal, command code, simulator, optical communication channel, signal carrier frequency, laser radiation*

I. INTRODUCTION

With the development of scientific and technological progress, the number of tasks related to the transmission of information in a complex electromagnetic environment is constantly increasing. The most optimal solution in this situation is the use of fiber-optic communication lines (FOCL). To solve these problems, a large number of FOCL designs have been developed, both digital and analog. Particular difficulties arise when transmitting information from multifunctional environmental monitoring complexes located on high-voltage power lines (PFL) [1, 2].

The operation of the developed prototype optical communication channel revealed a number of shortcomings in its performance. One of them is related to the formation of an analog signal for modulating laser radiation, which contains information about the measured parameters of the environment. Therefore, a new concept for constructing an optical communication channel was developed using a subcarrier frequency signal and a new method for forming and processing optical signals to transmit information about environmental parameters. Since this method is being used for the first time and it is extremely difficult to implement a 500 km long optical fiber communication line in laboratory conditions, it is necessary to develop a simulator to test the principles of operation of the optical fiber communication line and the method of forming and processing the optical signal. It should be noted that simulators are often used to test various equipment in radar and other complex systems.

The simulator for information transmission also uses laser radiation with a narrow beam pattern. [3, 4]

II. AN ANALOG OPTICAL COMMUNICATION CHANNEL FOR TRANSMITTING INFORMATION OVER DISTANCES OF UP TO 500 KM IN CONDITIONS OF HIGH LEVELS OF ELECTROMAGNETIC INTERFERENCE

To determine the principles for developing a simulator for testing the method of forming and recording analog signals in the form of special (command) codes, let us consider one of the options we have proposed for analog optical channels in which information is transmitted. Fig. 1 shows a block diagram of a newly developed optical communication channel (OCC) for transmitting information about the state of the environment over long distances from sensors located on power lines to a central computer [2-5].

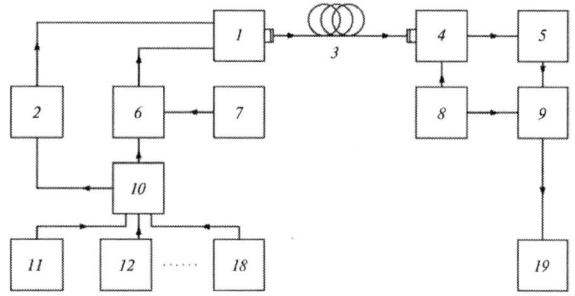

Fig. 1. Block diagram of an analog fiber optic communication system: 1 – semiconductor laser with λ = 1550 nm, 2 – laser power supply with direct modulation, 3 – optical fiber, 4 – photodetector, 5 – tunable LC filter, 6 – electronic key, 7 – high-stability quartz generator with frequency tuning, 8 – multifunctional power supply unit, 9 – information processing device, 10 – control device, 11–18 sensors for measuring various physical quantities in the environment, 19 – central computer.

In the developed AOS design, it is necessary to check the formation of information about physical quantities characterizing the state of the environment in the form of a sequence of pulses, the parameters of which change depending on the measurement results. It is also necessary to check the operation of key elements for supplying the carrier signal to the laser power supply unit (the pulses in the command code sequence are envelope signals for the carrier signal). When transmitting analog signals over fiber optic lines in the classical version, the carrier signal (pulses in the

979-8-3315-4784-4/26 $31.00 © 2026 IEEE

command code sequence are envelopes for the carrier signal). When transmitting analog signals over an optical fiber communication line in the classic version, the carrier signal is the envelope of the information signals. It is also necessary to check the operation of the transmitted information processing device 9 (Fig. 1).

Let us consider the algorithm for forming and transmitting data. Since an analog optical signal is used to transmit information, all the information formed in them is related to changes in the amplitude of pulses, their duration and duty cycle. In some cases, when transmitting information, it is necessary to take into account the phase shift between the transmitted signal and the signal registered on the photoreceptor device.

Further, the entire algorithm for forming a sequence of command codes can be divided into three stages. The first is the conversion of data into a convenient form for transmission over a communication channel and for reading on a photoreceptor device. The second is the formation of pulses in laser radiation containing information and their registration on a photoreceptor device. The third is the decoding of the recorded messages and the output of information to the display.

For a specific example (environmental monitoring system), let us consider the algorithm we propose, which can be further transformed to other simulation cases, with changes that take into account the specifics of information transmission. There can be up to 20 measuring sensors, and the signal for the measured environmental parameter from the sensor is output in analog or digital format. At the polling time (information from the sensor must be transmitted every 2 seconds – this is the maximum polling interval), there may be restrictions on the number of "unique" characters in the information packet about the parameters of the physical quantity.

When using a large number in the environmental monitoring complex, the number of characters can be increased. In the example given in the article for air flow velocity, the static part corresponds to symbol 2. In information encoding algorithms for information transmission, various symbols are used for separation. In the algorithm we propose with an analog signal, we suggest using the colon symbol ":" for separation. Then, in Arabic transcription, the information about the speed υ will be represented as follows: 6.5:2. After reading the separator, the receiving device decodes the number 2 (into a physical quantity and its unit of measurement).

In the design of the simulator, when registering an analog optical signal, it is necessary to ensure the start of counting the transmitted message and the moment of completion of data transmission about the measured physical quantity. In communication systems, the terms start bit and stop bit have been introduced. The start and end of counting are placed at the beginning and end of the packet with the transmitted information.

Fig. 2 shows an example of encoding airflow velocity information in the form of pulses (command codes) that are read by a photodetector.

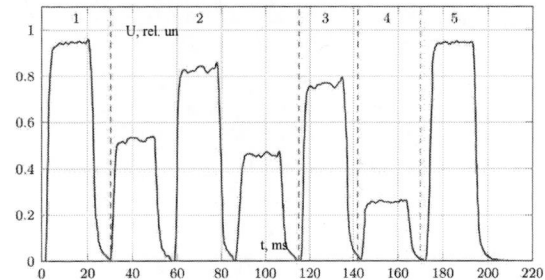

Fig. 2. Sequence of pulses received from the receiver for data with a value of $\upsilon = 6.5$ m/s.

III. DESIGN OF THE IMITATOR AND ITS COMPONENTS

Fig. 3 shows a block diagram of the developed simulator design. The simulator design uses a semiconductor laser diode with a wavelength of $\lambda = 613$ nm to transmit optical radiation, which receives rectangular pulses of varying amplitude (their duration can be changed if necessary).

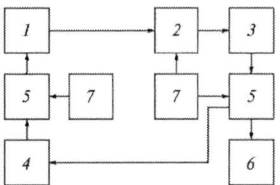

Fig. 3. Block diagram of the simulator: 1 – semiconductor laser module with $\lambda = 613$ nm with current control, 2 – photodetector, 3 – processing device, 4 – personal computer, 5 – microcontroller, 6 – display device (LDC display), 7 – power supply.

A photodetector 2 is located 15 cm away from 1. A FD-24K photodiode (with a diameter of more than 10 mm) was used in the photodetector. The radiation is transmitted through the air. The visible range of the laser diode radiation was chosen for the reason of convenience of adjusting its optical axis of radiation to the photosensitive layer of the photodetector when they are spaced at a distance of up to 20 m or more in the laboratory. When using IR radiation, a laser semiconductor diode is required as a reference point for introducing IR radiation to the photosensitive layer of the PD (an InGaAs-based avalanche photodiode is used to register radiation at $\lambda = 1550$ nm). It should be noted that the principle of forming the transmitted information and its decryption with a change in the wavelength of the optical signal radiation does not change. In addition, when testing developments for optical air communication channels, the visible red range is used, as these are the most extreme conditions of its operation compared to the IR range.

Processing device 3 includes additional filters for suppressing noise signals during the conversion of the optical signal to analog.

Next, the signal from the photodetector is sent to the signal processing device 3 for recoding into digital code, followed by transmission to the microcontroller 5. The microcontroller processes the incoming information, outputs it to the display device 6, and sends it to the input of the personal computer 4. Fig. 4 shows the schematic diagrams of the command signal generation blocks 3 and their processing after registration on the photodetector 6 (Fig. 3).

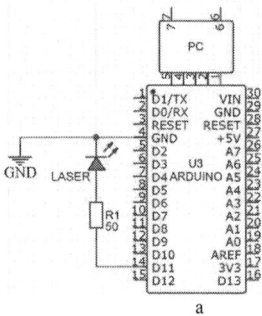 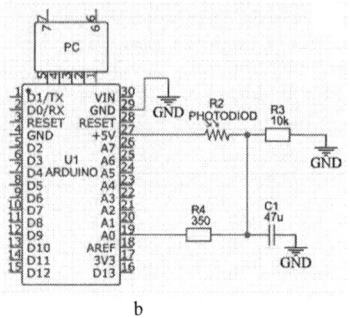

a   b

Fig. 4. Diagrams of information formation (a) and processing (b) blocks.

An Arduino debugging board with an atmega328p microcontroller is used as the control system for the transmitting laser module in this simulator design (Fig. 4a). This microcontroller in the Arduino platform is small in size and has low power consumption.

The size of the program memory (Flash) is 32 KB, the size of the RAM is 2 KB, and the operating frequency is 16 MHz, which will also be sufficient for implementing the program. The size of the program memory (Flash) is 32 KB, the size of the RAM is 2 KB, and the operating frequency is 16 MHz, which will also be sufficient for implementing an analog WDM simulator. In the developed design (Fig. 4a), a semiconductor laser diode with a wavelength of 613 nm is connected to digital pin D11 through a resistor R1 = 50 ohms. A PWM signal with specified parameters (frequency, duration and amplitude of pulses, which is tied by absolute value to U = 5 V (microcontroller supply voltage)) is applied to the semiconductor laser diode.

This is done to determine the functional capabilities of the developed simulator (in the air channel, it is sufficient to use pulses). When using the simulator with optical signal transmission via an optical fiber, the pulse amplitude is increased by a factor of 10. This is done to determine the functional capabilities of the developed simulator (in an air channel, it is sufficient to use pulses). When using a simulator with optical signal transmission via fiber optic cable, a subcarrier frequency is used. In this case, it is more expedient to use a PWM signal.

A personal computer (PC) is connected to the transmitting device, which simulates the operation of sensors for measuring various physical quantities in the environment. The processing recording circuit (Fig. 4b) also uses an Arduino debugging board with similar characteristics. For more accurate decoding of pulse amplitude values, a voltage divider is installed in the recording device, consisting of a photodiode with resistance varying with the power of the incoming radiation (maximum resistance R2 = 1 kOhm) and a resistor with resistance R3 = 10 kOhm. In this case, everything works as follows. When radiation in the form of a pulse arrives at the photodiode, the resistance R2 decreases and the voltage at pin A0 (Fig. 4b) increases. At the same time, resistor R3 is connected between pin A0 and ground, forming a divider.

The debug board reads the voltage value on pin A0 and converts it to a digital value between 0 and 1023 using the analogRead function. This is then used to decode the information into laser pulses.

IV. RESEARCH INTO THE FUNCTIONING OF THE SIMULATOR AND VERIFICATION OF COMMAND CODE FORMATION

Fig. 5 shows an example of the simulator operating mode using a PC to transmit and decode information about the absorbed dose in air from a Geiger-Müller counter. The PC display shows the sequence of pulses received by the photodetector and the result of decoding the information, shown on the LD display (this information is shown more clearly in Fig. 6).

Fig. 5. Photographs of pulses during simulation of radiation sensor operation and optical fibre optic cables as a possible communication line option and Photo of the data obtained from the radiation measurement sensor

Analysis of the information obtained in Figure 5 shows that the simulation of the transmission of information about the absorbed dose of radiation = 24 µR/h by the simulator was successfully processed. This shows that our proposed design with a voltage divider for the Arduino debugging board with an atmega328p microcontroller is justified and gives adequate results during operation.

Another important element in the design of the receiving device, which is not found in classical photodetectors for fiber-optic communication lines, is the use of a capacitor C1 (Fig. 4b) to "smooth out" parasitic noise in pulses (filtering (RC filter)), which arises for various reasons, especially in an air optical communication channel. Fig. 6 shows the result of noise filtering using C1 = 47 µF and R4 = 350 Ω for a cutoff frequency f_c = 1 kHz (noise is mainly generated at a frequency of about 500 Hz, so the cutoff frequency for the filter is taken with a margin).

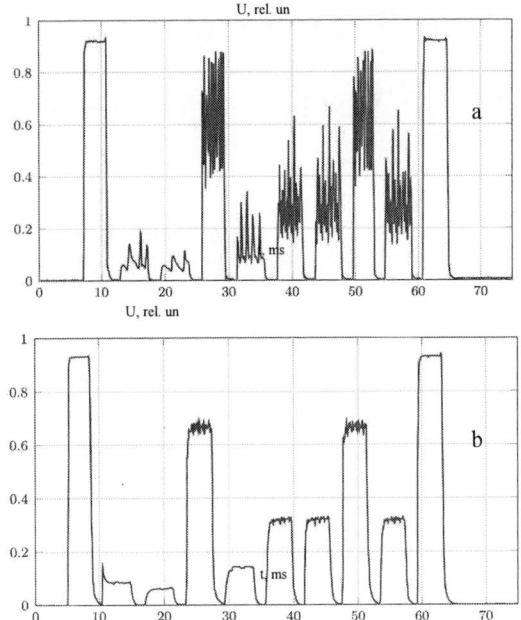

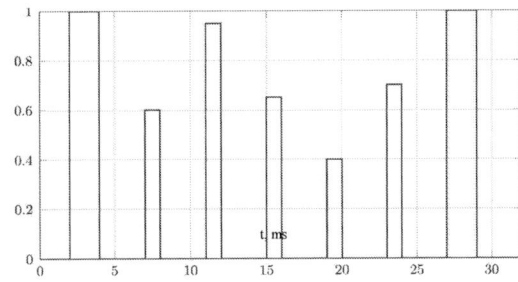

Fig. 6. Sequence of recorded pulses: without filtering (a), with filtering RC-filter (b).

When using the developed simulator to test an extended fiber-optic line with radiation input through a convector into an optical fiber and its subsequent output to a photodetector, filtering, as shown by experiments, does not need to be implemented (noise is insignificant). Figure 7 shows an example of the formed sequence in the form of command codes (rectangular pulses) in which data on the temperature T = 294.3 K is encrypted for transmission over the fiber optic line.

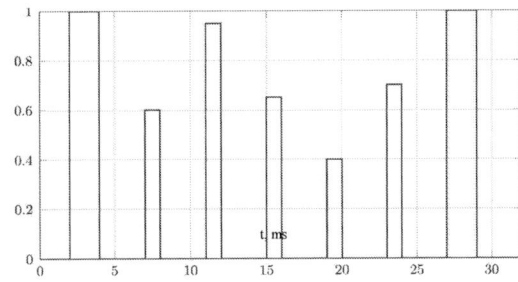

Fig. 7. Sequence of command codes in the form of rectangular pulses for transmitting the temperature value T = 294.3 K in analogue form via fibre optic communication lines.

Fig. 8 shows the test results for the analog fiber-optic communication line (decrypted data on the T value on the display, transmitted using a sequence of command codes).

Fig. 8. Result of transferring data on temperature value T using a simulator.

The result obtained once again confirmed the validity of the method we developed for transmitting information using a sequence of command codes (the decrypted value of temperature T from the received optical signal corresponds to a temperature of 294.3 K).

V. CONCLUSION

The results obtained demonstrated the validity of the proposed concept for constructing the simulator and the technical solutions used for its practical implementation. The developed simulator fully coped with the tasks set for testing the new method of forming information in analog form (sequence of command codes), its transmission and decryption for various types of communication lines (air optical communication channel, optical fiber, fiber-optic communication line).

For other variants of encoding and transmission of information in communication and telecommunication systems [2, 4, 7, 8], the proposed design of the simulator can be improved by using a more powerful microcontroller based on another debugging board with antenna devices included in it.

REFERENCES

[1] Kuzmin M.S., Rogov S.A. // Computer Optics. 2019. V. 43. № 3. P. 391.

[2] Podstrigaev A.S., Lukiyanov A.S., Smolyakov A.V. et. al. // J. Phys.: Conf. Ser. 2019. V. 1368. № 2. P. 022027.

[3] Dmitriev P.S., Kovalev A.V., Viktorov E.A. et. al. // Optics Lett. 2020. V. 45. № 22. P. 6150. 7. Reznikov B.K., Stepanenkov G.V., Logvinova E.A. et. al. // St. Petersburg Polytechnic Univ. J. Physics and Mathematics. 2023. T. 16. № 3.2. C. 143.

[4] Isaenko D., Rodin S., Stepanenkov G. et. al. // IEEE Int. Conf. on Electrical Engineering and Photonics, EExPolytech 2022. Saint-Petersburg, 2022. V. 2022. P. 316.

[5] Davydov V., Reznikov B., Dudkin V. // Energies. 2023. T. 16. № 3. C. 1040. 10. Borodkin A.I., Kovalev A.V., Verschelde A. et. al. // IEEE Photonics Technology Lett. 2022. V. 34. № 18. P. 989.

[6] Podstrigaev A.S., Lukiyanov A.S., Smolyakov A.V. et. al. // J. Phys.: Conf. Ser. 2019. V. 1410. № 1. P. 012155.

[7] Davydov R., Antonov V., Angelina M. // IEEE Int. Conf. on Electrical Engineering and Photonics, EExPolytech 2019. Saint-Petersburg, 2019. V. 8906791. P. 42.

[8] Petrov A.A., Shabanov V.E., Zalyotov D.V. et. al. // IEEE Int. Conf. on Electrical Engineering and Photonics, EExPolytech 2018. Saint-Petersburg, 2018. V. 8564389. P. 52.

A Mobile Multifunctional Sensor for Monitoring the Spectral Composition of Radiation Sources

Andrey Novikov
Peter the Great Saint-Petersburg Polytechnic University
Saint Petersburg, Russia
aanovikov2004@yandex.ru

Gleb Sozykin
Peter the Great Saint-Petersburg Polytechnic University
Saint Petersburg, Russia
sonnerwelt@gmail.com

Natalya Burenkova
J.-S. Company "CRI "Electron"
Saint Petersburg, Russia
n.burenkova@niielectron.ru

Vladimir V. Skryabin
J.-S. Company "CRI "Electron"
Saint Petersburg, Russia
v.skryabin@niielectron.ru

Anna S. Klyuchnik
J.-S. Company "CRI "Electron"
Saint Petersburg, Russia
a.klyuchnik@niielectron.ru

Abstract— The need to develop a mobile multifunctional sensor for monitoring the spectral composition of radiation sources is substantiated. Currently used mobile spectrometers have a number of disadvantages associated with limitations in the measured wavelength range and the possibility of failure of photodetector elements when exposed to high-power radiation. The newly developed design uses a silicon photodiode that is resistant to high-power radiation over a wide spectral range. When emitting high power this photoreceiving element enters saturation, and the user has time to remove the photoreceiving sensor from the active area of radiation exposure. To determine the spectral composition of radiation we use a set of small-sized filters developed by us with a range from 10 to 20 nm depending on the recorded wavelength of the spectrum. This ensures the determination of the spectral composition of radiation and the power of spectral components from the UV range including near IR. The experimental studies conducted confirmed the operability of our device.

Keywords— radiation, sensor, power, spectrum, wavelength, filter, photodetector, measurement error, reliability.

I. INTRODUCTION

Nowadays, with the rapid development of technology and the global digitalization of all kinds of information, the problem of deteriorating human health is acute. [1-3]. In addition to heart disease, an increasing number of people are experiencing progressive vision deterioration [4, 5]. There are many causes for this. One particularly dangerous problem for vision is the use of artificial light sources, including for lighting and various backlights. White light sources from LEDs, which are used in most lighting fixtures, unlike sunlight and incandescent lamps, produce a blue light spectrum [6, 7]. This blue light spectrum in a standard white LED produces 160% of the green spectrum's violet output.

In sunlight, the blue spectrum accounts for 60% or less of the radiant power, depending on the time of day [8, 9]. A violation of the proportions in the spectral composition, especially with prolonged exposure (more than an hour), affects visual functions: an excess of blue and violet light in the range of 390 - 450 nm has a negative effect on the cornea and retina of the eye [10]. In such light, the ciliary muscle, which controls the lens, is under increased tension. All of this leads to faster vision deterioration. Systems that attenuate radiation in this spectral range are not always installed on LEDs. LEDs without these systems are used in the manufacture of various lighting fixtures. During mass production of LEDs, minor defects in the spectral range are allowed, which, taken together, can increase the negative impact of blue light on the visual organs.

Powerful radiation sources with an imbalanced spectral power distribution or high levels of radiation in the violet-blue wavelength range pose a significant risk to human vision. In some cases, after a short exposure to this radiation, a person may temporarily lose visual acuity or may need to seek medical attention for vision problems.

Moreover, monitoring the spectral composition of such radiation sources at their work sites is primarily necessary for those who work with them (for example, medical personnel, pharmacists, water treatment plant workers, and many other professions), as their vision is constantly exposed to radiation from these sources. Personal protective equipment is designed with a certain power reserve for specific wavelengths of radiation. Blue light is attenuated only in extreme cases. Filtering blue light will distort the eye's ability to form a contrast image—the eye, like any video system (e.g., a television), operates in RGB mode [11, 12]. Distorted image perception can lead to poor performance. Therefore, it is crucial to maintain high-level visual function for as long as possible by monitoring the spectral composition of light, especially the violet-blue range, where it is needed. Red light overload, which sometimes predominates in some radiation sources, is also unpleasant. Its power ratio to other parts of the spectrum also needs to be monitored.

Industrial mobile spectrometers currently available operate in various wavelength ranges from 350 to 800 nm. This range is insufficient for monitoring illumination from various types of sources (part of the NIR spectrum is not captured). However, these devices are highly sensitive to radiation (using a matrix photodetector to detect it). In the case of high radiation power, this element burns out, and the spectrometer ceases operation. In some cases, the electronics also burn out (especially if there is high power in the NIR range, which the device barely detects), due to saturation mode [13, 14]. It is extremely difficult to implement blocking at high sensitivity of photodetectors in a mobile device [15, 16]. These devices do not provide a system of replaceable filters and polarizing elements to reduce radiation power.

Therefore, the aim of our work is to develop mobile sensors for monitoring the composition and power of radiation with an error of less than 2% in the wavelength range from 300 to 1100 nm, resistant to overloads of powerful radiation.

II. DESIGN OF THE MOBILE SENSOR

Despite the wide selection of CCD matrices for radiation detection, it was decided to use a classic silicon photodiode with a detection bandwidth of 350 to 1100 nm. This

979-8-3315-4784-4/26 $31.00 © 2026 IEEE

photodiode is resistant to short-term exposure to high-power radiation. A new technical solution in the developed sensor is the use of two photodiodes for radiation detection. A filter is installed in front of one of them. Figure 1 shows the front panel of the developed sensor with two photodiodes and a red-light filter. Figure 2 shows the sensor housing, the circuit board with electronic components, and the control unit with a silicon photodiode and ammeter.

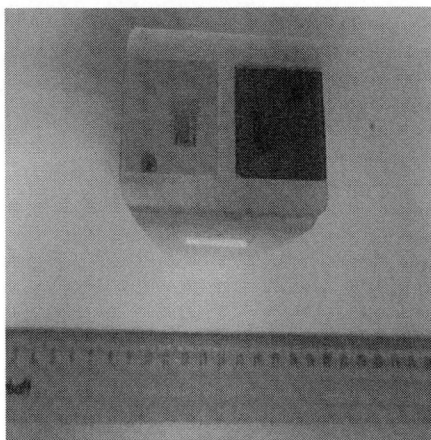

Fig. 1.   Front panel of the sensor.

The current from each photodiode is recorded. By comparing these currents, it is possible to determine the ratio of the radiation power of the studied range, for example, red, to the power of the entire range. The percentage of the radiation power of a specific light (wavelength) is calculated from the recorded radiation power perceived by the human visual system. Figure 3 shows the schematic diagram of the developed sensor. The measured radiation powers are compared using the STM32M microcontroller. The measurement result is displayed on the screen as a percentage (Figure 4).

Fig. 2.   External view of the sensor and electronic board with monitor.

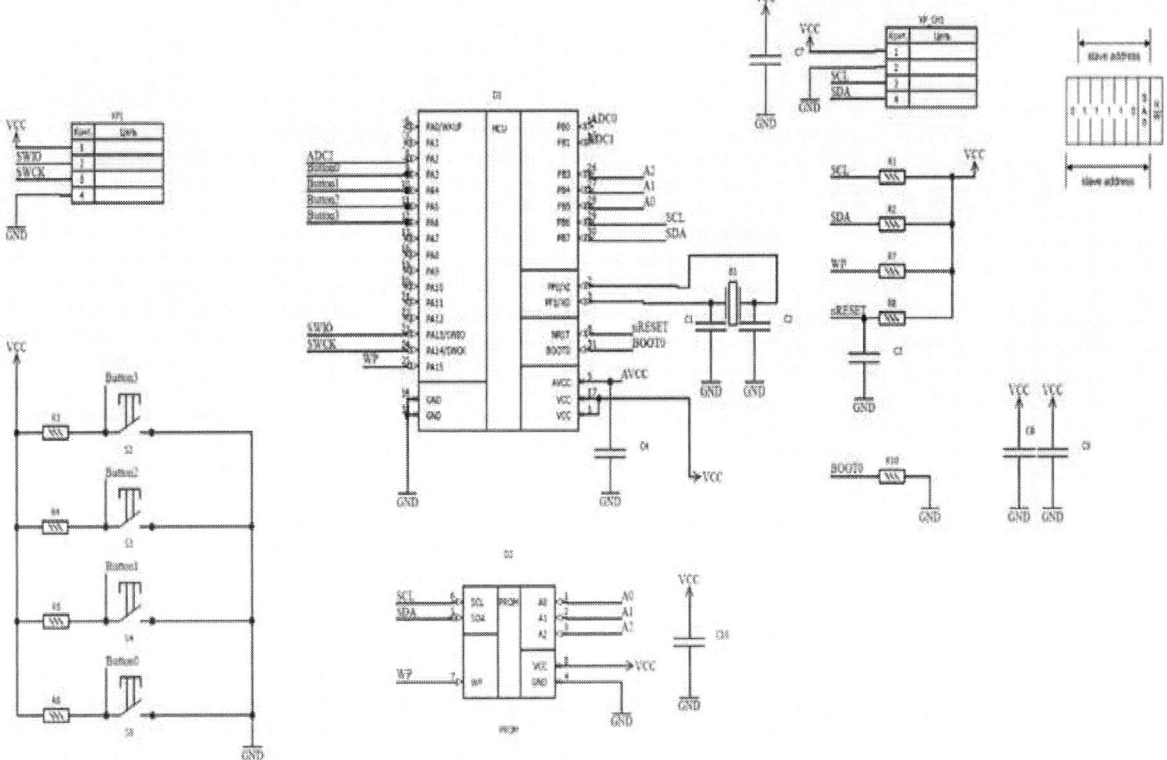

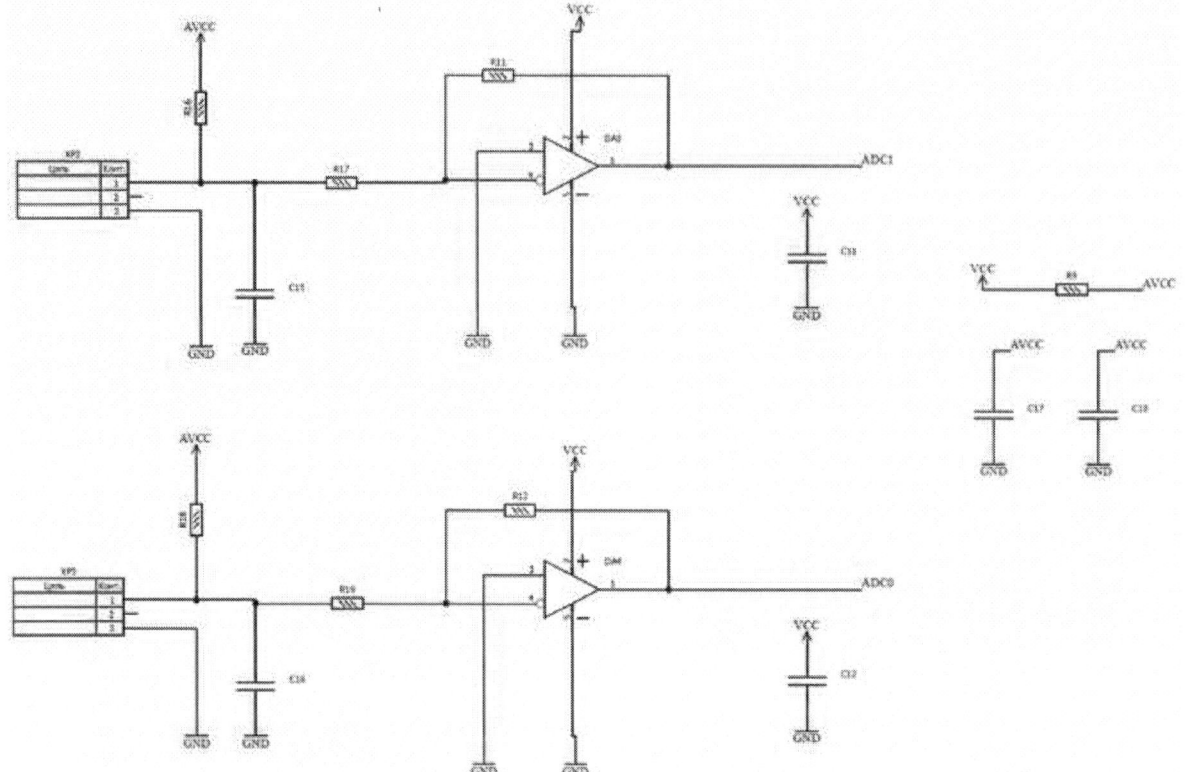

Fig. 3. Schematic diagram of the electronic part of the developed sensor.

Fig. 4. The operating mode of the developed sensor for determining the percentage of red light in the radiation.

. If necessary, the results of radiation power measurements for two channels (with and without a filter) can be displayed on the screen.

Fig. 5 shows an experimental setup for measuring the spectral characteristics of various radiation sources using the developed sensor.

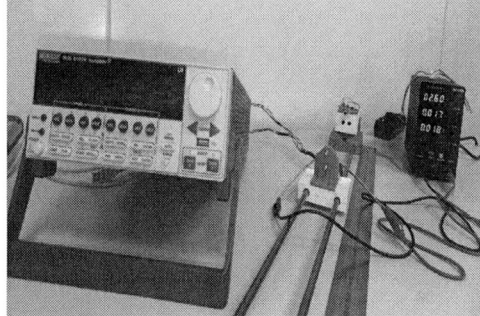

Fig. 5. External view of the experimental setup with control of the distance between the radiation source and the sensor.

Fig. 6 shows as an example the appearance of a high-power ARPL-3W-BCX45 white LED, which is often used in lighting sources.

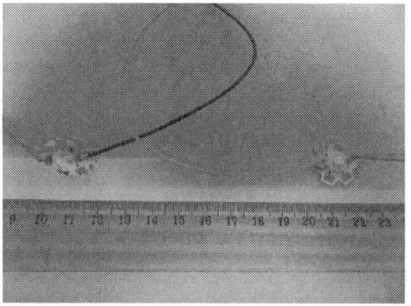

Fig. 6. Appearance of white LEDs at different thermal temperatures T_T: (a) - 6000 K, (b) – 4500 K.

979-8-3315-4784-4/26 $31.00 © 2026 IEEE

The remaining radiation sources, the models of which will be given in the studies of the sensor operation, do not require such a representation.

III. RESULTS OF EXPERIMENTAL RESEARCH THE SPECTRAL PARAMETERS OF DIFFERENT LIGHT SOURCES

Figure 7 shows a check of the parallel placement of filters in front of silicon photodiodes. Parallel placement of filters helps reduce errors in radiant power measurement, which are associated with the refraction of light rays at two interfaces (air-glass and glass-air) before they reach the photodiode.

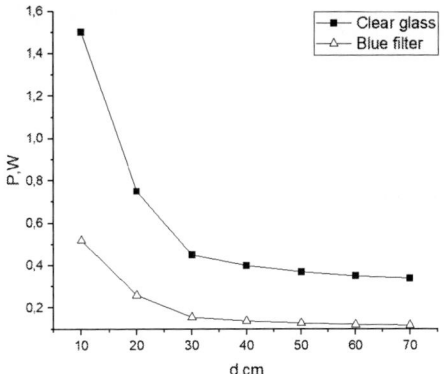

Fig. 7. Changes in the recorded radiant power of a white LED and photodiodes as a function of the distance between them. Graph 1 corresponds to the presence of a blue filter, graph 2 to transparent glass..

Analysis of the obtained results shows that the difference in recorded power does not change significantly with distance d from the radiation source (white LED (Fig. 6)) with a large divergence angle of the primary power (more than 60 degrees). This indicates that the difference in refraction of the rays arriving at the photodiodes is insignificant, confirming the parallel installation of the filters relative to each other and the photosensitive layer of the photodiodes.

An ATP2002 industrial high-resolution spectrometer was used to monitor the spectral composition of various radiation sources. Figure 8 shows the spectrum of a 40-watt incandescent lamp manufactured by Cosmos Electro LLC. Measurements of the red component of this lamp's spectrum are presented in Figure 4.

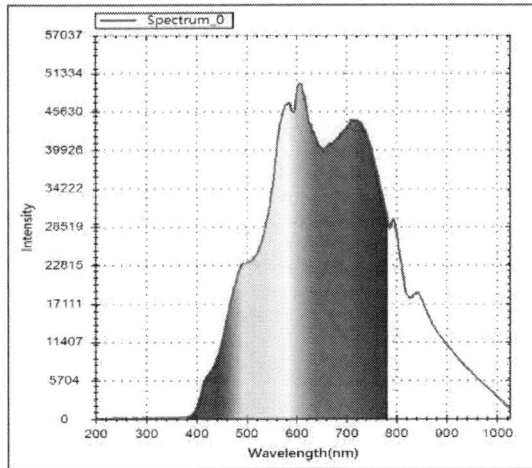

Fig. 8. Radiation spectrum of an incandescent lamp.

The relative portion of the red and other components of the spectrum by power using the obtained dependence (Fig. 8) is determined by calculating the area under the curve. A comparison of the obtained results on the contribution of the red part of the spectrum to the total radiation of the lamp showed that they coincide within the limits of measurement error.

To verify the performance of the developed sensor, the results of determining the red (Table 1) and blue (Table 2) spectral power were compared to the total radiated power of various light sources. The devices used in the measurements are presented in Tables 1 and 2.

TABLE I. THE RATIO BETWEEN THE POWERS OF THE RED PART OF THE RADIATION SPECTRUM TO THE TOTAL RADIATION POWER RECORDED WITHOUT FILTERING

Type of radiation source	The developed sensor	Silicon photodiode and current device	Spectrometer ATP2002
LED ARPL-3W-BCX45 with T_T = 6000 K	0,196±0,014	0,188±0,026	0,192
LED ARPL-3W-BCX45 with T_T = 4500 K	0,250±0,014	0,240±0,026	0,245
LED ARPL-3W-BCX45 with T_T = 3000 K	0,351±0,014	0,341±0,026	0,346
Incandescent lamp from Cosmos Electro LLC	0,344±0,014	0,341±0,026	0,356
White light lantern	0,240±0,014	0,231±0,026	0,235
Lighting in the laboratory	0,268±0,014	0,257±0,026	0,262
Sunlight on a clear day (outdoors)	0,344±0,014	0,332±0,026	0,337

TABLE II. THE RATIO BETWEEN THE POWERS OF THE BLUE PART OF THE RADIATION SPECTRUM TO THE TOTAL RADIATION POWER RECORDED WITHOUT FILTERING.

Type of radiation source	The developed sensor	Silicon photodiode and current device	Spectrometer ATP2002
LED ARPL-3W-BCX45 with T_T = 6000 K	0,346±0,014	0,336±0,026	0,341
LED ARPL-3W-BCX45 with T_T = 4500 K	0,339±0,014	0,329±0,026	0,334
LED ARPL-3W-BCX45 with T_T = 3000 K	0,249±0,014	0,241±0,026	0,245
Incandescent lamp from Cosmos Electro LLC	0,057±0,014	0,052±0,026	0,054
White light lantern	0,376±0,014	0,365±0,026	0,371
Lighting in the laboratory	0,272±0,014	0,263±0,026	0,267
Sunlight on a clear day (outdoors)	0,076±0,014	0,069±0,026	0,072

Analysis of the obtained results shows that certain parts of the spectra match each other in power within the measurement error. Since the contribution of spectral components to the total radiation power for spectra recorded using ATR2002 was determined based on the areas under the graphs, the error in determining the ratio in this column is not included (any accuracy can be achieved with such calculations on a computer).

IV. CONCLUSION

An analysis of the spectral results from various sources demonstrated the reliability of the developed sensor. A replaceable set of filters will allow, if necessary, the contribution of the NIR component, which is present in small amounts in sunlight, to light sources (such as incandescent lamps). The filter slots provided in the design can be used to accommodate optical attenuators, preserving the lifespan of the photodiode, which is significantly reduced when exposed to high-power radiation (the diode enters saturation mode and overheats). The presence of attenuators allows for the selection of the angle of radiation exposure to the photodiodes (without initially overloading them), at which measurements can be performed.

REFERENCES

[1] M. S. Mazing, and et. al. "Monitoring of oxygen supply of human tissues using a noninvasive optical system based on a multi-channel integrated spectrum analyzer," International Journal of Pharmaceutical Research, vol. 12, pp. 1974 – 1978. December 2020.

[2] E. Porfirieva, and et. al. "Features of the use of esCCO technology for the diagnosis of human condition," Proceedings - 9th IEEE International Conference on Information Technology and Nanotechnology, ITNT 2023, vol. 245732, pp. 234-239. April 2023.

[3] R. Davydov, and et. al. "New Methodology of Human Health Express Diagnostics Based on Pulse Wave Measurements and Occlusion Test" Journal of Personalized Medicine, vol. 13(3), pp. 443. March 2023.

[4] T. Noro, and et. al. "Age-related decline in retinal function in marmosets," Scientific Reports, vol. 15(1), pp. 22374. January 2025.

[5] C. L. Sturiale, and et. al. "The importance of the optic nerves unlocking during the resection of anterior skull base meningiomas for visual function preservation: surgical nuances and clinical outcome," Neurosurgical Review, vol. 48(1), pp. 31. January 2025.

[6] S. Kim, and et. al. "Evaluation of cytokinin types and LED light spectra for enhanced production of diarylheptanoids in Alnus incana subsp. incana," Scientific Reports, vol. 15(1), pp. 36564. January 2025.

[7] H. Farhangi, and et. al. "Optimizing LED lighting spectra for enhanced growth in controlled-environment vertical farms," Scientific Reports, vol. 15(1), pp. 30152. January 2025.

[8] J. Chen, and et. al. "Nanoscale heterophase regulation enables sunlight-like full-spectrum white electroluminescence," Nature Communications, vol. 16(1), pp. 3621. January 2025.

[9] I. Gureeva, and et. al. "Development of an optical system for lighting rooms with biologically safe and environmentally friendly sunlight," Journal of Physics Conference Series, vol. 1942(1), pp. 012087. December 2021.

[10] B. Yang, and et. al. "Exposure of A2E to blue light promotes ferroptosis in the retinal pigment epithelium," Cellular and Molecular Biology Letters, vol. 30(1), pp. 22. January 2025.

[11] L. Kivelä, and et. al. "Mitigating the light pollution problem via spectral adjustment: color-biased phototaxis in male glow-worms," Oecologia, vol. 207(8), pp. 133. August 2025.

[12] V. Dyumin, and et. al. "Charge-coupled Device with Integrated Electron Multiplication for Low Light Level Imaging," Proceedings of the 2019 IEEE International Conference on Electrical Engineering and Photonics Eexpolytech 2019 , vol. 8906868, pp. 308–310. October 2019.

[13] K. J. Smirnov, and et. al. "Temperature investigations of inp/ingaas based photocathodes," Proceedings of the 2018 IEEE International Conference on Electrical Engineering and Photonics Eexpolytech 2018, vol. 8564416, pp. 209–211. October 2018.

[14] V. V. Davydov, and et. al. "An Optical Method of Monitoring the State of Flowing Media with Low Transparency That Contain Large Inclusions," Measurement Techniques, vol. 62(6), pp. 519–526. June 2019.

[15] M. R. Ainbund, and et. al. "Hybrid multi-channel photodetector for 1-1.6 µm spectral range," Applied Physics, vol. 2018(6), pp. 54–59. June 2018.

[16] A. E. Zhukov, and et. al. "Model for Speed Performance of Quantum-Dot Waveguide Photodiode," Semiconductors, vol. 57(13), pp. 632–637. July 2023.

979-8-3315-4784-4/26 $31.00 © 2026 IEEE

Fractal Antenna Array for a Compact Active Electronically Scanned Array

Stanislav S. Sidorenko
MIREA – Russian Technological University
Moscow, Russia
stas.sidorenko.96@mail.ru

Aleksander A. Borisov
MIREA – Russian Technological University
Moscow, Russia

Valeriy V. Demshevsky
MIREA – Russian Technological University
Fryazino, Russia

Ilya A. Tsygitsa
MAI – Moscow Aviation Institute
Moscow, Russia

Grigory S. Anikin
MAI – Moscow Aviation Institute
Moscow, Russia

Vyacheslav V. Lobodin
MIREA – Russian Technological University
Fryazino, Russia

Abstract— **The urgency of the research issue is due to the fact that aviation radar systems significantly enlarge their functional possibilities when using dual-frequency antenna systems with various combinations in S, C, X, Ku frequency bands. The foundation of these directions is based on the principles and technical solutions for building radiating antenna systems with a beam electronic control.**

The paper considers the structure of a fractal antenna array of the Ku-band frequencies, such as a Sierpinski carpet, as well as the possibility of integrating a second X-band into this structure. In addition, the paper analyzes the possibility of reducing the T/R-modules by sparse antenna array.

Keywords—AESA, PAA, fractal, sparse antenna array, clustered antenna array, combined PAA

I. INTRODUCTION

The urgency of the research is due to the expansion of AESA functionalities through the use of dual-frequency antenna systems with various combinations in S, C, X, Ku frequency bands.

The main objective of this work is to synthesize a fractal phased antenna array (PAA) with overlapping bands on a single curtain allowing for electronic beam steering over a wide range of solid angles.

Fractals were first described in the 1970s by B. Mandelbrot and characterized by self-similar geometry and recursive repetition of elements. Formally it can be expressed through the ratio (1):

$$F_{n+1}=T(F_n),\qquad(1)$$

where F_n – fractal in the n-th iteration level, T – the transformative function.

In the field of antennas the mention of fractals is most often found in the context of designing single emitters, however, some studies, for example [1, 2] use the property of fractals to construct PAA. Fractal antenna arrays are well suited for combining several frequency bands, as they are already sparse and typically provide electronic scanning capabilities. By placing the elements of the second frequency band to the places, where the elements are already absent, it is possible to obtain the second electronically scanned PAA.

Paper [2] shows the optimization of a fractal PAA based on the Sierpinski carpet principle using the differential evolution method to minimize the side lobes level (SLL). As a result the authors of the article obtained a fractal-like phased array configuration, with a 27x27 element dimension and a SLL of less than -18 dB.

The paper mentioned above was used as a basis for developing a fractal-like PAA with frequency bands overlap.

II. FRACTAL-LIKE PAA

One of the ways to combine frequency ranges on the antenna array (AA) of a fractal-like shape is to implement an antenna array based on two types of independent antenna elements operating at separated frequency ranges. This method is flexible when combining two bands with a high frequency ratio ?[3 – 6]?. As mentioned above , the results of a fractal-like AA calculations [2] served as the basis for further research on the combined PAA of X- and Ku-frequency ranges. The appearance of the original and modified fractal-like PAA for placing elements in the Ku-frequency range is shown in Fig.1.

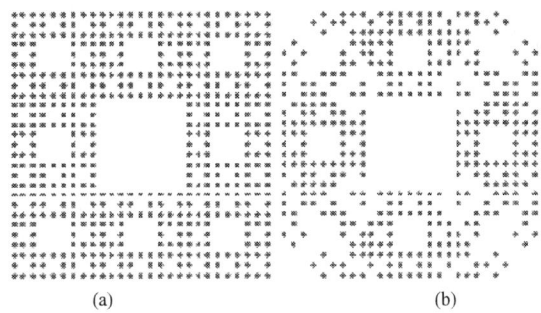

(a)   (b)

Fig. 1. Four The appearance of fractal-like PAA in Ku- frequency range. (a) original fractal-like PAA; (b) modified fractal-like PAA.

Formula (2) is used to calculate the radiation pattern for fractal-like AAs:

$$f_{AP}\,(\theta,\varphi) = f_{из}\,(\theta,\varphi)\cdot f_{реш}\,(\theta,\varphi)$$

where $f_{из}\,(\theta,\varphi)$ – patch-antenna radiation pattern, calculated in CAD-system; $f_{реш}\,(\theta,\varphi)$ – AA multiplier

$$f_{реш}(\theta,\varphi) = \frac{1}{N_x N_y}\sum_{m=0}^{N_x-1}\sum_{n=0}^{N_y-1} A_{mn}e^{j\,\phi_{mn}(\theta,\varphi)}$$

979-8-3315-4784-4/26 $31.00 © 2026 IEEE

where N_x N_y – number of elements in x and y; A_{mn} – amplitude of elements excitation m in x, n in y; $\phi_{mn}(\theta,\varphi)$ – phase shift on the elements in the direction of θ,φ.

The initial phase shift is calculated based on the phasing equation:

$$\phi_{mn}(\theta,\varphi) = -k\left(m\,d_x\cos\varphi + n\,d_y\sin\varphi\right)\sin\theta$$

where – wave number at the operating wavelength; d_x, d_y – grid pitch along x and y axes.

Fig. 2 shows the radiation pattern of fractal antenna array of Sierpinski carpet type. The graphs show that the arrangement of the elements according to the classical Sierpinski carpet structure has a high SLL relative to the main radiation pattern beam and does not provide a wide scanning range, unlike the modified fractal-like PAA.

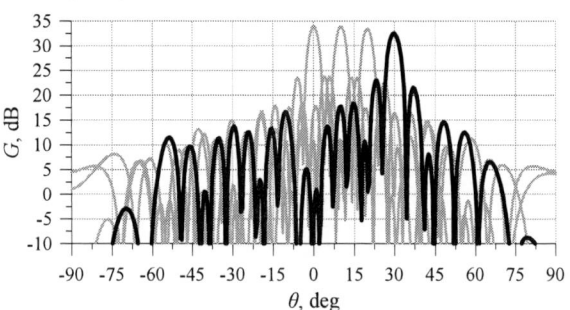

(a)

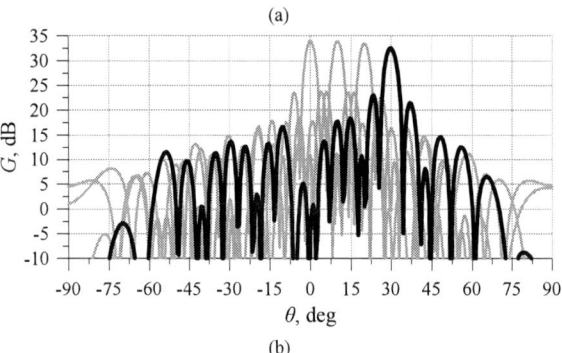

(b)

Fig. 2. The radiation pattern of a fractal antenna array of Sierpinski carpet type of Ku-frequency range. (a) azimuthal plane. (b) and elevation plane.

Fig. 3 shows the radiation pattern graphs of a modified fractal-like PAA when the beam is deflected in azimuthal and elevation planes with $0,5\lambda_{ku0}$ step of the arrangement of the elements. The graphs show that the radiation pattern beam deviation of antenna array in azimuthal and elevation planes reaches ±60°, while the SLL does not exceed -10 dB relative to the main beam.

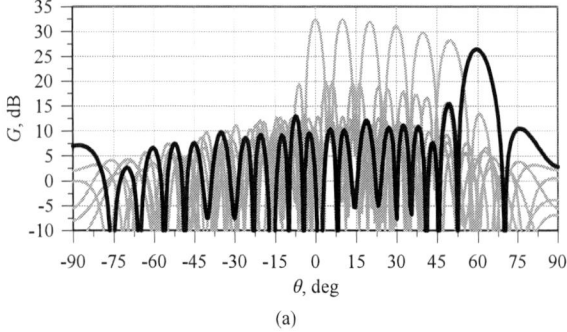

(a)

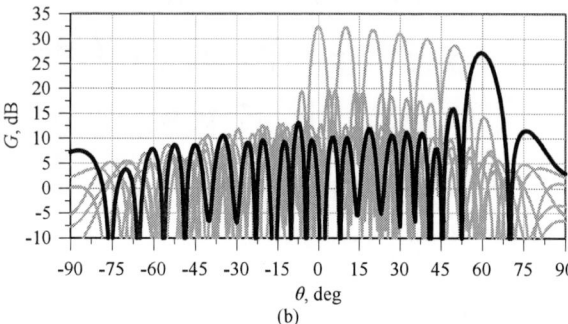

(b)

Fig. 3. The radiation pattern of a modified fractal-like PAA of Ku-frequency range. (a) azimuthal plane. (b) and elevation plane.

By placing the elements of the second range in places of the missing elements on the modified fractal-like PAA (Fig. 4), it was also possible to achieve ±60° electronic deviation of antenna array beam in the main regions, as in the case with the first range. The radiation pattern of antenna array X-band frequency response radiation patterns are presented in Fig.5, SLL does not exceed -10 dB relative to the main beam. The uneven formation of the side lobes is caused by a relatively small spacing of the X-band elements which is equal to $0,4\lambda_{x0}$.

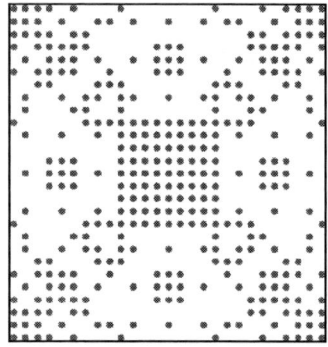

Fig. 4. The appearance of fractal-like PAA in X- frequency range

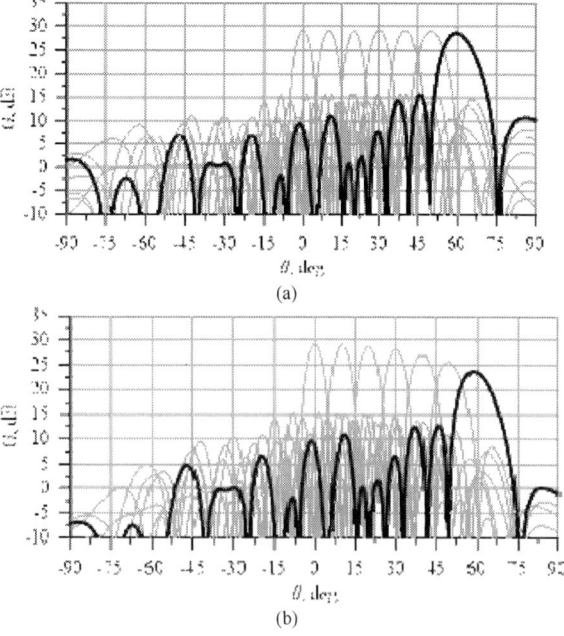

(a)

(b)

Fig. 5. The radiation pattern of a fractal-like PAA of X-frequency range. (a) azimuthal plane. (b) and elevation plane.

Combining two frequency bands according to the principle described above , will result in a dense arrangement of T/R-modules, which makes it impossible to install a heat dissipation system from the heated T/R-modules due to the lack of space in the AESA system. The operation of the T/R-modules in conditions of elevated temperature [7] (the maximum operating temperature of T/R-modules is usually between +85°C − +150°C depending on the structure of the module electronic components) can cause them to malfunction resulting in deterioration of antenna array operation and the overall characteristics of the radio system. The solution of this problem is an extra sparseness of fractal-like PAA , which will reduce the number of T/R-modules, thereby increasing the available space for installing the heat dissipation system and significantly reducing the economic costs for AESA production.

III. PAA SPARSENESS . METHODS OF SPARSENESS

Along with the classical method of PAA sparseness, which reduces the number of radiating elements on the antenna curtain (AC), the AA clustering method can be considered [8 − 13] as a special case of PAA sparseness. Clustering is a combination of several radiating antenna array elements into one group (cluster cell), which is excited by one T/R-modules. In this case the excitation amplitude and phase of the emitting elements in the cluster cell will be the same. Fig. 6 shows functional graphs of a regular Fig. 6(a), clustered Fig. 6(b) and sparse Fig. 6(c) PAA.

Keep your text and graphic files separate until after the text has been formatted and styled. Do not use hard tabs, and limit use of hard returns to only one return at the end of a paragraph. Do not add any kind of pagination anywhere in the paper. Do not number text heads-the template will do that for you.

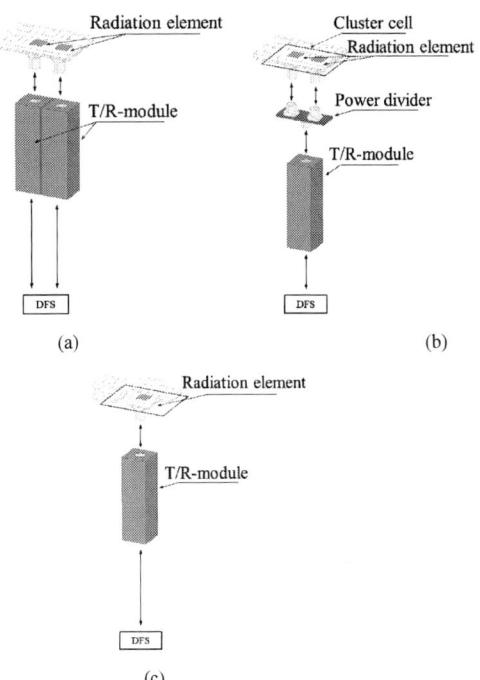

Fig. 6. PAA functional circuits, where DFC − a diagram forming system. a) regular PAA b) clusterized PAA c); sparse PAA

Hereinafter a clustered antenna array will refer to antenna array, constructed according to Fig. 6 (b), and a sparse antenna arRAY will refer to Fig. 6 (c). To determine the

advantage of one sparseness method over the other, the antenna array of the Ku-band (Fig. 1) и antenna array of the X-band (Fig. 4) were made sparse in stages, as it is shown in Fig 7 − 10.

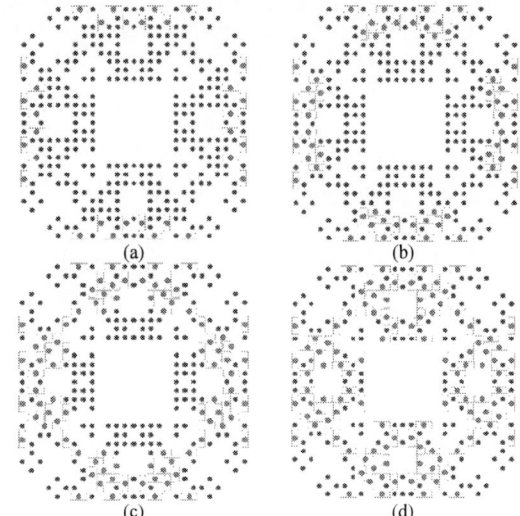

Fig. 7. Step-by-step process of the antenna array sparseness of Ku-frequency band using the classical method. (a) Iteration №1; (b) Iteration №2; (c) Iteration №3; (d) Iteration №4.

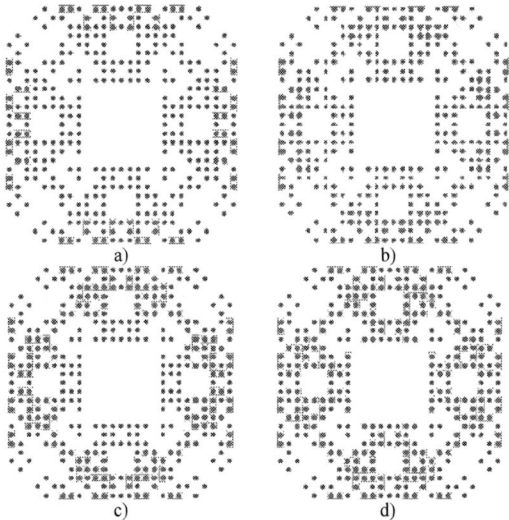

Fig. 8. Step-by-step process of the antenna array sparseness of Ku-frequency band using the clustering method. (a) Iteration №1; (b) Iteration №2; (c) Iteration №3; (d) Iteration №4.

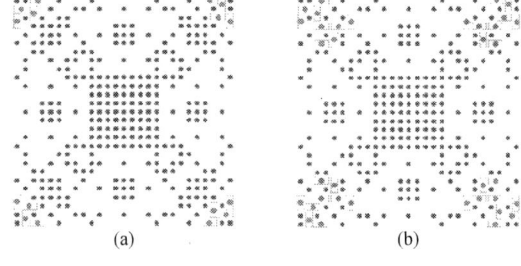

979-8-3315-4784-4/26 $31.00 © 2026 IEEE

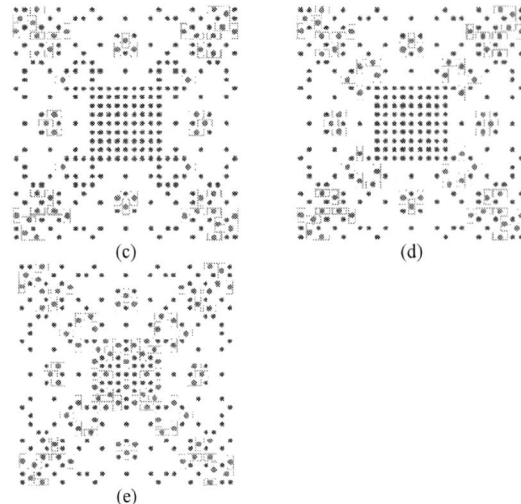

(c)   (d)

(e)

Fig. 9. Step-by-step process of the antenna array sparseness of X-frequency band using the classical method. (a) Iteration №1; (b) Iteration №2; (c) Iteration №3; (d) Iiteration №4; (e) Iiteration №5.

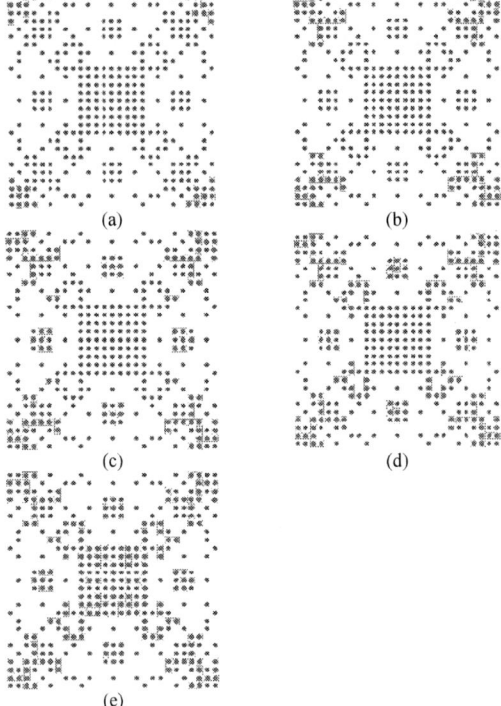

(a)   (b)

(c)   (d)

(e)

Fig. 10. Step-by-step process of the AC sparseness of X-frequency band using the clustering method. Iteration. a) Iteration №1; (b) Iteration №2; (c) Iteration №3; (d) Iteration №4; (e) Iiteration №5.

The results of PAAs of Ku- и X-frequency bands using step-by-step method of sparseness and clustering are shown in Fig. 11, 12.

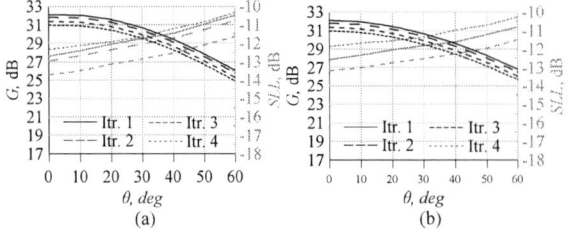

(a)   (b)

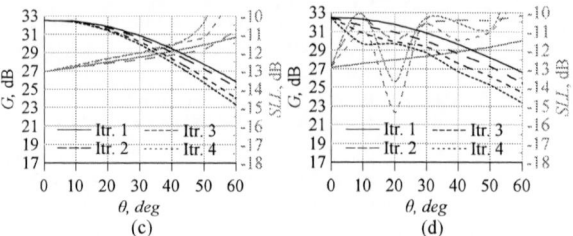

(c)   (d)

Fig. 11. Graphs of the PAA gain factor and SLL of the Ku-band dependences on the scanning angle, sparsed by the classical method in the (a) azimuthal plane; (b) in the elevation plane. Sparsed by clustering method in the azimuthal plane (c); in the elevation plane (d)

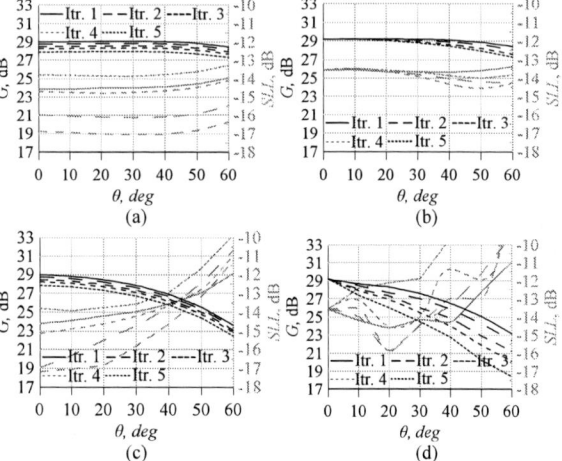

(a)   (b)

(c)   (d)

Fig. 12. Graphs of the PAA gain factor and SLL of the X-band dependences on the scanning angle, sparsed by the classical method in the (a) azimuthal plane; (b) in the elevation plane. Sparsed by clustering method in the azimuthal plane (c); in the elevation plane (d)

Based on the analysis of the graphs of dependences of the gain factor and SLL on the scanning angle of the sparse and clustered PAAs, as shown in Fig. 11 and 12, it can be concluded, that PAA clustering is a viable method for sparseness of antenna arrays with electronic angle scanning of 0° - 20°, while maintaining a higher level of the gain factor than sparsed PAAs.

IV. CONCLUSIONS

The paper studies the methods of sparse and clustered PAAs on the example of fractal-like PAA with a combination of Ku- and X- frequency bands on a single antenna array in order to reduce the number of T/R-modules and integrate the heat dissipation system of the heated modules when there is little space in the system with AESA. The modulation and simulation results showed that the clustered PAAs maintain a higher level of the gain factor with electronic angle scanning of 0° - 20°, than classic sparse PAAs. The value of gain factor difference in the cluster and sparse values may vary depending on the number of the sparse/clustered elements.

REFERENCES

[1] G. Thomas Spence and Douglas H. Werner, "Generalized Space-Filling Gosper Curves and Their Application to the Design of Wideband Modular Planar Antenna Arrays," IEEE transactions on antennas and propagation, vol. 58, no. 12, pp. 3931–3940, December 2010.

[2] A. Karmakar, R. Ghatak, R.K. Mishra and D.R. Poddar, "Sierpinski carpet fractal-based planar array optimization based on differential evolution algorithm," Journal of Electromagnetic Waves and Applications, 2015, DOI:10.1080/09205071.2014.997837.

[3] R. Pokuls, J. Uher, D. M. Pozar, "Dual-frequency and dual-polarization microstrip antennas for SAR applications," IEEE Trans. Antennas Propag. 1998.V.46, № 9. P. 1289–1296.

[4] S. S. Zhong, Z. Sun, L. B. Kong, C. Gao, W. Wang, M. P. Jin, "Tri-band dual-polarization shared-aperture microstrip array for SAR applications," IEEE Trans. Antennas Propagat. 2012. Vol. 60. №. 8. P. 4157–4165.

[5] C.-X. Mao, S. Gao, Q. Luo, T. Rommel, Q.-X. Chu, "Low-Cost X/Ku/Ka-Band Dual-Polarized Array with Shared Aperture," IEEE Trans. Antennas Propagat. 2017. Vol. 65. №. 7. P. 3520–3527.

[6] X. Xu, W.-Y. Yin, G. Xu, R. Chen, "Dual-band aperture-shared circular polarized array antenna for X-/Ku-band satellite communications» IEEE MTT-S International Conference on Numerical Electromagnetic and Multiphysics Modeling Optimization, 2020.

[7] V. V. Demshevsky, S. S. Sidorenko, A. D. Bazhenov, " Investigation of the operating characteristics of an antenna array under temperature influence," Conference of Young Researchers in Electrical and Electronics Engineering (2025 ElCon), January 28 – 30, 2025

[8] S. S. Sidorenko, V. V. Demshevsky, G. S. Anikin, D. V. Bagno, E. V. Iliin, A. E. Zaikin, "Clustering of the antenna array," Conference of Young Researchers in Electrical and Electronic Engineering (ElCon25), pp. 1205 - 1207. DOI:

[9] N. Anselmi, P. Rocca, M. Salucci, A. Massa, "Irregular phased array tiling by means of analytic schemata-driven optimization" IEEE Transaction on Antennas and Propagations. 2017. Vol. 65. №. 9. P. 4495–4510.

[10] Francesco Alessio Dicandia, Simone Genovesi, "Analysis of Performance Enhancement of Clustered-Based Phased Arrays Employing Mixed Antenna Element Factor" IEEE Trans. Antennas Propag. 2024. V. 72. No. 2. P. 1439-1448.

[11] P. Rocca, N. Anselmi, A. Polo, A. Massa, "Pareto-Optimal Domino-Tiling of Orthogonal Polygon Phased Arrays" IEEE Transaction on Antennas and Propagations. 2022. Vol. 70. №. 5. P. 3329–3341.

[12] Mailloux R. J., Santarelli S. G., Roberts T. M., and D. Luu, "Irregular Polyomino-Shaped Subarrays for Space-Based Active Arrays, "Hindawi Publishing Corporation International Journal of Antennas and Propagation Volume 2009, 9 pages, doi:10.1155/2009/956524.

[13] Jiang H., Gong Y., Zhang J., Dun S., "Irregular Modular Subarrayed Phased Array Tiling by Algorithm X and Differential Evolution Algorithm," in IEEE antennas and wireless propagation letters, vol. 22, no. 7, july 2023, p. 1532 – 1536, doi:10.1109/LAWP.2023.3250260.

Optimization of a Grow Light Design for Uniform Plant Illumination

Gulnaz Galina
Department of Photonics
Saint Petersburg Electrotechnical
University "LETI"
St. Petersburg, Russian Federation
galina-gulnaz0@mail.ru

Roman Kurenkov
Department of Photonics
Saint Petersburg Electrotechnical
University "LETI"
St. Petersburg, Russian Federation
rkurenkov7@gmail.com

Mariya Degtereva
Department of Photonics
Saint Petersburg Electrotechnical
University "LETI"
St. Petersburg, Russian Federation
mmromanovich@etu.ru

Yevgeniy Levin
Department of Photonics
Saint Petersburg Electrotechnical
University "LETI"
St. Petersburg, Russian Federation
e_levin@etu.ru

Alexander Degterev
Department of Photonics
Saint Petersburg Electrotechnical
University "LETI"
St. Petersburg, Russian Federation
aedegterev@etu.ru

Ivan Lamkin
Department of Photonics
Saint Petersburg Electrotechnical
University "LETI"
St. Petersburg, Russian Federation
ialamkin@etu.ru

Abstract — **A current area of lighting technology development for the crop production needs is the improvement and artificial lighting sources optimization. Special attention is devoted to the development of lamps for plants growth with unique and thoughtful shapes that maximize the luminous flux use, as well as provide uniform and comprehensive illumination of plants at all stages of their development. Such advanced technologies create optimal conditions for growth, contributing to higher yields and crop quality, which makes these plants indispensable in modern agriculture and home cultivation. This article discusses the specifics of the phytolamp enclosures development designed specifically for growing plants at home, as well as an overview of phytolamps for growing plants, which allows us to identify key advantages and identify current trends in this area.**

Keywords—lighting device, lighting control, agrophotonics, plant growth, LED.

I. INTRODUCTION

Rapidly developing technologies are being introduced into all areas of human life. The agricultural industry is no exception. In order to optimize the cultivation of various crops, the field of agrophotonics is increasingly being applied. Agro-photonic is based on the use of visible light radiation for growing plants in order to increase their productivity. By using artificial lighting, it is possible not only to compensate for the lack of natural light, but also to concentrate the radiation of the required spectrum depending on the growth phase of the plant.

Light is not only a source of energy that controls photosynthesis. Different parts of the spectrum are perceived by plants as signals that influence many aspects of plant life. (Fig. 1). Photosynthetically active radiation is the range of radiation used by plants for photosynthesis, ranging from 400 nm to 700 nm. The most significant radiation is in the red (600-720 nm) and blue (400-500 nm) ranges of the spectrum. These light waves are necessary for the formation of chlorophyll, and they also affect the rate of plant growth [1, 2]. The response of plants to the light conditions in which they are grown affects their growth and development in complex ways. The quality and quantity of light trigger a signaling cascade of specific photoreceptors, such as phytochromes, cryptochromes, and phototropins, which alter the expression of a large number of genes. When perceiving inductive wavelengths, activated phytochromes, together with blue (phototropins, cryptochromes) and ultraviolet light receptors, control plant physiology and development. During germination, the morphology changes, characterized by a fast-growing hypocotyl and a closed apical hook, which maximizes the chance of reaching the surface (Fig. 2).

Blue light also improves the photosynthetic efficiency of plants. According to research [3], the optimal ratio between red and blue light contributes to increased photosynthetic activity and crop yields (Fig. 3).

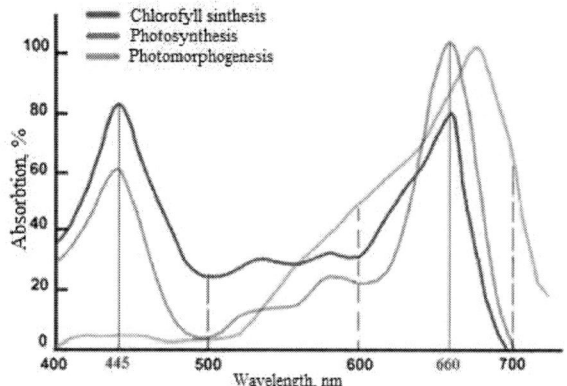

Fig. 1. Graph of dependence of regulatory processes in plants

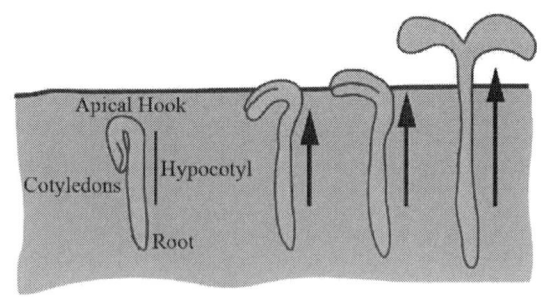

Fig. 2. The initial stages of plant development

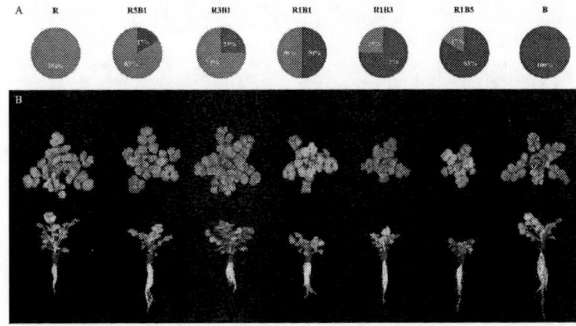

Fig. 3. Seven different settings for the ratio of red to blue illumination (R – monochromatic red, B – monochromatic blue, R5B1-R1B5 – various ratios of red to blue)

II. DESIGN DEVELOPMENT

KOMPAS-3D software was used to develop the design of LED luminaires. This CAD system allows you to create accurate 3D models that take into account all design features, ensuring high detail and ease of subsequent refinement. KOMPAS-3D tools ensure efficient design engineering.

A. Development of a cross-shaped design

The structure consists of four or more vertical posts 350 mm high and 80 mm wide, connected by a top crossbar to form a "cross" inside which the plant is placed. The inner surfaces of the posts have 50 mm wide mounting slots for installing linear LED modules, oriented towards the crown, which form multi-point ring lighting around the plant. The lower edge of the light-emitting elements is located at a height of 50 mm above the substrate surface, which minimizes direct illumination of the lower part of the stem and at the same time provides illumination of the lower tier of lea (Fig. 4).

Distributing light sources vertically and around the perimeter reduces local peaks of illumination on the top leaves and simultaneously increases the proportion of light reaching the lower and side tiers. Studies on optimizing the spatial arrangement of LEDs in greenhouses and vertical farms show that inter-crown and lateral lighting evens out photosynthetically active radiation, reduces light stress on the upper leaves, and increases the light efficiency of the entire system [4-6].

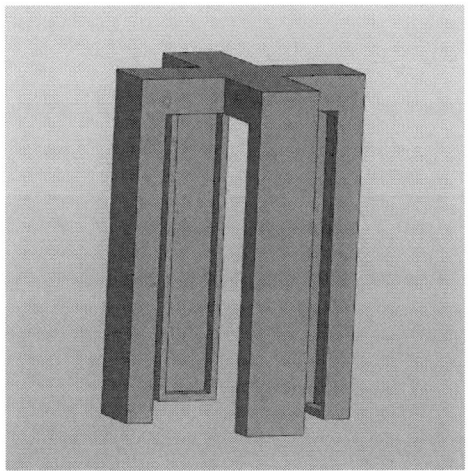

Fig. 4. Body construction

B. Development of a spiral design

The body of the LED luminaire is designed as a spatial cylindrical frame formed by a system of ring segments aligned along a common axis and connected in a spiral. The design forms a vertical light tunnel around the plant and is designed to create the most uniform radial and vertical distribution of light across the surface of the crown (Fig. 5).

The height of the entire structure is 500 mm, which provides coverage for the main volume of the crown of low-growing crops (lettuce, strawberries, herbs). The diameter of the cylindrical frame is 300 mm.

The width of the ring segment is 60 mm and is selected to provide sufficient space for mounting LED modules. Thanks to the spiral geometry, the LEDs can be evenly distributed along the height of the housing, which contributes to more uniform lighting of plants at all levels. Ring bases at the top and bottom provide rigidity and the ability to attach to support elements. This shape also facilitates the installation of the device and its integration into various plant growing systems. The spiral design not only increases the LED installation area compared to traditional straight housings, but also allows the light to evenly cover the plants from all sides, minimizing shading and increasing photosynthesis efficiency.

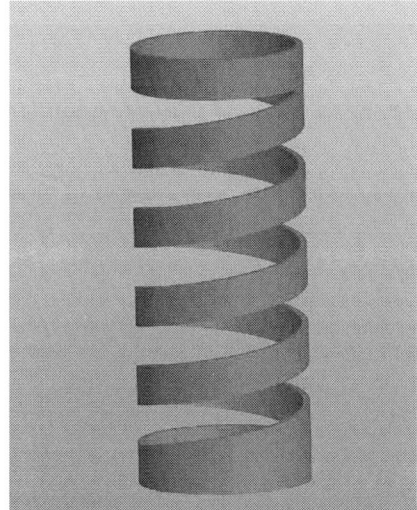

Fig. 5. Body construction

C. LEDs characteristics

LEDs with wavelengths of 450 and 660 nm were used as sources of photosynthetically active radiation. The main characteristics of LEDs are the radiation spectrum and watt-ampere characteristic, volt-ampere characteristic, and efficiency. To select the optimal radiation parameters, the radiation spectrum and watt-ampere characteristics of the LEDs were studied (Fig. 6-8)

The following parameters were taken into account when selecting LEDs: spectral characteristics corresponding to the photosynthetic requirements of plants, as well as nominal voltage and current to ensure an optimal ratio of red and blue light to stimulate growth and flowering [7-9].

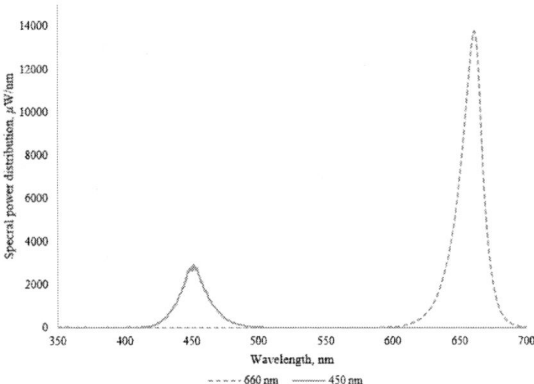

Fig. 6. Emission spectrum of LEDs

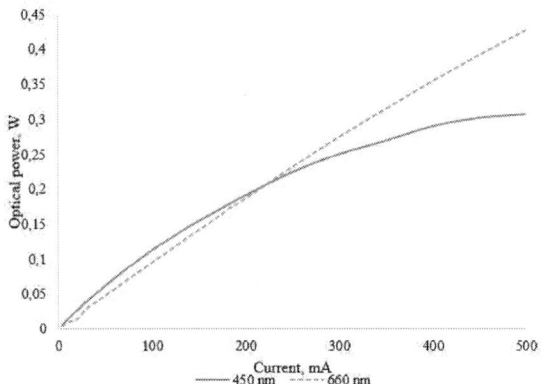

Fig. 7. Watt-ampere characteristics

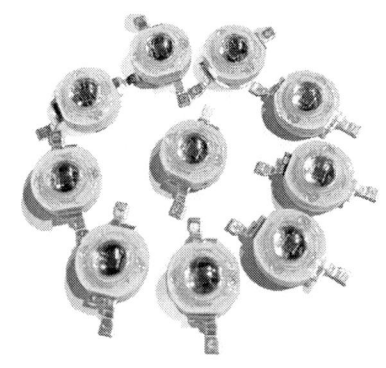

Fig. 8. External appearance of LEDs

III. CONCLUSIONS

The traditional configuration of artificial lighting in indoor plant cultivation, characterized by the use of an artificial light source positioned directly above a plant or group of plants, generates a significantly uneven spatial distribution of photosynthetically active radiation (PAR). This manifests itself in the formation of a pronounced central peak of illuminance under the source, followed by an exponential attenuation of intensity towards the peripheral areas of the plant canopy.

The cross-shaped design is structurally simpler, provides good access to the plant, and creates pronounced side lighting on four sides, making it convenient for plants or small groups of plants. The spiral housing, in turn, provides more complete surround lighting for the plant and can provide higher radial uniformity, which is especially important for compact crops grown in limited space.

Thus, the simulation results indicate that both geometries are promising for light field equalization tasks, while the choice of a specific design may be determined by requirements for manufacturability, crop type, and cultivation format (home installations, rack farms).

The developed designs can be used to study the effect of this light distribution on plants, to compare it with classic overhead lighting, and to study the density of the photosynthetic photon flux at different leaf levels. It is also possible to use different spectral combinations in volumetric lighting conditions to evaluate the effect of the spectrum on the biometric parameters of plant development.

REFERENCES

[1] State Standart 58461-2019. Plant lighting in protected ground structures. Terms and definitions. Moscow, Standartinform Publ., 2019, 19 p. (In Russian).

[2] V. N. Nurminsky, Y. B. Zaharov, "The effect of spectral composition and light intensity on plant photosynthesis," Collection of materials of the Annual Meeting of the Society of Plant Physiologists of Russia, the All-Russian Scientific Conference with international participation and the school of young scientists. Irkutsk, July 10-15, 2018 (In Russian).

[3] Qi Gao, Qiuhong Liao, Qingming Li et al, "Effects of LED Red and Blue Light Component on Growth and Photosynthetic Characteristics of Coriander in Plant Factory," Horticulturae. 2022, 8(12), p. 1165; https://doi.org/10.3390/horticulturae8121165.

[4] Yang X, Wang S, Liu W, Huang S, Xie Y, Meng X, Li Z, Jin N, Jin L, Lyu J, Yu J, "Different Spatial Configurations of LED Light Sources Enhance Growth in Tomato Seedlings by Influencing Photosynthesis, CO2 Assimilation, and Endogenous Hormones," Plants. 2025, vol. 14, 1369. DOI: 10.3390/plants14091369.

[5] Slattery RA, Ort DR, "Perspectives on improving light distribution and light use efficiency in crop canopies," Plant Physiol. 2021, vol. 185 p. 34-48. DOI: 10.1093/plphys/kiaa006.

[6] O.V. Avercheva, Yu.A. Berkovich, I.O. Konovalova, S.G. Radchenko, S.N. Lapach, E.M. Bassarskaya, G.V. Kochetova, T.V. Zhigalova, O.S. Yakovleva, I.G. Tarakanov, "Optimizing LED lighting for space plant growth unit: Joint effects of photon flux density, red to white ratios and intermittent light pulses," Life Sciences in Space Research. 2016, vol. 11, p. 29-42. https://doi.org/10.1016/j.lssr.2016.12.001.

[7] A. A. Gubina, M. M. Romanovich, A. E. Degterev, I. A. Lamkin, and S. A. Tarasov, "Development of Phytolamp and Study the Influence of LED Radiation on Tomatoes Growth," 2021 IEEE Conference of Russian Young Researchers in Electrical and Electronic Engineering (ElConRus), pp. 1310-1313, 2021. DOI: 10.1109/ElConRus51938.2021.9396515

[8] M. M. Romanovich, I. A. Lamkin, and S. A. Tarasov, "Accounting the quantum-confined Stark effect on the determination of the active LED region temperature," Journal of Physics Conference Series, vol. 1400, p. 0660467, 2019. DOI: 10.1088/1742-6596/1400/6/066046

[9] M. Degtereva, Y. Levin, A. Gubina, A. Degterev, I. Lamkin and et al. "Influence of the Spectral Composition of Illuminating Light Sources on Biometric and Phytochemical Characteristics of Ocimum basilicum L," Photonics 10, no. 12: 1369, 2023. DOI: 10.3390/photonics10121369

Optical Properties Study of Metal Halide Perovskites after Polymer Modification

Yuriy E. Isaev
Department of Photonics
Saint Petersburg Electrotechnical
University "LETI"
St.-Petersburg, Russia
yura.isaev.97@bk.ru

Maksim A. Balutin
Department of Photonics
Saint Petersburg Electrotechnical
University "LETI"
St.-Petersburg, Russia
mabalutin@stud.etu.ru

Aleksandr S. Tarasov
Department of Photonics
Saint Petersburg Electrotechnical
University "LETI"
St.-Petersburg, Russia
astarasov@etu.ru

Aleksander E. Degterev
Department of Photonics
Saint Petersburg Electrotechnical
University "LETI"
St.-Petersburg, Russia
aedegterev@etu.ru

Marina D. Pavlova
Department of Photonics
Saint Petersburg Electrotechnical
University "LETI"
St.-Petersburg, Russia
mdpavlova@etu.ru

Sergey A. Tarasov
Department of Photonics
Saint Petersburg Electrotechnical
University "LETI"
St.-Petersburg, Russia
satarasov@etu.ru

Abstract— Increasing the thermal stability and durability of metal halide perovskite thin films with a mixed anion is a critical and relevant task in photovoltaics. In this work, we investigated the effect of polyvinylpyrrolidone addition on perovskites containing the mixed anions $FAPbBr_2Cl$ and $FA_xCs_{1-x}PbBr_2Cl$. Specifically, the study examines how the stabilizing additive interacts with both FA and Cs cations. To assess the stability of the optical properties, the temperature and time dependencies of the samples' photoluminescence were investigated. The results indicate that PVP addition improves the uniformity of thin films and significantly enhances long-term and thermal stability. In the case of the cation synergy composition and polymer stabilization, a photobrightening effect is observed.

Keywords—mixed halide perovskite, chloride perovskite, photobrightening, photoluminescence spectra

I. INTRODUCTION

Lead halide perovskite semiconductors are promising materials for a broad range of optoelectronic devices. Their key advantage lies in the ease of tuning their properties through chemical substitution of components within the ABX_3 structure. For a long time, research focus was on iodide (I) and bromide (Br)-based perovskites. However, chloride (Cl) and mixed-halide perovskites constitute an important material class with unique benefits.

Beyond their large bandgap and outstanding optical properties, they exhibit enhanced moisture stability [1]. This opens a pathway for creating efficient emitters in the blue spectral region. Furthermore, the increased bandgap makes these materials ideal candidates for the top sub-cells in tandem solar cells paired with silicon or narrow-bandgap perovskites.

Nevertheless, the fabrication of mixed-halide perovskite films faces several technological challenges. The most significant among them is uncontrolled halide segregation during crystallization and under light exposure. This phenomenon is driven by differences in the ionic radius, solubility, and migration energy of Br^- and Cl^- ions. Consequently, it leads to the formation of inhomogeneous, defective films with phase impurities, which drastically increase the density of non-radiative recombination centers and degrade device efficiency [2].

An effective strategy to suppress segregation and enhance stability is cation engineering, for instance, the partial substitution of the organic FA^+ cation with inorganic Cs^+. The perovskite stability is further increased through defect passivation using polymeric additives. This work investigates the synergetic effect of the polymer polyvinylpyrrolidone (PVP) and the cesium cation for stabilizing the $FAPbBr_2Cl$ perovskite, synthesized in an ambient air environment under high humidity. The resulting films hold potential for application in efficient and cost-effective photosensitive and light-emitting elements.

II. EXPERIMENTAL SECTION

A. Preparation of precursor solutions

1) FAPbBr₂Cl with and without PVP

To create samples of thin films of $FAPbBr_2Cl$, 0,5 M solution was produced by mixing 0,5 M FABr, 0,25 M $PbBr_2$, 0,25 M $PbCl_2$ in solvent system with dimethylformamide (DMF) and dimethyl sulfoxide (DMSO), DMF:DMSO = 4:1. Then solution was stirred for 1 hour. After that we added 12,5 mg/ml of PVP and stirred until completely dissolved.

2) CsₓFA₁₋ₓPbBr₂Cl with and without PVP

To create samples of thin films of $Cs_xFA_{1-x}PbBr_2Cl$, 0,5 M solution was produced by mixing 0,25 M CsBr, 0,25 M FABr, 0,25 M $PbBr_2$, 0,25 M $PbCl_2$ in mixed solvent system with DMF:DMSO = 4:1. After that we added 12,5 mg/ml of PVP. Then solution was stirred for 24 hours. After that we filtered it with 0,22 μm PTFE filter.

B. Sample preparation

Glass substrates were washed with detergent and subsequently ultrasonically cleaned with distilled water and isopropyl alcohol for 15 min. The as-cleaned substrates were treated with ultraviolet radiation for 30 min. Substrates were pre-heated after drying. In this work, a one-step spin-coating method was used to create thin films of the perovskite. Films were deposited under ambient condition and 40% RH. Precursors were applied to hot substrates at 2000 rpm for 30 s. Perovskite films then were annealed on a hotplate at 120°C.

III. RESULTS

A. Effect of adding PVP in precursor solution of FAPbBr₂Cl

The research was funded by project № FSEE-2025-0013.

The influence of polyvinylpyrrolidone (PVP) as a polymer additive on the formation and properties of FAPbBr$_2$Cl perovskite films was investigated first. The morphological impact is drastic, as evidenced by the photographs (Fig. 1).

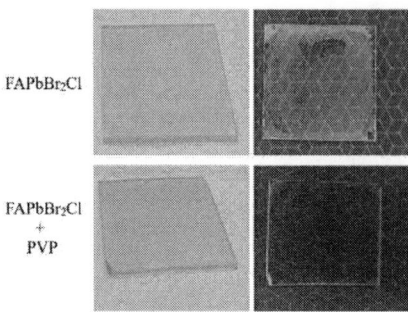

FAPbBr$_2$Cl

FAPbBr$_2$Cl + PVP

Fig. 1. Photographs of FAPbBr$_2$Cl films

The pristine FAPbBr$_2$Cl film exhibits severe inhomogeneity, haze, and incomplete substrate coverage, which are characteristic of rapid, uncontrolled crystallization leading to a high density of grain boundaries and structural defects. In contrast, the introduction of 1 wt% PVP into the precursor solution yields a smooth, pinhole-free, and highly uniform film (Fig. 2b). This is caused by passivation effect of PVP, described in previous works [3, 4]. The photoluminescence (PL) spectra at different time for these films are shown in Fig. 2.

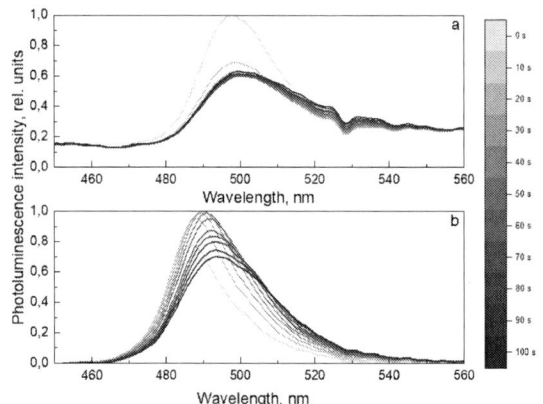

Fig. 2. Photoluminescence spectra of FAPbBr$_2$Cl (a) and FAPbBr$_2$Cl + PVP (b) films

Figure 2 shows the emergence of the photobrightening effect over 30 seconds in the presence of PVP. This indicates that in the presence of PVP, recombination centers can be filled or deactivated upon illumination. In contrast, the pristine film exhibits immediate photodarkening, indicative of dominant defect-activated recombination pathways.

The addition of PVP also leads to a blue shift of the PL peak. This suggests that the polymer suppresses phase segregation during the crystallization process and achieves the target stoichiometry. However, light-induced heating can subsequently promote phase segregation and the formation of a bromide-rich phase, resulting in a red shift of the PL peak. Additionally, this process generates new recombination centers, which cause the observed photodarkening effect [5].

Furthermore, the initial photoluminescence intensity is significantly higher in the PVP-containing composition. Specifically, the ratio of their initial PL peak intensities is

2,3. High photoluminescence intensity is a critical efficiency metric for optoelectronic devices. Its importance originates from its correlation with a high radiative quantum yield, which enhances the open-circuit voltage by ensuring the dominance of radiative recombination channels [6].

B. Studying the Incorporation of Cesium into FAPbBr$_2$Cl

It has been established that cation mixing in the perovskite structure slows down ion migration [7]. Photographs of samples with FA and Cs$_x$FA$_{1-x}$ cations are shown in Fig. 3.

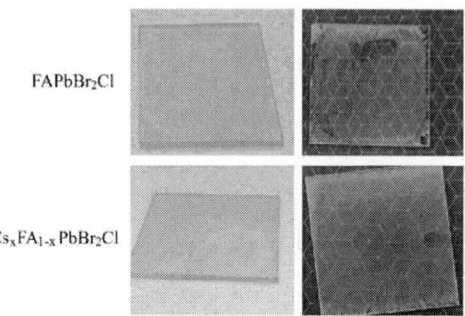

FAPbBr$_2$Cl

Cs$_x$FA$_{1-x}$PbBr$_2$Cl

Fig. 3. Photographs of FAPbBr$_2$Cl and Cs$_x$FA$_{1-x}$PbBr$_2$Cl films.

Figure 3 indicates higher uniformity of the films containing cesium cations. However, these films remain hazy due to a high density of scattering centers, which is likely a result of their fabrication under ambient conditions with high relative humidity. The time dependence of photoluminescence, depicted in Figure 4, also reveals a high density of recombination centers. This high defect density causes the observed photodarkening effect.

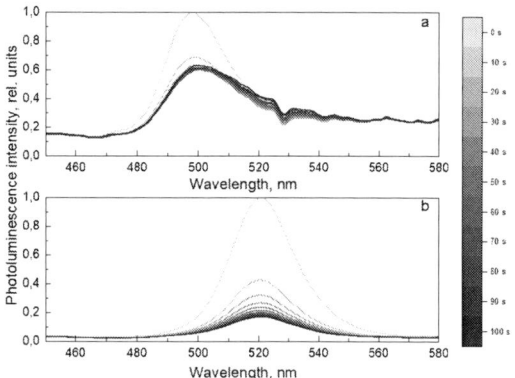

Fig. 4. Photoluminescence spectra of FAPbBr$_2$Cl (a) and Cs$_x$FA$_{1-x}$PbBr$_2$Cl (b) films

The incorporation of cesium cations was found to induce a red shift of the PL peak to 520 nm. It suggests a substantial alteration of the halide composition, likely favoring the formation of a bromide-rich perovskite phase, possibly due to the exclusion of chloride ions during crystallization in the presence of CsBr. Despite the presence of photoinduced recombination centers that cause photodarkening, the initial PL peak intensity of the Cs-containing sample is more than 20 times higher than that of the reference sample without Cs. This indicates a generally lower initial defect density within the crystalline lattice.

We propose that combining cation engineering with PVP addition could lead to enhanced crystallinity and PL stability.

C. Synergistic Effect of PVP and Cation Engineering

The effect of PVP addition on the $Cs_xFA_{1-x}PbBr_2Cl$ perovskite is illustrated in Figure 5.

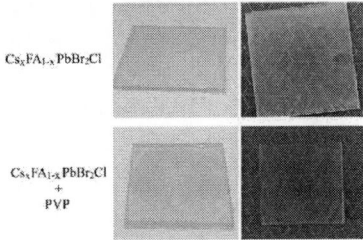

Fig. 5. Photographs of $Cs_xFA_{1-x}PbBr_2Cl$ films

A similar morphology is observed, yielding a uniform and transparent film, as in Figure 1. This confirms that the passivating effect of PVP is also effective for this composition. The corresponding photoluminescence spectra are presented in Figure 6.

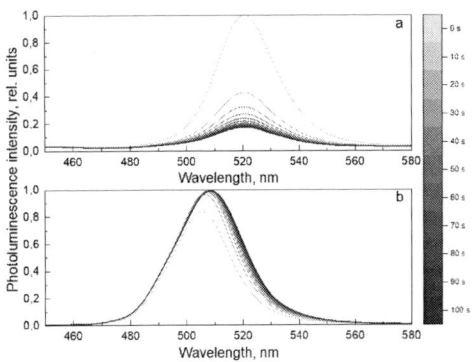

Fig. 6. Photoluminescence spectra of $Cs_xFA_{1-x}PbBr_2Cl$ (a) and PVP + $Cs_xFA_{1-x}PbBr_2Cl$ (b) films

As expected, the co-introduction of cesium cations and PVP enhances stability and induces a photo-brightening effect. Concurrently, the photoluminescence peak shifts to 505 nm. This shift confirms PVP's ability to suppress halide phase segregation during crystallization and facilitates the incorporation of chloride ions into the perovskite lattice. Furthermore, after an initial brief brightening period (15 seconds), the PL intensity stabilizes and remains constant. This PL stabilization indicates a significant reduction in ion migration and effective defect passivation.

Time dependence of PL peak intensity for all samples are presented in Fig. 7.

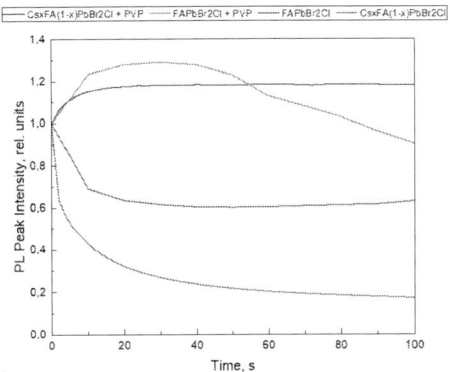

Fig. 7. Photoluminescence decay of $FAPbBr_2Cl$ and $Cs_xFA_{1-x}PbBr_2Cl$ films

The optimal properties are exhibited by the film where PVP and cesium cations act synergistically. Following a brief initial brightening period, the photoluminescence signal remains stable.

Excitation parameters, atmospheric composition, and ambient conditions can strongly influence photo-brightening and photodarkening effects [7]. In this work, all measurements were conducted in open (ambient) air at 40% relative humidity, using continuous-wave laser excitation at a wavelength of 400 nm and a power of 100 mW.

The initial PL peak wavelengths and the corresponding intensity ratios (normalized to pure $FAPbBr_2Cl$) for all investigated samples are summarized in Table 1.

TABLE I CHARACTERISTICS OF DIFFERENT PEROVSKITE FILMS

Perovskite Composition	Peak Intensity ratio, rel. units	Initial Peak Wavelength, nm
$FAPbBr_2Cl$	1	497
$FAPbBr_2Cl$ + PVP	2,3	486
$Cs_xFA_{1-x}PbBr_2Cl$	20,2	520
$Cs_xFA_{1-x}PbBr_2Cl$ + PVP	45,8	505

Thus, the maximum photoluminescence intensity and its temporal stability are achieved through the synergistic action of the cesium cation and the PVP polymer. Adding PVP alone to the $FAPbBr_2Cl$ composition results in a 2,3 increase in PL intensity, indicating a substantial improvement in film morphology. The most significant enhancement – a 45,8-fold increase relative to the pristine sample – is observed when Cs^+ and PVP are used in combination. This suggests that the key roles are played by the suppression of lattice point defect formation and the blocking of photoinduced ion migration.

It is important to note that PVP and CsBr differently influence the final phase composition of the film, which is directly reflected in the PL peak shift. The addition of PVP alone induces a blue shift (486 nm), evidencing the formation of a chloride-enriched phase due to the selective interaction of the polymer's functional groups with lead ions and altered crystallization kinetics. Introducing only CsBr causes a red shift (520 nm), pointing to the dominance of a bromide-rich phase. Only their combined application yields a homogeneous film of the target composition with an emission peak at 505 nm, confirming the role of synergism in precise stoichiometry control.

The obtained high initial PL intensity values and the improved stability for the synergistic composition make this material a promising candidate for application in optoelectronic devices, such as light-emitting diodes (LEDs) and photodetectors.

D. Thermal stability

Thermal stability is critical for perovskite thin-film devices, as their operating temperature – particularly in solar cells – can reach up to 50°C [7]. The temperature dependence of the photoluminescence peak is presented in Figure 8.

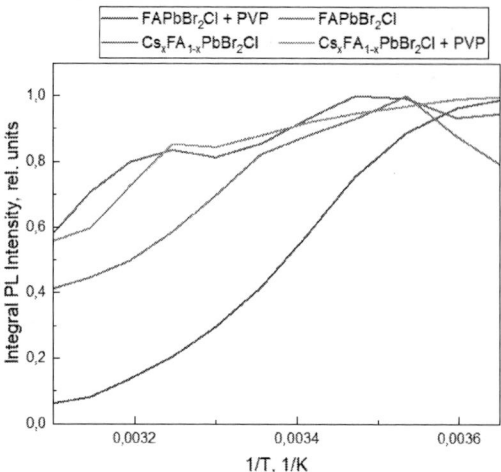

Fig. 8. Temperature dependence of integral PL intensity

The temperature dependence of photoluminescence for the studied samples (Fig. 8) reveals a fundamental difference in their thermal stabilization mechanisms. The mixed-cation $Cs_xFA_{1-x}PbBr_2Cl$ compositions exhibit high stability, retaining 60% of the initial PL intensity at 50 °C. In contrast, the PVP-modified $FAPbBr_2Cl$ film undergoes rapid degradation, losing over 90% of its signal. This difference arises from the nature of cation stabilization. Cs^+ ions suppress halide ion migration and raise the energy barrier for thermal decomposition.

Conversely, in the pure formamidinium system, stability is limited by the weaker bonding between the organic FA^+ cation and the halide framework. In this case, while PVP incorporation improves room-temperature crystallinity, it appears to facilitate the formation of a thermodynamically unstable phase (Cl-enriched, as inferred from the PL blue shift) and introduces a polymer matrix susceptible to degradation at elevated temperatures. This explains the sharp PL intensity drop.

IV. CONCLUSIONS

This work investigates the influence of the polymer additive polyvinylpyrrolidone and cation mixing (FA^+/Cs^+) on the morphological and optical properties of mixed-halide perovskite films.

It was demonstrated that PVP is a powerful agent for controlling film morphology. Its introduction leads to the formation of uniform, pinhole-free films and provides an increase in photoluminescence intensity by passivating surface defects. A key observation is the ability of PVP to influence halide stoichiometry, manifested by a blue shift of the PL peak.

Incorporating the cesium cation (Cs^+) into the perovskite structure induces an opposite, red shift of the PL peak (520 nm), pointing to the formation of a bromide-rich phase. Despite their higher initial PL intensity, Cs-containing films without PVP retain a significant density of recombination centers that become activated under illumination.

Maximum efficiency is achieved only through the synergistic combination of both strategies. The use of Cs^+ and PVP enables the fabrication of homogeneous, transparent films of the target composition with an emission peak at 505 nm. In this system, the initial PL intensity increases by a factor of 45 compared to the reference sample, and the signal stabilizes after a brief photobrightening effect. This is a direct consequence of suppressing both point defects and ion migration at grain boundaries.

Furthermore, the Cs^+ ion contributes to the thermal stability of the perovskite, while PVP is critical for achieving high morphological uniformity and photostability at room temperature.

Thus, cation doping combined with polymer-modified crystallization paves the way for creating high-quality mixed-halide perovskite films. This makes them promising candidates for application in optoelectronic devices such as light-emitting diodes (LEDs) and photodetectors.

REFERENCES

[1] Kontos A. G. et al. "Halogen–NH2+ interaction, temperature-induced phase transition, and ordering in (NH2CHNH2) PbX3 (X= Cl, Br, I) hybrid perovskites", The Journal of Physical Chemistry C, vol. 124, no. 16., pp. 8479-8487, March 2020, doi: https://doi.org/10.1021/acs.jpcc.9b11334.

[2] Xu F. et al. "Challenges and perspectives toward future wide-bandgap mixed-halide perovskite photovoltaics", Advanced Energy Materials, vol. 13, no. 13, p. 2203911, Feb. 2023, doi: https://doi.org/10.1002/aenm.202203911.

[3] B. Li et al., "Surface passivation engineering strategy to fully-inorganic cubic CsPbI3 perovskites for high-performance solar cells," Nature communications, vol. 9., no. 1., p. 1076, Mar. 2018, doi: 10.1038/s41467-018-03169-0.

[4] T. Ye et al., "Below 200° C fabrication strategy of black-phase CsPbI3 film for ambient-air-stable solar cells," Solar RRL, vol. 4., no. 5., Jan. 2020, doi: 10.1002/solr.202000014.

[5] Mahon N. S. et al. "Photoluminescence kinetics for monitoring photoinduced processes in perovskite solar cells", Solar Energy, vol. 195, pp. 114-120, Jan. 2020, doi: https://doi.org/10.1016/j.solener.2019.11.050.

[6] Campanari V. et al. "Reevaluation of photoluminescence intensity as an indicator of efficiency in perovskite solar cells", Solar RRL, vol. 6, no. 8, p. 2200049, May 2022, doi: https://doi.org/10.1002/solr.202200049.

[7] Goetz K. P. et al. "Shining light on the photoluminescence properties of metal halide perovskites", Advanced Functional Materials, vol. 30, no. 23, p. 1910004, March 2020, doi: https://doi.org/10.1002/adfm.201910004.

Study of Technology for Forming Arrays of Polymer Microlenses for Fluorimetric Biosensor Systems

Diana I. Khasanova
Department of micro- and nanoelectronics
Saint-Petersburg Electrotechnical University ETU "LETI"
St. Petersburg, Russia
d.i.a.n.a.2109@mail.ru

Anastasia D. Tarasenko
Department of micro- and nanoelectronics
Saint-Petersburg Electrotechnical University ETU "LETI"
St. Petersburg, Russia
tarasankon@mail.ru

Nikita O. Sitkov
Department of micro- and nanoelectronics
Saint-Petersburg Electrotechnical University ETU "LETI"
St. Petersburg, Russia
sitkov93@yandex.ru

Abstract—This paper explores ways to increase the sensitivity of optical biosensors through the use of polymer microlens arrays. The optical properties of the microlenses were modeled using the COMSOL Multiphysics software package, and the potential for signal amplification with their application was investigated. Three fabrication technologies for polymer microlenses were analyzed: injection molding, photopolymer 3D printing, and casting from silicone master molds. It was found that the choice of fabrication method significantly affects the quality and optical properties of the microlenses, which directly influences the amplification coefficient of the fluorescence signal. Experimental results demonstrated that integrating microlenses into biosensor systems can considerably increase their sensitivity in detecting low concentrations of protein markers. The results confirm the feasibility and potential of using polymer microlenses in the development of compact and highly sensitive systems for biomedical diagnostic application.

Keywords—*biosensors, microfabrication, microptics, 3D printing, polymer microlenses*

I. INTRODUCTION

With the increasing prevalence of diseases, the need for analytical tools that provide accurate, rapid, and reproducible analysis of biological samples is growing [1]. Biosensors are integrated analytical devices that convert the result of a specific interaction into a measurable physical signal, most often electrical, which determines their significant potential for the diagnosis of a wide range of biologically significant analytes [2]. The most critical element determining the sensitivity, detection limit, dynamic range, and reproducibility of measurements is the transducer. Depending on the physical nature of the effect being recorded, electrochemical, optical, piezoelectric, and other detection approaches are used, each of which is characterized by its own advantages and limitations [3, 4]. Among the listed methods, optical biosensors occupy a leading place due to their high sensitivity and specificity, as well as the ability to record signals directly related to molecular recognition processes [5, 6]. Optical systems utilize both label-free modes, based on detecting changes in optical parameters upon analyte binding, and label-based approaches, which utilize specialized fluorophores or other markers to provide signal amplification and selectivity [5, 6].

This paper examines an optical fluorimetric biosensor system designed for detecting protein markers in a microfluidic format. The system includes an excitation light source, a microfluidic biochip with integrated biorecognition elements based on peptide aptamers, an optical path with a phosphor and polymer microlenses, and a photodetector based on a digital camera's CMOS matrix [7]. The detection principle is based on recording the intrinsic fluorescence of proteins, caused by the presence of aromatic amino acid residues (phenylalanine, tyrosine, and tryptophan). Excitation is achieved by ultraviolet radiation with a wavelength of 280 nm, which is then converted by a ZnS:Cu phosphor, allowing the signal to be transferred to a longer wavelength and recorded using a standard CMOS matrix instead of specialized ultraviolet photodetectors [8].

Increasing the sensitivity of this approach is fundamental for the quantitative assessment of low concentrations of protein markers in real samples. One promising approach to signal amplification is the use of polymer microlenses, which provide localized focusing and increased radiation density at the photodetector through refraction and redistribution of the light flux. The aim of this study is to investigate the possibility of increasing the sensitivity of fluorimetric detection of a protein marker using polymer microlenses.

II. THE OPERATING PRINCIPLE OF MICROLENSES AND COMPUTER MODELING OF RAY TRAJECTORIES

The principles of lens operation are based on the laws of geometric optics, particularly the law of refraction. Refraction is the deflection of light by a certain amount upon entering or exiting a medium. The deflection caused by refraction is a function of the refractive index of the medium and the angle formed by the light relative to the surface normal. This deflection is described by Snell's law of refraction, which describes the relationship between the angles of incidence and refraction as a beam passes through n media. Snell's law is expressed as follows [5]:

$$Sin\theta_1/sin\theta_2 = n_2/n_1, \qquad (1)$$

where θ_1 is the angle of incidence, θ_2 is the angle of reflection, n_1 is the absolute refractive index of the first medium, n_2 is the absolute refractive index of the second medium. An optical system is defined as a set of refractive and reflective surfaces that enclose optically homogeneous media. Depending on the shape of the lens, a distinction is made between diverging and converging lenses. All optical systems obey Snell's law of refraction; therefore, the geometry of the optical lens determines the behavior of light as it passes through the optical system [5].

The microlenses studied in this paper are designed to focus the re-emitted signal, allowing all the light power to be concentrated at a single point. Therefore, focusing the fluorescence signal can be achieved using a converging lens, specifically a plano-convex lens.

Before fabricating microlenses, it is important to study the behavior of the rays on a model; this will allow an assessment of the potential for amplification by the microlenses. This paper examines the results obtained by

---

The reported study was funded by Russian Science Foundation, project number 25-79-10055.

979-8-3315-4784-4/26 $31.00 © 2026 IEEE

modeling the ray trajectory for two types of microlenses whose radius of curvature is the same over the entire microlens area: a conventional microlens and an elongated plano-convex microlens. Microlens models were developed in KOMPAS 3D, and then the resulting microlens models were added to COMSOL Multiphysics, a finite element analysis program.

The problem solved by tracing rays passing through an optical system is one of geometric optics, addressing the wave vector and position of individual rays. According to the law of rectilinear light propagation, when light propagates through a homogeneous medium, the rays move in a straight line at a velocity of $c = 10^8$ m/s. However, when passing through a medium with a gradient refractive index, the rays can follow a curved trajectory determined by integrating first-order ordinary coupled differential equations with respect to time. Expressions are calculated along each ray that include certain variables, such as the optical wavelength, wavelength, or intensity [5].

The ray trajectories obtained from the simulations correspond to theory: when rays propagating parallel to the optical axis strike a microlens, the rays passing through it converge at a single point, which corresponds to the microlens' focus. The simulation results for two types of microlenses, spherical and prolate, are shown in Fig. 1.

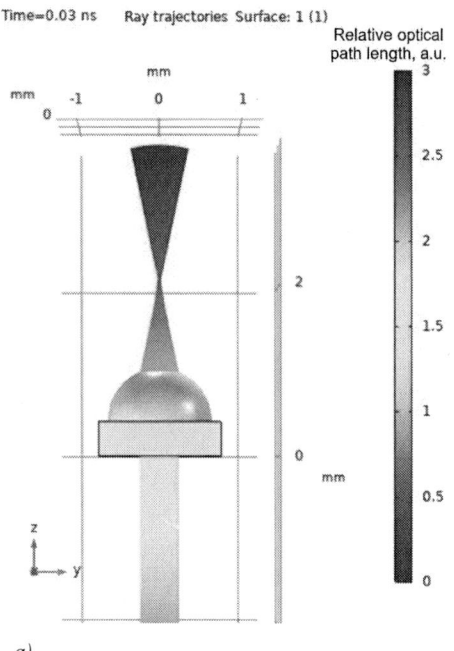
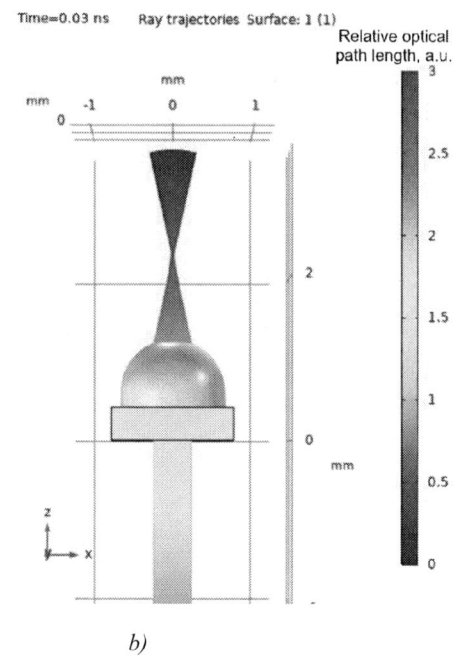

a)   b)

Fig. 1. Results of microlens amplification simulation: (a) – Hemispherical lens with a diameter of 1.3 mm, (b) – Elongated hemispherical lens with a diameter of 1.3 mm

Comparison of axisymmetric and elongated geometries allows us to evaluate how changes in aperture shape and curvature distribution affect the spatial localization of the focal region and, consequently, the amplification efficiency of the recorded fluorescence signal. Thus, using modeling, it is possible to visualize the path of rays through a microlens and determine the efficiency of radiation focusing.

III. TECHNOLOGIES FOR FORMING POLYMER MICROLENSES

A. Injection molding

One of the most commonly used methods for producing polymer microlenses is injection molding. It is a relatively inexpensive and simple method that produces high-quality microlenses.

PMMA, which has high dimensional stability and transparency, can be used as a material for injection molding microlenses. It is also important to select the right material for the master molds for microlens casting. Materials such as PCB and polypropylene are well suited for this purpose due to their resistance to high temperatures.

Injection molding of microlenses can be easily accomplished using a heat press heated to 130°C. The entire process takes approximately one hour, with about 30 minutes for the injection molding itself and about the same amount of time required for the heat press to cool slightly so that the parts can be safely removed. This time is sufficient for the PMMA lenses to form. The PCB master mold and the resulting microlenses are shown in Fig. 2.

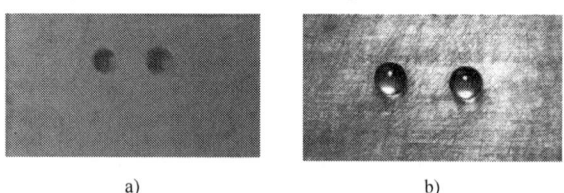

a)   b)

Fig. 2. Injection-molded microlenses: (a) – master mold made of PCB, (b) – microlenses made of PMMA

B. 3D printing technology

3D printing technology is a promising method for creating structured components, with particular attention given to the additive manufacturing of components for optical applications [6].

Microlenses were 3D printed as follows: first, the 3D lenses designed in KOMPAS 3D were printed on a Saturn 4 Ultra 12K 3D printer (Elegoo, China) using transparent Epoxy Resin (Elegoo, China). The printer operates by projecting light from an LED UV source through an LCD screen onto the photopolymer, which cures the entire model layer. It is important to prepare the model for printing, which requires selecting the orientation of the model relative to the printer's work surface. We decided to position the model at a 45° angle to the work surface. This orientation significantly increases print time, but also improves print quality. Furthermore, it is important to carefully position the supports, allowing the models to be printed at a certain height above the printer's work surface. Printing the simulated microlenses on a 3D printer took approximately 12 hours. After printing, the resulting microlenses must be washed in isopropanol to remove any remaining resin. After washing the printed models, remove the supports and, if necessary, sand the back of the supports with a sanding sheet. After printing, the resulting microlenses must be washed in isopropanol to remove any remaining resin. The resulting microlenses with supports are shown in Fig. 3.

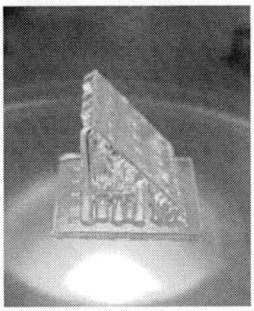

Fig. 3. 3D printed microlenses with supports

After washing the printed models, remove the supports and, if necessary, sand the back of the supports with a sanding sheet. The resulting microlenses are shown in Fig. 4.

Fig. 4. 3D printed microlenses

The advantage of 3D printing technology is its relatively high reproducibility, unlike injection molding, which requires controlled casting conditions. However, 3D printing does have some limitations, specifically in the precision of microlens shape. Due to the layer-by-layer printing process, roughness can occur, which degrades the optical performance of the microlenses.

C. Molding from Silicone Master Molds

Another method for producing polymer microlenses is casting from silicone molds. Silicone master molds are made from 3D-printed silicone microlenses (Epoxy Master, Russia) with a platinum-based material with a hardness of 30A Shore. The resulting silicone master mold is shown in Fig. 5.

Fig. 5. Silicone master mold for casting microlenses

The master mold can be filled with various photopolymers. In this work, UV resin (Epoxy Master, Russia) was used. Exposure to UV resin takes approximately 5–7 minutes, which is sufficient time for the resin to harden. If necessary, the object can be further cured for 2 minutes after removal from the mold. Microlenses cast from UV resin are shown in Fig. 6.

Fig. 6. UV resin molded micro lenses

It's important to note that cast microlenses also have a rough surface. To achieve a smooth surface, the microlenses can be additionally coated with a thin layer of photopolymer varnish. The resulting microlenses before and after coating with photopolymer varnish are shown in Fig. 7.

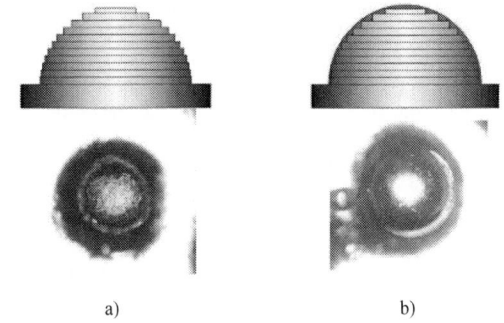

a)   b)

Fig. 7. UV resin microlenses: (a) – before coating with photopolymer varnish, (b) – after coating with photopolymer varnish

IV. STUDY OF THE EFFICIENCY OF POLYMER MICROLENSES

An experiment to evaluate the microlens amplification efficiency was performed using a setup that allows microlenses to be studied by coupling them to a cover glass coated with a phosphor layer to convert the LED's ultraviolet radiation to long-wavelength wavelengths, and a CMOS matrix [7]. The gain of the microlenses was then estimated by calculating the gain as the ratio of the intensity of the re-emitted radiation measured at the output of the microlens to the intensity of the radiation incident on the microlens:

$$K = I_{out}/I_{in}, \qquad (2)$$

where I_{out} is the intensity of the radiation emitted from the microlens, I_{in} is the intensity of the re-emitted radiation incident on the microlens. The graph of the dependence of the signal amplification on their diameter is shown in Fig. 8.

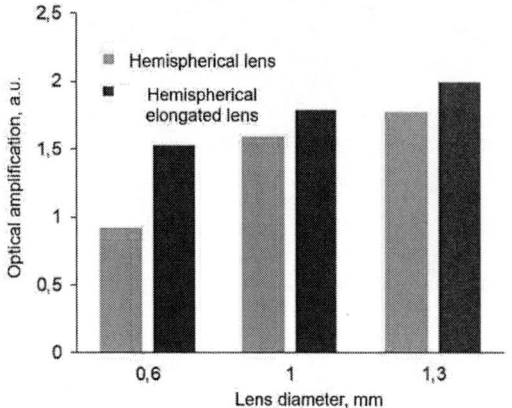

Fig. 8. Graph of the amplification as a function of the diameter of polymer microlenses

Thus, the dependence of the gain on the microlens diameter was determined; the best results were shown by microlenses with a diameter of 1.3 mm, elongated microlenses with a diameter of 1.3 mm make it possible to amplify the signal almost twice as much.

The practical significance of using microlenses was further confirmed by recording the concentration dependence of the signal in a microfluidic chip (Fig. 9).

Comparable series of measurements were obtained in two modes (with and without microlenses) while passing human lactoferrin solutions in phosphate buffer through the chip at concentrations ranging from 2×10^{-10} to 2×10^{-5} g/mL. A comparison of the concentration dependences showed that the use of microlenses leads to an increase in the relative emission intensity of the recognition pads across the entire studied concentration range, which corresponds to the expected effect of optical concentration of the re-emitted signal and improves the conditions for its recording on the CMOS matrix.

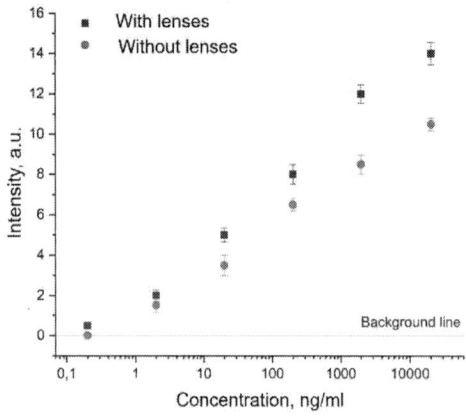

Fig. 9. Concentration dependence of the relative intensity of the emission from the recognition pads of microfluidic chips, obtained with and without microlenses, when passing a lactoferrin solution through them.

Taken together, these results confirm that the integration of polymer microlenses into the recording path is an effective approach to increasing the sensitivity of fluorimetric detection in microfluidic biosensor systems.

V. CONCLUSIONS.

Increasing the sensitivity of optical detection methods is a critical challenge for quantitatively assessing low concentrations of protein markers contained in small sample volumes. One solution to this problem is the use of polymer microlenses, which can amplify the signal re-emitted by the intrinsic fluorescence of proteins. The main technologies for manufacturing polymer microlenses were considered: injection molding, 3D printing, and casting from silicone master molds. The best results were achieved with microlenses produced using silicone master molds. Microlenses produced by this method are characterized by relatively high optical properties and shape reproducibility, and also provide a twofold amplification of the re-emitted signal.

Overall, the results demonstrate the feasibility of using polymer microlenses as a simple and technologically feasible means of increasing the sensitivity of fluorimetric detection in microfluidic biosensor systems. This approach expands the possibilities for creating compact, inexpensive, and highly sensitive biomedical diagnostic devices designed for analyzing small sample volumes. Further research could focus on optimizing microlens parameters (curvature, aperture, and material), standardizing surface finishing procedures, and assessing the impact of microlenses on the detection limit and metrological performance of the system when working with various protein markers and matrices of real biological samples.

ACKNOWLEDGMENT

The study was supported by a grant from the Russian Science Foundation No. 25-79-10055.

REFERENCES

[1] systems," Micromachines, vol. 12, no. 6, pp. 691–691, Jun. 2021

[2] V. Naresh and N. Lee, "A Review on Biosensors and Recent Development of Nanostructured Materials-Enabled Biosensors," Sensors, vol. 21, no. 4, p. 1109, Feb. 2021

[3] A. Uniyal, G. Srivastava, A. Pal, S. A. Taya, and Arjuna Muduli, "Recent Advances in Optical Biosensors for Sensing Applications: a Review," Plasmonics, vol. 18, no. 2, pp. 735–750, Feb. 2023

[4] M. Ramesh, R. Janani, C. Deepa, and L. Rajeshkumar, "Nanotechnology-Enabled Biosensors: A Review of Fundamentals, Design Principles, Materials, and Applications," Biosensors, vol. 13, no. 1, p. 40, Dec. 2022

[5] M. B. Kulkarni, N. H. Ayachit, and T. M. Aminabhavi, "Biosensors and Microfluidic Biosensors: From Fabrication to Application," Biosensors, vol. 12, no. 7, p. 543, Jul. 2022

[6] M. Y. Azab, M. F. O. Hameed, and S. S. A. Obayya, "Overview of Optical Biosensors for Early Cancer Detection: Fundamentals, Applications and Future Perspectives," Biology, vol. 12, no. 2, p. 232, Feb. 2023

[7] L. M. Bonanno and L. A. DeLouise, "Whole blood optical biosensor," Biosensors and Bioelectronics, vol. 23, no. 3, pp. 444–448, Oct. 2007

[8] N. O. Sitkov et al., "Study of the Fabrication Technology of Hybrid Microfluidic Biochips for Label-Free Detection of Proteins," Micromachines, vol. 13, no. 1, pp. 20–20, Dec. 2021

[9] N. O. Sitkov et al., "Toward Development of a Label-Free Detection Technique for Microfluidic Fluorometric Peptide-Based Biosensor Systems," Micromachines, vol. 12, no. 6, pp. 691–691, Jun. 2021

[10] N. Lindlein, "Simulation of micro-optical systems including microlens arrays," Journal of Optics A: Pure and Applied Optics, vol. 4, no. 4, pp. S1–S9, Jul. 2002

[11] L. M. Rooney et al., "Printing, Characterizing, and Assessing Transparent 3D Printed Lenses for Optical Imaging," Advanced Materials Technologies, vol. 9, no. 15, May 2024

Development of an Energy-Efficient Phytotron

Roman A. Kurenkov
Department of Photonics
Saint Petersburg Electrotechnical
University "LETI"
St. Petersburg, Russian Federation
rkurenkov7@gmail.com

Gulnaz M. Galina
Department of Photonics
Saint Petersburg Electrotechnical
University "LETI"
St. Petersburg, Russian Federation
galina-gulnaz0@mail.ru

Mariya M. Degtereva
Department of Photonics
Saint Petersburg Electrotechnical
University "LETI"
St. Petersburg, Russian Federation
mmromanovich@etu.ru

Yevgeniy Levin
Department of Photonics
Saint Petersburg Electrotechnical
University "LETI"
St. Petersburg, Russian Federation
e_levin@etu.ru

Alexander E. Degterev
Department of Photonics
Saint Petersburg Electrotechnical
University "LETI"
St. Petersburg, Russian Federation
aedegterev@etu.ru

Ivan A. Lamkin
Department of Photonics
Saint Petersburg Electrotechnical
University "LETI"
St. Petersburg, Russian Federation
ialamkin@etu.ru

Abstract—**Agrophotonics is a field of science that uses radiation to control the agricultural organisms vital activity. The most important factor influencing the growth and development of plants is the spectral composition and radiation source power. The article presents the development of an energy-efficient phytotron for growing plants in controlled environmental conditions. A radiation source design is shown, which includes three-watt LEDs at the required wavelengths (370 nm, 445 nm, 520 nm, 660 nm, 730 nm). A control system consisting of a microcontroller and the necessary sensors connected to an external device (tablet) via a Wi-Fi module is presented.**

Keywords—agrophotonics, LEDs, phytotron, controlled environment, photosynthesis.

I. Introduction

Plant growth is carried out through a sequence of chemical reactions – organic chains are formed from simple molecules, forming plant organs. The process of combining atoms from simple mineral molecules into much more complex organic ones, which occurs in plants under the influence of light rays, is called photosynthesis.

It was believed that for growing plants, lamps were needed, designed to create lighting close to natural. However, at the present stage of development of cultivation technology, it has become known that photosynthetically active radiation (PAR) is necessary for the vital activity of plants [1]. This is radiation in the visible spectrum range from 400 to 700 nm, which is most effective when absorbed by plants [2]. Inside the headlamps, the highest-energy blue (410-480 nm) and low-energy red (630-680 nm) spectral ranges have the greatest impact on plant development [3].

The main sources of lighting that are used nowadays are high pressure sodium (HPS) lamps and LED fixtures. The study [4] compares these types of radiators. The authors conducted a comparison of physiological parameters in the cultivation of tomato plants. The work notes the advantages of LED emitters over HPS in many physiological parameters: the dry mass and dry matter content of vegetative organs, the content of photosynthetic pigments, the functional activity of the photosynthetic apparatus, the number of flower brushes and the timing of their formation. The main advantages of LED radiation sources are the ability to specifically select the spectral composition and a longer service life [5]. However, despite the advantages of LED emitters, most of them still use LED lamps due to the economic component due to the high cost of switching to new LED lighting technologies.

II. Development of an Energy-Efficient Phytotron

A. The housing of an energy-efficent phytotron

First of all, when developing an energy-efficient phytotron, it was necessary to select the device body. For optimized cultivation under controlled conditions, it was decided to use a structure isolated from external influences. In this case, a metal cabinet (Fig. 1) may be the most successful, ensuring independence from external lighting, as well as being able to withstand loads on shelves and allow for convenient work with plants. Also, the white color of the inner walls will ensure the re-reflection of radiation from the walls, which will increase the efficiency of lighting. Dimensions of this cabinet 853*503*2000 mm (width*depth*height).

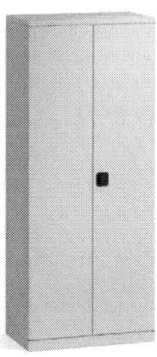

Fig. 1. An example of a metal cabinet as a housing for an energy-efficient phytotron

For optimal plant cultivation, there will be 2 tiers inside the installation, each with a height of 70 cm, which will ensure the cultivation of a large number of plant varieties. A container for an aqueous solution will be located on the lower tier. A plastic tank suitable for drinking water will be used as a water tank. To provide plants with the necessary micro- and macronutrients, a two-component Hoagland-Arnon solution containing the main mineral components necessary for plant growth and development will be added to the water.

To ensure that air enters the installation, fans equipped with dust filters will be installed in the amount of 1 piece per injection and blowout on each tier. This will ensure air circulation inside the installation.

979-8-3315-4784-4/26 $31.00 © 2026 IEEE

B. Hydroponics

After working through the device body, the next step was to develop an irrigation system. The modern method of irrigation in closed systems for growing plants is chosen by hydroponics, since it allows you to grow practically any plants (except trees, mushrooms, nodules of plants), and also ensures the rapid growth of herbs (for example, basil, mint, rosemary, sage, etc.) [6-8].

Polystyrene pallets were chosen for the device, as it does not emit a pungent odor and does not emit toxic substances that could harm plants during cultivation. Its size allows you to place up to 40 plants on one tier, depending on the type of plant. Each plant is grown in an individual pot in a special substrate designed for growing in hydroponic systems. The pots (Fig. 2, a) were selected in such a way that their height was not greater than the height of the pallet, and the root system of the plant freely occupied the necessary space inside the pallet. The substrate (Fig. 2, b) is a mineral wool that is safe for growing plants.

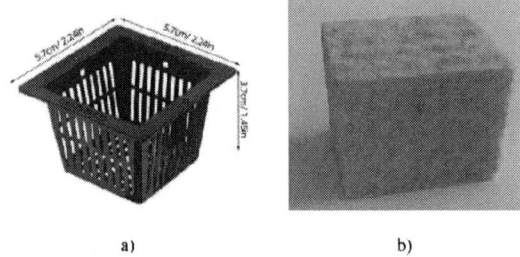

a)   b)

Fig. 2.   Growing devices (a – pot, b – substrate)

To ensure autonomy from the city's water supply, we will place a 70-liter box inside the installation, which will not require frequent replenishment of the tank. To ensure irrigation, it was necessary to choose a pump capable of raising water to a height of about 1.5 meters. A small pump shown in Fig. 3 is sufficient for these needs. It is capable of lifting water to a height of up to 3 meters, and also has a capacity of 5 liters /min, which will allow irrigation.

Fig. 3.   Pump example

In hydroponic systems, watering is carried out every 2 hours for a duration of 2-3 minutes. To ensure the frequency of watering, an Arduino controller will be used along with a set of sensors, which will be discussed further.

C. Electronic devices

To ensure the control of the ecosystem, it is necessary to select appropriate sensors, thanks to which it will be possible to monitor the state inside the phytotron. The main parameters that require close attention when growing plants are humidity, temperature, illumination, as well as the pH level and electrical conductivity of an aqueous mineral solution. These parameters will be monitored by appropriate sensors: humidity – a hygrometer, temperature – a thermometer combined in one device, illumination – a light sensor, which is a photoresistor with a voltage divider circuit, TDS and pH sensors, which are a control board connected to the controller, with a sensor in a waterproof housing with an output by contact. The device images are shown in Fig. 4.

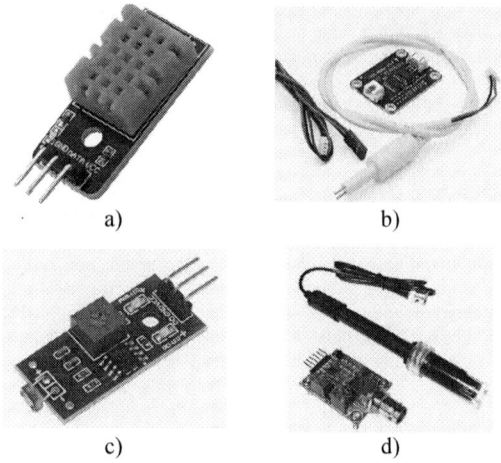

a)   b)

c)   d)

Fig. 4.   Ecosystem monitoring sensors (a – thermohygrometer, b – TDS sensor, c – light sensor, d – pH sensor)

To process the information received from the sensors, an Arduino Uno controller with 32 MB of memory will be used, which will be enough to complete the tasks. To connect to external devices and output information received from sensors, a Wi-Fi module connected to the Arduino controller will be used, and an application installed on the tablet will be written. The app on the tablet also includes the ability to adjust the illumination.

D. Lightning system

The basis of an energy-efficient phytotron is a lighting system. Three-watt LEDs with wavelengths of 370, 445, 520, 660, and 730 nm were selected as energy-efficient radiation sources (Fig. 5). These LEDs have a wide illumination angle of 140°, as well as a long service life of up to 50,000 hours. This lighting system will operate in a cycloadaptive mode, providing photoperiodism inside the phytotron. Thus, the lighting will work for 14 hours, providing a "day" inside the installation, and at 10 hours it will turn off the lighting, providing a "night".

Fig. 5.   Example of an LEDs housing

The most significant characteristics for the main LED emitters are shown in Fig. 6. LEDs with wavelengths of 370 and 730 nm will be used in small quantities (about 1% each), since high radiation in these spectral regions adversely affects the morphological characteristics of plants. LEDs with wavelengths of 370 and 730 nm will be used in small quantities (about 1% each), since high radiation in these spectral regions adversely affects the morphological characteristics of plants.

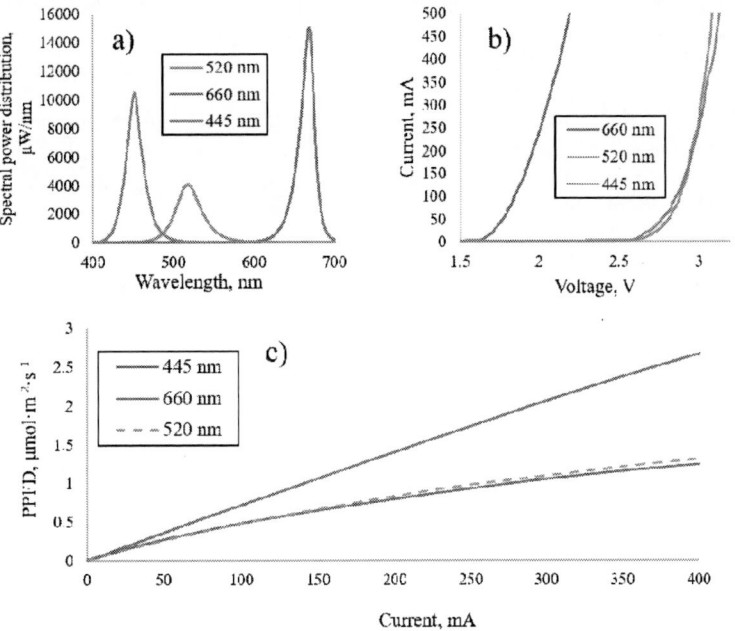

Fig. 6. Characteristics of the main LEDs (a – spectral characteristic, b – VAC, c – PPFD of one LED at a distance of 50 cm from the emitter)

However, it was also necessary to measure what photosynthetic photon flux density (PPFD) value the LEDs would produce when several radiators were turned on at the same time, since theoretically blue and red radiators should show approximately the same PPFD level, and Figure 6,b shows that blue and green LEDs produce similar PPFD values. For this purpose, an experimental plate was printed on a 3D printer (Fig. 7), on which up to 5 LEDs can be placed simultaneously.

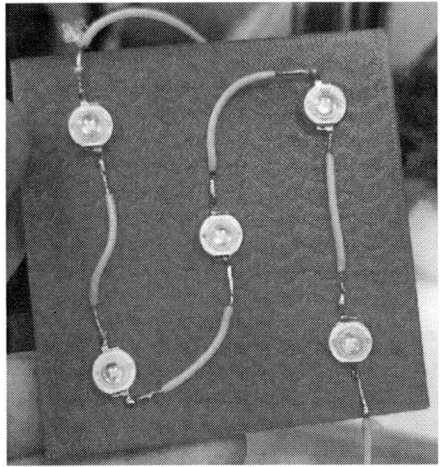

Fig. 7. A plate with LEDs for conducting an experiment

After that, the PPFD from 5 radiators was measured at a distance of 13 cm from the radiation source to the measuring device in two ways: perpendicular to the radiator and at a distance of 25 cm to the left of the radiator. The resulting dependencies are shown in Fig. 8.

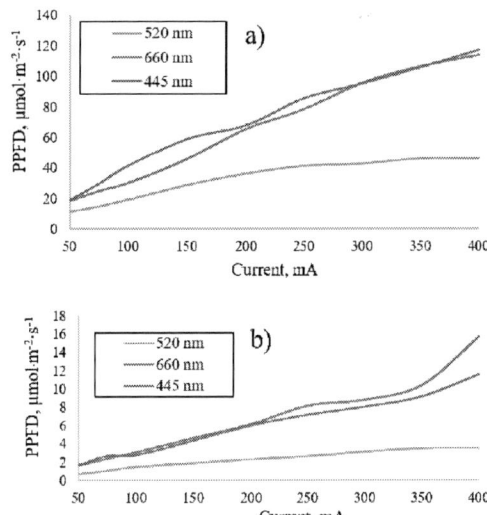

Fig. 8. PPFD measurement from 5 LEDs (a – perpendicular to the emitter, b – at a distance of 25 cm to the left of the emitter)

The measurements suggest that when the radiation from several LEDs is added together, the blue and red LEDs show approximately the same PPFD values, which are higher than those of the green emitters. It can be concluded that the radiation from red LEDs is distributed unevenly over the lighting area and the intensity in the center will be higher than at the edges, while for blue LEDs the intensity is distributed more evenly over the illuminated area.

III. Conclusion

During the development, a project of an energy-efficient phytotron for growing plants under controlled conditions was presented. The body of the device with the implemented hydroponics system for watering plants was thought out.

A sensor system was selected to monitor the internal ecosystem, including a temperature and humidity sensor, light, as well as pH and TDS sensors. All devices will be connected to an Arduino controller, and information will be output to an app on the tablet using Wi-Fi.

Studies of the selected radiation sources were also conducted, the characteristics of the main LEDs used were investigated, and experiments were conducted to study radiation from 5 simultaneously switched-on LEDs. All the results obtained will be used in the future in the implementation of an energy-efficient phytotron.

References

[1] N. V. Konyaev and S. E. Mamonova, "Universal lamp-irradiator of plants", *Effektivnost primeneniya innovatsionnykh tekhnologiy i tekhniki v selskom i vodnom khozyaystve* [Efficiency of application of innovative technologies and equipment in agriculture and water management], 2020, pp. 187-189 (in Russian).

[2] State Standart 58461-2019. Plant lighting in protected ground structures. Terms and definitions. Moscow, Standartinform Publ., 2019, 19 p. (In Russian).

[3] A. A. Bogdanov and M. M. Romanovich, "Advantages of using LED phytolents to improve plant growth", *78-ya nauchno-tekhnicheskaya konferentsiya SPb NTO RES im. A. S. Popova. posvyashchennaya Dnyu radio* [78th Scientific and Technical Conference of St. Petersburg NTO RES named after A. S. Popov, dedicated to Radio Day], 2023, pp. 347-349 (in Russian).

[4] O. V. Molchan, L. V. Obukhovskaya, T. N. Kudelina et al., "Activation of photosynthetic and growth processes of tomato plants by LED lighting in the conditions of a pilot production site", Proceedings of the National Academy of Science of Belarus, vol. 68, no 4, pp. 282-292 (in Russian). DOI: 10.29235/1029-8940-2023-68-4-282-292.

[5] I. M. Dovlatov, A. A. Smirnov, A.V. Sokolov et al., "Choosing the type of lighting for growing plants by the hydroponic method on racks", *Innovatsii v selskom khozyaystve* [Innovations in agriculture], i. 2, 2020, pp. 111-118 (in Russian).

[6] P. S. Yamov, "Hydroponics", *Aktualnyye voprosy nauki i khozyaystva: novyye vyzovy i resheniya* [Actual issues of science and economy: New Challenges and Solutions], 2021, pp. 743-746.

[7] A. A. Gubina, M. M. Romanovich, A. E. Degterev et al., Development of Phytolamp and Study the Influence of LED Radiation on Tomatoes Growth. Proceedings of the 2021 IEEE Conference of Russian Young Researchers in Electrical and Electronic Engineering, ElConRus 2021, p. 1310-1313. DOI:10.1109/ElConRus51938.2021.9396515.

[8] M. M. Degtereva, Y. Levin, A. E. Degterev, A.A. Bogdanov, I.A. Lamkin, S.A.Tarasov, P.A. Sergeev, Assessment Procedure for the Advantages of LED Phyto-Strip Application in the Industrial Greenhouse Complexes. Photonics Russia, 2023, 17(7), pp. 566–578. DOI:10.22184/1993-7296.FRos.2023.17.7.566.578.

Laser Polishing for Creating Microfluidic Chips on Stainless Steel

Valery V. Lavrinenko
Department of Photonics
Saint-Petersburg Electrotechnical University "LETI"
Saint-Petersburg, Russia
0009-0001-8930-5438

Abstract — **When creating microfluidic chips, which are used, for example, in microelectronics to create cooling systems or in medicine to create new drugs or point drug delivery, it is necessary to adjust the surface roughness. The low roughness parameter provides low resistance for the liquid inside the channels, helps to avoid additional turbulence and turbulence of the flow. In addition, with a low roughness of the docking surface, a more reliable and stable contact of the components of the microfluidic chip is observed. The paper shows a study of the process of creating polishing modes of laser treatment of stainless steel during the formation of microtopology. The results of measuring the roughness parameter of a polished surface and their application to a finished microfluidic topology are presented.**

Keywords — *microfluidic, laser polishing, microfluidic topology, laser processing, roughness*

I. INTRODUCTION

Microfluidics is a field of science and technology that deals with the control and manipulation of small volumes of liquids in miniature devices, the so—called microfluidic chips, simulating natural physiological processes for high-precision selection, sorting or analysis of biological material. This technology has been mainly developed in the biomedical field, where these devices are used for clinical analysis (rapid blood analysis), toxicological analysis, drug quality control, and analysis of physiological samples [1-3].

The main requirements for such systems are high sensitivity and high performance, cost-effectiveness, low weight, compactness and miniaturization. The methods of obtaining microfluidic systems (MFS) are diverse and are mainly determined by the material used. Microfluidic chips can be created from various materials, for example, glass [4], polymers [5] and many other materials [6]. The main methods of obtaining MFS are standard operations of the lithographic process [7], methods of formation (imprinting) [8], laser ablation [9, 10], soft lithography [11] and LIGA technologies [12]. The general scheme of MFS manufacturing is as follows: preparation of the substrate, production of a photolithographic template for the selected MFS topology, formation of a channel system and their sealing.

Photolithography and etching are combined technological processes where, with the help of the former, topology is transferred to the material by applying a photoresist to a substrate and exposing it to ultraviolet or other radiation through a photomask, and etching consists in removing an unnecessary surface layer of the material under the action of chemicals. These methods are among the most common in micro-manufacturing, but they have significant drawbacks, such as limited resolution, the use of consumables, and the creation of a mask for photolithography is quite an expensive process.

Due to the disadvantages of the above production methods, in this paper it is proposed to use a laser as a tool for creating a chip that allows you to create channels of any topology on the workpiece quickly and without the use of photo masks. In addition, the use of metal, namely stainless steel, as a substrate material reduces the cost of manufacturing these products, which makes them more accessible for a wide range of research and tasks. However, the process of interaction of laser radiation with metals is quite complex and represents a combination of such physical phenomena as heating of the material, formation of a liquid phase, metal evaporation and formation of plasma vapors. All this leads to the need to solve complex technological problems in laser processing of materials, in particular, to give channels a low roughness value.

The requirements for the channel surface vary depending on the application. So, for example, in order to preserve the laminarity of the flows, a small surface roughness and uniformity of the channel are necessary throughout its entire length, and for increased mixing of solutions, a sufficiently rough surface is needed to ensure the appearance of turbulent flows [13].

The MFS is usually a multi-layered system. a structure consisting of several parts. Moreover, it is advisable to form through-holes in the "sandwich structures", providing the possibility of introducing / removing reagents. Thus, another important technological process in the MFS is the sealing of the chip, which should ensure the tightness of the channel with minimal changes in its profile. Therefore, in order to use a metal plate as a blank for a microfluidic chip, a low surface roughness is required, since in our case a smooth and even surface is required, which will be glued between the parts of the microfluidic chip to create a more durable and airtight structure.

II. MATERIALS AND METHODS

A. Material

The research was carried out on a 2 mm thick stainless steel plate. The percentage of chemical elements (Table 1) was detected using X-ray fluorescence analysis (XRF). The data obtained have very approximate values, because XRF cannot register all chemical elements, for example, oxygen, which is precisely present in the material due to direct contact with atmospheric oxygen.

TABLE I.	PERCENTAGE CONTENT OF CHEMICAL ELEMENTS		
Elements	**Content, %**	**Elements**	**Content, %**
Iron (Fe)	71.738 ± 0.348	Silicon (Si)	0.424 ± 0.115
Chromium (Cr)	18.233 ± 0.147	Cobalt (Co)	0.259 ± 0.020
Nickel (Ni)	8.076 ± 0.138	Vanadium (V)	0.077 ± 0.011
Manganese (Mn)	1.165 ± 0.040	Titanium (Ti)	0.021 ± 0.006

According to the data obtained, we can see a fairly high content of chromium (Cr) and nickel (Ni), so we can classify this material as austenitic stainless steel, which has high strength, ductility and corrosion resistance in most liquids.

B. Laser system

The stainless steel plate was exposed to a precision laser marker "MiniMarker2" (Laser Center LLC, Russia) powered by an ytterbium fiber laser (Table 2). The topological elements on which the roughness parameter was measured were designed using the specialized MaxiGraf software supplied with the laser marker. The finished microfluidic topology was designed using 3D modeling software, which was then imported into the laser marker software.

TABLE II.	THE MAIN CHARACTERISTICS OF THE LASER MARKER
Characteristic	**Value**
Wavelength	1.07 μm
Pulse duration	From 1 to 350 ns
Pulse repetition rate	Up to 4000 kHz
Scanning speed	Up to 8000 mm/sec
Processing field	110 × 110 mm
Permission	0.1 mm

C. Profilometer

The roughness of the structures was assessed using Industrial NSRT-100 profilometer (NORGAU, Russia). Before evaluating the roughness of the model microchannels, the profilometer was calibrated on a reference sample with known roughness. The measurement error was calculated based on the specified measurement error of the device itself and the practical number of measurements with certain values. In our case, we performed 5 measurements each, and then calculated the relative measurement error based on these values. The measurement error of the parameter R_a (the arithmetic mean deviation of the surface profile) was about 5%.

III. EXPERIMENTAL RESULTS

There are 3 main processes of laser polishing (Fig. 1): removal of large surface areas, removal of local (small) surface areas, and polishing by remelting the surface layer of the material [14].

When polishing large areas by ablation, the material evaporates over the entire surface, while ablation of local areas leads to the removal of material from the tops of the surface (a complex and expensive measuring system is needed to determine the position of the peaks). When polishing by remelting, the thin surface layer melts, and the surface tension leads to the alignment of the material by melting the peaks and pouring the molten material into the recesses of the surface.

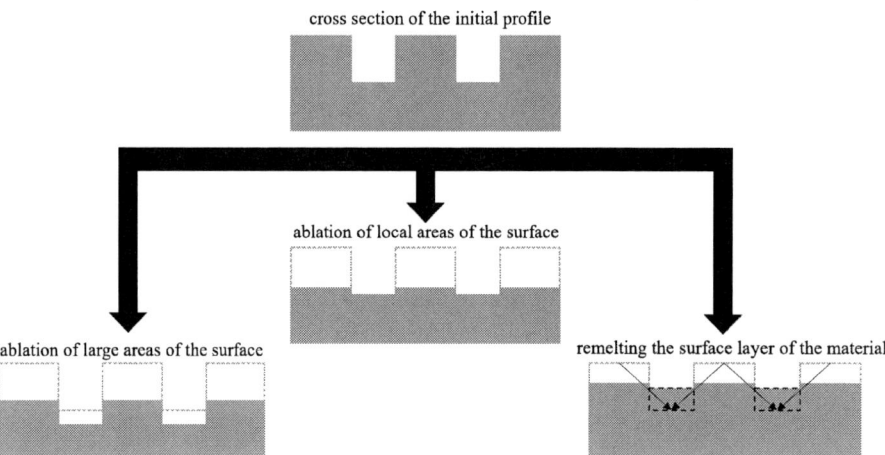

Fig. 1. Schematic representation of various laser polishing processes

For this work, a laser treatment process was chosen based on the 3rd method of polishing metal surfaces described above.

A. Polishing Modes

Before starting to create polishing modes, indentation was simulated on the surface using laser treatment. The surface roughness parameter R_a after this type of treatment was 0.754 microns. The polished surfaces (Fig. 2) were obtained by processing with a laser beam with a diameter of 50 microns and a pulse duration of 120 to 350 ns. Some of the polishing modes were created in 4 processing stages, which varied the power density (from 26.7 to 68.8 kW/cm²) sufficient to reach the melting point of stainless steel, pulse repetition rate (from 40 to 60 kHz), scanning speed (from 500 to 3500 mm/second), as well as filling density (from 100 to 300 lines/mm) and the filling path (from 0 to 270 degrees).

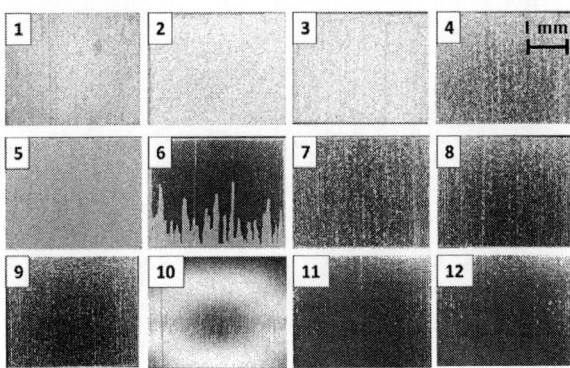

Fig. 2. Photos of polishing treatment modes

In the photos above, small furrows are visible on all the treated areas, which correspond to the directions of the laser scanning lines. Mode №6 has unique properties, which consist in the presence of an oxide film on the surface (the appearance of color in the lower part of the treated area). This effect is observed as a result of laser heating, during which atmospheric oxygen reacts with the material of the heated plate and forms oxide layers of iron and other impurities on its surface. The formation of an oxide film, as was found in one of the earlier papers [15], is a negative effect due to an increase in the surface roughness parameter.

The roughness of the treated areas was measured with a profilometer (Table 3).

TABLE III.   ROUGHNESS OF THE TREATED AREAS

№	R_a, μm	№	R_a, μm
1	0.420	7	0.232
2	0.370	8	0.188
3	0.447	9	0.180
4	0.270	10	0.173
5	0.423	11	0.115
6	0.621	12	0.088

Modes with lower roughness are seen darker in photographs due to the fact that they reflect the lens of the microscope during photography. As can be seen from the results above, processing mode № 12 gives the best value of the roughness parameter, reducing the initial value by 88%.

Using mode № 12, the workpiece with the finished microfluidic topology will be polished.

B. Polishing of a Workpiece With a Wicrofluidic Topology

For the presented microfluidic chip model (Fig. 3), a blank was developed (an inverted model where channels are turned into convex tracks) using 3D software.

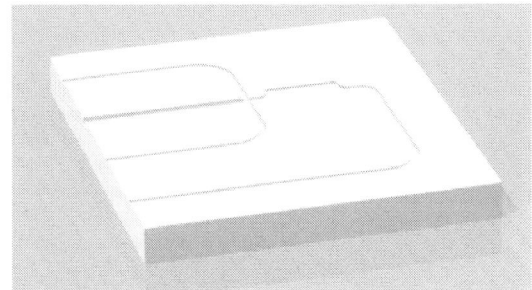

Fig. 3. Microfluidic chip model

Next, this file was imported into the MaxiGraf software and recreated on a stainless steel plate using laser micro-processing (Fig. 4).

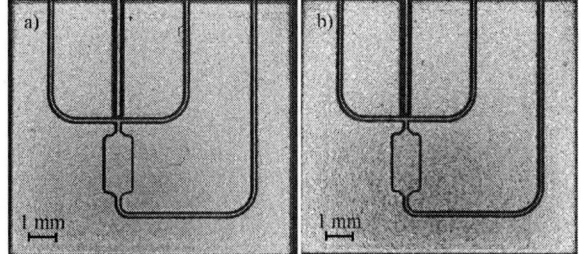

Fig. 4. Photographs of a metal billet with a microfluidic topology: a) Before polishing, b) After polishing

As can be seen from the figures above, carbon deposits are visible around the topological elements of the workpiece, which were formed during laser micro-processing. This deposit appears due to additional impact on the surface due to the rereflection of the laser beam from topological elements. The appearance of the treated surface after polishing does not correspond to what was obtained in mode 12 above, which indicates an increased level of roughness in this case. Indeed, thanks to the profilometer data, it turned out that the roughness parameter of the molding area before polishing was 0.654 microns, and after polishing - 0.457. The roughness value decreased after polishing, but became significantly higher than on the test samples. These results mean that the quality of polishing depends on the geometric shape and size of the area and requires significant refinement.

IV. CONCLUSION

With the help of IR laser treatment, polishing modes for the surface of stainless steel were achieved. The average number of processing iterations is 4. The best reduction in the roughness parameter is approximately 88% and is 0.088 microns in absolute value, which is acceptable for microfluidic tasks.

The application of this polishing mode to the created workpiece with a microfluidic plate showed significantly worse results due to its strong dependence on the geometry of the treated area.

Further stages of the work will be the modernization of existing processing modes to obtain the required level of roughness.

REFERENCES

[1] M.L. Zaveskin , A.A. Mironova , A.M. Popov, "Microfluidics and its prospects in medicine", Molekulyarnaya meditsina [Molecular medicine], vol. 5, 2012, 8 p. (in Russian).

[2] H. Jiang, X. Weng, D. Li et al. "Microfluidic whole-blood immunoassays", Microfluidics and Nanofluidics, vol. 10, 2011, pp. 941–964.

[3] Xu Y., Jang K., Yamashita T. et al. "Microchipbased cellular biochemical systems for practical applications and fundamental research: from microfluidics to nanofluidics", Springer – Analytical and Bioanalytical Chemistry, vol. 402, 2012, pp. 99–107.

[4] A. A. Evstrapov, T. A. Lukashenko, G. E. Rudnitskaya et al., "Microfluidic chips made of glass materials", Nauchnoe priborostroenie [Scientific Instrument Engineering], vol. 22, i. 2, pp. 27-43 (in Russian).

[5] A. A. Evstrapov, T. A. Lukashenko, S. G. Gorny, K. V. Yudin, "Microfluidic chips made of polymethylmethacrylate: a method of laser ablation and thermal bindings", Nauchnoe priborostroenie [Scientific Instrument Engineering], vol. 15, i. 2, pp. 72-81 (in Russian).

[6] K.L.Wlodarczyk, D.P. Hand, M.M. Maroto-Valer, "Maskless, Rapid manufacturing of glass microfuidic devices using a picosecond pulsed laser", Scientific Reports, 2019, no. 9, pp. 1-13.

[7] Nadim M. "An introduction to microelectromechanical systems engineering", Artech house MEMS library, 2004, p. 304.

[8] Martynova L. et.al. "Fabrication of plastic microfluid channels by imprinting methods", Analytical Chemistry, 1997, vol. 69, pp. 4783–4789.

[9] Snakenborg D., Klank H. "Microstructure Fabrication with a CO2 Laser System", Journal of Micromechanics and Microengineering, 2004, vol. 14, pp. 182–189.

[10] K. L. Wlodarczyk, R. M. Carter, A. Jahanbakhsh et al. "Rapid Laser Manufacturing of Microfluidic Devices from Glass Sub strates", Micromachines, 2018, vol. 9 (409), 14 p.

[11] Xia Y., George M., "Soft Lithography", Angewandte Chemie International Edition, 1998, vol. 37, pp. 550–575.

[12] Chen Y., Pepin A., "Nanofabrication: Conventional and nonconventional methods", J. Electrophoresis, 2001, vol. 22, pp. 187–207.

[13] Stroock A., Stephan K, "Chaotic Mixer for Microchannels", Science, 2002, vol. 295, pp. 647–651.

[14] Chesnokov D. V., Shapran E. V., "Scientific and technical aspects of the development of laser glass polishing methods", Interexpo Geo-Siberia, 2016, pp. 76–81 (in Russian).

[15] Lavrinenko V.V., Vasilieva A.V., Parfenov V.A., Novikov I.A., "Investigation of microfluidic topology formation with the use of IR pulse laser", St. Petersburg State Polytechnical University Journal. Physics and Mathematics, vol. 17 (3.2), 2024, pp. 173–176.

Laser Cleaning of Parchment: Assessment of Treatment Results

Taniana K. Lepekhina
Laser Measurement and Navigation Systems Department
Saint-Petersburg Electrotechnical University "LETI"
Saint Petersburg, Russia
tanialep@mail.ru

Angelina D. Neelova
Laser Measurement and Navigation Systems Department
Saint-Petersburg Electrotechnical University "LETI"
Saint Petersburg, Russia
angelina.neelova@gmail.com

Yan O. Guttovskiy
Electronic Devices and Systems Department
Saint-Petersburg Electrotechnical University "LETI"
Saint Petersburg, Russia
yanpheonix426@yandex.ru

Abstract—**This study addresses the application of laser cleaning to historical parchment – a highly sensitive organic substrate widely used in manuscripts and archival documents. Despite the growing adoption of laser-based methods in cultural heritage conservation, their use on parchment remains underexplored due to risks of thermal and structural damage. In this work, we evaluate the efficacy and safety of laser cleaning using analytical approaches such as scanning electron microscopy (SEM) and spectrophotometry. These techniques enable comprehensive assessment of surface morphology and colorimetric changes before and after laser treatment. Experiments were conducted on both model parchment samples and historical artifacts, with systematic variation of laser parameters. The results provide critical insights into threshold fluences that preserve parchment integrity while effectively removing contaminants. This research contributes to establishing evidence-based protocols for the laser cleaning of parchment, supporting its integration into conservation practice.**

Keywords—*laser cleaning, parchment, cultural and historical heritage, multi-pulse laser radiation, SEM, spectrophotometer*

I. INTRODUCTION

In recent years, laser technologies have found increasing application in the preservation of cultural and historical heritage, encompassing both laser cleaning technology and a range of analytical methods, including spectroscopy, interferometry, holography, and 3D scanning, which enable the determination of chemical composition and the structural diagnostics, documentation, and monitoring of monuments [1-5].

Laser cleaning technology is used in restoration to remove anthropogenic and natural contaminants from the surface of monuments and is an integral part of any restoration work. Currently, cleaning the surface of artifacts is a very labor-intensive and time-consuming process. Furthermore, existing methods have a number of significant drawbacks, including the risk of damaging the original material, low efficiency, and the toxicity of some of the reagents used, which can pose a health risk to restorers. Compared to traditional restoration methods, laser cleaning offers several advantages: high efficiency, precision, and speed of contaminant removal, as well as a gentle effect on the material being treated [6]. Laser cleaning of objects made of inorganic materials, primarily metals and stone, is currently the most studied and well-developed method [1-3].

However, the use of lasers for processing organic materials, such as leather and parchment, remains understudied. Among organic materials, paper has received the most attention in the literature, while publications devoted to laser processing of other organic materials are extremely rare [7, 8]. This is due to the high sensitivity of organic materials, including parchment, to external influences, which requires a thorough study of the mechanisms of their interaction with laser radiation.

Therefore, research in the field of laser cleaning of leather materials, and parchment in particular, represents a pressing scientific challenge with high practical significance. In this work, the effectiveness and safety of laser cleaning of parchment has been evaluated when varying the energy parameters of radiation using analytical methods based on SEM and spectrophotometry.

II. EXPERIMENTS

A. Samples

The experiment used modern parchment and a fragment of a historical parchment manuscript. The objects of study are shown in Figure 1 and Figure 2. The modern parchment samples were prepared for conservation purposes. The parchment manuscript was written in Europe and may date back to the XVII century or earlier. The surface structure of the historical parchment has significantly deteriorated due to its age and environmental exposure. It has ink writings on its front side. The fragment has traces of worn-in dust. The back side of the manuscript is more contaminated than the front side.

To simulate parchment contamination on model samples, graphite and carbon dust were pre-applied to the surface of the study objects, simulating dust and other surface contaminants.

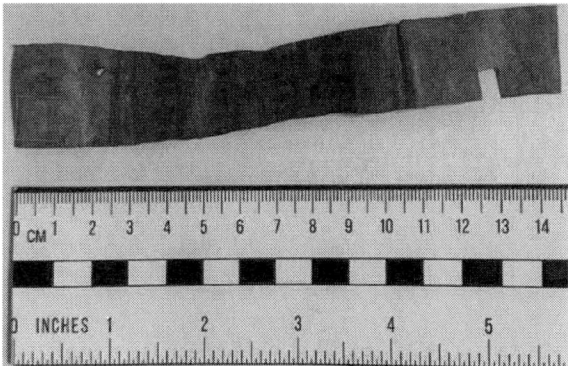

Fig. 1. Parchments samples: fragment of the manuscript from front side. Laser treated areas are marked with red squares (size 10 by 10 mm).

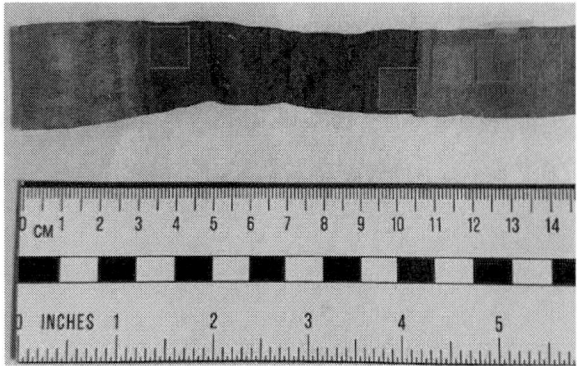

Fig. 2. Parchments samples: fragment of the manuscript from the back side (b). Laser treated areas are marked with red squares (size 10 by 10 mm).

B. Equipment

A multi-pulse laser was used for laser cleaning of the parchment surface, primarily because it minimizes thermal impact on the sensitive material. Unlike continuous exposure, pulsed mode delivers energy in short periods of time (in this case, 100 ns) and in high-intensity pulses, allowing for the effective removal of contaminants without transferring them to the bulk of the material. This is especially important for fragile materials like parchment, whose structure is easily damaged by localised overheating. It is known from scientific literature that the use of multiple pulsed irradiation combined with rapid scanning of the laser emitter creates a so-called multi-pulse microprocessing mode, which removes contaminants layer by layer, while maintaining minimal heat buildup in the material [9]. Thus, it is the pulsed mode, with its adjustable parameters (pulse duration, energy, frequency, conversion coefficient), that ensures a balance between cleaning efficiency and preservation of the original material, making it an indispensable factor in the careful restoration of organic cultural heritage objects.

Thus, in this study laser cleaning of samples from contaminants was performed using a multi-pulse ytterbium fiber laser "MiniMarker" (manufactured by "Laser Center", Ltd., Russia). The "MiniMarker" system is an industrial scanning laser for marking and engraving. The main parameters of the laser system are presented in Table 1. The average power density of laser radiation varied during the experiments.

TABLE I. OUTCOME IRRADIATION PARAMETERS OF THE USED LASER

Parameter	Value
Emission wavelength, nm	1064
Pulse duration, ns	100
Pulse repetition frequency, kHz	20
Maximum pulse energy, mJ	1
Radius of the radiation spot in the focal zone, μm	50
Average power density, W/cm2	$0.06 \cdot 10^3 - 0.3 \cdot 10^3$

Before conducting the laser parchment cleaning experiment, the reflectivity was estimated using an integrating sphere («AvaShpere» produced by «Avantes», Netherlands). The integrating sphere is an optical device designed to measure diffuse and specular reflections, as well as the total reflection coefficient of samples due to multiple scattering of radiation inside a cavity with a highly reflective inner coating. This type of device provides spatial averaging of reflected light, which makes it possible to obtain representative and reproducible data on the spectral reflective characteristics of the material under study. The results are shown in Figure 3. The spectra show that near 1000 nm, the reflectivity exceeds 75% for modern parchment samples and 50% for older ones. Based on this data, it can be assumed that a 1064 nm laser can be safely used for cleaning parchment.

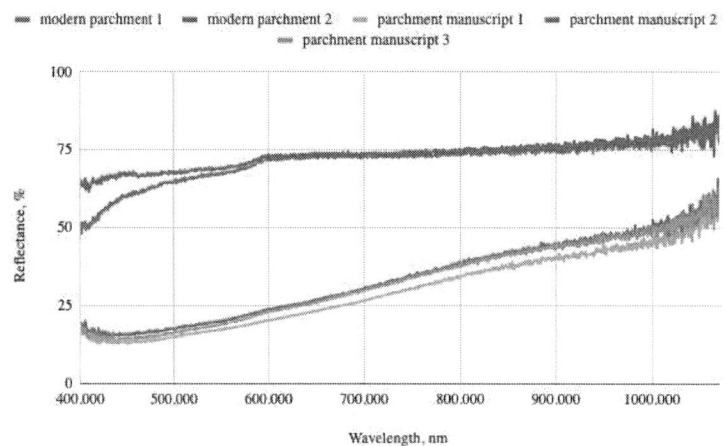

Fig. 3. Reflectance spectra of parchment samples and parchment manuscript. Model samples were evaluated in two points, manuscript in three point in different areas of the fragment.

III. RESULTS AND DISCUSSION

During the experiment, parchment samples were cleaned from surface contaminations using multi-pulse laser radiation. Laser treated areas were 10 by 10 mm. The best results were obtained with the following laser parameters: power density of $0{,}2 \cdot 10^3$ W/cm², the pulse repetition frequency of 20 kHz. Scanning parameters: beam scanning speed of 800 lines per mm with the filling of 25 lines per mm. Areas with the ink writings were excluded from the treatment.

Based on the visual examination laser cleaning of parchment manuscript (Figure 1, Figure 2) was a success, parchment in the cleaned areas notably lightened in tone.

979-8-3315-4784-4/26 $31.00 © 2026 IEEE

A. Evaluation of Laser Cleaning Results Using Optical Microscopy

More profound evaluation of the results of laser cleaning of the parchment were performed by optical microscope "Levenhuk" (USA), scanning electron microscope "Hitachi" (Japan) along with the spectrophotometry device "Exact" ("X-rite", USA).

Images by optical microscope (Figure 4) show no visible damage done to the surface of the parchment artefact by the laser irradiation.

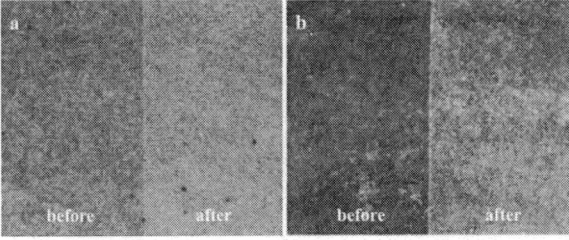

Fig. 4. Optical microscopy images of the parchment manuscript after laser cleaning in two areas: (a) light in tone area of the fragment; (b) – before. Magnification 20x.

B. Evaluation of Laser Cleaning Results Using Spectrophotometry

At the initial stage of the experiment, the effectiveness of laser cleaning was evaluated by visual inspection and optical microscopy.

At the next stage, an instrumental approach was used to objectively confirm visual observations, in which changes in the color coordinates L*, a* and b*, were calculated for areas of the parchment manuscript that were subjected to laser treatment and without laser exposure, which made it possible to quantify the removal of surface contamination from parchment samples.

Data of the visual examination of the laser cleaning of parchment from contaminations was confirmed by the instrumental method using a portable spectrophotometer. Evaluation was performed in the untreated areas and in the areas after laser cleaning. Evaluation aperture was 1.5 mm. The results of spectrophotometry are presented in Table 2. In Table 2 and Figure 5 letter "w" marks the area without cleaning and "a" - after cleaning.

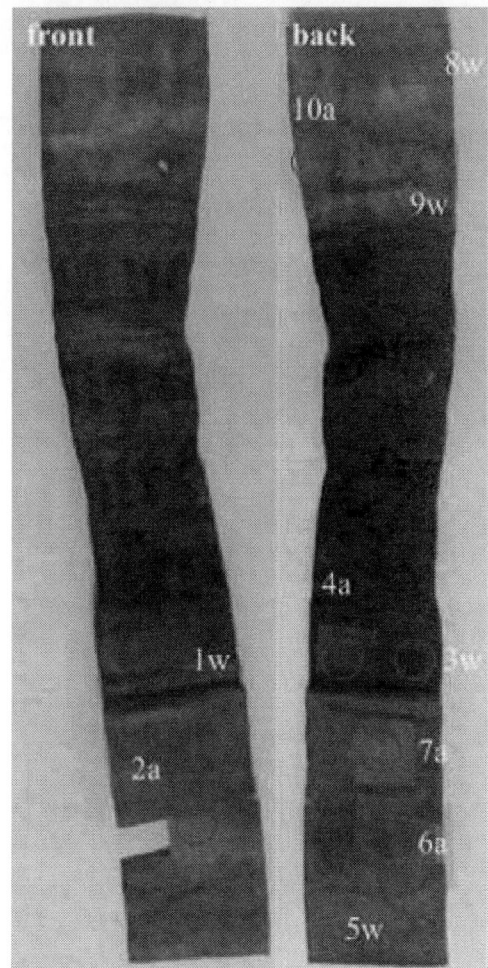

Fig. 5. of the points of spectrophotometric analysis of parchment manuscript, front and back sides. Evaluation areas are marked with red.

TABLE II. COLORIMETRIC PARAMETERS OF PARCHMENT IN THE STUDY AREAS

Sample №	L*	a*	b*
1w	51	5	17
2a	55	6	19
3w	26	5	5
4a	39	7	13
5w	53	5	18
6a	56	6	19
7a	58	5	19
8w	50	8	21
9w	43	6	14
10 a	47	7	16

Having the values of the color coefficients indicated in Table 3, it is possible to calculate the magnitude of the total color difference [10]. The formula for calculating the magnitude of the color difference is as follows:

$$(\Delta E)^2 = (\Delta L^*)^2 + (\Delta a^*)^2 + (\Delta b^*)^2 \qquad (1)$$

The calculation data is shown in Table 3. The darkest area of the parchment manuscript corresponds to point 3, so the magnitude of the total color difference between this point and point 4, where laser cleaning was performed (15.4) is significantly more important than in other dimensions. The remaining treated areas have a total color difference in the range from 3.3 to 5.9. Based on these values, it can be concluded that the processing areas of the parchment manuscript sample have become significantly lighter after laser cleaning.

TABLE III. EVALUATION OF THE EFFECTIVENESS OF LASER REMOVAL OF CONTAMINANTS FROM THE SURFACE OF PARCHMENT

Sample №	The magnitude of the overall color difference ΔE
1-2	4.6
3-4	15.4
5-6	3.3
5-7	5.1
8-10	5.9
9-10	3.3

Thus, a comparison of the values of neighboring points on contaminated and cleaned areas shows the effectiveness of laser cleaning, since the treated areas become lighter in tone.

C. Evaluation of Laser Cleaning Results Using Scanning Electron Microscopy Method

Scanning electron microscopy (Figure 6) proved that there were no structural damages to the parchment fibres and pores.

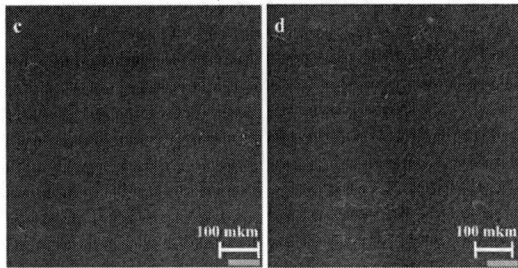

Fig. 6. Scanning electron microscopy of modern parchment after laser cleaning: (a) – parchment sample without cleaning; (b) – after laser cleaning, no damages; (c) – before cleaning; (d) – after cleaning, no damages.

It is important to note that the parchment manuscript has a deteriorated surface structure as a consequence of its age and prominence. Due to this fact the fragment is especially fragile, yet SEM-images of the fragment after laser treatment (Figure 7 a, c) do not indicate the presence of any damages in comparison to the images of untreated areas of the fragment (Figure 7 b, d).

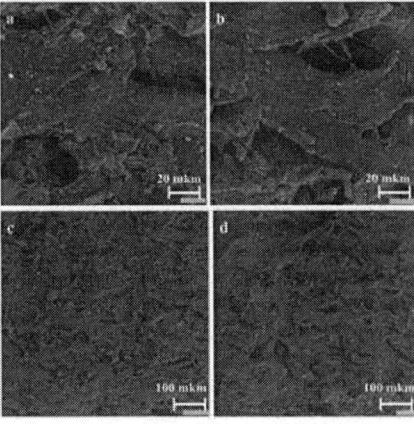

Fig. 7. Scanning electron microscopy of parchment manuscript before and after laser cleaning: (a) – manuscript before cleaning; (b) – after cleaning; (c) – before cleaning; (d) – after cleaning, no damages.

IV. CONCLUSION

This study demonstrates that laser cleaning using a multi-pulse 1064 nm ytterbium-doped fiber laser can be effectively and safely applied to both modern and historical parchment, provided that appropriate energy density thresholds are met, which correspond to the following values: power density of $0{,}2 \cdot 10^3$ W/cm^2, beam scanning speed of 800 lines per mm with the filling of 25 lines per mm.

Using an analytical approach combining spectrophotometry, optical microscopy and scanning electron microscopy (SEM), the effect of laser irradiation on the colorimetric properties of the surface of the parchment samples, as well as their surface integrity and morphology, was assessed. The results indicate that controlled laser processing successfully removes carbon-containing contaminants, such as graphite and dust, without visible or structural damage to the parchment base. Spectrophotometric data confirm change in tone after cleaning which indicates the removal of contamination substance.

SEM images demonstrate the preservation of the surface microstructure in both model and historical samples when using optimal laser parameters. These results highlight the potential of laser cleaning as a precise, contactless, and non-toxic alternative to traditional conservation methods for organic cultural heritage materials. However, due to the inherent variability and fragility of historical parchment, processing protocols must be individually adapted and rigorously tested. This work contributes to the development of scientifically sound guidelines for the safe implementation of laser technologies in parchment conservation practices, paving the way for wider application in the field of cultural heritage conservation.

ACKNOWLEDGMENT

The authors would like to express their gratitude to V. A. Parfenov (St. Petersburg Institute of History of the Russian Academy of Sciences, St. Petersburg Electrotechnical University "LETI") for his scientific guidance and constant support throughout this study.

SEM studies were carried out in the electron microscopy laboratory of the Department of Electronic Instruments and Devices of St. Petersburg Electrotechnical University with the support of the Department of Physical Chemistry and a strategic partnership with the Federal state unitary enterprise "Alexandrov Research Institute of Technology". The authors would like to thank G.A.Konoplev (St. Petersburg Electrotechnical University "LETI") for his help in conducting research on skin reflection spectra.

REFERENCES

[1] Lasers in the preservation of cultural heritage. Principles and applications / Fotakis C., Anglos D., Zafiropulos V. et al. (USA): CRC Press, Taylor &Francis Group, 2007. 364 p.

[2] Siano S., Grazzi F., Parfenov V. A. Laser cleaning of gilded bronze surfaces // Journal of Optical. 2008. Vol. 75, No. 7. P. 18-29.

[3] 2. Siano S., Giamello M., Bartoli L., Mencaglia A., Parfenov V., Salimbeni R. Laser cleaning of stone by different laser pulse duration and wavelength // Laser Physics. 2008. Vol.18, No.1. P. 1-10.

[4] Melita, L.N.; Węgłowska, K.; Tamburini, D.; Korenberg, C. Investigating the Potential of the Er:YAG Laser for the Removal of Cemented Dust from Limestone and Painted Plaster. *Coatings*, 2020, *10*, 1099.

[5] Bertasa M., Korenberg C. Successes and challenges in laser cleaning of metal artefacts: A review // Journal of Cultural Heritage. 2022. V. 53. P. 100–117.

[6] Cooper M. Laser Cleaning in Conservation: An Introduction. Oxford: Butterworth-Heinemann, 1998.

[7] Abdel-Maksouda, G.; Emamb, H.; Ragab, N.M. From Traditional to Laser Cleaning Techniques of Parchment Manuscripts: A Review Advanced Research in Conservation Science. Adv. Res. Conserv. Sci. 2020, 1, 52–76.

[8] Součková M., Sokolová, M., Neoralová, J., Vávrová, P., Mašková, L., Jandová, V., & Smolík, J. Evaluation methods of effect of cleaning techniques on library collagen materials //Optics for Arts, Architecture, and Archaeology VII. SPIE, 2019. V. 11058. pp. 244-251.

[9] Parfenov V., Galushkin, A., Tkachenko, T., Aseev, V. Laser Cleaning as Novel Approach to Preservation of Historical Books and Documents on a Paper Basis //Quantum Beam Science. 2022. 2022, 6, 23.

[10] Dobrusina S.A., Parfenov V.A., Podgornaya N.I., Samsygina N.D., Titov S.V., Petrov A.A., Aseev V.A. Laser removal of foxings from pages and paper-based documents // Optical Journal. 2023. Vol. 90. N° 10. Pp. 93-107.

Peculiarities of the Influence of Laser Radiation Parameters on the Measurement Error of the Refractive Index of Liquid Media in a Differential Refractometer

Daniil S. Provodin
Peter the Great Saint Petersburg Polytechnic University
Saint Petersburg, Russia
provodindanya@gmail.com

Alexandra D. Kurkova
The Bonch-Bruevich Saint Petersburg State University of Telecommunications
Saint Petersburg, Russia
alexsa99k@gmail.com

Abstract During refractive index measurements of liquid media in the express-control mode using a differential refractometer operating in the refractive index range from 1.23 to 2.63, a number of laser radiation parameters have been identified that affect the measurement uncertainty to varying degrees. In particular, the divergence of the laser beam focused onto the photodiode array plays a significant role, since the refractive index value is determined based on the position of the intensity maximum registered on the array pixels. The wavelength of the laser radiation also influences the measurement error, as different liquid media may exhibit different absorption characteristics at a given wavelength. In addition, adjustment and stabilization of the laser output power affect the measurement accuracy. A series of studies was carried out to investigate the influence of these laser radiation parameters on the measurement error, with the aim of improving the functional capabilities of the differential refractometer.

Keywords — measurement error, refraction, liquid, refractive index, Anderson cuvette, laser radiation, photodiode array.

I. INTRODUCTION

In the context of rapid technological development and increasing anthropogenic impact on the environment, liquid media are exposed to a growing number of adverse factors [1–3]. These factors manifest themselves in a wide range of fields, from scientific research, including specialized facilities, and industrial production to environmental monitoring and control of technological processes [4–7]. Humans consume large amounts of water and products containing it, and the quality of this water, as well as that of other liquid media used in production processes, has a direct impact on human health [8, 9]. The increasing variability of conditions and compositions of the liquids under study necessitates the development of devices capable of providing rapid and reliable analysis in an express-control mode.

To address the challenges of rapid diagnostics, a large number of mobile measuring instruments based on various physical principles have been developed [3, 10–13]. However, modern express-control methods must satisfy several mandatory requirements: the measurement method should not alter the structure of the investigated liquid or change its chemical composition, and measurements must be performed at the temperature of the sample. In practice, only two types of portable devices fully meet these requirements—refractometers and nuclear magnetic resonance spectrometers [2, 10–12, 14–16].

Despite meeting operational requirements, most refractometers, including mobile ones, have limitations that significantly reduce their functionality when working with a wide range of liquids [10, 17–21]. Among the key challenges is the lack of safe methods for analyzing volatile and potentially hazardous compounds, especially when measurements must be carried out over a wide temperature range [18, 19, 21–23]. Additional difficulties arise in the study of hydrocarbon mixtures and other liquid media that do not chemically interact with each other: determining the composition and concentration of components in such systems becomes problematic, particularly during long-term measurements accompanied by temperature variations [19, 21–24].

Previous attempts to address these challenges, including those based on the phenomenon of total internal reflection, have demonstrated the potential of refractometric methods for assessing the state of complex media [19, 21, 22–27]. However, the mobile instrument designs used in these studies are characterized by a limited refractive index measurement range and relatively high measurement uncertainty (up to 0.0002), which is insufficient for reliable analysis of complex mixtures [25, 27–30].

In this regard, an important and timely task is to expand the functional capabilities of a differential refractometer in order to improve measurement accuracy and informational content. This approach makes it possible to determine the composition and concentrations of components in mixtures formed by liquid media that do not undergo chemical interaction upon mixing, based on refractive index measurements.

II. MOBILE DIFFERENTIAL REFRACTOMETER AND ITS OPERATING PRINCIPLE

When measuring the refractive index of liquid media in the express-control mode using a differential refractometer operating in the refractive index range from 1.23 to 2.63, a number of laser radiation parameters have a significant influence on the resulting measurement uncertainty. Each of these parameters determines the accuracy of selecting the pixel used to calculate the refractive index value. Figure 1 shows a block diagram of the differential refractometer prototype incorporating an Anderson cuvette.

979-8-3315-4784-4/26 $31.00 © 2026 IEEE

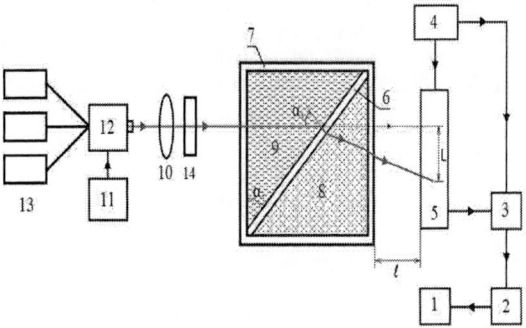

Fig. 1. Block diagram of the laboratory prototype of the mobile differential refractometer: 1 – data display system, 2 – data processing and storage unit, 3 – ADC, 4 – multifunctional power supply, 5 – photodiode array with a set of sensors, 6 – partition in the differential cuvette, 7 – side wall of the differential cuvette, 8 – test liquid medium with varying transparency, 9 – reference liquid, 10 – focusing lens, 11 – tunable power supply unit, 12 – multiplexer, 13 – semiconductor lasers of different wavelengths, 14 – collimator.

A laser beam with a flat, parallel wavefront passes through the cuvette wall at normal incidence and reaches partition 6 (Fig. 1), where the first refraction occurs. The beam then undergoes several successive refractions and exits the cuvette. A linear photodiode array 5 is positioned at a distance 1 from the exit wall of the cuvette, where the position of the optical axis of the laser radiation is recorded. When an Anderson cuvette filled with a liquid is placed in the path of the laser beam, the position of the optical axis shifts along the array by a distance L. A cuvette model was developed and its parameters were calculated (Fig. 2).

Fig. 2. Render of an Anderson cuvette with a rectangular base.

Several Anderson cuvettes were fabricated for the experiments. One of these cuvettes is shown in Fig. 3. For studies involving hazardous liquids or acids, an Anderson cuvette made of quartz should be used. Glass cuvettes can be employed for all other types of liquids. In this respect, the quartz cuvette is more versatile.

Fig. 3. External view of a quartz Anderson cuvette.

The derivation of the relationship for determining the refractive index based on the displacement measured on the photodiode array is described in detail in previous studies [18, 20, 28].

In our measurements aimed at determining the refractive index, a significant factor is the occurrence of total internal reflection at the interface between two media [11, 17, 19, 21]. When this phenomenon occurs, the laser beam does not reach the photodiode array, making signal registration impossible. To avoid such situations and to enable real-time monitoring of the process, we use only visible-range laser radiation for the study of liquid media, allowing the beam path to be visually tracked. When infrared radiation is used, detecting total internal reflection becomes considerably more difficult, which can lead to incorrect assumptions regarding the source of measurement errors. In this regard, Anderson cuvettes of various geometrical dimensions were fabricated for the experiments.

One of the key parameters is the beam cross-section (spot size) of the laser, which is focused onto the photodiode array. The beam size determines the width and shape of the light intensity distribution. If the beam is too large, the intensity maximum becomes "spread out," making it difficult to accurately determine the peak position. Conversely, a very small beam increases sensitivity to mechanical vibrations, surface inhomogeneities, and diffraction effects. Therefore, optimizing the beam size is critical for minimizing the uncertainty in selecting the pixel corresponding to the refractive index of the studied medium.

Equally important is the wavelength of the laser radiation. Since different liquids exhibit different absorption spectra, the choice of wavelength affects the transmission of the medium. If the wavelength falls within a region of high absorption, the signal on the photodiode array decreases, leading to increased noise and reduced contrast of the intensity distribution. This, in turn, increases the uncertainty in determining the position of the laser intensity maximum on the photodiode array. Moreover, the refractive index itself depends on the wavelength (dispersion), imposing additional requirements on the stability and monochromaticity of the laser.

Another significant parameter is the laser output power, which must be stabilized throughout the measurement cycle. Insufficient power reduces the signal-to-noise ratio and increases the probability of errors in determining the peak position on the array. Excessive power can lead to photodiode saturation, nonlinearity in response, and distortion of the intensity peak profile. Furthermore, power fluctuations caused by temperature variations or laser drift also negatively affect measurement stability.

III. INVESTIGATION OF LIQUID MEDIA AND THEIR MIXTURES AT DIFFERENT WAVELENGTHS AND DISCUSSION

Figures 4 and 5 show the results of the study of AI-92 gasoline and a mixture of three gasolines—AI-98, AI-95, and AI-92—at specific concentrations. In the first case, AI-92 is a commonly used gasoline, which is often stored under conditions that do not meet recommended standards. The second case occurs when expired AI-95 and small amounts of AI-92 are added to AI-98 gasoline. Such a mixture can cause accelerated wear of pistons and cylinders in certain classes of vehicles and also increases the time required for the vehicle to reach its operating speed during engine start-up.

Fuel monitoring for detecting such cases requires high accuracy and broad functionality. Therefore, the measurements were conducted at three different wavelengths.

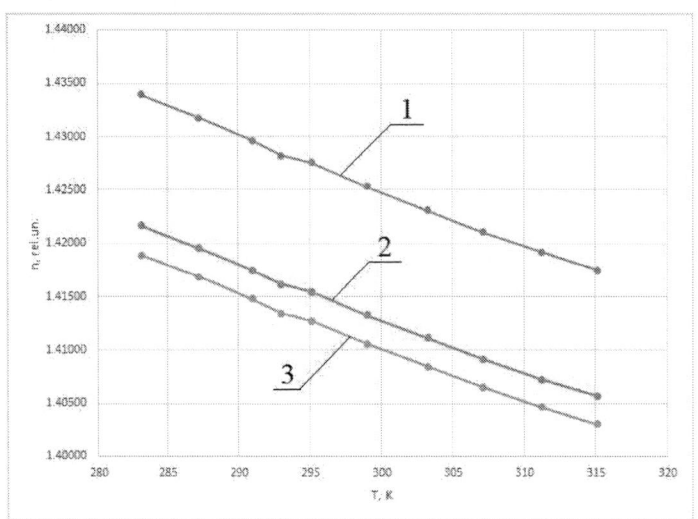

Fig. 4. Study of the refractive index n of AI-92 gasoline as a function of temperature T. Curves 1, 2, and 3 correspond to laser wavelengths λ (nm) of 436.4, 589.3, and 657.2, respectively.

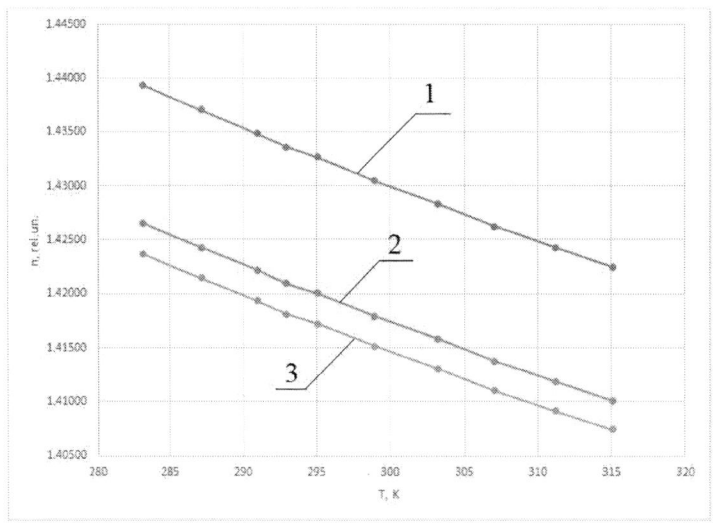

Fig. 5. Study of the refractive index n of a gasoline mixture (AI-98:AI-95:AI-92 in an 8:1:1 ratio) as a function of temperature T. Curves 1, 2, and 3 correspond to laser wavelengths λ (nm) of 436.4, 589.3, and 657.2, respectively.

The analysis of the obtained data indicates that the behavior of the n(T) dependencies for different wavelengths is consistent with results reported by other researchers [11, 19, 21, 27, 29]. This confirms the correct implementation of measurements using three laser wavelengths and the adjustment of laser power to achieve maximum pixel resolution on the photodiode array.

To further validate the reliability of the refractive index measurements, considering the noted features in the design of the differential refractometer, AI-92 gasoline stored under substandard conditions and a mixture of three gasolines were investigated. Measurements were performed using both the differential refractometer and the industrial Abbemat WR/MW refractometer. The results of the comparison between the two measurement methods are presented in Tables I and II.

TABLE I.   INVESTIGATION OF THE REFRACTIVE INDEX OF AI-92 GASOLINE AS A FUNCTION OF TEMPERATURE T.

T, K	Mobile differential refractometer			Industrial Abbemat WR/MW refractometer		
	λ, nm			λ, nm		
	436.4	589.3	657.2	436.4	589.3	657.2
283.1 ± 0.1	1.4393 ± 0.0001	1.4269 ± 0.0001	1.4241 ± 0.0001	1.43924 ± 0.00005	1.42682 ± 0.00005	1.42406 ± 0.00005
287.2 ± 0.1	1.4371 ± 0.0001	1.4247 ± 0.0001	1.4219 ± 0.0001	1.43709 ± 0.00005	1.42466 ± 0.00005	1.42187 ± 0.00005
291.0 ± 0.1	1.4349 ± 0.0001	1.4225 ± 0.0001	1.4198 ± 0.0001	1.43489 ± 0.00005	1.42246 ± 0.00005	1.41970 ± 0.00005
293.0 ± 0.1	1.4337 ± 0.0001	1.4214 ± 0.0001	1.4187 ± 0.0001	1.43368 ± 0.00005	1.42131 ± 0.00005	1.41864 ± 0.00005
295.1 ± 0.1	1.4326 ± 0.0001	1.4204 ± 0.0001	1.4177 ± 0.0001	1.43257 ± 0.00005	1.42034 ± 0.00005	1.41765 ± 0.00005
299.0 ± 0.1	1.4305 ± 0.0001	1.4183 ± 0.0001	1.4156 ± 0.0001	1.433045 ± 0.00005	1.41826 ± 0.00005	1.41554 ± 0.00005
303.2 ± 0.1	1.4283 ± 0.0001	1.4162 ± 0.0001	1.4135 ± 0.0001	1.42828 ± 0.00005	1.41615 ± 0.00005	1.41347 ± 0.00005
307.1 ± 0.1	1.4263 ± 0.0001	1.4142 ± 0.0001	1.4115 ± 0.0001	1.42626 ± 0.00005	1.41427 ± 0.00005	1.41147 ± 0.00005
311.2 ± 0.2	1.4243 ± 0.0001	1.4122 ± 0.0001	1.4096 ± 0.0001	1.42425 ± 0.00005	1.41215 ± 0.00005	1.40954 ± 0.00005
315.1 ± 0.1	1.4225 ± 0.0001	1.4105 ± 0.0001	1.4079 ± 0.0001	1.42246 ± 0.00005	1.41045 ± 0.00005	1.40786 ± 0.00005

TABLE II.   INVESTIGATION OF THE REFRACTIVE INDEX OF A GASOLINE MIXTURE (AI-100, AI-98, AND AI-95 IN AN 8:1:1 RATIO) AS A FUNCTION OF TEMPERATURE T.

T, K	Mobile differential refractometer			Industrial Abbemat WR/MW refractometer		
	λ, nm			λ, nm		
	436.4	589.3	657.2	436.4	589.3	657.2
283.1 ± 0.1	1.4365 ± 0.0001	1.4240 ± 0.0001	1.4212 ± 0.0001	1.43645 ± 0.00005	1.42394 ± 0.00005	1.42116 ± 0.00005
287.2 ± 0.1	1.4344 ± 0.0001	1.42209 ± 0.0001	1.4193 ± 0,0001	1.43436 ± 0.00005	1.42087 ± 0.00005	1.41926 ± 0.00005
291.0 ± 0.1	1.4323 ± 0.0001	1.4200 ± 0.0001	1.4172 ± 0.0001	1.43225 ± 0.00005	1.41995 ± 0.00005	1.41718 ± 0.00005
293.0 ± 0.1	1.4307 ± 0.0001	1.4184 ± 0.0001	1.4157 ± 0.0001	1.43067 ± 0,00005	1.41834 ± 0.00005	1.41564 ± 0.00005
295.1 ± 0.1	1.4302 ± 0.0001	1.4178 ± 0.0001	1.4152 ± 0.0001	1.43015 ± 0.00005	1.41776 ± 0.00005	1.41513 ± 0.00005
299.0 ± 0.1	1.4279 ± 0.0001	1.4157 ± 0.0001	1.4130 ± 0.0001	1.42786 ± 0.00005	1.41567 ± 0.00005	1.41297 ± 0.00005
303.2 ± 0.1	1.4258 ± 0.0001	1.4137 ± 0.0001	1.4110 ± 0.0001	1.42578 ± 0.00005	1.41367 ± 0.00005	1.41097 ± 0.00005
307.1 ± 0.1	1.4238 ± 0.0001	1.4118 ± 0.0001	1.4091 ± 0.0001	1.42373 ± 0.00005	1.41176 ± 0.00005	1.40908 ± 0.00005
311.2 ± 0.2	1.4220 ± 0.0001	1.4099 ± 0.0001	1.4073 ± 0.0001	1.42198 ± 0.00005	1.40987 ± 0.00005	1.40723 ± 0.00005
315.1 ± 0.1	1.4203 ± 0.0001	1.4082 ± 0.0001	1.4056 ± 0.0001	1.42025 ± 0.00005	1.40812 ± 0.00005	1.40556 ± 0.00005

The analysis of the data from the two sets of refractive index measurements showed that the obtained values were consistent within the respective measurement uncertainties.

IV. CONCLUSION

The results of the study confirmed the necessity of taking the laser wavelength into account when measuring the refractive index. In some cases, switching from one wavelength to another allows measurements to be conducted, particularly in mixtures, without adjusting the laser power. This is especially critical for hydrocarbons, where the media do not chemically interact with each other.

As a result of the conducted studies, an important feature for differential refractometry was identified. The use of a collimator to form a flat wavefront largely eliminates the effect of laser beam divergence. Measurements are performed based on the position of the optical axis on the photodiode array. In a flat wavefront, all rays are refracted uniformly, and the influence on the optical axis power is negligible, unlike in other refractometers that use a diverging laser beam [11, 17, 19, 21, 25, 30].

It should be noted that all functional capabilities of the mobile differential refractometer are preserved when using multiple wavelengths in the design developed in this work. Laser power, which affects beam divergence when increased, is an important parameter for measurements of poorly transparent media. The conducted studies showed that it is necessary to provide a power adjustment range from 0.5 to 25 mW by varying the current supplied to the semiconductor structure. This range is sufficient for investigating most media and their mixtures (while considering the dimensions of the refractometer). In cases of high turbidity of the studied medium, a smaller Anderson cuvette is used (reducing the optical path of the laser beam in the medium). All these measures ensure a refractive index measurement uncertainty of less than 0.0001 within the measurement range of 1.23 to 2.63, which is sufficient for express-control applications.

REFERENCES

[1] D. Nikolaev, and et. al. "Determining the location of an object during environmental monitoring in conditions of limited possibilities for the use of satellite positioning," IOP Conference Series: Earth and Environmental Science, vol. 578(1), pp. 012052. *July 2020*.

[2] E. Rukin, and et. al. "The development of a new method for making justified decisions by municipal authorities in the management of territories on the basis of the results of the environmental express-control of the state of various media," MATEC Web of Conferences, vol. 2455, pp. 12002. December 2018.

979-8-3315-4784-4/26 $31.00 © 2026 IEEE

[3] A. Karseev, and V. Vologdin, "Features of nuclear magnetic resonance signals registration in weak magnetic fields for express - Control of biological solutions and liquid medium by nuclear magnetic spectroscopy method," Journal of Physics: Conference Series, vol. 643(1), pp. 012108. December 2015.

[4] V. I. Antonov, and et. al. "Simulation of water flow management by the flood control facilities in the adjacent river basins," Journal of Physics: Conference Series, vol. 1400(7), pp. 077049, November 2019.

[5] V. V. Davydov, "Some specific features of the NMR study of fluid flows Optics and Spectroscopy (English translation of Optika i Spektroskopiya)," vol. 121(1), pp. 18–24, Jule 2016.

[6] N. Myazin, and et. al. "On the need for express control of the quality of consumer goods within the concept 'Internet of things,'" IOP Conference Series: Materials Science and Engineering, vol. 497, pp. 012111. April 2019.

[7] A. Moroz, and et. al. "On the possibility of growing vegetables and fruits on the lunar base," IOP Conference Series: Earth and Environmental Science, vol. 578(1), pp. 012006, November 2020.

[8] M. S. Mazing, and et. al. "Monitoring of oxygen supply of human tissues using a noninvasive optical system based on a multi-channel integrated spectrum analyzer," International Journal of Pharmaceutical Research, vol. 12, pp. 1974 – 1978. December 2020.

[9] R. Davydov, and et. al. "New Methodology of Human Health Express Diagnostics Based on Pulse Wave Measurements and Occlusion Test" Journal of Personalized Medicine, vol. 13(3), pp. 443. March 2023.

[10] A. Cheremiskina, and et. al. "Express-control of biological solutions by portable nuclear-magnetic spectrometer," Journal of Physics: Conference Series, vol. 541, pp. 12014. March 2014.

[11] G. V. Stepanenkov, D. V. Vakorina, and B. K. Reznikov, "Features of Control of Condensed Media in Visible Light," Proceedings of the Seminar on Fields, Waves, Photonics and Electro-Optics: Theory and Practical Applications, FWPE 2023, vol. 2023, pp. 139–142. November 2023.

[12] V. V. Davydov, and et. al. "On the Possibility of Express Recording of Nuclear Magnetic Resonance Spectra of Liquid Media in Weak Fields," Technical Physics, vol. 63(12), pp. 1845–1850. December 2018.

[13] V. Dyumin, and et. al. "Charge-coupled Device with Integrated Electron Multiplication for Low Light Level Imaging," Proceedings of the 2019 IEEE International Conference on Electrical Engineering and Photonics, EExPolytech 2019, vol. 8906868, pp. 308–310. October 2019.

[14] M. A. Sadovnikova, and et. al. "HYSCORE Spectroscopy to Resolve Electron–Nuclear Structure of Vanadyl Porphyrins in Asphaltenes from the Athabasca Oil Sands In Situ Conditions," Energies, vol. 15(17), pp. 6204. September 2022.

[15] R. Davydov, and et. al. "The Nuclear Magnetic Flowmeter for Monitoring the Consumption and Composition of Oil and Its Complex Mixtures in Real-Time," Energies, vol. 15(9), pp. 3259. May 2022.

[16] V. V. Davydov, V. I. Dudkin, and A. Yu. Karseev, "A Compact Nuclear Magnetic Relaxometer for the Express Monitoring of the State of Liquid and Viscous Media," Measurement Techniques, vol. 57(8), pp. 912–918. September 2014.

[17] N. M. Grebenikova, R. V. Davydov, and V. Y. Rud, "Features of the signal registration and processing in the study of liquid flow medium by the refraction method," Journal of Physics: Conference Series, vol. 1326(1), pp. 012012. December 2019.

[18] D. S. Provodin, and M. A. Yakusheva, "Development a Mobile Differential Refractometer for Express Control of Liquid Medium State with Refractive Index Measuring Range from 1.23 to 2.63," Proceedings of the 2024 Conference of Young Researchers in Electrical and Electronic Engineering, ElCon 2024, vol. 2024, pp. 745–749. January 2024.

[19] G. V. Stepanenkov, D. V. Vakorina, and B. K. Reznikov, "A New Method of Express Control in Fuel Industry by Using Small-Sized Mobile Refractometer," Proceedings of the 2024 Conference of Young Researchers in Electrical and Electronic Engineering, ElCon 2024, vol. 2024, pp. 773–776. January 2024.

[20] V. V. Davydov, and et. al. "A new method for describing the change of the laser beam axis trajectory in the Anderson differential cuvette to measure the refractive index of liquids," Computer Optics, vol. 48(2), pp. 217–224. April 2024.

[21] M. A. Karabegov, "A method of increasing the accuracy of analytical instruments by structural correction," Measurement Techniques, vol. 52(10), pp. 1126–1133, October 2009.

[22] N. M. Grebenikova, and et. al. "Features of monitoring the state of the liquid medium by refractometer," Journal of Physics: Conference Series, vol. 1135(1), pp. 012055, December 2018.

[23] V. V. Davydov, D. I. Nikolaev, and A. V. Moroz, "Design of a Flow-Through Refractometer for Monitoring the State of Transparent Media with a Cylindrical Insert in the Form of a Vertical Section of a Pipeline," Measurement Techniques, vol. 64(4), pp. 305–313. May 2021.

[24] V. V. Davydov, N. M. Grebenikova, and K. Y. Smirnov, "An Optical Method of Monitoring the State of Flowing Media with Low Transparency That Contain Large Inclusions," Measurement Techniques, vol. 62(6), pp. 519–526. June 2019.

[25] M. A. Karabegov, "Metrological and technical characteristics of total internal reflection refractometers," Measurement Techniques, vol. 47(11), pp. 1106-1112, December 2004.

[26] N. M. Grebenikova, and et. al. "The optical method for condition control of flowing medium," Journal of Physics: Conference Series, vol. 1124(4), pp. 041011. December 2018.

[27] V. V. Davydov, D. V. Vakorina, and G. V. Stepanenkov, "A new optical method for control in visible light of volatile hydrocarbon media and their mixtures using data from light-shadow boundary images," Computer Optics , vol. 48(1), pp. 93–101. January 2024.

[28] D. S. Provodin, "New technique for control of liquid media state by optical method in express mode," St. Petersburg State Polytechnical University Journal: Physics and Mathematics, vol. 15(3.2), pp. 124-129. December 2022.

[29] V. Davydov, and et. al. "New Method for State Express Control of Unstable Hydrocarbon Media and Its Mixtures," Energies, vol. 16(6), pp. 2529. May 2023.

[30] G. Morales-Luna, and et. al. "Plasmonic biosensor based on an effective medium theory as a simple tool to predict and analyze refractive index changes," Optics and Laser Technology, vol. 131, pp. 106332. October 2020.

Compact Cross-Pumped 2-μm Laser with Passive Q-Switching by Glass:PbS Nanocrystals

Mhmad Salhab
Department of Photonics
Saint Petersburg Electrotechnical
University "LETI"
Saint Petersburg, Russia
mhmadsalhab@gmail.com

Tatiana Zotova
JSC «Scientific Production Association
State Optical Institute Named after
Vavilov S.I.»
Saint Petersburg, Russia
bizina0802@mail.ru

Vladimir Ivanov
Department of Photonics
Saint Petersburg Electrotechnical
University "LETI"
Saint Petersburg, Russia
vnivan@mail.ru

Aleksei Onushchenko
JSC «Scientific Production Association
State Optical Institute Named after
Vavilov S.I.»
Saint Petersburg, Russia
alarkon@mail.ru

Aleksandr Titov
JSC «Scientific Production Association
State Optical Institute Named after
Vavilov S.I.»
Saint Petersburg, Russia
alextitov57@mail.ru

Anastasia Vasilieva
Department of Photonics
Saint Petersburg Electrotechnical
University "LETI"
Saint Petersburg, Russia
anastasiastru@mail.ru

Abstract— **This work presents an investigation of effective passive Q-switching of a 2-μm laser using glass doped with PbS nanocrystals and compares its performance with active Q-switching modes. The laser gain media were based on Tm, Ho-doped KY(WO₄)₂ and KLu (WO₄)₂ crystals. An electro-optical Q-switcher employing an RKTP crystal was used for active modulation, while passive Q-switching was achieved using glass: PbS nanocrystal samples. The compact laser system was pumped in a cross-configuration by laser diode modules emitting at 802–805 nm. The Experimental results demonstrate that PbS nanocrystal-doped glass enables stable passive Q-switching, confirming its potential for compact, efficient mid-infrared laser systems.**

Keywords—Q-switching, PbS quantum dots, 2.1 μm laser, active Q-switching, passive Q-switching, Tm,Ho-doped tungstate crystals, RKTP, electro-optic modulator, mid-infrared laser, cross-pumped laser,

I. INTRODUCTION

Laser sources operating near 2.0–2.1 μm are of significant interest due to their relevance in eye-safe applications, including rangefinding, LIDAR, atmospheric sensing, medical diagnostics, and free-space optical communication. The 2.1 μm spectral region lies within a favorable atmospheric transmission window and exhibits strong absorption in water, making it both biologically safe and well-suited for environmental sensing [1].

A widely adopted approach for generating radiation in this region involves co-doping host crystals with thulium (Tm³⁺) and holmium (Ho³⁺) ions. Tm³⁺ ions exhibit efficient absorption in the 801–808 nm range, enabling direct pumping by commercially available laser diodes. Due to a cross-relaxation mechanism, one absorbed photon can excite up to 1.6 Tm³⁺ ions, thereby improving quantum efficiency [2]. The excitation energy is subsequently transferred to Ho³⁺ ions, which emit in the 2.05–2.1 μm range—an ideal band for atmospheric propagation [3].

Crystals such as KY(WO₄)₂ and KLu(WO₄)₂ are frequently used as laser hosts due to their high rare-earth solubility, broad emission bandwidths, and favorable thermal and optical properties [4]. These hosts support efficient cross-pumped diode geometries, enabling compact and thermally stable laser configurations.

Short-pulse generation in such systems is often accomplished through active Q-switching using electro-optic

devices such as RKTP-based modulators. While this method remains effective, passive Q-switching is increasingly favored for compact and cost-sensitive designs due to its simpler setup, lower power requirements, and elimination of driver electronics [5].

Among the materials investigated for passive Q-switching at 2 μm, PbS quantum dot-doped silicate glass has shown particular promise. These glasses exhibit excitonic absorption near the laser wavelength and have been successfully implemented in passively Q-switched Ho³⁺-doped and Tm³⁺,Ho³⁺ co-doped lasers [6,7].

Passive Q-switches offer additional advantages: they are more compact, cost-effective, and free from induced electrical noise. However, there is currently a lack of commercially available passive optical shutters that operate efficiently near the 2.1 μm wavelength and match the performance of electro-optic modulators. This gap motivated a comparative investigation into the performance of electro-optic Q-switches versus passive Q-switching using custom-fabricated potassium silicate glass samples embedded with PbS quantum dots.

This work investigates the use of PbS-doped silicate glass as a passive Q-switch in cross-pumped Tm,Ho-doped lasers near 2.1 μm, with its performance compared to active Q-switching modes.

II. MATERIALS AND METHODS

Active Q-switching was implemented using a quarter-wave electro-optic modulator (EOM) composed of two RKTP crystals. A quartz glass plate, positioned at the Brewster angle within the resonator, served as a polarizer. The resonator consisted of two mirrors. The output mirror had a reflectivity of 80% and was flat, while the high-reflectivity (HR) mirror was spherical with a radius of curvature of 2.4 m.

For the experimental studies, laser crystals of KY(WO₄)₂:Tm,Ho (KYW:Tm,Ho) and KLu(WO₄)₂:Tm,Ho (KLuW:Tm,Ho) were grown. Laser elements were fabricated from these crystals with dimensions of Ø1.6 × 15 mm for KYW:Tm,Ho and Ø2.2 × 20 mm for KLuW:Tm,Ho. Each element was enclosed in a leucosapphire tube coated with a silver reflective layer. An immersion liquid layer, approximately 0.1 mm thick, was introduced between the tube and the active element to enhance thermal conductivity from the gain medium to the surrounding laser housing.

In both configurations, side-pumping was employed using three laser diode modules operating at wavelengths of 802–805 nm, arranged symmetrically at 120° intervals around the tube containing the active element. The maximum output power of each diode module was 1.6 kW for KYW:Tm,Ho and 2.1 kW for KLuW:Tm,Ho. The pump pulse duration was adjustable within the range of 50 to 450 µs.

The pump delivery system was designed to accommodate both the anisotropic absorption characteristics of the laser crystals and the polarization of the diode laser emission. Two optical schemes were tested for delivering pump radiation from the laser diodes to the active elements. In the first configuration, mirror-based concentrators were employed. For the KLuW:Tm,Ho laser, a hybrid scheme was implemented: cylindrical lenses were used to manage the lower divergence axis of the diode modules, while parallel plates with reflective coatings addressed the higher divergence axis. The length of the KLuW:Tm,Ho active element was matched to the length of the pump module. The illumination slits on the silver-coated tube were dimensioned to match the pump radiation stripe formed by the concentrator or cylindrical lens and were approximately equal to the diameter of the active element. The pump energy was varied in the range of 1 to 2.5 J. Experimental results showed that the hybrid pumping configuration offered superior efficiency.

The 20 mm length of the KLuW:Tm,Ho active element was chosen specifically to ensure stable Q-switched operation at high power densities of the generated radiation. Double tungstate crystals, such as KYW and KLuW, are known to have low Raman scattering gain and exhibit favorable resistance to optical loading within the resonator [8]. Increasing the length of the active element from 15 mm (KYW:Tm,Ho) to 20 mm (KLuW:Tm,Ho) made it possible to completely suppress optical damage phenomena, such as surface burnouts, across all resonator elements, including the electro-optic modulator (EOM).

For passive Q-switching, glass samples were used with an absorption maximum at 2.06 µm, corresponding to the first excitonic absorption peak of PbS quantum dots. Polished plates of 0.8 mm thickness with varying initial transmittance levels were fabricated from this glass.

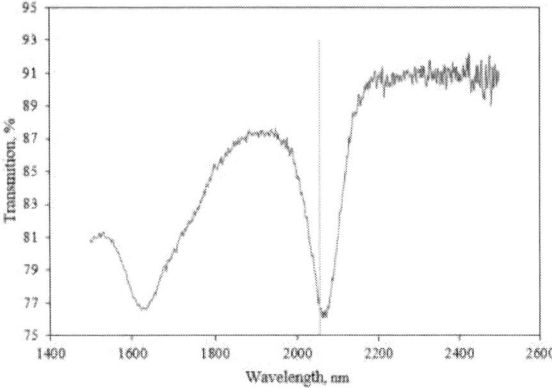

Fig. 1. Transmission spectrum of the glass sample containing PbS nanocrystals. The red line indicates the laser emission wavelength at λ = 2056 nm.

Due to the absence of anti-reflection coatings on the surfaces of the passive Q-switch plates, they were positioned in the resonator at the Brewster angle with respect to the laser's optical axis. Under normal incidence, Fresnel reflection losses from both surfaces amounted to 9.29%, corresponding to a refractive index of n = 1.558. This relatively high refractive index is attributed to the elevated lead content in the glass. The initial transmittance of the passive shutter was $T_0 = 84.6\%$ when accounting for Fresnel losses; excluding these losses, the intrinsic transmittance was $T_0 = 76.7\%$.

III. RESULTS AND DISCUSSION

A. Active Q-Switching in KLu(WO₄)₂:Tm,Ho at 2.061 µm

Under single-pulse operation with active Q-switching using an RKTP electro-optic modulator, the KLuW:Tm,Ho laser produced pulse energies exceeding 40 mJ at a central wavelength of 2.061 µm, with a full-width at half-maximum (FWHM) pulse duration of 22 ns This level of pulse energy is consistent with high-performance active Q-switched solid-state lasers operating near 2 µm, where active modulation enables efficient extraction of stored energy in the gain medium. The high output energy observed in this work demonstrates both effective cross-pumping and superior resonator design.

B. Single-Mode Generation in KY(WO₄)₂:Tm,Ho at 2.056 µm

For the KYW:Tm,Ho laser operated in single-longitudinal-mode (SLM) at a wavelength of 2.056 µm, pulse energies up to 15 mJ were obtained with a 25 ns (FWHM) pulse duration. The SLM operation ensures spectral purity and stability, which are important for applications such as coherent LIDAR and high-resolution sensing, and confirms that the cross-pumped configuration with carefully managed resonator losses supports efficient energy extraction in this crystal host.

C. Passive vs. Active Q-Switching in KYW:Tm,Ho

Comparative experiments were performed on the KYW:Tm,Ho laser in passive and active Q-switching configurations under identical pumping conditions (equal pump energy and pump-pulse duration). In the passive regime, a PbS–quantum-dot–doped glass saturable absorber enabled pulse energies up to 6 mJ with a 22 ns FWHM duration. Under active Q-switching with the RKTP-based electro-optic modulator, the same laser produced 7–8 mJ pulses with a comparable duration of approximately 25 ns.

When placed in the context of prior work, the achieved mJ-level passive Q-switching is notably higher than most compact 2-µm passively Q-switched systems reported in the literature. Early demonstrations of PbS-doped glass as a saturable absorber in diode-pumped Tm:KYW lasers reported pulse energies on the order of tens of µJ (for example, 44 µJ at 2.5 kHz) [9] rather than millijoules. Likewise, passively Q-switched double-tungstate microchip Tm,Ho lasers using established mid-IR saturable absorbers such as Cr:ZnS typically operate in the µJ pulse-energy range (single-digit to ~10 µJ, with nanosecond durations) due to the short cavities and limited stored energy [10].

At the same time, mJ-level energies in passively Q-switched operation have been demonstrated in higher-power bulk architectures and different gain media. For instance, a Cr²⁺:ZnS-passively-Q-switched Ho:YAG ceramic laser reported up to 0.6 mJ pulse energy with a 29 ns pulse width at 24.4 kHz [11]. In general, published side-by-side comparisons frequently demonstrate a clear difference between passive and active Q-switching, where passive Q-switching stays at the µJ level while active Q-switching reaches the mJ regime (e.g., 10.3 mJ active versus 23.31 µJ passive in a Tm,Ho:YAP system using an AO Q-switch and a MoS_2 saturable absorber mirror) [12].

IV. Conclusion

This study demonstrates the successful development of a compact laser system operating near 2.1 µm using cross-pumped Tm,Ho-doped tungstate crystals. Both active and passive Q-switching modulations were evaluated. Notably, the implementation of a passive Q-switch based on PbS quantum dot-doped silicate glass enabled pulse energies of up to 6 mJ with a 22 ns (FWHM) duration—performance that approaches that of active electro-optic modulation in the same cavity configuration.

In contrast to many previously reported passively Q-switched 2 µm lasers—where output energies often remain in the microjoule range—this work demonstrates that PbS quantum dot-doped glass, when employed in a properly engineered KYW:Tm,Ho cavity, can achieve significantly higher pulse energies, bridging the gap toward active Q-switching levels. These findings are especially important for the development of compact mid-infrared pulsed lasers, where passive Q-switches are favored for their simplicity, cost-efficiency, and ease of integration.

However, it should be noted that direct comparison across different studies requires careful consideration of parameters such as resonator length, output coupling, mode area on the saturable absorber, repetition rate, and total intracavity losses—all of which critically influence achievable pulse energy.

Acknowledgment

The authors would like to thank Ph.D. Volinkin V. M. for providing the immersion liquid used to fix the active element and for the valuable consultations.

References

[1] Y. Y. Kalisky, "Two Micron Lasers: Holmium and Thulium Doped Crystals," Bellingham, WA, USA: International Society for Optics and Photonics, 2006, pp. 125–140, doi:10.1117/3.660249.CH9.

[2] P. Koopmann, R. Peters, K. Petermann, and G. Huber, "Highly efficient, broadly tunable Tm:Lu$_2$O$_3$ Laser at 2 µm," European Quantum Electronics Conference, p. 1, Jun. 2009, doi: 10.1109/CLEOE-EQEC.2009.5196613.

[3] G. A. Newburgh and M. Dubinskii, "Resonantly diode pumped Ho³⁺:YVO₄ 2.1 µm laser," Proceedings of SPIE - The International Society for Optical Engineering, vol. 8039, p. 803905, May 2011, doi: 10.1117/12.886571.

[4] V.N. Ivanov et al., "2 µm KYW:Tm,Ho Laser with Transverse Diode Pumping for Passive Q-Switching." 2022 International Conference Laser Optics (ICLO), pp. 1–1, Jun. 2022, doi: 10.1109/iclo54117.2022.9840131.

[5] Y. Zhao et al., "The passive Q-switching Tm, Ho co-doped GAGG state laser operating near 2.1µm," Optical Materials, vol. 129, p. 112499, Jul. 2022, doi: 10.1016/j.optmat.2022.112499.

[6] M. S. Gaponenko et al., "Passive Q-switching of 2-µm holmium lasers with PbS-quantum dot-doped glass," Proceedings of SPIE, vol. 6054, p. 605403, Dec. 2005, doi: 10.1117/12.660492.

[7] P. Loiko et al., "Ho:KLu(WO₄)₂ Microchip Laser Q-Switched by a PbS Quantum-Dot-Doped Glass," IEEE Photonics Technology Letters, vol. 27, no. 17, pp. 1795–1798, Sep. 2015, doi: 10.1109/LPT.2015.2439576.

[8] I. F. Balashov, B. G. Berezin, V. N. Ivanov, A. V. Lukin, V. D. Belyaev, O. B. Storoshchuk, and L. V. Osolikhin, "Pulsed solid-state laser," USSR State Patent SU 1819489, Sep. 25, 1990.

[9] M. S. Gaponenko et al., "Diode-pumped Tm:KY(WO₄)₂ laser passively Q-switched with PbS-doped glass," Applied Physics B, vol. 93, no. 4, pp. 787–791, Nov. 2008, doi: 10.1007/s00340-008-3266-1.

[10] J. M. Serres et al., "Passive Q-switching of a Tm,Ho:KLu(WO₄)₂ microchip laser by a Cr:ZnS saturable absorber," Applied Optics, vol. 55, no. 14, pp. 3757–3757, May 2016, doi: 10.1364/ao.55.003757.

[11] B. Yao et al., "High-power Cr²⁺:ZnS saturable absorber passively Q-switched Ho:YAG ceramic laser and its application to pumping of a mid-IR OPO," Optics Letters, vol. 40, no. 3, pp. 348–351, Jan. 2015, doi:10.1364/OL.40.000348.

[12] L. Li et al., "Active/passive Q switching operation of 2 µm Tm,Ho:YAP laser with an acousto optical Q switch/MoS₂ saturable absorber mirror," Photonics Research, vol. 6, no. 6, pp. 614–619, May 2018, doi:10.1364/PRJ.6.000614.

Functional Automatic Laser Beam Alignment System Capabilities

Sergey A. Shagako
The Bonch-Bruevich Saint Petersburg State University of Telecommunications
St. Petersburg, Russia
s.shagako15@yandex.ru

Diana D. Shagako
The Bonch-Bruevich Saint Petersburg State University of Telecommunications
St. Petersburg, Russia
diana.diana.tsyganova@mail.ru

Vyacheslav G. Nesterov
The Bonch-Bruevich Saint Petersburg State University of Telecommunications
St. Petersburg, Russia
nesterov.vyacheslav7@gmail.com

Abstract — This study addresses the core problem in Free-Space Optics (FSO): the misalignment of the laser beam and photodetector axes during signal transmission. Optimal parameters were determined for the plates in a developed beam steering system that employs an electro-optic mechanism, where control voltages modulate the refractive index of the optical material. The functional capabilities of the developed system for the precise alignment of the laser beam axis via the Pockels effect are described. The influence of plate temperature on the developed system performance of has been determined and compensation methods for thermal effects are proposed.

Keywords — automatic laser radiation adjustment system atmospheric optical communication channel, free-space optic communication lines, laser radiation.

I. INTRODUCTION

In today's world, communication networks play a crucial role in the development of infrastructure [1-3]. However, due to high traffic loads, networks face significant operational challenges. The increase in traffic exacerbates this problem, and current data transmission methods are unable to completely solve it [4-7]. The only viable solution remains the construction of new networks, but in densely populated urban areas, this is accompanied by a number of challenges.

The main challenge is that the existing underground utility infrastructure is already overloaded (and often in a poor condition). Any further work in this area would increase the risk of further damage. Therefore, installing new cables would require a large-scale reconstruction of the infrastructure, which would be extremely costly and time-consuming.

Simultaneously, the use of aerial cables with optical fiber is not feasible in urban areas due to their negative impact on the aesthetic appeal, environmental impact, and high probability of failure in adverse weather conditions. Free-Space Optics (FSO), which transmit data using laser radiation through the air over distances up to 200 kilometers or more, could be a promising alternative for urban environments [8-11]. FSO are also suitable for mountainous areas where installing traditional cables is difficult, as the communication range in such areas can extend up to 2,000 meters.

Atmospheric optical channels have been actively implemented at industrial enterprises, airports, and seaports [12, 13], where laying additional cables involves high costs and technical challenges.

During the operation of FSO lines, several negative factors inevitably occur, as noted in [14-16], the impact of which can be partially mitigated but not fully eliminated. These factors present a significant challenge for the stable functioning of the system and require special attention during the design and operation of FSO.

The most significant issues are:

Low-frequency vibrations in structures:

- Rocking of buildings and racks housing AOLS equipment due to wind loads, vibrations from heavy vehicles or industrial facilities;
- Microseismic oscillations of the earth's surface;
- Resonance phenomena in building structures.

Deformation processes in building structures:

- Gradual settlement of building foundations, particularly noticeable in the early years after construction;
- Seasonal ground movements (heaving in winter);
- Temperature deformations in load-bearing components.

These issues lead to a critical consequence for system operation: displacement of the optical axis of the laser beam from the center of the photodetector's photosensitive area. This results in:

- Partial or complete loss of transmitted data;
- Reduced stability of the communication link;
- Deterioration of main transmission parameters.

This issue is particularly acute during the autumn-winter months and during periods of rainfall, when the amplitude of structural vibrations increases due to increased wind loads, more pronounced temperature-induced deformations occur, and corrosion processes and fastener weakening increase.

A new solution to the problem of optical axis stabilization in atmospheric communication lines was proposed in [14, 15]. The developed system of automatic adjustment of the laser beam position is described in detail in [14, 15].

The system shown in fig. 1 demonstrates an approach to compensate for optical axis deviations caused by external influences. The system is based on four optical glass plates of a special design, which, when installed in a certain configuration, allow precise adjustment of the direction of laser radiation.

979-8-3315-4784-4/26 $31.00 © 2026 IEEE

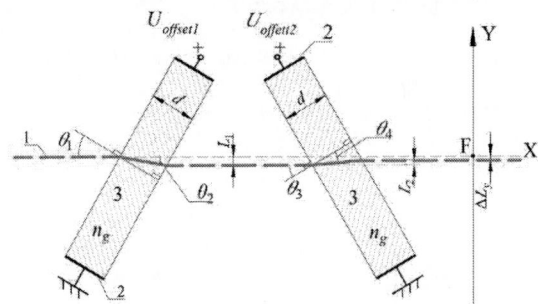

Fig. 1. Schematic diagram of the developed automatic electro-optical system: 1 – Laser radiation; 2 – control flat electrodes with a connected constant voltage source; 3 – a plate made of optical material.

II. DIMENSIONS OF THE GLASS PLATES

A. Plate width

To determine the required adjustment range for the optical axis position of laser radiation, we will consider the most critical operational case where the receiving equipment is mounted (Fig. 2) atop an antenna-mast structure (AMS).

In accordance with the current Russian regulatory requirements, specifically SP 70.13330.2012 "Load-Bearing and Enclosing Structures" (an updated version of SNiP 3.03.01-87), the maximum allowable vertical deviation of the top of these structures is regulated at 1/1000 of their total height.

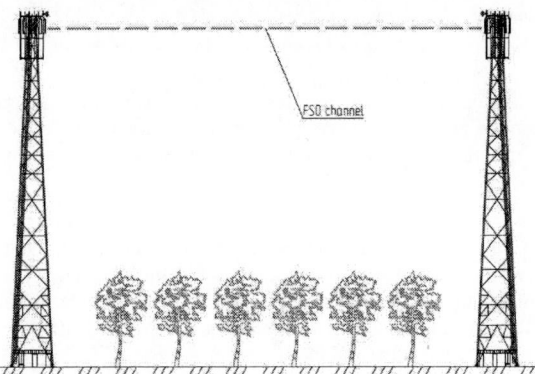

Fig. 2. Atmospheric optical channel between two AMS

This implies that for a standard AMC with a height of 120 meters, which can be utilized as part of the infrastructure for atmospheric optical communication links, the maximum permissible peak deviation would be 12 centimeters. This standard additionally considers the dynamic factors resulting from wind loading, temperature deformations, and other external influences. It is also noteworthy that the distance between the transmitter and receiver is irrelevant in this context.

To determine the required plate width, we will use the ratio that we derived in [14]:

$$\Delta L_y = L_1 - L_2 = d \left(\frac{\sin\left(\theta_1 - \arcsin\left(\sin(\theta_1)\frac{n_a}{n_g}\right)\right)}{\sqrt{\frac{n_g^2 - n_a^2 + n_a^2 \cos^2(\theta_1)}{n_g^2}}} - \frac{\sin\left(\theta_3 - \arcsin\left(\sin(\theta_3)\frac{n_a}{n_g}\right)\right)}{\sqrt{\frac{n_g^2 - n_a^2 + n_a^2 \cos^2(\theta_3)}{n_g^2}}} \right) \quad (1)$$

where n_a is the refractive index of air; n_g is the refractive index of the material from which the plate is made; θ_1, θ_3 are the angle of incidence of radiation on plates 1 and 2; θ_2, θ_4 are the angle of refraction of plates 1 and 2.

In [14], the maximum angles of inclination of the plates were determined – 80° for the first and 10° for the second. Using (1), it is easy to determine that the plate width required to ensure a total deviation of 12 mm is 15 mm.

B. Plate length

The key parameter is the minimum required length of the correction plate, which is directly dependent on the chosen method of its inclination.

We can consider a simplified scenario where the plate is positioned in such that its pivot point remains fixed at the point where the laser beam strikes the surface (Fig. 3a). In this ideal scenario, the length of the plate is essentially irrelevant, as it can be arbitrarily small since the beam would always fall within the operational area of the optical component, regardless of the rotation angle.

However, practical implementation of this mechanism presents significant technical challenges due to the need for a sophisticated system for dynamically positioning the axis of rotation. This makes this approach impractical for real-world applications.

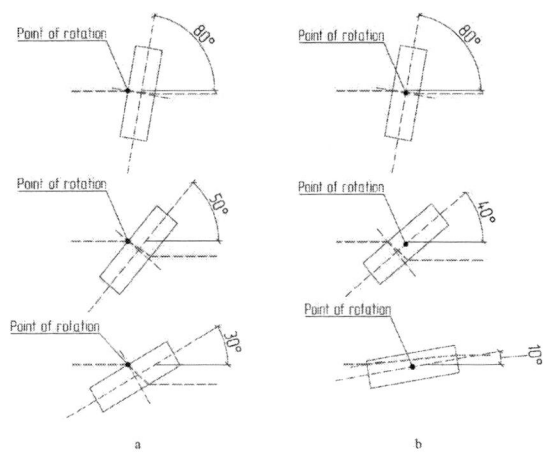

Fig. 3. Comparison of glass plates rotation principles: (a) an idealized scheme with a dynamically displaced axis of rotation, ensuring that the beam constantly hits the center of the plate; (b) a practical implementation with a fixed axis of rotation, requiring increased plate dimensions to avoid radiation hitting the end face at maximum deflection angles

In practice, the most practical and technically viable option is for the axis of plate rotation to remain fixed relative to the body of the device, maintaining its perpendicular orientation (Fig. 3b). However, this approach does have a limitation – at significant inclination angles of the plate, there is a risk that the laser beam will strike the end of the optical element, which would be unacceptable for the proper operation of the system.

979-8-3315-4784-4/26 $31.00 © 2026 IEEE

This issue becomes especially critical when the maximum operating angle of the plate is 80 degrees (corresponding to an incident beam angle of 10 degrees). Geometric calculations show that in order to ensure radiation does not reach the end of the plate at these extreme inclination angles, its minimum length must be at least 95 millimeters.

This value is derived by taking into account the plate thickness, the refractive index of the material, and the desired operating range of correction angles. The specified length allows for the necessary clearance for the beam to pass through the plate's working surface, even at maximum deflection angles, thereby eliminating the risk of undesired optical effects and signal loss.

Although the example of a 120-meter-tall structure may seem extreme, it places maximum demands on the correction system, it is important to note that most antenna mast structures are significantly shorter, typically not exceeding 70 meters in height. This significantly reduces the requirements for the adjustment system as the allowable vertical deviations at the top of these structures are proportionately smaller.

Even more favorable conditions occur when equipment is installed directly on buildings, as is typical in urban communication networks. In such cases, the requirements for compensation systems become significantly less stringent, as buildings, due to their structural characteristics (large mass, rigid framework, developed foundation), exhibit a magnitude of order lower vertical deviations in comparison to towers and communication masts.

Fig. 4 shows the relationship between the width of the correction plate and the maximum achievable beam deflection for specific system parameters. The calculations took into account the following key characteristics: the plate material used is lithium niobate with a refractive index of n = 2.286, as well as the optimal plate tilt angles of 80 and 10 degrees, established in previous studies.

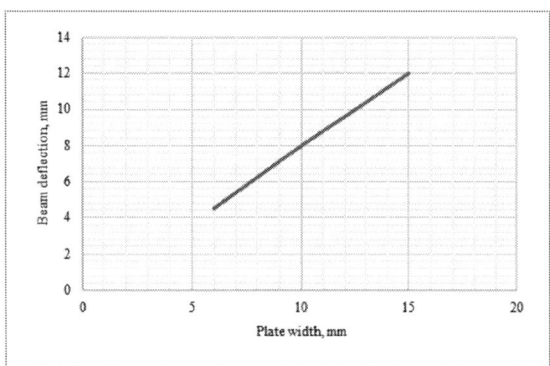

Fig. 4. The relation between the width of the plate and the possible beam deflection.

III. SYSTEM'S CAPABILITIES TO PRECISELY ADJUST USING THE POCKELS EFFECT

It was found in [14] that the choice of lithium niobate as an optical material for FSO systems is the optimal solution, provided there are no significant restrictions on the project budget. The key advantages of lithium niobate are: firstly, its unique electro-optical properties, which provide a wide dynamic range of adjustment of optical parameters; secondly, low values of control voltages not exceeding 30 V, necessary for effective control of the refractive index. Alternative optical materials require the use of significantly higher control voltages - on the order of several kilovolts, which creates serious problems during operation.

Taking into account the optimal plate sizes obtained, the actual dependence of L on the plate width d for various materials was obtained. The relationship is shown in Fig. 5.

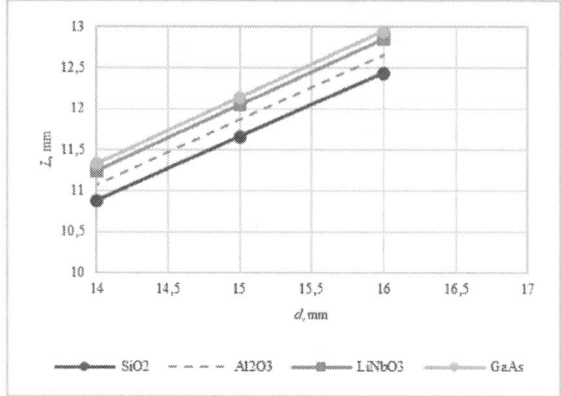

Fig. 5. Dependence of L on the width of plate d for various materials

For a system consisting of two lithium niobate plates, each 15 mm wide, the dependence Δn on the control voltage U_{offset} (Fig. 6) and the dependence of the resulting displacement L on the control voltage U_{offset} (Fig. 7) were obtained.

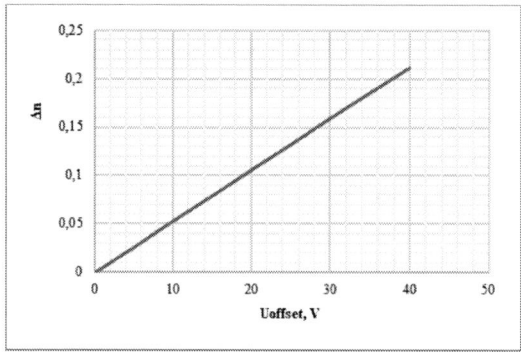

Fig. 6. Dependence of Δn on the control voltage U_{offset}

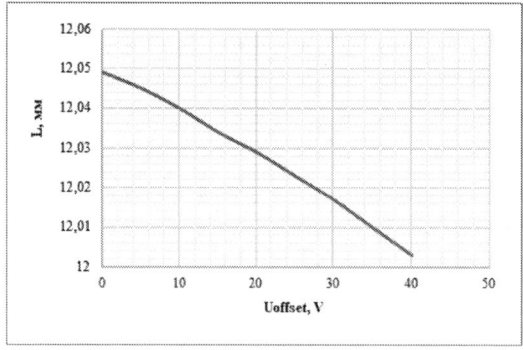

Fig. 7. Dependence of the resulting displacement L on the control voltage U_{offset}

An analysis of the results shows that the applied control voltage of 40 volts makes it possible to achieve an accurate displacement of the optical axis of the laser radiation by 0.05 millimeters. This high positioning accuracy is particularly valuable when performing precision optical system adjustments. The resulting offset value is optimal for fine equipment alignment when minimal but extremely accurate beam position adjustments are required. It is especially important that such precise control is achieved at relatively

979-8-3315-4784-4/26 $31.00 © 2026 IEEE   214

low control voltages, which indicates the effectiveness of the chosen approach and the correctness of the use of lithium niobate as a working material.

IV. THE IMPACT OF TEMPERATURE ON SYSTEM PERFORMANCE

When the temperature changes, there is thermal expansion of the plate material, which results in a change in its refractive index. This, in turn, can cause uncontrolled deviations of the laser beam and reduce the accuracy of the positioning process. Fig. 8 shows the dependence of the total laser radiation axis displacement on the plates' temperature. It is proposed to use the electro-optical Pockels effect to compensate for temperature distortions. The applied control voltage allows to dynamically adjust the refractive index of the material, neutralizing temperature changes in optical characteristics.

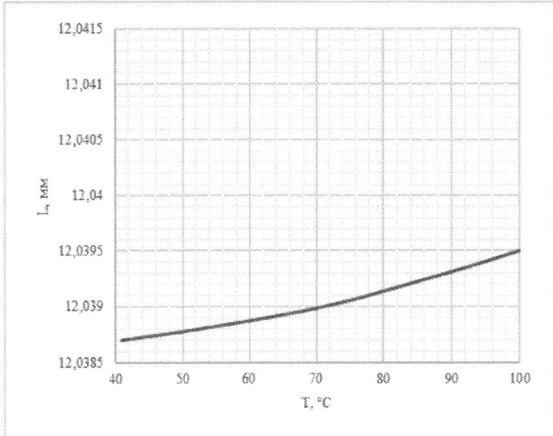

Fig. 8. Dependence of the resulting displacement L on the plates' temperature

Additional active cooling is not required – passive thermoregulation methods are sufficient. The heat sink through natural convection ensures that the temperature of the plates is maintained within acceptable limits.

V. CONCLUSION

The analysis of the results obtained has shown the effectiveness of the developed system for automatically adjusting the optical axis of the FSO systems using electro-optical elements made of lithium niobate. It has been found that the system can compensate for both static and dynamic deviations caused by external influences with minimal control voltages.

The optimal geometric parameters of the plates (15×95 mm) and the tilt angles ensure stable operation even under extreme operating conditions.

The possibilities for precise adjustment of the laser radiation axis are considered. This adjustment makes it possible to compensate for even minor temperature drifts and mechanical vibrations that inevitably occur in real-world operating conditions.

The practical significance of the solution lies in its adaptability to various infrastructure facilities, from high—rise masts to buildings, where the requirements for offset compensation vary significantly. Further research may be aimed at optimizing the design for operation over extremely long distances and in extreme temperatures.

REFERENCES

[1] I. A. Alimi, P. P. Monteiro, "High-capacity optical transmission: A review of advanced modulation formats", Optics and Laser Technology, vol. 176, p. 110917, September 2024.

[2] E. E. Elsayed, "Distributed HAPS-assisted communications in FSO/RF space-air-ground integrated network", Optical and Quantum Electronics, vol. 56, no. 6, p. 1020, May 2024.

[3] V. V. Davydov, B. K. Reznikov,V. I. Dudkin, "New Optical System for Long Distance Control of Electrical Energy Flows", Energies, vol. 16, no. 3, p. 1040, January 2023.

[4] A. N. Ermolaev, and et. al. "Compensation of chromatic and polarization mode dispersion in fiber-optic communication lines in microwave signals transmittion", Journal of Physics: Conference Series, vol. 741, no. 1, p. 012171, August 2016.

[5] E. E. Elsayed, "Investigations on OFDM UAV-based free-space optical transmission system with scintillation mitigation for optical wireless communication-to-ground links in atmospheric turbulence", Optical and Quantum Electronics, vol. 56, no. 5, p. 837, March 2024.

[6] V. V. Davydov, and et. al., "Investigation of speckle structures formed by the optical vortices of fiber lightguides", Journal of Optical Technology, vol. 82, no. 3, pp. 132–135, March 2015.

[7] N. I. Popovskiy, V. V. Davydov, I. M. Gureeva, Development to high-rate fiber optic communication line with code division multiplexing", in Proc. ACM International Conference Proceeding Series, 2021, pp. 527–531.

[8] D. Paul, D. Nandi, "Prediction of transmittance for a free space quantum channel and improving quantum Keyrate in adverse atmospheric condition", Optical and Quantum Electronics, vol. 56, no. 6, p. 100, June 2024.

[9] L. Li, and et. al., "Design and Experimental Demonstration of an Atmospheric Turbulence Simulation System for Free-Space Optical Communication", Photonics, vol. 11, no. 4, p. 334, April 2024.

[10] Y. Chen, and et. al., All-in-One BPSK/QPSK Switchable Transmission and Reception for Adaptive Free-Space Optical Communication Links", Photonics, vol. 11, no. 4, p. 326, March 2024.

[11] Y.-X. Cheng et al., "Time transfer over 113 km free space laser communication channel ", Opt. Express, vol. 32, no. 7, pp. 12645–12655, March 2024.

[12] J. Wang et al., "Performance analysis and verification of FSO based inter-ship communication systems on sea", Opt. Commun., vol. 557, p. 130271, April 2024.

[13] S. L. Sathiya Narayanan et al., "Implementation of forward error correction for improved performance of free space optical communication channel in adverse atmospheric conditions" Results Opt., vol. 16, p. 100689, July 2024.

[14] S. A. Shagako, V. V. Davydov, and V. V. Naumova, "Features of controlling the position of the laser radiation axis using controlled electro-optical systems in atmospheric optical communication lines", Uchenye Zapiski Fiz. Fak. Mosk. Univ., no. 3, pp. 2431111–1–2431111–6, 2024.

[15] K. S. Nazarova and V. V. Davydov, " Development of an automatic system for adjusting the position of the laser radiation axisfor an air communication channel", Uchenye Zapiski Fiz. Fak. Mosk. Univ., no. 4, pp. 2341112–1–2341112–5, September 2023.

[16] R. V. Davydov et al., " Fiber-Optic Transmission System for the Testing of Active Phased Antenna Arrays in an Anechoic Chamber", in Proc. Int. Conf. Comput. Sci., LNCS, vol. 10531. Cham: Springer, , pp. 177–183, September 2017.

Modeling of the GaAs Optical Waveguide Phase Shifter

Alexander Shevtsov
Department of Physical Electronics and Technology
Saint-Petersburg Electrotechnical University "LETI"
Saint-Petersburg, Russia
aashevtsovleti@gmail.com

Vitalii Vitko
Department of Physical Electronics and Technology
Saint-Petersburg Electrotechnical University "LETI"
Saint-Petersburg, Russia
Vitaly.vitko@gmail.com

Alexey Ustinov
Department of Physical Electronics and Technology
Saint-Petersburg Electrotechnical University "LETI"
Saint-Petersburg, Russia
ustinov_rus@yahoo.com

Abstract— **Over the last two decades, the development of integrated microwave photonic devices has grown rapidly. Significant interest in this field is driven by the potential for compact, energy-efficient components with low optical losses. The GaAs platform, with its direct bandgap and high refractive index contrast, represents a promising technology for such devices. Operating at a wavelength of 1.55 μm, device control is achieved via the free carrier dispersion effect. Applying a reverse bias voltage (V_{bias}) increases the depletion region at the p-n junction interface, enabling the tuning of the effective refractive index (n_{eff}). Numerical calculations were performed for a symmetric p-n junction with a doping concentration of $2.5 \cdot 10^{17}$ cm^{-3}, with V_{bias} swept from -1 to -10 V. An n_{eff} change of $1.4 \cdot 10^{-4}$ is achieved at -10 V, resulting in a π phase shift in a 5-mm-long waveguide.**

Keywords—GaAs, free-carrier dispersion, phase shifter, p-n junction.

I. Introduction

Microwave photonics (MWP) is an interdisciplinary science field focused on the generation, conversion and analysis of the microwave signals on the optical carrier [1]. Significant research advancements in this field over the past two decades have been directed towards tunable devices including tunable lasers [2], optical filters [2], true time delay lines [3], modulators [4], and phase shifters [5].

One of the general active elements of the integration MWP is a high-speed tunable phase-shifters. Such phase-shifters are used in the advanced MWP systems applied for quantum communication, radio monitoring, navigation for the aerospace and maritime industries [6]. The dominant material platform for implementation of passive MWP components is silicon-on-insulator (SOI). However, the SOI has a lack of opportunity to the monolithic integration of active and passive elements on a single chip due to the silicon indirect bandgap. One potential solution is the use of direct bandgap semiconductors from the A$_3$B$_5$ group. In particular, GaAs represents a mature and established technological platform for MWP [7].

In order to realize the high-speed tuning of optical signal phase in the integral GaAs rib waveguide the p-n junction have to be placed into its cross-section. The underlying physical principle of control is the free-carrier dispersion (FCD) effect [5]. The application of an external reverse bias to the p-n-junction leads to a depletion region formation. Thus, the tuning of effective refractive index and absorption coefficient of the rib waveguide are carried out.

The aim of this work is to model the performance characteristics of a phase-shifter based on a rib GaAs-waveguide with a p-n junction. The geometrical parameters of the rib waveguide and doping profiles are chosen for

reason of eigenmode waveguide regime and achieving complete depletion of the rib.

II. Modeling

A. Free-Carrier Dispersion

The investigated GaAs phase-shifter on a rib waveguide with a vertical p-n junction is shown in Fig.1(a). Preliminary modal analysis was carried out to optimize the geometry of rib waveguide to ensure its eigenmode regime at a wavelength of $\lambda = 1.55$ μm. Thus, the parameters of the rib waveguide for further calculations are 100-nm-thick slab, and 450-nm-width and 200-nm-thick rib. The calculated electric field intensity $|E|^2$ distribution in the cross-section of the investigated waveguide for the E_x^0 mode is shown in Fig. 1(b).

One can see, that the electric field of the propagated wave is localized in the rib of the waveguide. The mode area covers whole area of the rib. Therefore, the change in the area of the depletion region due to the reverse displacement of the p-n junction produces the change in the refractive index.

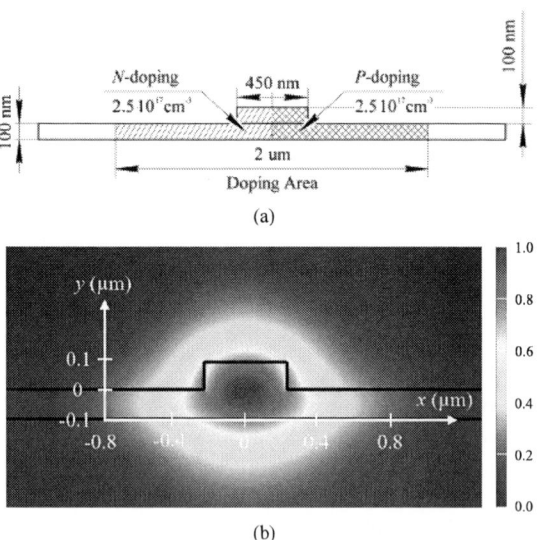

(a)

(b)

Fig. 1. Schematic of the cross-section GaAs phase shifter (a) and simulation of electric field intensity $|E|^2$ distribution for the E_x^0 mode in it (b).

In order to calculate the change in the refractive index due to the FCD effect several theoretical models was applied. The classical Drude model [8] provides a general dependence of the refractive index on free carrier density:

979-8-3315-4784-4/26 $31.00 © 2026 IEEE

$$\Delta n = -\frac{e^2 \lambda^2}{8\pi^2 c^2 \varepsilon_0 n} \left[\frac{N}{m_e} + P \frac{m_{hh}^{1/2} + m_{lh}^{1/2}}{m_{hh}^{3/2} + m_{lh}^{3/2}} \right], \qquad (1)$$

where $e = 1.602 \cdot 10^{-19}$ C is the electron charge, $c = 299792458$ m/s is the speed of light, $\varepsilon_0 = 8.854 \cdot 10^{-12}$ F/m is the dielectric constant, $m_e = 9.109 \cdot 10^{-31}$ kg is the electron mass, $m_{hh} = 0.62m_e$ kg is a heavy hole mass, $m_{lh} = 0.087m_e$ kg is a light hole mass, N and P are electron and hole concentration, respectively, $n = 3.377$ [9] is a refractive index of GaAs at a wavelength of 1.55 μm. The result of numerical simulation by (1) is shown by solid purple line in Fig. 2. In order to check the adequacy of applied model we compare the results with the experimental data from [8], which is shown by the punctured dots in Fig. 2. One can see the insufficient agreement between Drude model and experimental data.

Therefore, we use an empirical formula that takes into account the effects of band gap shift, intervalence band absorption and scattering by free carrier [10]:

$$\Delta n_{FC} = D_n N + D_p P \qquad (2)$$

with following coefficients

$$D_n = -\frac{1.09 \cdot 10^{-20}}{n E_\gamma^2} - \frac{2.03 \cdot 10^{-21}}{E_g^2 - E_\gamma^2} (eV^2 cm^3), \qquad (3)$$

$$D_p = -\frac{2.9 \cdot 10^{-22}}{E_\gamma^2} - \frac{1.35 \cdot 10^{-21}}{n E_\gamma^2} - \frac{1.5 \cdot 10^{-21}}{E_g^2 - E_\gamma^2} (eV^2 cm^3), \qquad (4)$$

where $E_g = 1.42$ eV is a band gap of GaAs at 300K, $E_\gamma = 0.8$ eV is a photon energy at $\lambda = 1.55$ μm. The result is shown by blue line in Fig. 2. Unfortunately, the comparison shows better but still insufficient agreement between theory and experiment.

As we know from [11] the best agreement with the experimental data for SOI technology was using the Richard Soref's power dependence approximation. It describes both the change in the refractive index and the absorption coefficient by power functions:

$$-\Delta n = p \Delta N_e^q + r \Delta N_h^s, \qquad (5)$$

$$\Delta \alpha = a \Delta N_e^b + c \Delta N_h^d, \qquad (6)$$

where ΔN_e and ΔN_h are changes in electron and hole concentration, respectively: a, b, c, d, p, q, r, s are empirical coefficients.

To determine the empirical coefficients, we used the experimental dependence of the refractive index on free carrier concentration from [8], [12-13] and the absorption coefficient dependences from [14-16]. The best agreement with the experimental data was obtained for following expressions:

$$-\Delta n = 10^{-24} \Delta N_e^{1.197} + 10^{-26} \Delta N_h^{1.3}, \qquad (7)$$

$$\Delta \alpha = 9.69 \cdot 10^{-19} \Delta N_e^{1.04} + 2.38 \cdot 10^{-16} \Delta N_h^1, \qquad (8)$$

where $\Delta N_e = \Delta N_h$ varies from 10^{17} cm^{-3} to 10^{18} cm^{-3}. The results for refractive index change are shown by black line in Fig. 2. One can see a perfect fitting of the theory curve with the experimental points.

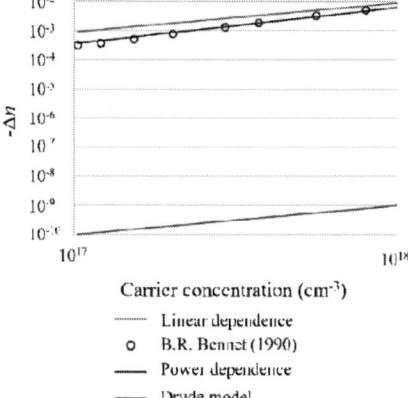

Fig. 2. Dependence of the refractive index change on the carrier concentration.

B. Phase-shifting

The phase shift tuning $\Delta\varphi$ is determined by the relation $\Delta\varphi = (2\pi / \lambda) \Delta n_{eff} L$, where Δn_{eff} is the induced change in the effective index and L is the length of a phase-shifter.

The variation of free carrier concentration within the rib waveguide is achieved by applying a reverse bias to the p-n junction. In the absence of external voltage, a depletion region is formed in the center of the waveguide structure due to the contact potential difference V_{cont}. When a reverse voltage is applied to the p-n junction, the depletion region expands. The dependence of the depletion region width on the voltage reverse can be obtained by solving Poisson's equation. For a simplified model considering uniform impurity distribution with $N_a = N_d$, this equation takes the following form [17]:

$$W = \sqrt{\frac{2\varepsilon_0 \varepsilon_s}{qN} (V_{bias} + V_{cont})}, \qquad (9)$$

where ε_s is a dielectric permittivity of GaAs, $V_{cont} = (kT/e) ln(N^2/n_i^2)$, $N = N_a = N_d$ is the dopant concentration, $T = 300$ K, $n_i^2 = 5.207 \cdot 10^8$ cm^{-3} is the intrinsic carrier concentration in GaAs, $k = 1.381 \cdot 10^{-23}$ J·K^{-1} is a Boltzmann constant.

The $V_{cont} = 1.105$ V for GaAs doped with $N = N_a = N_d = 2.5 \cdot 10^{17}$ cm^{-3}. Thus, the depletion region width in the absence of external voltage is $W = 54.74$ nm. The applied reverse voltage 10 V enhance the depletion region width to 178.6 nm. However, this estimate is only valid for one-dimensional models.

For a quantitative analysis of the electro-optical characteristics of the proposed structure at a wavelength of 1.55 μm, numerical simulations were performed using the Finite-Difference Time-Domain (FDTD) method. The depletion region width in the absence of external voltage is W

= 110 nm. The applied reverse voltage 10 V enhance the depletion region width to 500 nm. The calculated effective refractive index changes and absorption coefficients for different reverse voltages are shown in Fig. 3.

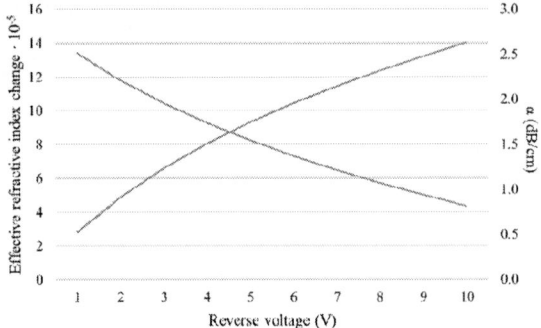

Fig. 3. The calculated effective refractive index changes and absorption coefficients for different reverse voltages.

The length of the phase-shifter at a fixed value of V_{bias}, at which a phase shift is equal to $\Delta\varphi$, calculated as follows [10]:

$$L_{\Delta\varphi} = \frac{\Delta\varphi}{360°} \frac{\lambda}{\Delta n_{eff}} . \qquad (10)$$

One of the key parameters characterizing the efficiency of a phase-shifter is the 180°-phase shift length (L_π) is the minimum structure length required to accumulate a phase shift of 180 degrees. The phase shift was calculated as a function of the phase-shifter length for various reverse voltages. The tuning of the phase-shifter through the applying voltage from -2 V to -10 V is shown in Fig.4.

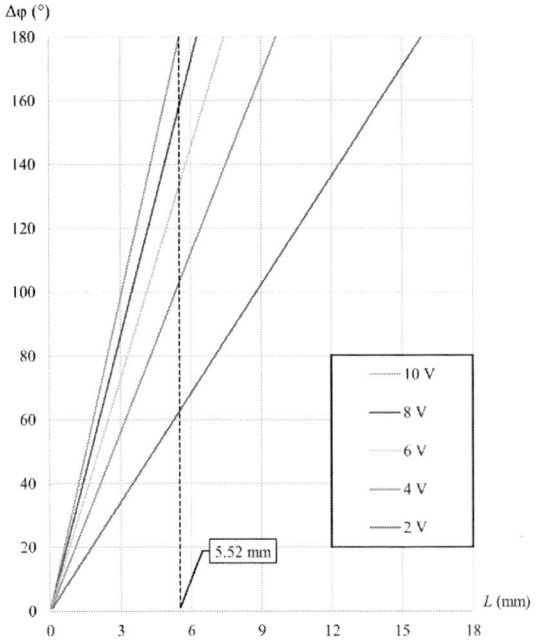

Fig. 4. Dependence of phase shift change on the length of the device for different reverse voltage.

Thus, the voltage required to induce a π-phase shift over a length of L_π will be denoted as V_π.

Total propagation losses calculated as follows:

$$\alpha_L = L_{\Delta\varphi}\alpha . \qquad (11)$$

According to (10), the phase-shifter lengths at which the phase shift is 180° for different reverse voltages are presented in Table I. Total propagation losses at the π-phase transition length are given in Table I for different reverse voltages.

TABLE I. PHASE-SHIFTER Π-LENGTHS AND TOTAL PROPAGATION LOSSES

V_π	L_π (mm)	α_{L_π} (dB)
1	27.59	6.95
2	15.83	3.50
3	11.74	2.30
4	9.63	1.67
5	8.32	1.28
6	7.43	1.02
7	6.77	0.82
8	6.27	0.67
9	5.86	0.55
10	5.52	0.45

The efficiency of an integrated phase-shifter is quantitatively assessed using figure of merit $V_\pi L_\pi$ [18]. The minimum of the $V_\pi L_\pi$ allows us to find a compromise between energy efficiency and the length of the phase shifter. In other words, a phase-shifter with length of 27.59 mm and 1 V supply voltage allows us to achieve a 180°-phase-shift. However, at this length the propagation loss is 6.95 dB, as shown in Table I. Often, a phase shifter has to introduce minimal signal losses. Therefore, it is necessary to find compromise between insertion losses and power efficiency. That's why a 5.52 mm phase-shifter appears more promising due to 0.45 dB losses.

III. CONCLUSION

This work presents the design and analysis of a GaAs phase-shifter whose operating principles relies on free-carrier dispersion electro-optic effect. The device is implemented as a rib waveguide with an integrated vertical p-n-junction. To form a symmetrical depletion region upon reverse bias application. The doping concentration in the n- and p-regions were chosen $N_a = N_d = 2.5 \cdot 10^{17}$ cm^{-3}. Based on the Soref numerical model the effective refractive index change of the investigated structure was performed using the FDTD method. The final device geometry was chosen considering a compromise between its length and propagation losses. As a result, the optimal configuration with a π-phase shift length of L_π = 5.52 mm was identified and recommended for implementation, as it provides the minimally achievable propagation losses level.

ACKNOWLEDGMENT

This work was supported by the Ministry of Science and Higher Education of the Russian Federation (grant number FSEE-2025-0014)

REFERENCES

[1] J. Yao, "Microwave Photonics", Journal of Lightwave Technology, vol.27, №3, pp. 314-335, February 2009.

979-8-3315-4784-4/26 $31.00 © 2026 IEEE

[2] J. Zhou et al., "Tunable Multi-Tap Bandpass Microwave Photonic Filter Using a Windowed Fabry-Perot Flter-Based Multi-Wavelength Tunable Laser", Journal of Lightwave Technology, vol.29, №22, pp. 3381-3386, November 2011.

[3] W. Xue, J. Mork, "Tunable true-time delay of a microwave photonic signal realized by cross gain modulation in a semiconductor waveguide", J. Appl. Phys., vol. 99, №23, December 2011.

[4] R.G. Walker et al., "Electro-Optic Modulators for Space Using Gallium Arsenide", in ICSO (2016), Vol. 10562, pp.369-377, September 2017.

[5] M. Nickerson et al., "Broadband Optical Phase Modulator with Low Residual Amplitude Modulation", in Optica Advanced Photonics Congress (2022), Paper IW4B.4 (Optica Publishing Group, 2022).

[6] G. Moody et al., "2022 Roadmap on integrated quantum photonics", Journal of Physics: Photonics, vol. 4, №1, P.012501, January 2022.

[7] E. Mobini et al., "AlGaAs Nonlinear Integrated Photonics", Micromachines, vol. 13, №7, P.991, June 2022.

[8] B. R. Bennett et al., "Carrier-Induced Change in Refractive Index of InP, GaAs, and InGaAsP", IEEE Journal of Quantum Electronics, vol.26, №1, pp.113-122, January 1990.

[9] K. Papatryfonos et al., Refractive indices of MBE-grown $Al_xGa_{1-x}As$ ternary alloys in the transparent wavelength region // AIP Advances. – 2021. – Vol. 11, №2. – P. 025327.

[10] M.J. Nickerson, "Gallium Arsenide Photonic Integrated Circuit Platform for Optical Phased Array Applications". Diss. University of California, Santa Barbara, 2023.

[11] M. Nedeljkovic, R. Soref, G.Z. Mashanovich, "Free-Carrier Electrorefraction and Electroabsorption Modulation Predictions for Silicon Over the 1-14-μm Infrared Wavelength Range", IEEE Photonics Journal, vol.3, №6, pp.1171-1180, December 2011.

[12] H.C. Huang and S. Yee, "Change in refractive index for ptype GaAs at λ=1.06, 1.3, and 1.55 μm due to free carriers", J. Appl. Phys., vol.70, №2, pp. 925-929, April 1991.

[13] H.C. Huang, S.Yee and M.Soma, "The carrier effects on the change of refractive index for n-type GaAs at λ=1.06, 1.3, and 1.55 μm", J. Appl. Phys., vol.67, №3, pp. 1497-1503, February 1990.

[14] W.G. Spitzer and J.M. Whelan, "Infrared Absorption and Electron Effective Mass in n-Type Gallium Arsenide", Phys. Rev., vol.114, №1, pp. 59-63, April 1959.

[15] D.I. Babic et al., "Design and analysis of double-fused 1.55-μm vertical cavity lasers", IEEE J. Quantum Electron, vol. 33, №8, pp. 1369-1383, August 1997.

[16] W.J. Turner and W.E. Reese, "Absorption Data of Laser-Type GaAs at 300° and 77°K", J. Appl. Phys., vol. 35, №2, pp. 350-352, February 1964.

[17] Chrostowski Lukas, and Michael Hochberg. Silicon Photonics Design: From Devices to Systems. – Cambridge: Cambridge University Press, 2015.

[18] M. Nickerson et al., "Broadband Low Residual Amplitude Modulation Phase Modulator Arrays for Optical Beamsteering Applications", in Conference on Lasers and Electro-Optics (2022), Paper SS1D.4 (Optica Publishing Group).

Development of Organic Photosensitive Structures with a Bulk Heterojunction Based on P3HT and PBDTT-DPP

Gordey A. Shutkin
Department of Photonics
Saint Petersburg Electrotechnical
University "LETI"
St. Petersburg, Russia
gashutkin@etu.ru

Marina D. Pavlova
Department of Photonics
Saint Petersburg Electrotechnical
University "LETI"
St. Petersburg, Russia
0000-0001-8815-8993

Nikita A. Khorshev
Department of Photonics
Saint Petersburg Electrotechnical
University "LETI"
St. Petersburg, Russia
0009-0007-4762-2164

Aleksandr E. Degterev
Department of Photonics
Saint Petersburg Electrotechnical
University "LETI"
St. Petersburg, Russia
0000-0002-6151- 6567

Aleksander S. Tarasov
Department of Photonics
Saint Petersburg Electrotechnical
University "LETI"
St. Petersburg, Russia
0000-0003-1360-3137

Ivan A. Lamkin
Department of Photonics
Saint Petersburg Electrotechnical
University "LETI"
St. Petersburg, Russia
0000-0002-3680-7725

Abstract— **Bulk heterojunction photosensitive structures for the visible and near-infrared ranges were fabricated and studied. The structures were based on the organic polymers Poly(3-hexylthiophene-2,5-diyl) (P3HT) and Poly{2,6´-4,8-di(5-ethylhexylthienyl)benzo[1,2-b;3,4-b]dithiophene-alt-5-dibutyloctyl-3,6-bis(5-thiophen-2-yl)pyrrolo[3,4-c]pyrrole-1,4-dione} (PBDTT-DPP), which were used as donors, and a fullerene derivative – [6,6]-Phenyl-C71-butyric acid methyl ester (PC71BM) which was employed as the acceptor. The influence of the substrate spin-coating speed on the uniformity of the fabricated active-layer films was studied. The optimal deposition rate was found to be 800 rpm for the P3HT:PC71BM blend and 1000 rpm for the PBDTT-DPP:PC71BM blend. The photoresponsivity spectra were investigated. The P3HT:PC71BM samples showed a spectral response from 400 to 800 nm, with a peak at 630 nm. The PBDTT-DPP:PC71BM structures were sensitive from 400 to 900 nm, exhibiting two maxima at 740 nm and 820 nm. Additionally, structures based on a ternary P3HT:PBDTT-DPP:PC71BM blend were developed. These demonstrated a broad spectral response across the 400 to 900 nm range.**

Keywords—Organic photosensitive structure, spin coating, absorption spectrum, P3HT, PBDTT-DPP, PC71BM

I. INTRODUCTION

Currently, photosensitive structures are finding an increasing number of new applications across diverse fields of optoelectronics [1]. For instance, they are used in optical communications, biomedical imaging, spectrometry, night vision devices, remote sensing, space exploration, and many other areas [2]. To date, photosensitive structures are typically fabricated based on inorganic semiconductors, which are produced through complex technological processes and require expensive equipment. This limits the widespread adoption of such devices due to their high cost [3].

For this reason, devices based on organic semiconductors as the active layer have seen significant development, employing simple and cost-effective deposition techniques such as spin coating or inkjet printing [4]. Furthermore, due to their light weight and high mechanical strength, the use of organic photosensitive structures (OPS) enables the application of flexible substrates, which further expands their

The study was carried out under project № FSEE-2025-0013.

potential fields of application. Another important advantage of organic semiconductors is the possibility of tuning certain material properties through modification of the molecular structure. This allows for the control and adjustment of the spectral absorption range, charge carrier mobility, and energy level positions [5-6].

The objective of this work is to develop photosensitive structures with a bulk heterojunction based on the organic donors Poly(3-hexylthiophene-2,5-diyl) (P3HT) and Poly{2,6´-4,8-di(5-ethylhexylthienyl)benzo[1,2-b;3,4-b]dithiophene-alt-5-dibutyloctyl-3,6-bis(5-thiophen-2-yl)pyrrolo[3,4-c]pyrrole-1,4-dione} (PBDTT-DPP) and to investigate their characteristics.

II. EXPERIMENTAL PART

The active layer materials used were P3HT and PBDTT-DPP, which serve as the donor components. P3HT is the most extensively studied donor material for the development of organic photosensitive structures due to its high absorption coefficient and charge carrier mobility of approximately 0.1 cm²/V·s [7]. The energy bandgap of P3HT is 2.0 eV [8]. PBDTT-DPP is a conjugated polymer consisting of donor benzodithiophene-thiophene (BDTT) blocks and acceptor diketopyrrolopyrrole (DPP) blocks linked via ring fusion [9]. PBDTT-DPP features a narrow bandgap of 1.67 eV and a high charge carrier mobility of 0.2 cm²/V·s, making it a promising material for organic photosensitive structures operating in the visible and near-infrared spectral ranges [10]. The acceptor material used was a fullerene derivative – [6,6]-Phenyl-C71-butyric acid methyl ester (PC71BM) – was used, which exhibits good solubility and a broad absorption spectrum in the visible range.

During the course of this work, OPS with a bulk heterojunction were fabricated and studied using the spin coating method. A schematic representation of the samples investigated in this study is shown in Fig. 1.

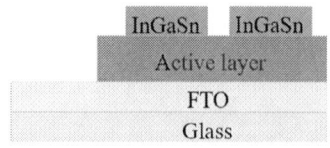

Fig. 1. Schematic representation of the investigated structures

Glass substrates coated with a fluorine-doped tin oxide (FTO) layer were used for sample fabrication. The substrates were cleaned in an ultrasonic bath for 20 min, then additionally rinsed with isopropyl alcohol and annealed at 80 °C for 10 min to remove residual solvents. Simultaneously with substrate preparation, the solution for the active layer deposition was prepared. For all samples, the donor-to-acceptor ratio in the active layer was 1:1. Chlorobenzene was used as the solvent, with a solution concentration of 40 mg/mL. The volume of solution deposited onto each substrate was 60 µL. After deposition, the samples were further annealed at 80 °C for 10 min to remove residual solvent and to complete the formation of the structure.

To achieve an efficient photosensitive structure, several conditions must be met: the combination and ratio of materials in the active layer must be optimized to ensure compatible energy level alignment, effective exciton dissociation at the heterojunction, and efficient charge carrier transport followed by collection at the electrodes. Additionally, high absorption of radiation within the specified spectral range is required. Optimizing both the composition and the thickness of the active layer is essential to meet these requirements.

A key parameter in the spin-coating process that influences the thickness and uniformity of the deposited film is the rotational speed of the substrate during active layer formation. This speed must be optimized to meet two critical requirements: the resulting films must have sufficient thickness to ensure high light absorption and exciton generation, while simultaneously exhibiting high uniformity to facilitate unimpeded charge carrier transport and minimize recombination losses.

Therefore, the initial stage of the research involved studying the influence of the substrate rotational speed on the parameters of the fabricated structures. For this purpose, samples were created at speeds ranging from 400 rpm to 1800 rpm in increments of 100 rpm. Optical methods were used to assess the quality of the layers, the primary one being the investigation of the spectral absorption coefficient characteristics, measured using a fast-scanning spectrometer. For each speed, a batch of 5 samples was prepared. Film quality was evaluated by comparing the mean deviation of the absorption intensity at the peak of the spectral dependence, measured at 25 points across the substrate surface. This was calculated as the arithmetic mean between the highest and lowest absorption intensities. The points on the sample were selected to cover the entire sample area, including the edges, during measurements. The results are presented in Fig. 2 and 3.

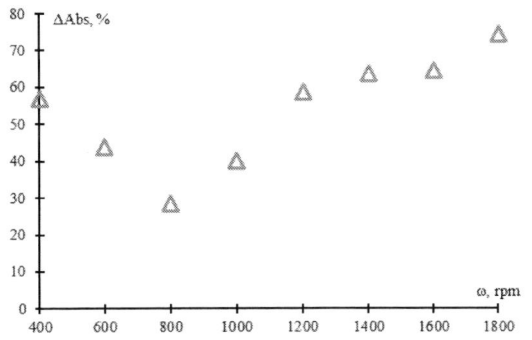

Fig. 2. Average absorption coefficient deviation of P3HT:PC71BM films versus substrate rotational speed

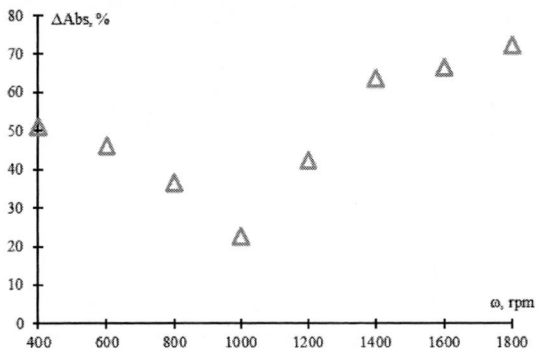

Fig. 3. Average absorption coefficient deviation of PBDTT-DPP:PC71BM films versus substrate rotational speed

It was revealed that the optimal substrate rotational speed for depositing films based on the P3HT:PC71BM blend is 800 rpm, while for the PBDTT-DPP:PC71BM blend it is 1000 rpm. At these speeds, the smallest deviation of the absorption coefficient was observed across the sample batch, while high absorption coefficient values were maintained. Speeds lower than optimal are insufficient for uniform distribution of the solution towards the substrate edges due to its high viscosity. Conversely, at speeds exceeding the optimum, the deposited film becomes thinner in the center compared to the edges.

Subsequent to determining the optimal spin-coating parameters, an investigation of the spectral dependencies of the absorption coefficient and photosensitivity was conducted. Fig. 4 presents the absorption spectra of films based on the P3HT:PC71BM blend, as well as its individual components. Fig. 5 shows the photosensitivity spectrum of the FTO/P3HT:PC71BM/InGaSn structure deposited at a substrate rotational speed of 800 rpm.

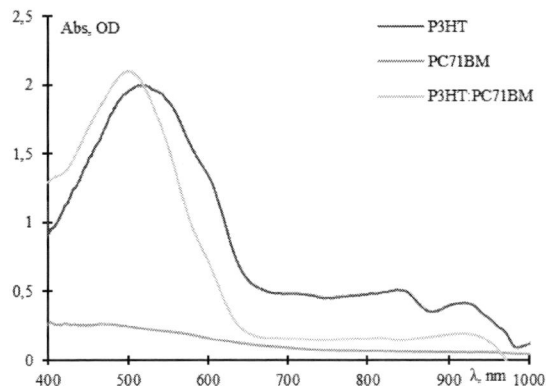

Fig. 4. Absorption spectra of films based on P3HT, PC71BM, and P3HT:PC71BM blend

The absorption of films based on P3HT covers the wavelength range from 400 to 950 nm. The highest absorption of PC71BM-based films is observed in the short-wavelength region. The absorption spectrum of the blend is a sum of the absorption curves of the pure materials, with the greatest absorption occurring in the wavelength range of 400 – 650 nm, peaking at around 500 nm.

979-8-3315-4784-4/26 $31.00 © 2026 IEEE

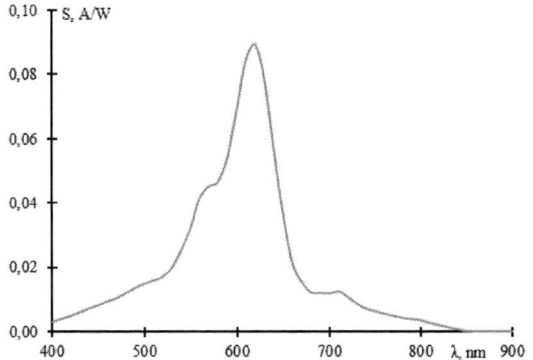

Fig. 5. Photosensitivity spectrum of the FTO/P3HT:PC71BM/InGaSn structure deposited at a substrate rotational speed of 800 rpm

The photosensitivity spectrum of the structure FTO/P3HT:PC71BM/InGaSn spans the wavelength range from 400 to 800 nm. The full width at half maximum (FWHM) of the spectrum is 80 nm. The maximum photosensitivity reaches 92 mA/W at a wavelength of 630 nm, corresponding to optical transitions with an energy of approximately 2 eV. These transitions are explained by the bandgap between the energy levels of the P3HT donor.

Fig. 6 presents the absorption spectra of films based on the PBDTT-DPP:PC71BM blend, as well as the individual components of the film. Fig. 7 shows the photosensitivity spectrum of the structure FTO/PBDTT-DPP:PC71BM/InGaSn, deposited at a substrate spin-coating speed of 1000 rpm.

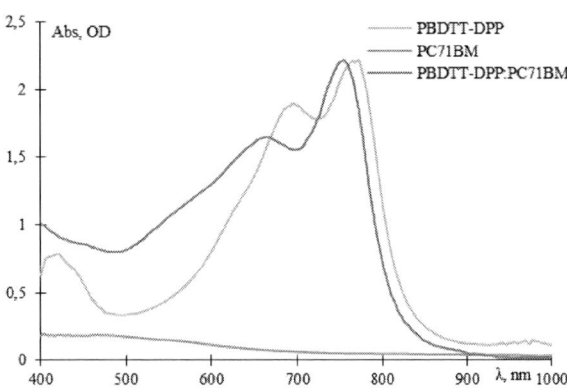

Fig. 6. Absorption spectra of films based on PBDTT-DPP, PC71BM, and PBDTT-DPP:PC71BM blend

Films based on PBDTT-DPP exhibit the highest absorption in the wavelength range of 500 to 850 nm. The absorption spectrum of the PBDTT-DPP:PC71BM blend film represents an envelope of the individual components' absorption spectra and covers the wavelength range from 400 to 820 nm, while maintaining the characteristic splitting of the absorption peak typical for PBDTT-DPP.

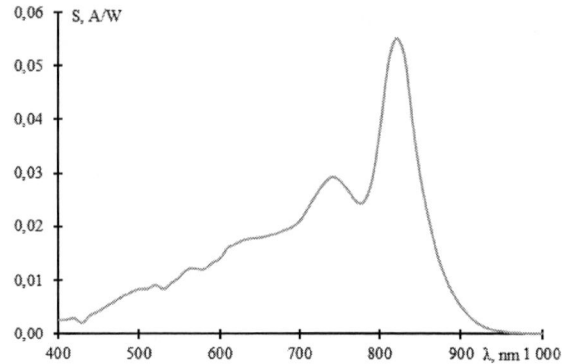

Fig. 7. Photosensitivity spectrum of the FTO/PBDTT-DPP:PC71BM/InGaSn structure deposited at a substrate rotational speed of 1000 rpm

The photosensitivity spectrum of the studied structure covers the wavelength range from 400 to 900 nm, featuring two peaks: 55 mA/W at 820 nm and 29 mA/W at 740 nm. The first peak can be explained by the proximity of the absorption maximum combined with a high quantum efficiency, while the second peak is attributed to $\pi-\pi^*$ transitions in PBDTT-DPP. The full width at half maximum (FWHM) of the spectrum is 100 nm.

Thus, the responsivity of structures based on the P3HT:PC71BM blend is concentrated in the visible range, and its signal is nearly twice as high as that for samples based on the PBDTT-DPP:PC71BM blend. The responsivity of the latter structure, in turn, is concentrated in the near-infrared range. To broaden the spectral responsivity range, samples based on a ternary blend comprising two donors and one acceptor, P3HT:PBDTT-DPP:PC71BM, were investigated. The material concentration ratio in the solution was 0.5:0.5:1. The solution was deposited at a substrate rotational speed of 1000 rpm. The absorption spectrum of the studied samples is presented in Fig. 8.

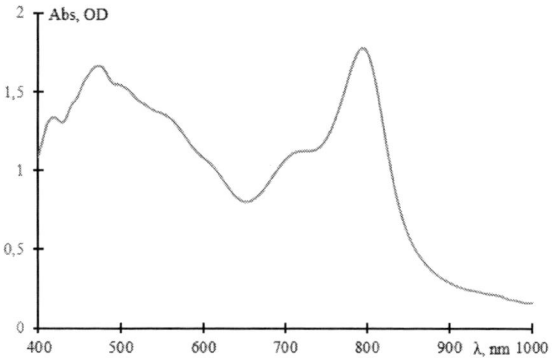

Fig. 8. Absorption spectrum of films based on P3HT:PBDTT-DPP:PC71BM blend

As expected, the absorption spectrum of these samples is a combination of the absorption spectra of structures based on P3HT:PC71BM and PBDTT-DPP:PC71BM blends. The absorption occurs over the wavelength range from 400 to 1000 nm, with the absorption edge of the samples at around 850 nm, corresponding to transitions in PBDTT-DPP. There are local maxima at wavelengths of 450 nm, 710 nm, and 800 nm, which are also characteristic of PBDTT-DPP but slightly red-shifted. The high absorption in the 400 – 600 nm range is attributed to P3HT absorption.

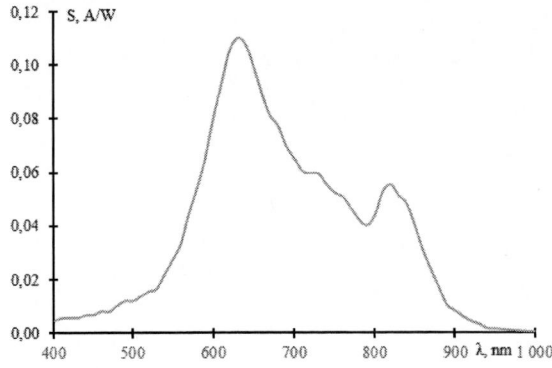

Fig. 9. Photosensitivity spectrum of the FTO/P3HT:PBDTT DPP:PC71BM/InGaSn structure deposited at a substrate rotational speed of 1000 rpm

Fig. 9 presents the photoresponsivity spectrum of the FTO/P3HT:PBDTT-DPP:PC71BM/GaInSn structure. The sample exhibits photoresponsivity across the wavelength range from 400 to 900 nm and combines the spectral characteristics of its constituent dual-donor systems. Specifically, the short-wavelength responsivity is attributed to the P3HT:PC71BM component, while the long-wavelength responsivity originates from the PBDTT-DPP:PC71BM component. This combination allows for significant spectral broadening, achieving a full width at half maximum (FWHM) of 180 nm, which is greater than that of the preceding binary structures. Furthermore, an increase in the photoresponsivity level up to 0.11 A/W is observed. This enhancement is achieved due to the relative alignment of the materials' energy levels, which facilitates a stepwise transfer of photogenerated charge carriers towards the contacts.

III. CONCLUSIONS

In this work, bulk heterojunction organic photosensitive structures based on the donor polymers P3HT and PBDTT-DPP were fabricated and characterized. An investigation was conducted on the influence of substrate rotational speed during active layer formation. It was established that the optimal spin-coating speeds are 800 rpm for layers based on the P3HT:PC71BM blend and 1000 rpm for samples based on the PBDTT-DPP:PC71BM blend. At these speeds, the minimum deviation of the absorption coefficient across the sample surface was observed. The study of the spectral dependence of photoresponsivity revealed that samples based on the P3HT:PC71BM blend exhibit a spectral response in the visible range, with a maximum of 92 mA/W at a wavelength of 630 nm. Samples based on the PBDTT-DPP:PC71BM blend demonstrate a response across the 400-900 nm wavelength range but show a lower signal (55 mA/W at 820 nm).

REFERENCES

[1] Y. Li, Y. Guo, Z. Liu, et al., "Recent Progress on Advanced Optical Structures for Emerging Photovoltaics and Photodetectors," Adv. Energy Sustainability Res., vol. 1, no. 2, p. 2000044, 2020

[2] T. Ma, N. Xue, A. Muhammad, G. Fang, J. Yan, R. Chen, J. Sun, and X. Sun, "Recent Progress in Photodetectors: From Materials to Structures and Applications," Micromachines, vol. 15, no. 10, p. 1249, 2024.

[3] A. A. Firoozi, A. A. Firoozi, M. R. Maghami, et al., "Harnessing photovoltaic innovation: Advancements, challenges, and strategic pathways for sustainable global development," Energy Convers. Manag.: X, vol. 27, p. 101058, 2024.

[4] L. A. Ruiz - Preciado, P. Pešek, C. Guerra-Yánez, Z. Ghassemlooy, S. Zvánovec, and G. Hernandez-Sosa, " Inkjet-printed high-performance and mechanically flexible organic photodiodes for optical wireless communication," Sci. Rep., vol. 14, Art. no. 3296, 2024.

[5] R. Lin, C.-H. Hsueh, S. Qin, et al., "The challenge and opportunity of organic semiconductors in photocatalysis," Microstruct., vol. 5, Art. no. 2025070, 2025.

[6] Z. Wu, Y. Yan, Y. Zhao, and Y. Liu, "Recent Advances in Realizing Highly Aligned Organic Semiconductors by Solution-Processing Approaches," Small Methods, vol. 6, no. 10, p. 2200752, 2022.

[7] H. Hoppe and N. S. Sariciftci, "Morphology of polymer/fullerene bulk heterojunction solar cells," J. Mater. Chem., vol. 16, pp. 45–61, 2006.

[8] Z. Ahmad, M. Awais, M. A. N. Nellikkal, F. Touati, et al., "Poly(3-Hexylthiophene) (P3HT), Poly(Gamma-Benzyl-l-Glutamate) (PBLG) and Poly(Methyl Methacrylate) (PMMA) as Energy Harvesting Materials," Springer International Publishing, p. 106, 2017.

[9] W. Zhuang, S. Wang, Q. Tao, et al., "Synthesis and Electronic Properties of Diketopyrrolopyrrole-Based Polymers with and without Ring-Fusion," Macromolecules, vol. 54, no. 2, pp. 970–980, 2021.

[10] M. Riede, C. Ulrich, J. Widmer, et al., "Efficient organic tandem solar cells based on small molecules," Adv. Funct. Mater., vol. 21, pp. 3019–3028, 2011.

Integrated Optical Delay Line with Electronic Control

Vitalii Vitko
Department of Physical Electronics and Technology
Saint-Petersburg Electrotechnical University "LETI"
Saint-Petersburg, Russia
Vitaliy.vitko@gmail.com

Alexander Shevtsov
Department of Physical Electronics and Technology
Saint-Petersburg Electrotechnical University "LETI"
Saint-Petersburg, Russia
aashevtsovleti@gmail.com

Andrey Nikitin
Department of Physical Electronics and Technology
Saint-Petersburg Electrotechnical University "LETI"
Saint-Petersburg, Russia
and.a.nikitin@gmail.com

Alexey Ustinov
Department of Physical Electronics and Technology
Saint-Petersburg Electrotechnical University "LETI"
Saint-Petersburg, Russia
Ustinov_rus@yahoo.com

Abstract— Compensation of phase distortions in fiber-optic links is critical for various applications such as radio interferometry, phased antenna arrays, and quantum communication systems. Existing thermo-optic solutions suffer from high power consumption and low speed (microseconds), which are incompatible with the demands of modern high-speed systems. This work proposes a fast and energy-efficient alternative: a voltage-tunable integrated optical delay line on a silicon-on-insulator platform. The device utilizes electronically controlled ring modulators, allowing rapid switching of the delay time. The modulator has an all-pass ring resonator configuration based on a rib waveguide with a vertical *p-n* junction. It is shown that applying a reverse voltage of 10 V produces a negative shift of the resonance frequency of 6 GHz, which is sufficient to switch the ring from the 'off' to 'on' state. The combination of a Y-splitter and two microring modulators as a double-throw switch routes the signal flow between one of two rib waveguides of different lengths. The switched-line delay line is designed for discrete binary delay times of $\tau = 5$ ps, 10 ps, 20 ps and 40 ps. Thus, a four-bit delay line circuit provides delay times from 0 to 75 ps with switching timestep of 5 ps. Numerical model predicts the total insertion loss of the delay line of less than 15 dB. The developed integrated optical delay line with electronic control is promising for the compensation of phase distortions that occur in real-life fiber-optic applications.

Keywords— Photonic Integrated Circuit, Delay line, Microring modulator, Silicon-on-Insulator, Y-splitter

I. Introduction

The modern development of radar and navigation systems, as well as radio monitoring systems for the aerospace and maritime industries, is characterized by increasing demands on size, weight, energy efficiency, and broadband capabilities. of the developed devices. Traditional semiconductor microwave electronics has reached its limits in terms of speed and operating frequencies. Therefore, the development of integrated microwave photonic devices, which combine the advantages of optical and microwave technologies, is becoming particularly relevant [1]. One of the critically important tasks is the compensation of phase

distortions that occur in fiber-optic communication lines under the influence of temperature, mechanical stress, and other external factors. These distortions are fundamentally significant for applications such as radio interferometry, phased array antennas, and quantum communication systems, where even minor phase shifts can significantly reduce measurement accuracy and signal transmission quality.

In a previously published work [2], a traditional approach to compensating for phase distortions based on photonic integrated circuits (PICs) fabricated using silicon-on-insulator (SOI) technology was presented. The SOI platform has provided a dramatic increase in the integration density of photonic elements on a single chip, as well as full compatibility with complementary metal-oxide-semiconductor (CMOS) technology. This direction aligns with global trends in the development of integrated microwave photonics [3]. A disadvantage of the proposed device is its relatively low switching speed due to the use of thermo-optic control elements, which is insufficient for modern high-speed systems. An alternative approach for tuning the SOI structures is the electro-optic effect realized in a *p-n* junction within a rib waveguide. This approach is utilized in the construction of microring and microdisk modulators operating in GHz frequency bandwidth [4]. Replacing the thermo-optic control elements with the proposed modulators in the design of an integrated optical delay line allows for a sufficient improvement in performance characteristics. Thus, the purpose of this work is to develop the design of a four-bit integrated optical delay line with microring modulators operating as electronically controlled switches.

II. Operating Principle and Architecture of the Tunable Delay Line

The schematic of the proposed four-bit delay line is shown in Fig. 1. The central element of the construction is a double-throw switch that routes the signal flow between one of two delay lines based on SOI rib waveguides. The switch consists of a combination of a Y-splitter and two microring modulators (MRMs) in all-pass configuration.

979-8-3315-4784-4/26 $31.00 © 2026 IEEE

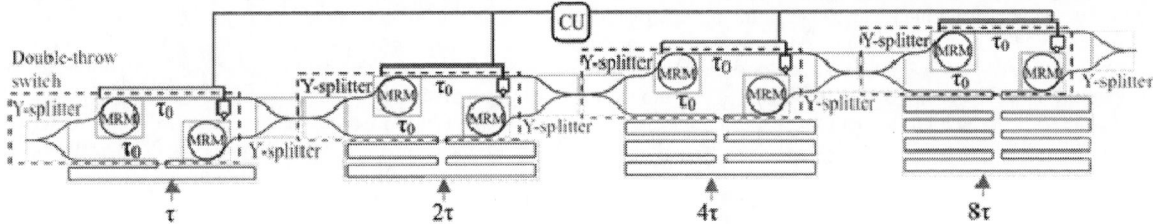

Fig. 1. General scheme of an electronic-control integrated-optical tunable delay line

Consider first the operating principle of the switch (see the region marked by a dashed line in Fig.1). An optical signal with frequency f is applied to the switch input. The signal is equally divided between two arms by the Y-splitter. Each arm is loaded with an identical MRM having the resonance frequency f_0 (the modulator's absorption frequency) at zero bias voltage. If the signal frequency f equals f_0, then the signal is switched off due to the MRM absorption. One should apply some reverse bias to the MRM p-n junction to shift the resonance frequency away from the operational frequency and switches the signal on. Applying a reverse bias to the MRM in one arm must occur synchronously with the removal of the reverse bias from the MRM in the other arm. Thus, an inverter is included in the control circuit as is shown in Fig.1. This synchronization of control pulses ensures signal propagation through only single arm of the switch. It gives a switched-line delay, providing two discrete delay times of τ_0 and $\tau_0+\tau$, where τ_0 is a reference delay time produced by waveguide interconnections and τ is the designed delay time. A cascade connection of four switches with delays of τ, 2τ, 4τ and 8τ forms a four-bit delay line circuit that provides $2^4=16$ states. Let's assume that $\tau = 5$ ps then this circuit offers an electrically controlled delay time varrying from 0 to 75 ps with a step of 5 ps. The core elements of the proposed delay line —which are designed in the following section—are: delay lines of 5 ps, 10 ps, 20 ps, and 40 ps; Y-splitters; and MRMs.

III. SIMULATION OF CORE ELEMENTS

Consider design of the core elements that are based on the SOI rib waveguide with following typical parameters: slab thickness $h = 90$ nm, rib thickness $s = 110$ nm, and width $w = 500$ nm. To demonstrate device operation, the central frequency f is set to 193.453 THz.

Let us start with the delay line. Fig. 2 shows the modeling results for the spatial distribution of the electric field component E_y. The calculations shows the predominant propagation of quasi-TE modes in this waveguide configuration, characterized by the following parameters: effective refractive index $n_{eff} = 2.51274$, group index $n_g = 3.70944$, and propagation loss $\alpha = 3$ dB/cm. The obtained group index is used to calculate the lengths of the delay lines for delay times of τ, 2τ, 4τ and 8τ, as is shown in Fig. 1. According to well-known equation $l = \tau c/n_g$, where c is the speed of light, the length of the first delay line is $l = 404.1$ µm. Therefore, the second, the third, and the fourth delay lines have lengths of $2l$, $4l$ and $8l$, respectively. These four delay lines introduce insertion losses of 0.12 dB, 0.24 dB, 0.48 dB, and 0.96 dB, respectively.

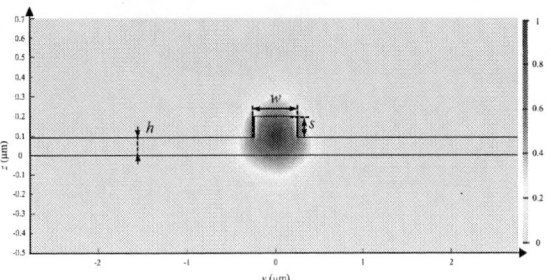

Fig. 2. Spatial distribution of the electric field component E_y in a rib waveguide.

The second element is the Y-splitter, which divides the signal into the arms of the switch and combine it back after. For equal routing, a symmetrical Y-splitter has been designed. Its schematic is shown in Fig. 3(a). Optimization of the Y-splitter's design and geometry is performed using a variational-iterative method. The dimensions of the designed Y-splitter do not exceed 15×7 µm². Simulation of signal propagation is performed for a Gaussian beam applied at the input of the Y-splitter, as is shown in Fig. 3(b). One can see that the optical power is divided equally between the two arms.

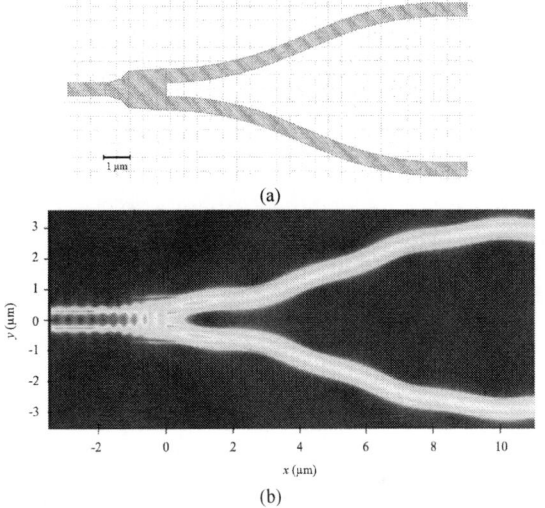

(a)

(b)

Fig. 3. Schematic of the Y-splitter (a) and simulation of optical signal propagation through it (b)

Due to reciprocity of the designed Y-splitter the optical signals applied to its output ports will be combined. The cascade of the double-throw switch and the Y-splitter as a combiner has a minimum loss of 3 dB. Therefore, the four-bit electronically controlled integrated-optical tunable delay line has a total loss of 12 dB for zero delay time.

The last element is the MRM. We propose an original MRM topology, schematically shown in Fig. 4. The red line indicates the rib of the integrated waveguide, shades of blue represent the p and $p+$ regions, and shades of purple represent the n and $n+$ regions on both sides of the waveguide slab. The active region of the MRM is assumed to constitute 67% of the total ring length.

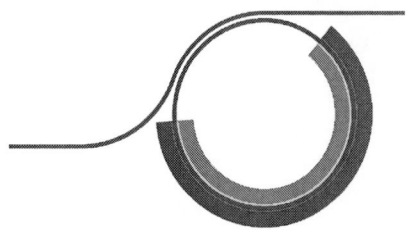

Fig. 4. The scheme of the original microring modulator

For the implementation of the MRM, a rib waveguide with a vertical p-n junction designed in work [4] is used. The waveguide cross-section of the MRM, applying three levels of dopants on each P and N regions, is shown in Fig. 5(a). An offset of 120 nm denotes the distance from the center of the rib to the center of the p-n junction. The distribution of dopant concentrations in this region is shown by color in Fig. 5(a).

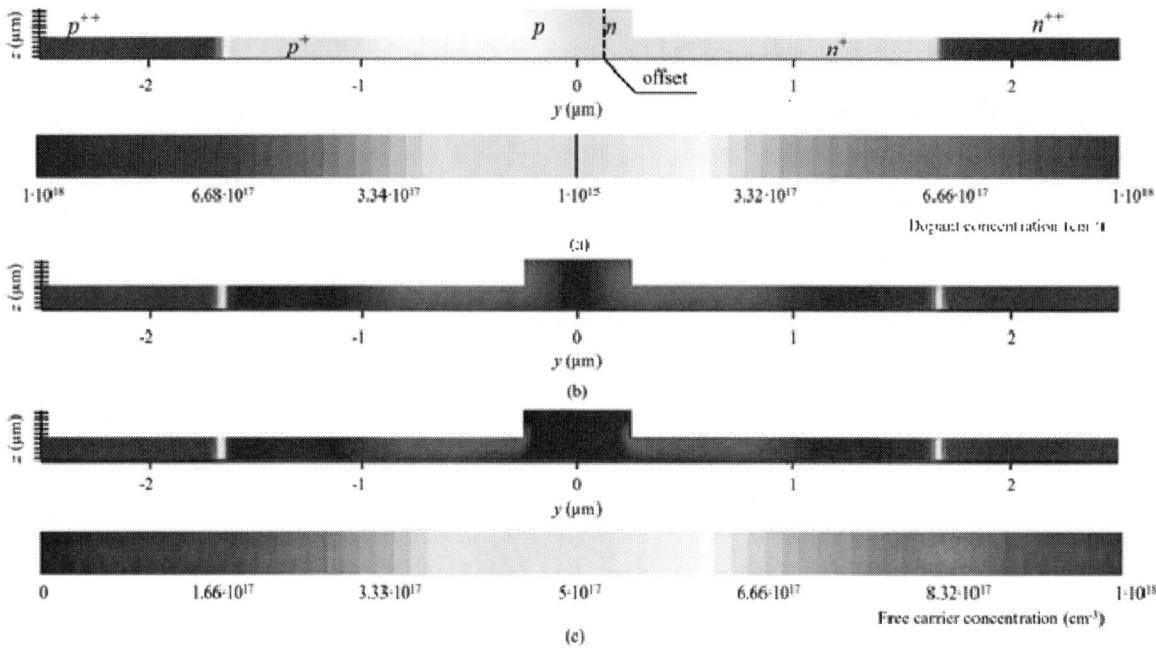

Fig. 5. Distribution of dopant in the cross-section of a rib waveguide with a vertical p-n-junction (a), simulation of the distribution of free carrier concentration with U_{rev}=0 V (b) and with U_{rev}=10 V (c)

For zero bias voltage U_{rev}=0, a depletion region forms in the middle of the rib waveguide due to the contact potential difference, as shown in Fig. 5 (b). Free carriers in the rib waveguide lead to a decrease in the refractive index due to the plasmonic electro-optic effect. The dependence of refractive index on both electron and hole densities through the electro-optic effect is described by Soref's theory [5]. Thus, the rib waveguide exhibits an inhomogeneous refractive index in its cross-section. Taking into account this distribution, the dispersion is calculated using the finite-difference time-domain method. The obtained parameters of the rib waveguide with a p-n junction are $n_{eff} = 2.51251$, $n_g = 3.70925$. These parameters allow us to proceed to the modeling of a ring resonator's transfer characteristic.

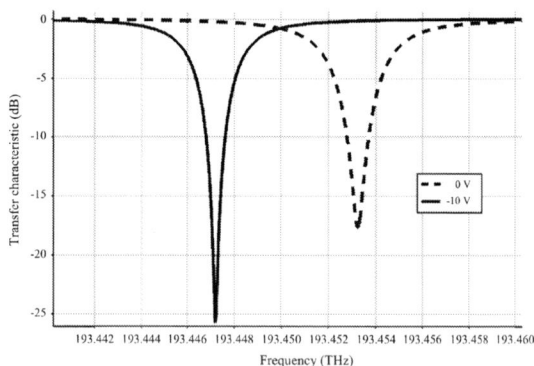

Fig. 6. The transfer characteristics of the original microring modulator at U_{rev} =0 V and U_{rev}=10 V

In order to obtain the resonant frequency directly at the operating frequency 193.453 THz, the diameter of the ring resonator is chosen as 25.7 μm. The numerical calculation of the transmission characteristic is shown by dashed line in Fig. 6. Optimization of the length of the microring resonator coupling region and the coupling gap is performed to achieve

979-8-3315-4784-4/26 $31.00 © 2026 IEEE

a transfer characteristic with an attenuation of -18 dB at the resonant frequency, which corresponds to switch insertion loss for the OFF state. The obtained coupling gap is 0.5 μm.

Applying U_{rev} leads to an increase in the depletion region. To deeper understand the physical limits of the proposed effect, the simulation of free carrier distribution for U_{rev} =

10 V is shown in Fig. 5 (c). One can see that this bias voltage is sufficient to remove free charge carriers from the rib region. It leads to an increase in the effective refractive index, which became close to the case of intrinsic rib waveguide. To demonstrate this, we calculate the dependence of the effective refractive index change Δn_{eff} on U_{rev}, which is shown in Fig. 7.

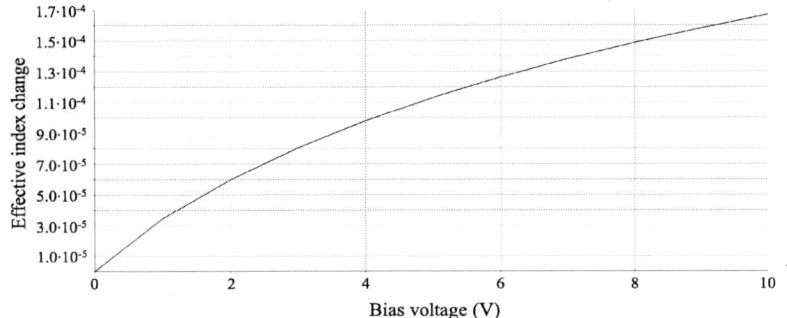

Fig. 7. The numerical simulation of the effective refractive index in the rib waveguide dependence on the applying reverse bias voltage U_{rev}

An increase in n_{eff} produces a negative shift of the MRM resonance frequency, which leads to a decrease in the insertion loss at the operating frequency of 193.453 THz. The numerical calculation of the dependence of the insertion loss on U_{rev} is shown in Fig 8. It can be seen that to achieve a minimal insertion loss of 0.2 dB in the ON state, a bias voltage of U_{rev} = 10 V is required. The transfer characteristics of the MRM calculated for U_{rev} = 10 V is shown by the solid line in Fig. 6. As it is seen the applied bias voltage is sufficient to shift the modulator's absorption frequency from 193.453 THz to 193.447 THz.

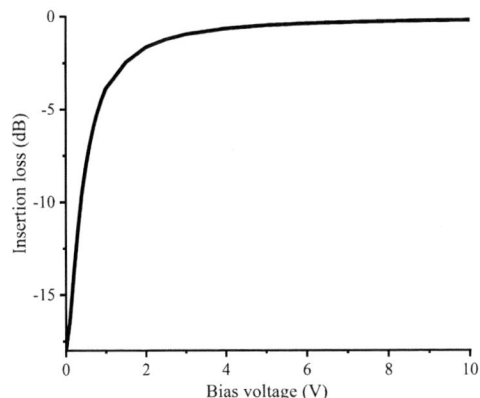

Fig. 8. The insertion loss as function of the bias voltage U_{rev}

Therefore, the proposed double-throw switch, which routes the optical signal into one of two arms via synchronized control of the MRMs, exhibits an insertion loss of no more than 3.2 dB for the selected ('ON') arm, while providing an extinction of more than 21 dB for the other ('OFF') arm.

IV. PERFORMANCE CHARACTERISTICS

The cascade connection of four switches with delay lines in the proposed schematic of four-bit delay line shown in Fig. 1 enables of 2^4=16 discrete states encoded in a binary format (see column 1 in the Table I).

TABLE I.   BINARY CODES AND THEIR CORRESPONDING PARAMETERS FOR THE INTEGRATED OPTICAL TUNABLE DELAY LINE.

Binary code	Length ΔL (μm)	Delay time Δτ (ps)	Total losses Δα (dB)
1	2	3	4
0000	0	0	0
0001	404.1	5	0.12
0010	808.2	10	0.24
0011	1212.3	15	0.36
0100	1616.4	20	0.48
0101	2020.5	25	0.60
0110	2424.6	30	0.72
0111	2828.7	35	0.84
1000	3232.8	40	0.96
1001	3636.9	45	1.08
1010	4041	50	1.2
1011	4445.1	55	1.32
1100	4849.2	60	1.44
1101	5253.3	65	1.56
1110	5657.4	70	1.68
1111	6061.5	75	1.8

A logical '1' corresponds to the OFF state of the MRM in the upper arm and to the ON state of the MRM in the lower arm. Conversely, a logical '0' level corresponds to the opposite states of the MRMs. Depending on the binary code, the optical path length changes with the time step of 5 ps. The binary code '0000' corresponds to the minimum delay line length, which serves as the reference. For other combinations, the optical path length increases by the amount ΔL shown in the second column of the Table I. The corresponding total change in delay time Δτ is presented in column 3 of the Table I. The total device loss consists of the losses in the Y-splitter (3 dB per cascade) and losses in the integrated delay lines Δα (see column 4 of the Table I). As shown in the table, the maximum propagation loss in the delay lines does not exceed 1.8 dB. When combined with the losses from the Y-splitters and MRMs, the total loss for the maximum delay time of 75 ps will be no more than 15 dB.

V. Conclusion

This work proposes a design of the four-bit electronically controlled integrated-optical delay line on a silicon-on-insulator platform. The device consists of four cascades. Each cascade includes a double-throw switch that routes the signal flow between one of two delay lines and the Y-splitter for recombine it after. All elements are fabricated using silicon-on-insulator rib waveguides. To overcome the limitations of conventional thermo-optic approaches, we propose the architecture replacing a slow thermo-optic elements with high-speed microring modulators based on vertical *p-n* junctions, enabling rapid switching in the GHz range. Through detailed simulation of the core elements—including SOI rib waveguides, Y-splitters, and microring modulators—we have validated the operational principles and performance of the device. The cascaded connection of four switches with delay line lengths of 404.1 µm, 808.2 µm, 1616.4 µm, and 3232.8 µm enables the design of an electronically controlled integrated-optical delay line. The device provides 16 discrete states with a time step of 5 ps in the range from 0 to 75 ps. The total optical loss remains below 15 dB, which is acceptable for integration into practical systems. The developed delay line represents a compact, CMOS-compatible solution that meets the demands for speed, low power consumption, and integration density in next-generation microwave photonic and optical communication systems.

Acknowledgment

This work was supported by the Ministry of Science and Higher Education of the Russian Federation (grant number FSEE-2025-0014).

References

[1] Unchenko I. V., Emelyanov A. A. Specific Features of Designing Microwave Photonic Receiving and Transmitting Channels of Onboard Systems for Communication, Radar and Radio Monitoring // Journal of the Russian Universities. – 2023. – Vol. 26, No. 1. – P. 59.

[2] Nikitin A.A., Vitko V.V., Emelyanov A.A., Ustinov A.B. Integrated-Optical Tunable Delay Line for Application in Fiber-Optic Communication Lines // Radiotekhnika. – 2023. – Vol. 87, No. 11. – P. 47-53.

[3] Kosolobov S. S. et al. Silicon integrated photonics //Uspekhi Fizicheskikh Nauk. – 2024. – T. 194. – №. 11. – C. 1223-1239.

[4] Dube-Demers R. et al. Analytical modeling of silicon microring and microdisk modulators with electrical and optical dynamics // Journal of Lightwave Technology. – 2015. – Vol. 33, No. 20. – P. 4240-4252.

[5] Soref R., Bennett B. Electrooptical effects in silicon // IEEE Journal of Quantum Electronics. – 1987. – Vol. 23, No. 1. – P. 123-129.

AUTHOR INDEX

Agafonov, Dmitry V. 100
Ahmad, Abdulhadi Haj 131
Alzhanov, Danyar 154
Amir, Sohail 61
Anikin, Grigoriy S. 145
Anikin, Grigory S. 175
Bagno, Dmitri V. 145
Bair Ts., Rakshaev 110
Balutin, Maksim A. 89, 183
Baranov, Artyom I. 57
Bazhenov, Andrey D. 141
Bobrovskaia, Valeriia 154
Bogachev, Igor A. 141
Borisov, Aleksander A. 175
Borisov, Alexander A. 141
Bui, Cong Doan 93
Burenkova, Natalya 170
Buzovkin, Sergey 96
Cheburkin, Yuri V. 5
Chemchem, Selemani A. 1
Degterev, Aleksander E. 89, 183
Degterev, Aleksandr E. 220
Degterev, Alexander E. 191
Degterev, Alexander 180
Degtereva, Mariya M. 89, 191
Degtereva, Mariya 180
Demshevsky, Valeriy V. 141, 145, 175
Dina, Snarskaya D. 110
Dmitriev, Kirill 154
Dudin, Anatoly L. 57
Epifanova, Elisavata I. 5
Galimov, Sergey R. 76
Galina, Gulnaz M. 191
Galina, Gulnaz 180
Glushets, Ekaterina 10
Gribovskaya, Olga S. 14
Gritskevich, Ivan Yu. 18
Gryzulev, Sergey S. 160, 166
Gurin, Sergey A. 100
Guttovskiy, Yan O. 199
Haponchyk, Roman 150
Hlaing, Ye Min 27
Iliin, Eugene V. 145
Isaenko, Dmitry I. 160, 166
Isaev, Yuriy E. 89, 183
Ivanov, Vladimir 1, 209
Kalugin, Viktor V. 27, 124
Karymsakov, Kamil 154

Khasanova, Diana I. 187
Khmelnitskiy, Ivan K. 5
Khorshev, Nikita A. 220
Klyuchnik, Anna S. 170
Kochurina, Elena S. 27
Kolobov, Alexey A. 5, 46
Kolybelnikov, Nikolay Yu. 160, 166
Kolybelnikov, Nikolay 33
Korotkevich, Ivan A. 124
Kotov, Timofey A. 160, 166
Kryukov, Roman S. 57
Kryukov, Roman 96
Kurenkov, Roman A. 191
Kurenkov, Roman 180
Kurkova, Alexandra D. 204
Kuznetsov, Dmitry M. 76
Lamkin, Ivan A. 89, 191, 220
Lamkin, Ivan 180
Lavasani, Laleh M. Moazzami 157
Lavrinenko, Valery V. 195
Lepekhina, Taniana K. 199
Levin, Yevgeniy 180, 191
Lobodin, Vyacheslav V. 141, 175
Losev, Vladimir 72
Mhmad, Salhab 1
Minin, Evgeny S. 160, 166
Moe, Satt Naing 27
Mohammed, Hayder J. 131
Moshnikov, Vyacheslav 96
Moskvin, Alexey V. 18
Muratova, Ekaterina N. 23
Naimi, Qusai K. Al 131
Nalimova, Svetlana S. 93, 115
Nalimova, Svetlana 96
Neelova, Angelina D. 199
Nesterov, Vyacheslav G. 106, 212
Nikitin, Andrey 224
Novichkov, Maksim D. 100
Novikov, Andrey 170
Onushchenko, Aleksei 209
Pavel, Konditerov B. 110
Pavlova, Marina D. 183, 220
Pecherskaya, Ekaterina A. 100
Perepelovsky, Vadim V. 57
Petrov, Pavel 33
Petruhanov, Alexander V. 76
Peyman, Fallahzade Z. 110
Potapov, Mikhail A. 43

Printseva, Alina S. 46

Provodin, Daniil S. 204

Radaykin, Dmitry G. 50

Repkin, Matvey 54

Reznikov, Bogdan K. 160, 166

Ryabko, Andrey A. 5, 46

Rybina, Arina 96

Salhab, Mhmad 209

Samsygin, Pavel F. 115

Sapozhnikov, Alexander V. 57

Sedegova, Tatyana 119

Shagako, Diana D. 106, 212

Shagako, Sergei A. 106

Shagako, Sergey A. 212

Shepeleva, Anastasia E. 100

Shevtsov, Alexander 216, 224

Shiryaev, Maksim E. 124

Shutkin, Gordey A. 220

Sidorenko, Stanislav S. 141, 145, 175

Simonov, Boris M. 38

Sitkov, Nikita O. 5, 46, 187

Skryabin, Vladimir V. 170

Solovyov, Ilya A. 64

Sozonov, Maxim V. 68

Sozykin, Gleb 170

Svetlana, Kononova V. 110

Tarasenko, Anastasia D. 187

Tarasov, Aleksander S. 220

Tarasov, Aleksandr S. 89, 183

Tarasov, Sergey A. 183

Tatiana, Zimina M. 110

Thu, Paing Soe 27

Timoshenko, Aleksandr 72

Timoshenkov, Sergey P. 38

Titov, Aleksandr 209

Titov, Andrey Yu. 64

Traktirshchikov, Viktor S. 124

Trushlyakova, Valentina V. 46

Tsepilova, Anastasiia 72

Tsygitsa, Ilya A. 145, 175

Tun, Phyo Win 38

Ustinov, Alexey 150, 216, 224

Vasilieva, Anastasia 1, 209

Vertyanov, Denis V. 64

Vitko, Vitalii 216, 224

Volkov, Vadim S. 100

Walo, Al 79, 83

Yaroshenko, Ivan Y. 76

Zakasovsky, Igor N. 127

Zamoshets, Vladislav E. 136

Zaytsev, Oleg E. 127

Zotova, Tatiana 209

Zubko, Svetlana 154

ISBN 979-8-3315-4784-4